Chevrolet & GMC Pick-ups Automotive Repair Manual

by Larry Warren
and John H Haynes
Member of the Guild of Motoring Writers

Models covered:

Chevrolet and GMC pick-ups - 1967 through 1987
Blazer, Jimmy and Suburban - 1967 through 1991
Two- and four-wheel drive versions

Does not include diesel engine, GMC 305 V6 gasoline engine or 1988 and later pick-up information

(7P29 - 24064)

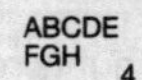

Haynes Publishing Group
Sparkford Nr Yeovil
Somerset BA22 7JJ England

Haynes North America, Inc
861 Lawrence Drive
Newbury Park
California 91320 USA

Acknowledgements
We are grateful for the help and cooperation of Tomco Industries, 1435 Woodson Road, St. Louis, Missouri 63132, for their assistance with technical information and illustrations. Technical writers who contributed to this project include Jeff Killingsworth and Arnold Sanchez.

A book in the Haynes Automotive Repair Manual Series

Printed in the U.S.A.

ISBN 1 85010 764 5

Library of Congress Catalog Card Number 91-71282

98-352

Contents

Introductory pages

About this manual	0-6
Introduction to the Chevrolet and GMC pick-ups	0-6
Vehicle identification numbers	0-7
Buying parts	0-9
Maintenance techniques, tools and working facilities	0-9
Booster battery (jump) starting	0-17
Jacking and towing	0-18
Automotive chemicals and lubricants	0-19
Conversion factors	0-20
Safety first!	0-21
Troubleshooting	0-22

Chapter 1
Tune-up and routine maintenance **1-1**

Chapter 2 Part A
V8 and V6 engines **2A-1**

Chapter 2 Part B
6-cylinder inline engines **2B-1**

Chapter 2 Part C
General engine overhaul procedures **2C-1**

Chapter 3
Cooling, heating and air conditioning systems **3-1**

Chapter 4
Fuel and exhaust systems **4-1**

Chapter 5
Engine electrical systems **5-1**

Chapter 6
Emissions control systems **6-1**

Chapter 7 Part A
Manual transmission **7A-1**

Chapter 7 Part B
Automatic transmission **7B-1**

Chapter 7 Part C
Transfer case **7C-1**

Chapter 8
Clutch and driveline **8-1**

Chapter 9
Brakes **9-1**

Chapter 10
Steering and suspension systems **10-1**

Chapter 11
Body **11-1**

Chapter 12
Chassis electrical system **12-1**

Wiring diagrams **12-11**

Index **IND-1**

Haynes mechanic, author and photographer with Chevrolet Silverado pick-up

1981 Chevrolet Silverado

About this manual

Its purpose

The purpose of this manual is to help you get the best value from your vehicle. It can do so in several ways. It can help you decide what work must be done, even if you choose to have it done by a dealer service department or a repair shop; it provides information and procedures for routine maintenance and servicing; and it offers diagnostic and repair procedures to follow when trouble occurs.

We hope you use the manual to tackle the work yourself. For many simpler jobs, doing it yourself may be quicker than arranging an appointment to get the vehicle into a shop and making the trips to leave it and pick it up. More importantly, a lot of money can be saved by avoiding the expense the shop must pass on to you to cover its labor and overhead costs. An added benefit is the sense of satisfaction and accomplishment that you feel after doing the job yourself.

Using the manual

The manual is divided into Chapters. Each Chapter is divided into numbered Sections, which are headed in bold type between horizontal lines. Each Section consists of consecutively numbered paragraphs.

At the beginning of each numbered Section you will be referred to any illustrations which apply to the procedures in that Section. The reference numbers used in illustration captions pinpoint the pertinent Section and the Step within that Section. That is, illustration 3.2 means the illustration refers to Section 3 and Step (or paragraph) 2 within that Section.

Procedures, once described in the text, are not normally repeated. When it's necessary to refer to another Chapter, the reference will be given as Chapter and Section number. Cross references given without use of the word "Chapter" apply to Sections and/or paragraphs in the same Chapter. For example, "see Section 8" means in the same Chapter.

References to the left or right side of the vehicle assume you are sitting in the driver's seat, facing forward.

Even though we have prepared this manual with extreme care, neither the publisher nor the author can accept responsibility for any errors in, or omissions from, the information given.

NOTE

A **Note** provides information necessary to properly complete a procedure or information which will make the procedure easier to understand.

CAUTION

A **Caution** provides a special procedure or special steps which must be taken while completing the procedure where the Caution is found. Not heeding a Caution can result in damage to the assembly being worked on.

WARNING

A **Warning** provides a special procedure or special steps which must be taken while completing the procedure where the Warning is found. Not heeding a Warning can result in personal injury.

Introduction to the Chevrolet and GMC pick-ups

Chevrolet and GMC pick-up trucks are of conventional front engine/rear wheel drive layout with 4-wheel drive (4WD) available on some models. Inline 6-cylinder, V6 and V8 engines have been used over the long life of these models, with fuel injection available on later models.

Power from the engine is transferred to either a 3 or 4-speed manual or 3 or 4-speed automatic transmission. A transfer case and driveshaft are used to drive the front wheels on 4-wheel drive models.

On 2-wheel drive models, suspension is by coil springs and shock absorbers at all four wheels on early models and coil springs and shock absorbers at the front with leaf springs and shock absorbers at the rear on later models. Four-wheel drive models use a solid front axle with suspension by leaf springs and shock absorbers.

The steering box is mounted to the left of the engine and is connected to the steering arms through a series of rods with power assist available as an option.

The brakes are drum-type on all four wheels on early models with disc-type at the front and drums at the rear on later models. Vacuum assist is available.

Vehicle identification numbers

Modifications are a continuing and unpublicized process in vehicle manufacturing. Since spare parts manuals and lists are compiled on a numerical basis, the individual vehicle numbers are essential to correctly identify the component required.

Vehicle Identification Number (VIN)

This very important identification number is located on a plate attached to the left side cowling just inside the windshield **(see illustration)**. The VIN also appears on the Vehicle Certificate of Title and Registration. It contains information such as where and when the vehicle was manufactured, the model year and the body style.

Service parts identification label

This label is located inside the glove box and should always be referred to when ordering parts. The Vehicle Service Parts Identification Label contains the VIN, wheelbase, paint and special equipment and option codes.

The Vehicle Identification Number (VIN) (arrow) is visible on the driver's side cowling inside the windshield

Engine code numbers

The engine code numbers can be found in a variety of locations, depending on engine type.

On the inline 6-cylinder engine the engine code number is found on a pad located on the right side of the block, to the rear of the distributor. On V6 engines, the code number is located on a pad at the front edge of the block, under the right cylinder head, adjacent to the water pump. On all V8 engines except the Mark IV (454 cu in) engine, the code can be found on a pad at the front edge of the block under the right cylinder head **(see illustration)**. On Mark IV engines, the code is located on a pad at the top center of the engine block adjacent to the front edge of the intake manifold.

Manual transmission unit number

The manual transmission unit number is located on the top left side of the case on Tremec 3-speed transmissions and on the lower left side of the case on other 3-speed transmissions. On 4-speed transmissions, the unit number can be found on the rear face of the case on most models **(see illustration)**.

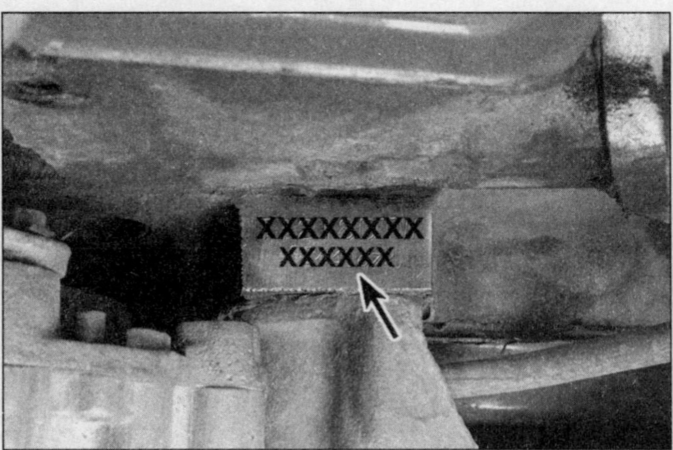

Typical V8 (except Mark IV) engine number locations

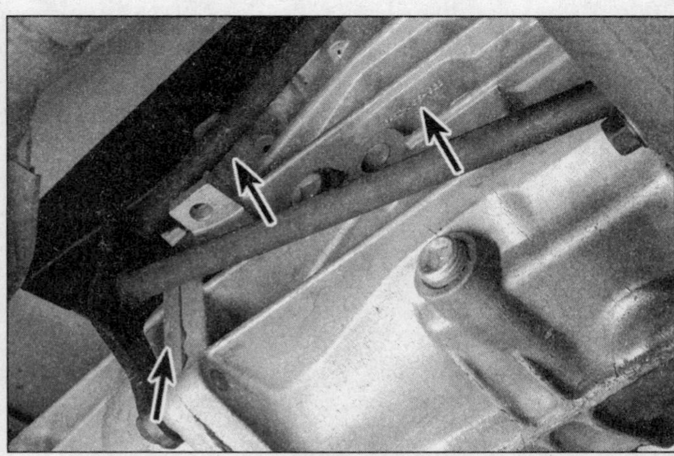

Typical 4-speed manual transmission number locations (arrows)

Automatic transmission numbers

On the THM 350 transmission, the unit number can be found the right rear vertical surface of the oil pan. On the THM 400 the serial number is located on a plate on the right side of the case on most models.

Transfer case build date number

The build date number is located on a tag on the front face of the case on most models.

Vehicle Emissions Control Information (VECI) label

On models so equipped, the emissions control information label is found under the hood (see Chapter 6 for an illustration of the label and its location).

Starter motor numbers

The serial number and production date are stamped on the rear of the case on most models.

Alternator numbers

The serial and part numbers are located on the drive (pulley) end of the end frame.

Axle code numbers

On front axles, the code is stamped on the top rear surface of the left axle tube and on rear axles on the top surface of the right axle tube.

Buying parts

Replacement parts are available from many sources, which generally fall into one of two categories - authorized dealer parts departments and independent retail auto parts stores. Our advice concerning these parts is as follows:

Retail auto parts stores: Good auto parts stores will stock frequently needed components which wear out relatively fast, such as clutch components, exhaust systems, brake parts, tune-up parts, etc. These stores often supply new or reconditioned parts on an exchange basis, which can save a considerable amount of money. Discount auto parts stores are often very good places to buy materials and parts needed for general vehicle maintenance such as oil, grease, filters, spark plugs, belts, touch-up paint, bulbs, etc. They also usually sell tools and general accessories, have convenient hours, charge lower prices and can often be found not far from home.

Authorized dealer parts department: This is the best source for parts which are unique to the vehicle and not generally available elsewhere (such as major engine parts, transmission parts, trim pieces, etc.).

Warranty information: If the vehicle is still covered under warranty, be sure that any replacement parts purchased - regardless of the source - do not invalidate the warranty!

To be sure of obtaining the correct parts, have engine and chassis numbers available and, if possible, take the old parts along for positive identification.

Maintenance techniques, tools and working facilities

Maintenance techniques

There are a number of techniques involved in maintenance and repair that will be referred to throughout this manual. Application of these techniques will enable the home mechanic to be more efficient, better organized and capable of performing the various tasks properly, which will ensure that the repair job is thorough and complete.

Fasteners

Fasteners are nuts, bolts, studs and screws used to hold two or more parts together. There are a few things to keep in mind when working with fasteners. Almost all of them use a locking device of some type, either a lockwasher, locknut, locking tab or thread adhesive. All threaded fasteners should be clean and straight, with undamaged threads and undamaged corners on the hex head where the wrench fits. Develop the habit of replacing all damaged nuts and bolts with new ones. Special locknuts with nylon or fiber inserts can only be used once. If they are removed, they lose their locking ability and must be replaced with new ones.

Rusted nuts and bolts should be treated with a penetrating fluid to ease removal and prevent breakage. Some mechanics use turpentine in a spout-type oil can, which works quite well. After applying the rust penetrant, let it work for a few minutes before trying to loosen the nut or bolt. Badly rusted fasteners may have to be chiseled or sawed off or removed with a special nut breaker, available at tool stores.

If a bolt or stud breaks off in an assembly, it can be drilled and removed with a special tool commonly available for this purpose. Most automotive machine shops can perform this task, as well as other repair procedures, such as the repair of threaded holes that have been stripped out.

Flat washers and lockwashers, when removed from an assembly, should always be replaced exactly as removed. Replace any damaged washers with new ones. Never use a lockwasher on any soft metal surface (such as aluminum), thin sheet metal or plastic.

Fastener sizes

For a number of reasons, automobile manufacturers are making wider and wider use of metric fasteners. Therefore, it is important to be able to tell the difference between standard (sometimes called U.S. or SAE) and metric hardware, since they cannot be interchanged.

All bolts, whether standard or metric, are sized according to diameter, thread pitch and length. For example, a standard 1/2 - 13 x 1 bolt is 1/2 inch in diameter, has 13 threads per inch and is 1 inch long. An M12 - 1.75 x 25 metric bolt is 12 mm in diameter, has a thread pitch of 1.75 mm (the distance between threads) and is 25 mm long. The two bolts are nearly identical, and easily confused, but they are not interchangeable.

In addition to the differences in diameter, thread pitch and length, metric and standard bolts can also be distinguished by examining the bolt heads. To begin with, the distance across the flats on a standard bolt head is measured in inches, while the same dimension on a metric bolt is sized in millimeters (the same is true for nuts). As a result, a standard wrench should not be used on a metric bolt and a metric wrench should not be used on a standard bolt. Also, most standard bolts have slashes radiating out from the center of the head to denote the grade or strength of the bolt, which is an indication of the amount of torque that can be applied to it. The greater the number of slashes, the greater the strength of the bolt. Grades 0 through 5 are commonly used on automobiles. Metric bolts have a property class (grade) number, rather than a slash, molded into their heads to indicate bolt strength. In this case, the higher the number, the stronger the bolt. Property class numbers 8.8, 9.8 and 10.9 are commonly used on automobiles.

Strength markings can also be used to distinguish standard hex nuts from metric hex nuts. Many standard nuts have dots stamped into one side, while metric nuts are marked with a number. The greater the number of dots, or the higher the number, the greater the strength of the nut.

Metric studs are also marked on their ends according to property class (grade). Larger studs are numbered (the same as metric bolts), while smaller studs carry a geometric code to denote grade.

It should be noted that many fasteners, especially Grades 0 through 2, have no distinguishing marks on them. When such is the case, the only way to determine whether it is standard or metric is to measure the thread pitch or compare it to a known fastener of the same size.

Standard fasteners are often referred to as SAE, as opposed to metric. However, it should be noted that SAE technically refers to a non-metric fine thread fastener only. Coarse thread non-metric fasteners are referred to as USS sizes.

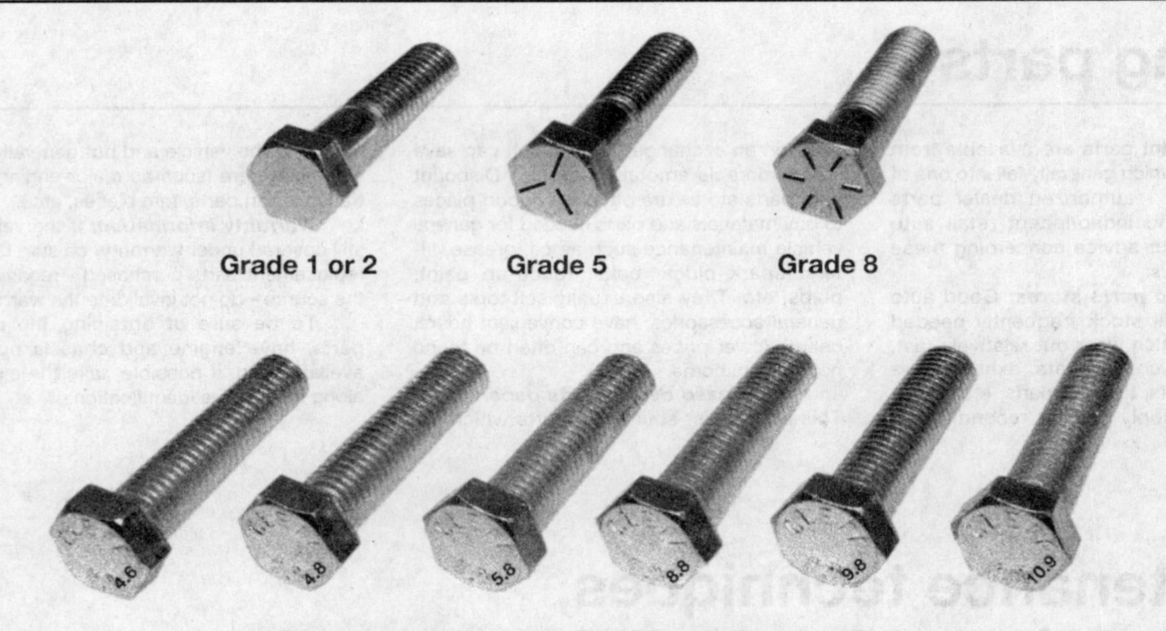

Bolt strength marking (standard/SAE/USS; bottom - metric)

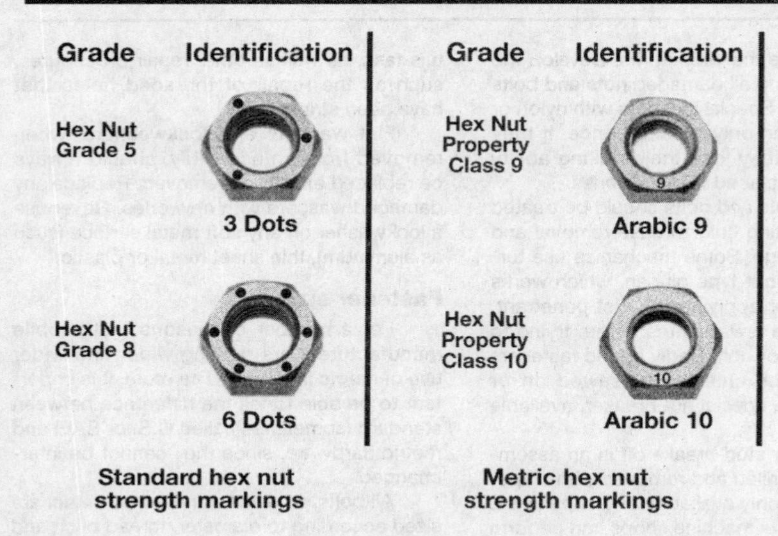

Standard hex nut strength markings

Metric hex nut strength markings

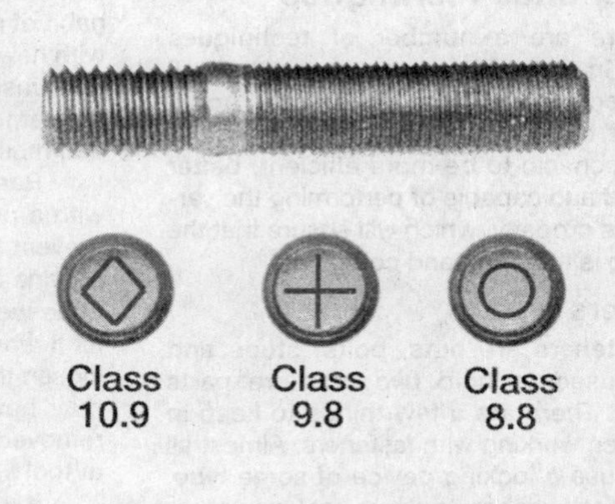

Metric stud strength markings

00-1 HAYNES

Since fasteners of the same size (both standard and metric) may have different strength ratings, be sure to reinstall any bolts, studs or nuts removed from your vehicle in their original locations. Also, when replacing a fastener with a new one, make sure that the new one has a strength rating equal to or greater than the original.

Tightening sequences and procedures

Most threaded fasteners should be tightened to a specific torque value (torque is the twisting force applied to a threaded component such as a nut or bolt). Overtightening the fastener can weaken it and cause it to break, while undertightening can cause it to eventually come loose. Bolts, screws and studs, depending on the material they are made of and their thread diameters, have specific torque values, many of which are noted in the Specifications at the beginning of each Chapter. Be sure to follow the torque recommendations closely. For fasteners not assigned a specific torque, a general torque value chart is presented here as a guide. These torque values are for dry (unlubricated) fasteners threaded into steel or cast iron (not aluminum). As was previously mentioned, the size and grade of a fastener determine the amount of torque that can safely be applied to it. The figures listed here are approximate for Grade 2 and Grade 3 fasteners. Higher grades can tolerate higher torque values.

Fasteners laid out in a pattern, such as cylinder head bolts, oil pan bolts, differential cover bolts, etc., must be loosened or tightened in sequence to avoid warping the component. This sequence will normally be shown in the appropriate Chapter. If a specific pattern is not given, the following procedures can be used to prevent warping.

Metric thread sizes	**Ft-lbs**	**Nm**
M-6	6 to 9	9 to 12
M-8	14 to 21	19 to 28
M-10	28 to 40	38 to 54
M-12	50 to 71	68 to 96
M-14	80 to 140	109 to 154
Pipe thread sizes		
1/8	5 to 8	7 to 10
1/4	12 to 18	17 to 24
3/8	22 to 33	30 to 44
1/2	25 to 35	34 to 47
U.S. thread sizes		
1/4 - 20	6 to 9	9 to 12
5/16 - 18	12 to 18	17 to 24
5/16 - 24	14 to 20	19 to 27
3/8 - 16	22 to 32	30 to 43
3/8 - 24	27 to 38	37 to 51
7/16 - 14	40 to 55	55 to 74
7/16 - 20	40 to 60	55 to 81
1/2 - 13	55 to 80	75 to 108

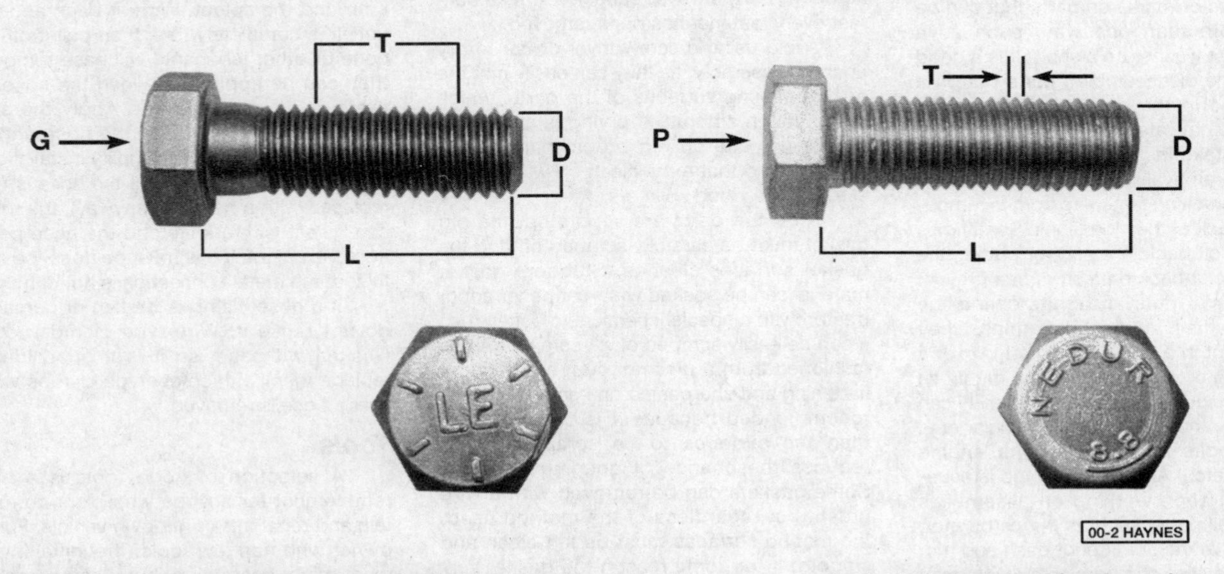

Standard (SAE and USS) bolt dimensions/grade marks

G *Grade marks (bolt strength)*
L *Length (in inches)*
T *Thread pitch (number of threads per inch)*
D *Nominal diameter (in inches)*

Metric bolt dimensions/grade marks

P *Property class (bolt strength)*
L *Length (in millimeters)*
T *Thread pitch (distance between threads in millimeters)*
D *Diameter*

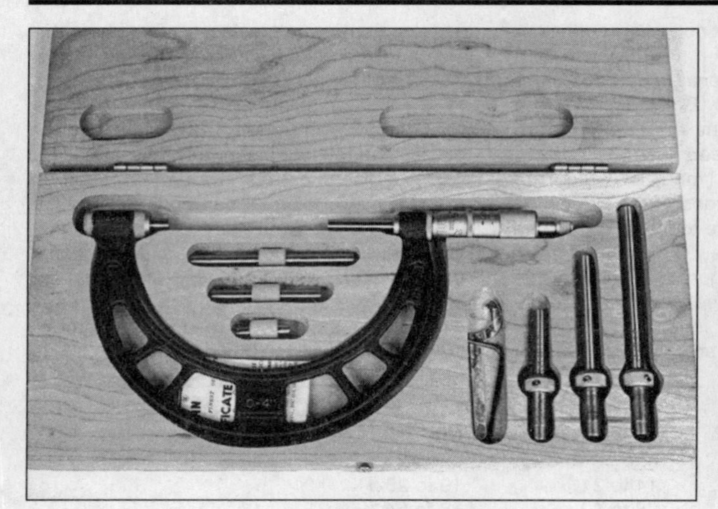

Micrometer set

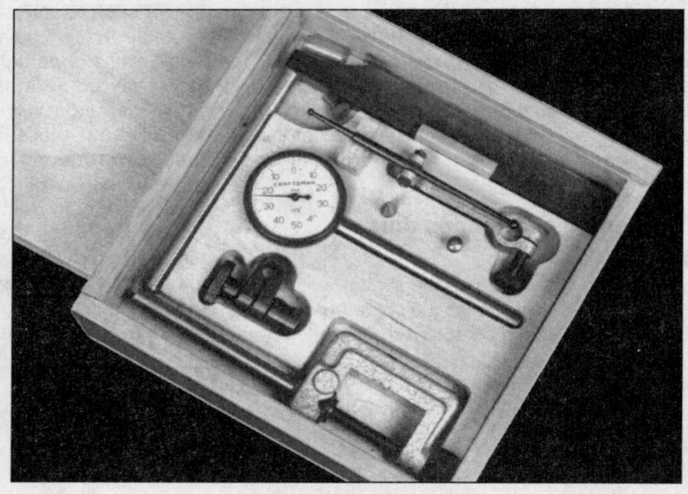

Dial indicator set

Initially, the bolts or nuts should be assembled finger-tight only. Next, they should be tightened one full turn each, in a criss-cross or diagonal pattern. After each one has been tightened one full turn, return to the first one and tighten them all one-half turn, following the same pattern. Finally, tighten each of them one-quarter turn at a time until each fastener has been tightened to the proper torque. To loosen and remove the fasteners, the procedure would be reversed.

Component disassembly

Component disassembly should be done with care and purpose to help ensure that the parts go back together properly. Always keep track of the sequence in which parts are removed. Make note of special characteristics or marks on parts that can be installed more than one way, such as a grooved thrust washer on a shaft. It is a good idea to lay the disassembled parts out on a clean surface in the order that they were removed. It may also be helpful to make sketches or take instant photos of components before removal.

When removing fasteners from a component, keep track of their locations. Sometimes threading a bolt back in a part, or putting the washers and nut back on a stud, can prevent mix-ups later. If nuts and bolts cannot be returned to their original locations, they should be kept in a compartmented box or a series of small boxes. A cupcake or muffin tin is ideal for this purpose, since each cavity can hold the bolts and nuts from a particular area (i.e. oil pan bolts, valve cover bolts, engine mount bolts, etc.). A pan of this type is especially helpful when working on assemblies with very small parts, such as the carburetor, alternator, valve train or interior dash and trim pieces. The cavities can be marked with paint or tape to identify the contents.

Whenever wiring looms, harnesses or connectors are separated, it is a good idea to identify the two halves with numbered pieces of masking tape so they can be easily reconnected.

Gasket sealing surfaces

Throughout any vehicle, gaskets are used to seal the mating surfaces between two parts and keep lubricants, fluids, vacuum or pressure contained in an assembly.

Many times these gaskets are coated with a liquid or paste-type gasket sealing compound before assembly. Age, heat and pressure can sometimes cause the two parts to stick together so tightly that they are very difficult to separate. Often, the assembly can be loosened by striking it with a soft-face hammer near the mating surfaces. A regular hammer can be used if a block of wood is placed between the hammer and the part. Do not hammer on cast parts or parts that could be easily damaged. With any particularly stubborn part, always recheck to make sure that every fastener has been removed.

Avoid using a screwdriver or bar to pry apart an assembly, as they can easily mar the gasket sealing surfaces of the parts, which must remain smooth. If prying is absolutely necessary, use an old broom handle, but keep in mind that extra clean up will be necessary if the wood splinters.

After the parts are separated, the old gasket must be carefully scraped off and the gasket surfaces cleaned. Stubborn gasket material can be soaked with rust penetrant or treated with a special chemical to soften it so it can be easily scraped off. A scraper can be fashioned from a piece of copper tubing by flattening and sharpening one end. Copper is recommended because it is usually softer than the surfaces to be scraped, which reduces the chance of gouging the part. Some gaskets can be removed with a wire brush, but regardless of the method used, the mating surfaces must be left clean and smooth. If for some reason the gasket surface is gouged, then a gasket sealer thick enough to fill scratches will have to be used during reassembly of the components. For most applications, a non-drying (or semi-drying) gasket sealer should be used.

Hose removal tips

Warning: *If the vehicle is equipped with air conditioning, do not disconnect any of the A/C hoses without first having the system depressurized by a dealer service department or a service station.*

Hose removal precautions closely parallel gasket removal precautions. Avoid scratching or gouging the surface that the hose mates against or the connection may leak. This is especially true for radiator hoses. Because of various chemical reactions, the rubber in hoses can bond itself to the metal spigot that the hose fits over. To remove a hose, first loosen the hose clamps that secure it to the spigot. Then, with slip-joint pliers, grab the hose at the clamp and rotate it around the spigot. Work it back and forth until it is completely free, then pull it off. Silicone or other lubricants will ease removal if they can be applied between the hose and the outside of the spigot. Apply the same lubricant to the inside of the hose and the outside of the spigot to simplify installation.

As a last resort (and if the hose is to be replaced with a new one anyway), the rubber can be slit with a knife and the hose peeled from the spigot. If this must be done, be careful that the metal connection is not damaged.

If a hose clamp is broken or damaged, do not reuse it. Wire-type clamps usually weaken with age, so it is a good idea to replace them with screw-type clamps whenever a hose is removed.

Tools

A selection of good tools is a basic requirement for anyone who plans to maintain and repair his or her own vehicle. For the owner who has few tools, the initial investment might seem high, but when compared to the spiraling costs of professional auto maintenance and repair, it is a wise one.

To help the owner decide which tools are needed to perform the tasks detailed in this manual, the following tool lists are offered: *Maintenance and minor repair,*

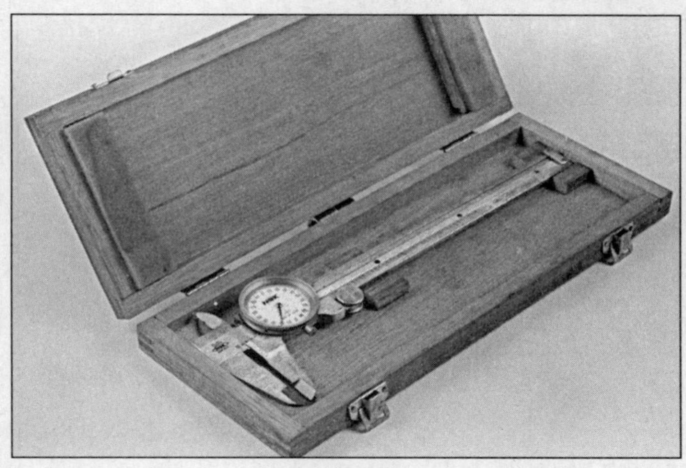
Dial caliper

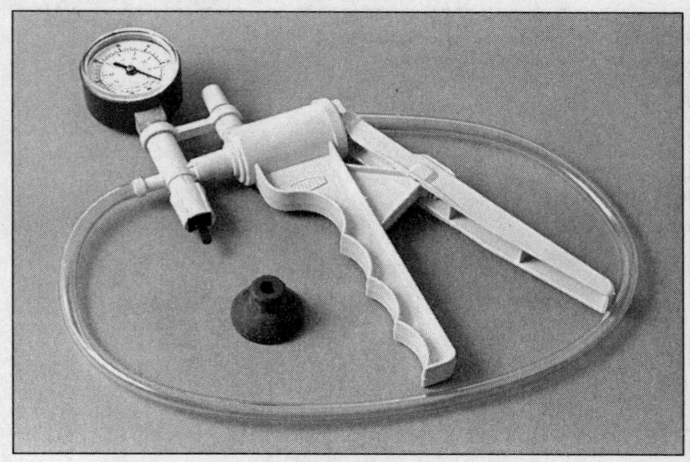
Hand-operated vacuum pump

Timing light

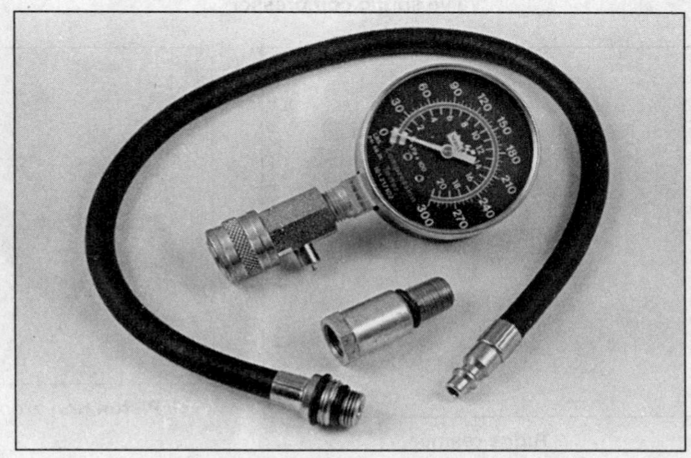
Compression gauge with spark plug hole adapter

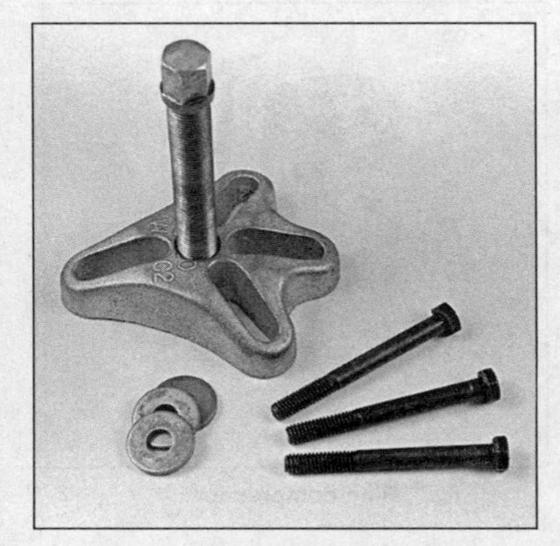
Damper/steering wheel puller

General purpose puller

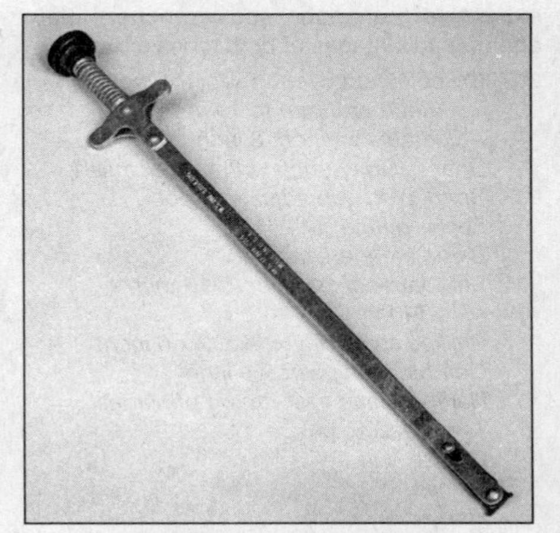
Hydraulic lifter removal tool

Repair/overhaul and *Special.*

The newcomer to practical mechanics should start off with the *maintenance and minor repair* tool kit, which is adequate for the simpler jobs performed on a vehicle. Then, as confidence and experience grow, the owner can tackle more difficult tasks, buying additional tools as they are needed. Eventually the basic kit will be expanded into the *repair and overhaul* tool set. Over a period of time, the experienced do-it-yourselfer will assemble a tool set complete enough for most repair and overhaul procedures and will add tools from the special category when it is felt that the expense is justified by the frequency of use.

Maintenance and minor repair tool kit

The tools in this list should be considered the minimum required for performance of routine maintenance, servicing and minor repair work. We recommend the purchase of combination wrenches (box-end and open-

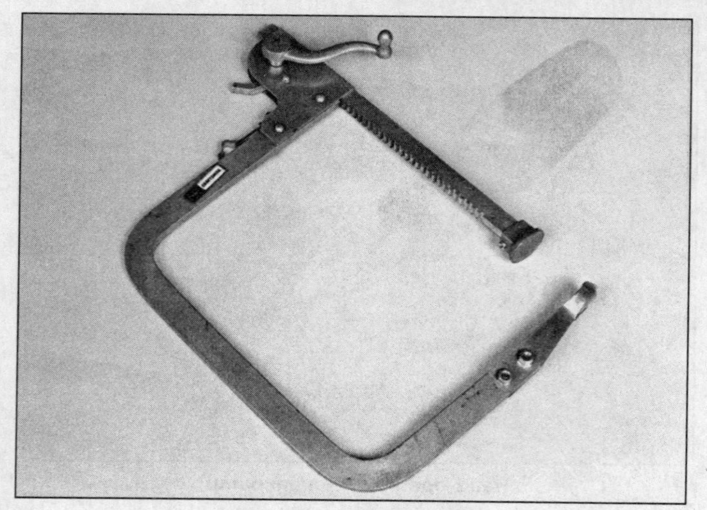

Valve spring compressor

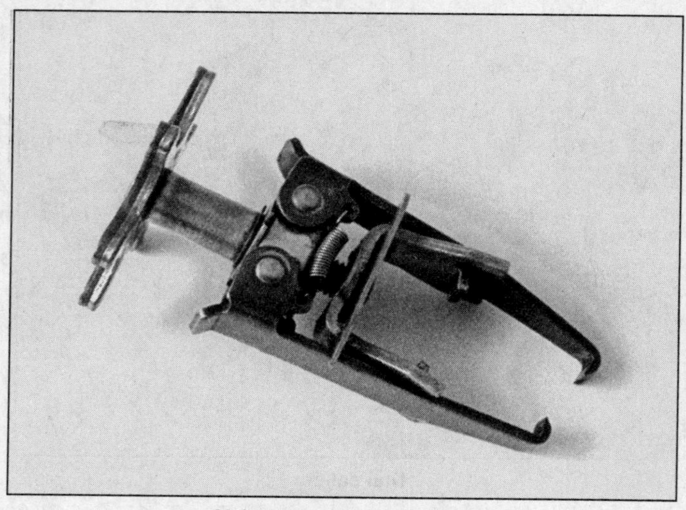

Valve spring compressor

Ridge reamer

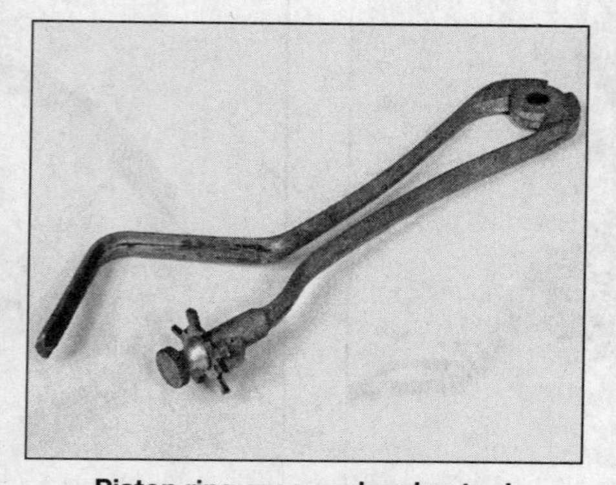

Piston ring groove cleaning tool

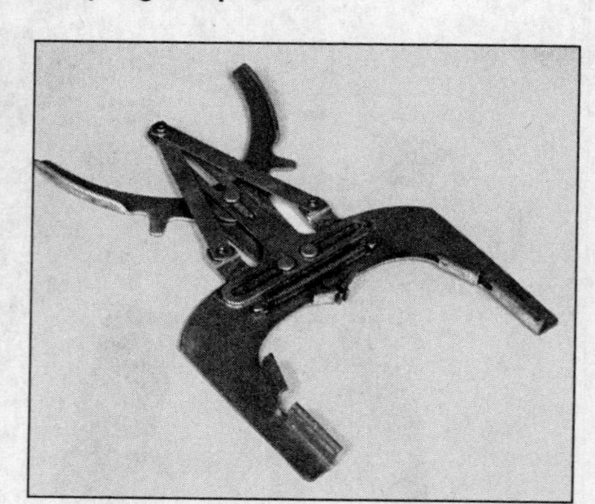

Ring removal/installation tool

end combined in one wrench). While more expensive than open end wrenches, they offer the advantages of both types of wrench.

Combination wrench set (1/4-inch to 1 inch or 6 mm to 19 mm)
Adjustable wrench, 8 inch
Spark plug wrench with rubber insert
Spark plug gap adjusting tool
Feeler gauge set
Brake bleeder wrench
Standard screwdriver (5/16-inch x 6 inch)
Phillips screwdriver (No. 2 x 6 inch)
Combination pliers - 6 inch
Hacksaw and assortment of blades
Tire pressure gauge
Grease gun
Oil can
Fine emery cloth
Wire brush
Battery post and cable cleaning tool
Oil filter wrench
Funnel (medium size)
Safety goggles
Jackstands (2)
Drain pan

Note: *If basic tune-ups are going to be part of routine maintenance, it will be necessary to purchase a good quality stroboscopic timing light and combination tachometer/dwell meter. Although they are included in the list of special tools, it is mentioned here because they are absolutely necessary for tuning most vehicles properly.*

Repair and overhaul tool set

These tools are essential for anyone who plans to perform major repairs and are in addition to those in the maintenance and minor repair tool kit. Included is a comprehensive set of sockets which, though expensive, are invaluable because of their versatility, especially when various extensions and drives are available. We recommend the 1/2-inch drive over the 3/8-inch drive. Although the larger drive is bulky and more expensive, it has the capacity of accepting a very wide range of large sockets. Ideally, however, the mechanic should have a 3/8-inch drive set and a 1/2-inch drive set.

Socket set(s)
Reversible ratchet
Extension - 10 inch
Universal joint
Torque wrench (same size drive as sockets)
Ball peen hammer - 8 ounce
Soft-face hammer (plastic/rubber)

Ring compressor

Standard screwdriver (1/4-inch x 6 inch)
Standard screwdriver (stubby - 5/16-inch)
Phillips screwdriver (No. 3 x 8 inch)
Phillips screwdriver (stubby - No. 2)
Pliers - vise grip
Pliers - lineman's
Pliers - needle nose
Pliers - snap-ring (internal and external)
Cold chisel - 1/2-inch

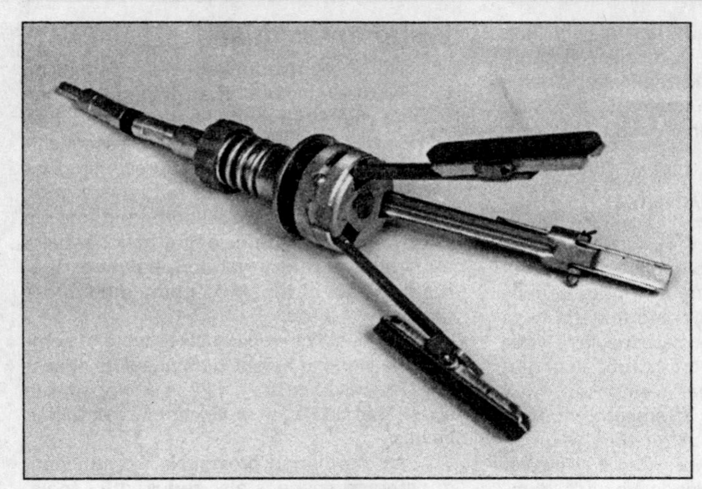
Cylinder hone

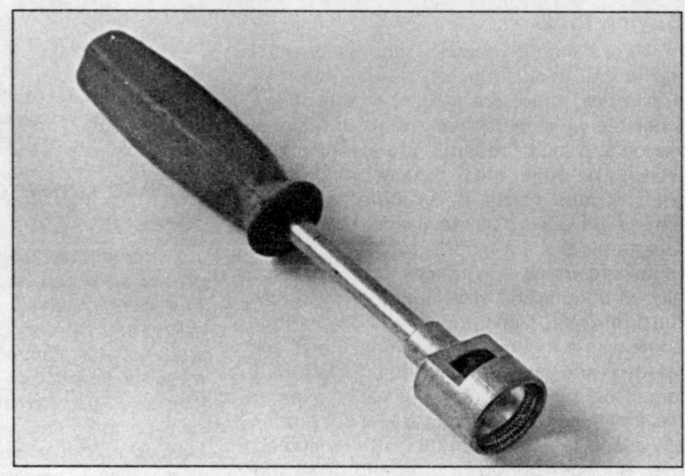
Brake hold-down spring tool

Scribe
Scraper (made from flattened copper tubing)
Centerpunch
Pin punches (1/16, 1/8, 3/16-inch)
Steel rule/straightedge - 12 inch
Allen wrench set (1/8 to 3/8-inch or 4 mm to 10 mm)
A selection of files

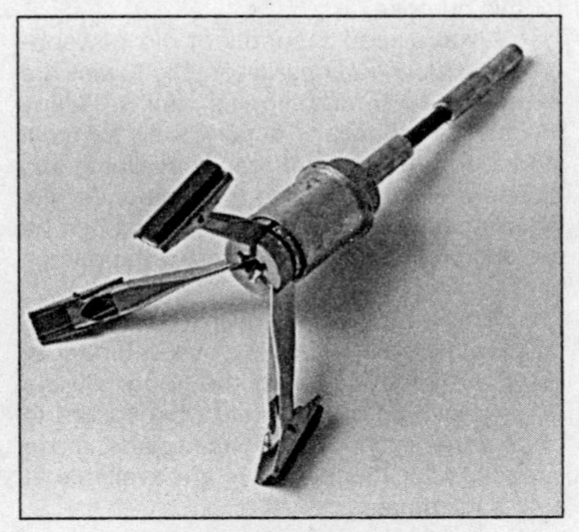
Brake cylinder hone

Wire brush (large)
Jackstands (second set)
Jack (scissor or hydraulic type)

Note: *Another tool which is often useful is an electric drill with a chuck capacity of 3/8-inch and a set of good quality drill bits.*

Special tools

The tools in this list include those which are not used regularly, are expensive to buy, or which need to be used in accordance with their manufacturer's instructions. Unless these tools will be used frequently, it is not very economical to purchase many of them. A consideration would be to split the cost and use between yourself and a friend or friends. In addition, most of these tools can be obtained from a tool rental shop on a temporary basis.

This list primarily contains only those tools and instruments widely available to the public, and not those special tools produced by the vehicle manufacturer for distribution to dealer service departments. Occasionally, references to the manufacturer's special tools are included in the text of this manual. Generally, an alternative method of doing the job without the special tool is offered. However, sometimes there is no alternative to their use. Where this is the case, and the tool cannot be purchased or borrowed, the work should be turned over to the dealer service department or an automotive repair shop.

Valve spring compressor
Piston ring groove cleaning tool
Piston ring compressor
Piston ring installation tool
Cylinder compression gauge
Cylinder ridge reamer
Cylinder surfacing hone
Cylinder bore gauge
Micrometers and/or dial calipers
Hydraulic lifter removal tool
Balljoint separator
Universal-type puller
Impact screwdriver
Dial indicator set
Stroboscopic timing light (inductive pick-up)
Hand operated vacuum/pressure pump
Tachometer/dwell meter
Universal electrical multimeter
Cable hoist
Brake spring removal and installation tools
Floor jack

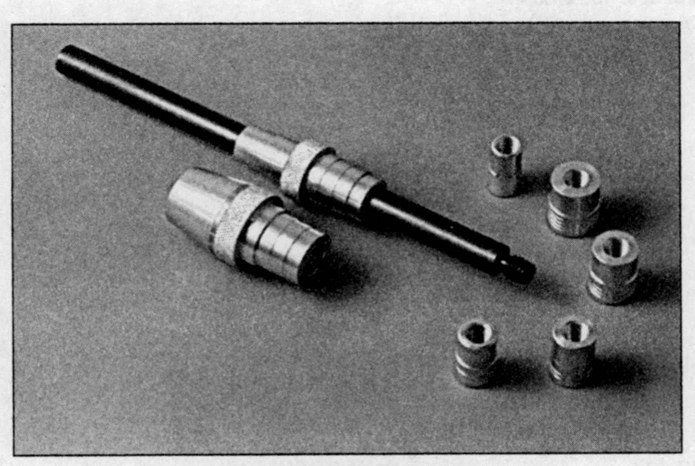
Clutch plate alignment tool

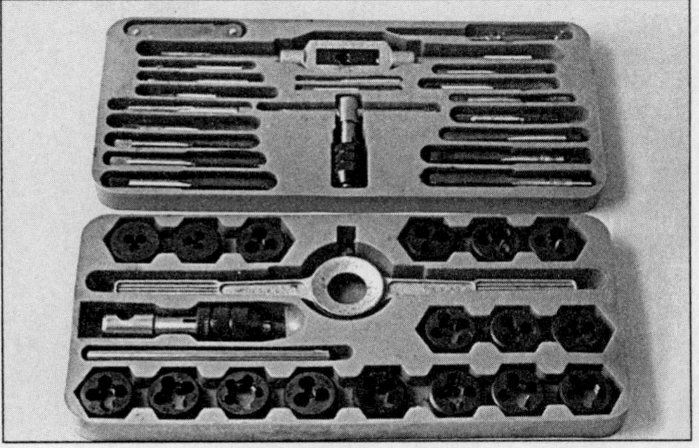
Tap and die set

Buying tools

For the do-it-yourselfer who is just starting to get involved in vehicle maintenance and repair, there are a number of options available when purchasing tools. If maintenance and minor repair is the extent of the work to be done, the purchase of individual tools is satisfactory. If, on the other hand, extensive work is planned, it would be a good idea to purchase a modest tool set from one of the large retail chain stores. A set can usually be bought at a substantial savings over the individual tool prices, and they often come with a tool box. As additional tools are needed, add-on sets, individual tools and a larger tool box can be purchased to expand the tool selection. Building a tool set gradually allows the cost of the tools to be spread over a longer period of time and gives the mechanic the freedom to choose only those tools that will actually be used.

Tool stores will often be the only source of some of the special tools that are needed, but regardless of where tools are bought, try to avoid cheap ones, especially when buying screwdrivers and sockets, because they won't last very long. The expense involved in replacing cheap tools will eventually be greater than the initial cost of quality tools.

Care and maintenance of tools

Good tools are expensive, so it makes sense to treat them with respect. Keep them clean and in usable condition and store them properly when not in use. Always wipe off any dirt, grease or metal chips before putting them away. Never leave tools lying around in the work area. Upon completion of a job, always check closely under the hood for tools that may have been left there so they won't get lost during a test drive.

Some tools, such as screwdrivers, pliers, wrenches and sockets, can be hung on a panel mounted on the garage or workshop wall, while others should be kept in a tool box or tray. Measuring instruments, gauges, meters, etc. must be carefully stored where they cannot be damaged by weather or impact from other tools.

When tools are used with care and stored properly, they will last a very long time. Even with the best of care, though, tools will wear out if used frequently. When a tool is damaged or worn out, replace it. Subsequent jobs will be safer and more enjoyable if you do.

How to repair damaged threads

Sometimes, the internal threads of a nut or bolt hole can become stripped, usually from overtightening. Stripping threads is an all-too-common occurrence, especially when working with aluminum parts, because aluminum is so soft that it easily strips out.

Usually, external or internal threads are only partially stripped. After they've been cleaned up with a tap or die, they'll still work. Sometimes, however, threads are badly damaged. When this happens, you've got three choices:

1) *Drill and tap the hole to the next suitable oversize and install a larger diameter bolt, screw or stud.*
2) *Drill and tap the hole to accept a threaded plug, then drill and tap the plug to the original screw size. You can also buy a plug already threaded to the original size. Then you simply drill a hole to the specified size, then run the threaded plug into the hole with a bolt and jam nut. Once the plug is fully seated, remove the jam nut and bolt.*
3) *The third method uses a patented thread repair kit like Heli-Coil or Slimsert. These easy-to-use kits are designed to repair damaged threads in straight-through holes and blind holes. Both are available as kits which can handle a variety of sizes and thread patterns. Drill the hole, then tap it with the special included tap. Install the Heli-Coil and the hole is back to its original diameter and thread pitch.*

Regardless of which method you use, be sure to proceed calmly and carefully. A little impatience or carelessness during one of these relatively simple procedures can ruin your whole day's work and cost you a bundle if you wreck an expensive part.

Working facilities

Not to be overlooked when discussing tools is the workshop. If anything more than routine maintenance is to be carried out, some sort of suitable work area is essential.

It is understood, and appreciated, that many home mechanics do not have a good workshop or garage available, and end up removing an engine or doing major repairs outside. It is recommended, however, that the overhaul or repair be completed under the cover of a roof.

A clean, flat workbench or table of comfortable working height is an absolute necessity. The workbench should be equipped with a vise that has a jaw opening of at least four inches.

As mentioned previously, some clean, dry storage space is also required for tools, as well as the lubricants, fluids, cleaning solvents, etc. which soon become necessary.

Sometimes waste oil and fluids, drained from the engine or cooling system during normal maintenance or repairs, present a disposal problem. To avoid pouring them on the ground or into a sewage system, pour the used fluids into large containers, seal them with caps and take them to an authorized disposal site or recycling center. Plastic jugs, such as old antifreeze containers, are ideal for this purpose.

Always keep a supply of old newspapers and clean rags available. Old towels are excellent for mopping up spills. Many mechanics use rolls of paper towels for most work because they are readily available and disposable. To help keep the area under the vehicle clean, a large cardboard box can be cut open and flattened to protect the garage or shop floor.

Whenever working over a painted surface, such as when leaning over a fender to service something under the hood, always cover it with an old blanket or bedspread to protect the finish. Vinyl covered pads, made especially for this purpose, are available at auto parts stores.

Booster battery (jump) starting

Certain precautions must be observed when using a booster battery to jump start a vehicle.

a) Before connecting the booster battery, make sure the ignition switch is in the Off position.
b) Turn off the lights, heater and other electrical accessories.
c) The eyes should be shielded. Safety goggles are a good idea.
d) Make sure the booster battery is the same voltage as the dead one in the vehicle.
e) The two vehicles must not touch each other.
f) Make sure the transmission is in Neutral (manual transmission) or Park (automatic transmission).
g) If the booster battery is not a maintenance-free type, remove the vent caps and lay a cloth over the vent holes.

Connect the red jumper cable to the positive (+) terminals of each battery.

Connect one end of the black jumper cable to the negative (-) terminal of the booster battery. The other end of this cable should be connected to a good ground on the vehicle to be started, such as a bolt or bracket on the engine block. Use caution to insure that the cable will not come into contact with the fan, drivebelts or other moving parts of the engine.

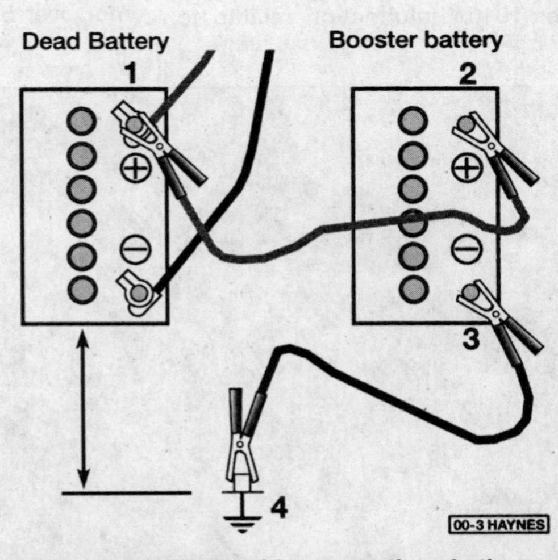

Make the booster battery cable connections in the numerical order shown (note that the negative cable of the booster battery is NOT attached to the negative terminal of the dead battery)

Jacking and towing

Jacking

The jack supplied with the vehicle should only be used for raising the vehicle when changing a tire or placing jackstands under the frame. Warning: Never work under the vehicle or start the engine while this jack is being used as the only means of support.

The vehicle should be on level ground with the wheels blocked and the transmission in Park (automatic) or Reverse (manual). If the wheel is being replaced, loosen the wheel nuts one-half turn and leave them in place until the wheel is raised off the ground. Refer to Chapter 10 for information related to removing and installing the tire.

Place the jack under the vehicle suspension in the indicated position (see illustration). Operate the jack with a slow, smooth motion until the wheel is raised off the ground.

Lower the vehicle, remove the jack and tighten the nuts (if loosened or removed) in a criss-cross sequence.

Towing

Vehicles can be towed with all four wheels on the ground, provided that speeds do not exceed 35 mph and the distance is not over 50 miles, otherwise transmission damage can result.

Towing equipment specifically designed for this purpose should be used and should be attached to the main structural members of the vehicle, not the bumper or brackets.

Safety is a major consideration when towing and all applicable state and local laws must be obeyed. A safety chain system must be used for all towing.

While towing, the parking brake should be released and the transmission must be in Neutral. The steering must be unlocked (ignition switch in the Off position). Remember that power steering and power brakes will not work with the engine off.

Typical front jacking location

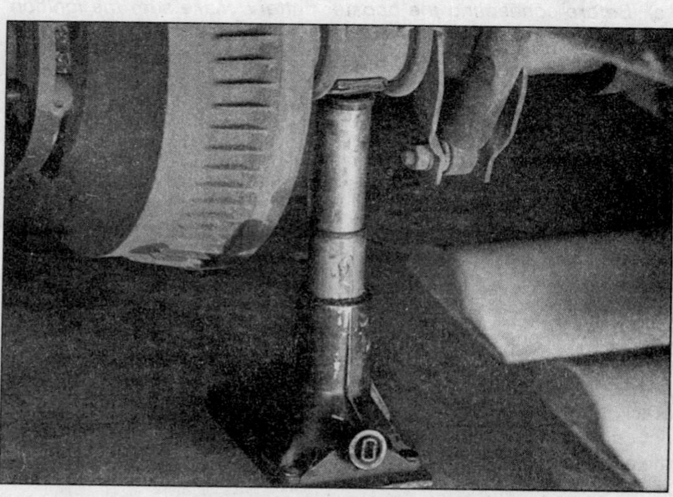

Typical rear jacking location

Automotive chemicals and lubricants

A number of automotive chemicals and lubricants are available for use during vehicle maintenance and repair. They include a wide variety of products ranging from cleaning solvents and degreasers to lubricants and protective sprays for rubber, plastic and vinyl.

Cleaners

Carburetor cleaner and choke cleaner is a strong solvent for gum, varnish and carbon. Most carburetor cleaners leave a dry-type lubricant film which will not harden or gum up. Because of this film it is not recommended for use on electrical components.

Brake system cleaner is used to remove grease and brake fluid from the brake system, where clean surfaces are absolutely necessary. It leaves no residue and often eliminates brake squeal caused by contaminants.

Electrical cleaner removes oxidation, corrosion and carbon deposits from electrical contacts, restoring full current flow. It can also be used to clean spark plugs, carburetor jets, voltage regulators and other parts where an oil-free surface is desired.

Demoisturants remove water and moisture from electrical components such as alternators, voltage regulators, electrical connectors and fuse blocks. They are non-conductive, non-corrosive and non-flammable.

Degreasers are heavy-duty solvents used to remove grease from the outside of the engine and from chassis components. They can be sprayed or brushed on and, depending on the type, are rinsed off either with water or solvent.

Lubricants

Motor oil is the lubricant formulated for use in engines. It normally contains a wide variety of additives to prevent corrosion and reduce foaming and wear. Motor oil comes in various weights (viscosity ratings) from 0 to 50. The recommended weight of the oil depends on the season, temperature and the demands on the engine. Light oil is used in cold climates and under light load conditions. Heavy oil is used in hot climates and where high loads are encountered. Multi-viscosity oils are designed to have characteristics of both light and heavy oils and are available in a number of weights from 5W-20 to 20W-50.

Gear oil is designed to be used in differentials, manual transmissions and other areas where high-temperature lubrication is required.

Chassis and wheel bearing grease is a heavy grease used where increased loads and friction are encountered, such as for wheel bearings, balljoints, tie-rod ends and universal joints.

High-temperature wheel bearing grease is designed to withstand the extreme temperatures encountered by wheel bearings in disc brake equipped vehicles. It usually contains molybdenum disulfide (moly), which is a dry-type lubricant.

White grease is a heavy grease for metal-to-metal applications where water is a problem. White grease stays soft under both low and high temperatures (usually from -100 to +190-degrees F), and will not wash off or dilute in the presence of water.

Assembly lube is a special extreme pressure lubricant, usually containing moly, used to lubricate high-load parts (such as main and rod bearings and cam lobes) for initial start-up of a new engine. The assembly lube lubricates the parts without being squeezed out or washed away until the engine oiling system begins to function.

Silicone lubricants are used to protect rubber, plastic, vinyl and nylon parts.

Graphite lubricants are used where oils cannot be used due to contamination problems, such as in locks. The dry graphite will lubricate metal parts while remaining uncontaminated by dirt, water, oil or acids. It is electrically conductive and will not foul electrical contacts in locks such as the ignition switch.

Moly penetrants loosen and lubricate frozen, rusted and corroded fasteners and prevent future rusting or freezing.

Heat-sink grease is a special electrically non-conductive grease that is used for mounting electronic ignition modules where it is essential that heat is transferred away from the module.

Sealants

RTV sealant is one of the most widely used gasket compounds. Made from silicone, RTV is air curing, it seals, bonds, waterproofs, fills surface irregularities, remains flexible, doesn't shrink, is relatively easy to remove, and is used as a supplementary sealer with almost all low and medium temperature gaskets.

Anaerobic sealant is much like RTV in that it can be used either to seal gaskets or to form gaskets by itself. It remains flexible, is solvent resistant and fills surface imperfections. The difference between an anaerobic sealant and an RTV-type sealant is in the curing. RTV cures when exposed to air, while an anaerobic sealant cures only in the absence of air. This means that an anaerobic sealant cures only after the assembly of parts, sealing them together.

Thread and pipe sealant is used for sealing hydraulic and pneumatic fittings and vacuum lines. It is usually made from a Teflon compound, and comes in a spray, a paint-on liquid and as a wrap-around tape.

Chemicals

Anti-seize compound prevents seizing, galling, cold welding, rust and corrosion in fasteners. High-temperature ant-seize, usually made with copper and graphite lubricants, is used for exhaust system and exhaust manifold bolts.

Anaerobic locking compounds are used to keep fasteners from vibrating or working loose and cure only after installation, in the absence of air. Medium strength locking compound is used for small nuts, bolts and screws that may be removed later. High-strength locking compound is for large nuts, bolts and studs which aren't removed on a regular basis.

Oil additives range from viscosity index improvers to chemical treatments that claim to reduce internal engine friction. It should be noted that most oil manufacturers caution against using additives with their oils.

Gas additives perform several functions, depending on their chemical makeup. They usually contain solvents that help dissolve gum and varnish that build up on carburetor, fuel injection and intake parts. They also serve to break down carbon deposits that form on the inside surfaces of the combustion chambers. Some additives contain upper cylinder lubricants for valves and piston rings, and others contain chemicals to remove condensation from the gas tank.

Miscellaneous

Brake fluid is specially formulated hydraulic fluid that can withstand the heat and pressure encountered in brake systems. Care must be taken so this fluid does not come in contact with painted surfaces or plastics. An opened container should always be resealed to prevent contamination by water or dirt.

Weatherstrip adhesive is used to bond weatherstripping around doors, windows and trunk lids. It is sometimes used to attach trim pieces.

Undercoating is a petroleum-based, tar-like substance that is designed to protect metal surfaces on the underside of the vehicle from corrosion. It also acts as a sound-deadening agent by insulating the bottom of the vehicle.

Waxes and polishes are used to help protect painted and plated surfaces from the weather. Different types of paint may require the use of different types of wax and polish. Some polishes utilize a chemical or abrasive cleaner to help remove the top layer of oxidized (dull) paint on older vehicles. In recent years many non-wax polishes that contain a wide variety of chemicals such as polymers and silicones have been introduced. These non-wax polishes are usually easier to apply and last longer than conventional waxes and polishes.

Conversion factors

Length (distance)

Inches (in)	X 25.4	= Millimetres (mm)	X 0.0394	= Inches (in)
Feet (ft)	X 0.305	= Metres (m)	X 3.281	= Feet (ft)
Miles	X 1.609	= Kilometres (km)	X 0.621	= Miles

Volume (capacity)

Cubic inches (cu in; in^3)	X 16.387	= Cubic centimetres (cc; cm^3)	X 0.061	= Cubic inches (cu in; in^3)
Imperial pints (Imp pt)	X 0.568	= Litres (l)	X 1.76	= Imperial pints (Imp pt)
Imperial quarts (Imp qt)	X 1.137	= Litres (l)	X 0.88	= Imperial quarts (Imp qt)
Imperial quarts (Imp qt)	X 1.201	= US quarts (US qt)	X 0.833	= Imperial quarts (Imp qt)
US quarts (US qt)	X 0.946	= Litres (l)	X 1.057	= US quarts (US qt)
Imperial gallons (Imp gal)	X 4.546	= Litres (l)	X 0.22	= Imperial gallons (Imp gal)
Imperial gallons (Imp gal)	X 1.201	= US gallons (US gal)	X 0.833	= Imperial gallons (Imp gal)
US gallons (US gal)	X 3.785	= Litres (l)	X 0.264	= US gallons (US gal)

Mass (weight)

Ounces (oz)	X 28.35	= Grams (g)	X 0.035	= Ounces (oz)
Pounds (lb)	X 0.454	= Kilograms (kg)	X 2.205	= Pounds (lb)

Force

Ounces-force (ozf; oz)	X 0.278	= Newtons (N)	X 3.6	= Ounces-force (ozf; oz)
Pounds-force (lbf; lb)	X 4.448	= Newtons (N)	X 0.225	= Pounds-force (lbf; lb)
Newtons (N)	X 0.1	= Kilograms-force (kgf; kg)	X 9.81	= Newtons (N)

Pressure

Pounds-force per square inch (psi; lbf/in^2; lb/in^2)	X 0.070	= Kilograms-force per square centimetre (kgf/cm^2; kg/cm^2)	X 14.223	= Pounds-force per square inch (psi; lbf/in^2; lb/in^2)
Pounds-force per square inch (psi; lbf/in^2; lb/in^2)	X 0.068	= Atmospheres (atm)	X 14.696	= Pounds-force per square inch (psi; lbf/in^2; lb/in^2)
Pounds-force per square inch (psi; lbf/in^2; lb/in^2)	X 0.069	= Bars	X 14.5	= Pounds-force per square inch (psi; lbf/in^2; lb/in^2)
Pounds-force per square inch (psi; lbf/in^2; lb/in^2)	X 6.895	= Kilopascals (kPa)	X 0.145	= Pounds-force per square inch (psi; lbf/in^2; lb/in^2)
Kilopascals (kPa)	X 0.01	= Kilograms-force per square centimetre (kgf/cm^2; kg/cm^2)	X 98.1	= Kilopascals (kPa)

Torque (moment of force)

Pounds-force inches (lbf in; lb in)	X 1.152	= Kilograms-force centimetre (kgf cm; kg cm)	X 0.868	= Pounds-force inches (lbf in; lb in)
Pounds-force inches (lbf in; lb in)	X 0.113	= Newton metres (Nm)	X 8.85	= Pounds-force inches (lbf in; lb in)
Pounds-force inches (lbf in; lb in)	X 0.083	= Pounds-force feet (lbf ft; lb ft)	X 12	= Pounds-force inches (lbf in; lb in)
Pounds-force feet (lbf ft; lb ft)	X 0.138	= Kilograms-force metres (kgf m; kg m)	X 7.233	= Pounds-force feet (lbf ft; lb ft)
Pounds-force feet (lbf ft; lb ft)	X 1.356	= Newton metres (Nm)	X 0.738	= Pounds-force feet (lbf ft; lb ft)
Newton metres (Nm)	X 0.102	= Kilograms-force metres (kgf m; kg m)	X 9.804	= Newton metres (Nm)

Vacuum

Inches mercury (in. Hg)	X 3.377	= Kilopascals (kPa)	X 0.2961	= Inches mercury
Inches mercury (in. Hg)	X 25.4	= Millimeters mercury (mm Hg)	X 0.0394	= Inches mercury

Power

Horsepower (hp)	X 745.7	= Watts (W)	X 0.0013	= Horsepower (hp)

Velocity (speed)

Miles per hour (miles/hr; mph)	X 1.609	= Kilometres per hour (km/hr; kph)	X 0.621	= Miles per hour (miles/hr; mph)

*Fuel consumption**

Miles per gallon, Imperial (mpg)	X 0.354	= Kilometres per litre (km/l)	X 2.825	= Miles per gallon, Imperial (mpg)
Miles per gallon, US (mpg)	X 0.425	= Kilometres per litre (km/l)	X 2.352	= Miles per gallon, US (mpg)

Temperature

Degrees Fahrenheit = (°C x 1.8) + 32

Degrees Celsius (Degrees Centigrade; °C) = (°F - 32) x 0.56

****It is common practice to convert from miles per gallon (mpg) to litres/100 kilometres (l/100km), where mpg (Imperial) x l/100 km = 282 and mpg (US) x l/100 km = 235***

Safety first!

Regardless of how enthusiastic you may be about getting on with the job at hand, take the time to ensure that your safety is not jeopardized. A moment's lack of attention can result in an accident, as can failure to observe certain simple safety precautions. The possibility of an accident will always exist, and the following points should not be considered a comprehensive list of all dangers. Rather, they are intended to make you aware of the risks and to encourage a safety conscious approach to all work you carry out on your vehicle.

Essential DOs and DON'Ts

DON'T rely on a jack when working under the vehicle. Always use approved jackstands to support the weight of the vehicle and place them under the recommended lift or support points.

DON'T attempt to loosen extremely tight fasteners (i.e. wheel lug nuts) while the vehicle is on a jack - it may fall.

DON'T start the engine without first making sure that the transmission is in Neutral (or Park where applicable) and the parking brake is set.

DON'T remove the radiator cap from a hot cooling system - let it cool or cover it with a cloth and release the pressure gradually.

DON'T attempt to drain the engine oil until you are sure it has cooled to the point that it will not burn you.

DON'T touch any part of the engine or exhaust system until it has cooled sufficiently to avoid burns.

DON'T siphon toxic liquids such as gasoline, antifreeze and brake fluid by mouth, or allow them to remain on your skin.

DON'T inhale brake lining dust - it is potentially hazardous (see *Asbestos* below).

DON'T allow spilled oil or grease to remain on the floor - wipe it up before someone slips on it.

DON'T use loose fitting wrenches or other tools which may slip and cause injury.

DON'T push on wrenches when loosening or tightening nuts or bolts. Always try to pull the wrench toward you. If the situation calls for pushing the wrench away, push with an open hand to avoid scraped knuckles if the wrench should slip.

DON'T attempt to lift a heavy component alone - get someone to help you.

DON'T rush or take unsafe shortcuts to finish a job.

DON'T allow children or animals in or around the vehicle while you are working on it.

DO wear eye protection when using power tools such as a drill, sander, bench grinder, etc. and when working under a vehicle.

DO keep loose clothing and long hair well out of the way of moving parts.

DO make sure that any hoist used has a safe working load rating adequate for the job.

DO get someone to check on you periodically when working alone on a vehicle.

DO carry out work in a logical sequence and make sure that everything is correctly assembled and tightened.

DO keep chemicals and fluids tightly capped and out of the reach of children and pets.

DO remember that your vehicle's safety affects that of yourself and others. If in doubt on any point, get professional advice.

Asbestos

Certain friction, insulating, sealing, and other products - such as brake linings, brake bands, clutch linings, torque converters, gaskets, etc. - may contain asbestos. Extreme care must be taken to avoid inhalation of dust from such products, since it is hazardous to health. If in doubt, assume that they do contain asbestos.

Fire

Remember at all times that gasoline is highly flammable. Never smoke or have any kind of open flame around when working on a vehicle. But the risk does not end there. A spark caused by an electrical short circuit, by two metal surfaces contacting each other, or even by static electricity built up in your body under certain conditions, can ignite gasoline vapors, which in a confined space are highly explosive. Do not, under any circumstances, use gasoline for cleaning parts. Use an approved safety solvent.

Always disconnect the battery ground (-) cable at the battery before working on any part of the fuel system or electrical system. Never risk spilling fuel on a hot engine or exhaust component. It is strongly recommended that a fire extinguisher suitable for use on fuel and electrical fires be kept handy in the garage or workshop at all times. Never try to extinguish a fuel or electrical fire with water.

Fumes

Certain fumes are highly toxic and can quickly cause unconsciousness and even death if inhaled to any extent. Gasoline vapor falls into this category, as do the vapors from some cleaning solvents. Any draining or pouring of such volatile fluids should be done in a well ventilated area.

When using cleaning fluids and solvents, read the instructions on the container carefully. Never use materials from unmarked containers.

Never run the engine in an enclosed space, such as a garage. Exhaust fumes contain carbon monoxide, which is extremely poisonous. If you need to run the engine, always do so in the open air, or at least have the rear of the vehicle outside the work area.

If you are fortunate enough to have the use of an inspection pit, never drain or pour gasoline and never run the engine while the vehicle is over the pit. The fumes, being heavier than air, will concentrate in the pit with possibly lethal results.

The battery

Never create a spark or allow a bare light bulb near a battery. They normally give off a certain amount of hydrogen gas, which is highly explosive.

Always disconnect the battery ground (-) cable at the battery before working on the fuel or electrical systems.

If possible, loosen the filler caps or cover when charging the battery from an external source (this does not apply to sealed or maintenance-free batteries). Do not charge at an excessive rate or the battery may burst.

Take care when adding water to a non maintenance-free battery and when carrying a battery. The electrolyte, even when diluted, is very corrosive and should not be allowed to contact clothing or skin.

Always wear eye protection when cleaning the battery to prevent the caustic deposits from entering your eyes.

Household current

When using an electric power tool, inspection light, etc., which operates on household current, always make sure that the tool is correctly connected to its plug and that, where necessary, it is properly grounded. Do not use such items in damp conditions and, again, do not create a spark or apply excessive heat in the vicinity of fuel or fuel vapor.

Secondary ignition system voltage

A severe electric shock can result from touching certain parts of the ignition system (such as the spark plug wires) when the engine is running or being cranked, particularly if components are damp or the insulation is defective. In the case of an electronic ignition system, the secondary system voltage is much higher and could prove fatal.

Troubleshooting

Contents

Symptom	Section
Engine	
Engine backfires	13
Engine diesels (continues to run) after switching off	15
Engine hard to start when cold	4
Engine hard to start when hot	5
Engine lacks power	12
Engine lopes while idling or idles erratically	8
Engine misses at idle speed	9
Engine misses throughout driving speed range	10
Engine rotates but will not start	2
Engine stalls	11
Engine starts but stops immediately	7
Engine will not rotate when attempting to start	1
Pinging or knocking engine sounds during acceleration or uphill	14
Starter motor noisy or excessively rough in engagement	6
Starter motor operates without rotating engine	3
Engine electrical system	
Battery will not hold a charge	16
CHECK ENGINE light comes on	19
Ignition light fails to come on when key is turned on	18
Ignition light fails to go out	17
Fuel system	
Excessive fuel consumption	20
Fuel leakage and/or fuel odor	21
Cooling system	
Coolant loss	26
External coolant leakage	24
Internal coolant leakage	25
Overcooling	23
Overheating	22
Poor coolant circulation	27
Clutch	
Clutch slips (engine speed increases with no increase in vehicle speed)	29
Clutch pedal stays on floor when disengaged	33
Fails to release (pedal pressed to the floor - shift lever does not move freely in and out of Reverse)	28
Grabbing (chattering) as clutch is engaged	30
Squeal or rumble with clutch fully disengaged (pedal depressed)	32
Squeal or rumble with clutch fully engaged (pedal released)	31
Manual transmission	
Difficulty in engaging gears	38
Noisy in all gears	35
Noisy in Neutral with engine running	34
Noisy in one particular gear	36
Oil leakage	39
Slips out of high gear	37
Automatic transmission	
Fluid leakage	43
General shift mechanism problems	40
Transmission slips, shifts rough, is noisy or has no drive in forward or reverse gears	42
Transmission will not downshift with accelerator pedal pressed to the floor	41
Transfer case	
Transfer case difficult to shift into the desired range	44
Transfer case noisy in all gears	45
Noisy or jumps out of 4-wheel drive Low range	46
Lubricant leaks from the vent or output shaft seals	47
Driveshaft	
Knock or clunk when the transmission is under initial load (just after transmission is put into gear)	49
Metallic grating sound consistent with vehicle speed	50
Oil leak at front of driveshaft	48
Vibration	51
Axles	
Noise	52
Vibration	53
Oil leakage	54
Brakes	
Brake pedal feels spongy when depressed	58
Brake pedal pulsates during brake application	61
Excessive brake pedal travel	57
Excessive effort required to stop vehicle	59
Noise (high-pitched squeal with the brakes applied)	56
Pedal travels to the floor with little resistance	60
Vehicle pulls to one side during braking	55
Suspension and steering systems	
Excessive pitching and/or rolling around corners or during braking	64
Excessive play in steering	66
Excessive tire wear (not specific to one area)	68
Excessive tire wear on inside edge	70
Excessive tire wear on outside edge	69
Excessively stiff steering	65
Lack of power assistance	67
Shimmy, shake or vibration	63
Tire tread worn in one place	71
Vehicle pulls to one side	62

This section provides an easy reference guide to the more common problems which may occur during the operation of your vehicle. These problems and possible causes are grouped under various components or systems, such as Engine, Cooling system, etc., and also refer to the Chapter and/or Section which deals with the problem.

Remember that successful troubleshooting is not a mysterious black art practiced only by professional mechanics. It's simply the result of a bit of knowledge combined with an intelligent, systematic approach to the problem. Always work by a process of elimination, starting with the simplest solution and working through to the most complex - and never overlook the obvious. Anyone can forget to fill the gas tank or leave the lights on overnight, so don't assume that you are above such oversights.

Finally, always get clear in your mind why a problem has occurred and take steps to ensure that it doesn't happen again. If the electrical system fails because of a poor connection, check all other connections in the system to make sure that they don't fail as well. If a particular fuse continues to blow, find out why - don't just go on replacing fuses. Remember, failure of a small component can often be indicative of potential failure or incorrect functioning of a more important component or system.

Engine

1 Engine will not rotate when attempting to start

1 Battery terminal connections loose or corroded. Check the cable terminals at the battery. Tighten the cable or remove corrosion as necessary.
2 Battery discharged or faulty. If the cable connections are clean and tight on the battery posts, turn the key to the On position and switch on the headlights and/or windshield wipers. If they fail to function, the battery is discharged.
3 Automatic transmission not completely engaged in Park or clutch not completely depressed.
4 Broken, loose or disconnected wiring in the starting circuit. Inspect all wiring and connectors at the battery, starter solenoid and ignition switch.
5 Starter motor pinion jammed in flywheel ring gear. If equipped with a manual transmission, place the transmission in gear and rock the vehicle to manually turn the engine. Remove the starter and inspect the pinion and flywheel at earliest convenience.
6 Starter solenoid faulty (Chapter 5).
7 Starter motor faulty (Chapter 5).
8 Ignition switch faulty (Chapter 12).

2 Engine rotates but will not start

1 Fuel tank empty.
2 Battery discharged (engine rotates slowly). Check the operation of electrical components as described in previous Section.
3 Battery terminal connections loose or corroded. See previous Section.
4 Carburetor flooded and/or fuel level in carburetor incorrect. This will usually be accompanied by a strong fuel odor from under the hood. Wait a few minutes, depress the accelerator pedal all the way to the floor and attempt to start the engine.
5 Choke control inoperative (Chapter 1).
6 Fuel not reaching carburetor or fuel injectors. With the ignition switch in the Off position, open the hood, remove the top plate of the air cleaner assembly and observe the top of the carburetor (manually move the choke plate back if necessary). Have an assistant depress the accelerator pedal and check that fuel spurts into the carburetor. If not, check the fuel filter (Chapter 1), fuel lines and fuel pump (Chapter 4).
7 Fuel injector or fuel pump faulty (fuel injected vehicles) (Chapter 4).
8 No power to fuel pump (Chapter 4).
9 Worn, faulty or incorrectly gapped spark plugs (Chapter 1).
10 Broken, loose or disconnected wiring in the starting circuit (see previous Section).
11 Distributor loose, causing ignition timing to change. Turn the distributor as necessary to start the engine, then set the ignition timing as soon as possible (Chapter 1).
12 Broken, loose or disconnected wires at the ignition coil or faulty coil (Chapter 5).

3 Starter motor operates without rotating engine

1 Starter pinion sticking. Remove the starter (Chapter 5) and inspect.
2 Starter pinion or flywheel teeth worn or broken. Remove the cover at the rear of the engine and inspect.

4 Engine hard to start when cold

1 Battery discharged or low. Check as described in Section 1.
2 Choke control inoperative or out of adjustment (Chapter 4).
3 Carburetor flooded (see Section 2).
4 Fuel supply not reaching the carburetor (see Section 2).
5 Carburetor/fuel injection system in need of overhaul (Chapter 4).
6 Distributor rotor carbon tracked and/or mechanical advance mechanism rusted (Chapter 5).
7 Fuel injection malfunction (Chapter 4).

5 Engine hard to start when hot

1 Air filter clogged (Chapter 1).
2 Fuel not reaching the injectors (see Section 2).
3 Corroded electrical leads at the battery (Chapter 1).
4 Bad engine ground (Chapter 1).
5 Starter worn (Chapter 5).
6 Corroded electrical leads at the fuel injection (Chapter 4).

6 Starter motor noisy or excessively rough in engagement

1 Pinion or flywheel gear teeth worn or broken. Remove the cover at the rear of the engine (if so equipped) and inspect.
2 Starter motor mounting bolts loose or missing.

7 Engine starts but stops immediately

1 Loose or faulty electrical connections at distributor, coil or alternator.
2 Insufficient fuel reaching the carburetor or fuel injector. Disconnect the fuel line. Place a container under the disconnected fuel line and observe the flow of fuel from the line. If little or none at all, check for blockage in the lines and/or replace the fuel pump (Chapter 4).
3 Vacuum leak at the gasket surfaces of the carburetor or fuel injection unit. Make sure that all mounting bolts/nuts are tightened securely and that all vacuum hoses connected to the carburetor or fuel injection unit and manifold are positioned properly and in good condition.

8 Engine lopes while idling or idles erratically

1 Vacuum leakage. Check mounting bolts/nuts at the carburetor/fuel injection unit and intake manifold for tightness. Make sure that all vacuum hoses are connected and in good condition. Use a stethoscope or a length of fuel hose held against your ear to listen for vacuum leaks while the engine is running. A hissing sound will be heard. Check the carburetor/fuel injector and intake manifold gasket surfaces.
2 Leaking EGR valve or plugged PCV valve (see Chapters 1 and 6).
3 Air filter clogged (Chapter 1).
4 Fuel pump not delivering sufficient fuel to the carburetor/fuel injector (see Section 7).
5 Carburetor out of adjustment (Chapter 4).
6 Leaking head gasket. If this is suspected, take the vehicle to a repair shop or

dealer where the engine can be pressure checked.
7 Timing chain and/or gears worn (Chapter 2).
8 Camshaft lobes worn (Chapter 2).

9 Engine misses at idle speed

1 Spark plugs worn or not gapped properly (Chapter 1).
2 Faulty spark plug wires (Chapter 1).
3 Choke not operating properly (Chapter 1).
4 Sticking or faulty emissions system components (Chapter 6).
5 Clogged fuel filter and/or foreign matter in fuel. Remove the fuel filter (Chapter 1) and inspect.
6 Vacuum leaks at the intake manifold or at hose connections. Check as described in Section 8.
7 Incorrect idle speed or idle mixture (Chapter 1).
8 Incorrect ignition timing (Chapter 1).
9 Uneven or low cylinder compression. Check compression as described in Chapter 1.

10 Engine misses throughout driving speed range

1 Fuel filter clogged and/or impurities in the fuel system (Chapter 1).
Also check fuel output at the carburetor/fuel injector (see Section 7).
2 Faulty or incorrectly gapped spark plugs (Chapter 1).
3 Incorrect ignition timing (Chapter 1).
4 Check for cracked distributor cap, disconnected distributor wires and damaged distributor components (Chapter 1).
5 Leaking spark plug wires (Chapter 1).
6 Faulty emissions system components (Chapter 6).
7 Low or uneven cylinder compression pressures. Remove the spark plugs and test the compression with gauge (Chapter 1).
8 Weak or faulty ignition system (Chapter 5).
9 Vacuum leaks at the carburetor/fuel injection unit or vacuum hoses (see Section 8).

11 Engine stalls

1 Idle speed incorrect (Chapter 1).
2 Fuel filter clogged and/or water and impurities in the fuel system (Chapter 1).
3 Choke improperly adjusted or sticking (Chapter 1).
4 Distributor components damp or damaged (Chapter 5).
5 Faulty emissions system components (Chapter 6).
6 Faulty or incorrectly gapped spark plugs (Chapter 1). Also check spark plug wires (Chapter 1).
7 Vacuum leak at the carburetor/fuel injection unit or vacuum hoses. Check as described in Section 8.

12 Engine lacks power

1 Incorrect ignition timing (Chapter 1).
2 Excessive play in distributor shaft. At the same time, check for worn rotor, faulty distributor cap, wires, etc. (Chapters 1 and 5).
3 Faulty or incorrectly gapped spark plugs (Chapter 1).
4 Fuel injection unit not adjusted properly or excessively worn (Chapter 4).
5 Faulty coil (Chapter 5).
6 Brakes binding (Chapter 1).
7 Automatic transmission fluid level incorrect (Chapter 1).
8 Clutch slipping (Chapter 8).
9 Fuel filter clogged and/or impurities in the fuel system (Chapter 1).
10 Emissions control system not functioning properly (Chapter 6).
11 Use of substandard fuel. Fill tank with proper octane fuel.
12 Low or uneven cylinder compression pressures. Test with compression tester, which will detect leaking valves and/or blown head gasket (Chapter 1).

13 Engine backfires

1 Emissions system not functioning properly (Chapter 6).
2 Ignition timing incorrect (Chapter 1).
3 Faulty secondary ignition system (cracked spark plug insulator, faulty plug wires, distributor cap and/or rotor) (Chapters 1 and 5).
4 Carburetor/fuel injection unit in need of adjustment or worn excessively (Chapter 4).
5 Vacuum leak at the fuel injection unit or vacuum hoses. Check as described in Section 8.
6 Valve clearances incorrectly set, and/or valves sticking in guides (Chapter 2).
7 Crossed plug wires (Chapter 1).

14 Pinging or knocking engine sounds during acceleration or uphill

1 Incorrect grade of fuel. Fill tank with fuel of the proper octane rating.
2 Ignition timing incorrect (Chapter 1).
3 Carburetor/fuel injection unit in need of adjustment or overhaul (Chapter 4).
4 Improper spark plugs. Check plug type against Emissions Control Information label located in engine compartment. Also check plugs and wires for damage (Chapter 1).
5 Worn or damaged distributor components (Chapter 5).
6 Faulty emissions system (Chapter 6).
7 Vacuum leak. Check as described in Section 8.
8 Carbon build-up in cylinders (Chapter 2).

15 Engine diesels (continues to run) after switching off

1 Idle speed too high (Chapter 1).
2 Electrical solenoid at side of carburetor not functioning properly (not all models, see Chapter 4).
3 Ignition timing incorrectly adjusted (Chapter 1).
4 Thermo-controlled air cleaner heat valve not opêrating properly (Chapter 6).
5 Excessive engine operating temperature. Probable causes of this are malfunctioning thermostat, clogged radiator, faulty water pump (Chapter 3).
6 Incorrect grade of fuel. Fill tank with fuel of the proper octane rating.

Engine electrical system

16 Battery will not hold a charge

1 Alternator drivebelt defective or not adjusted properly (Chapter 1).
2 Electrolyte level low or battery discharged (Chapter 1).
3 Battery terminals loose or corroded (Chapter 1).
4 Alternator not charging properly (Chapter 5).
5 Loose, broken or faulty wiring in the charging circuit (Chapter 5).
6 Short in the vehicle wiring causing a continual drain on the battery.
7 Battery defective internally.

17 Ignition light fails to go out

1 Fault in alternator or charging circuit (Chapter 5).
2 Alternator drivebelt defective or not properly adjusted (Chapter 1).

18 Ignition light fails to come on when key is turned on

1 Warning light bulb defective (Chapter 12).
2 Alternator faulty (Chapter 5).
3 Fault in the printed circuit, dash wiring or bulb holder (Chapter 12).

19 "Check engine' light comes on

See Chapter 6.

Fuel system

20 Excessive fuel consumption

1 Dirty or clogged air filter element (Chapter 1).
2 Incorrectly set ignition timing (Chapter 1).
3 Choke sticking or improperly adjusted (Chapter 1).
4 Emissions system not functioning properly (not all vehicles, see Chapter 6).
5 Carburetor idle speed and/or mixture not adjusted properly (Chapter 1).
6 Carburetor/fuel injection internal parts excessively worn or damaged (Chapter 4).
7 Low tire pressure or incorrect tire size (Chapter 1).

21 Fuel leakage and/or fuel odor

1 Leak in a fuel feed or vent line (Chapter 4).
2 Tank overfilled. Fill only to automatic shut-off.
3 Emissions system filter clogged (Chapter 1).
4 Vapor leaks from system lines (Chapter 4).
5 Carburetor/fuel injection internal parts excessively worn or out of adjustment (Chapter 4).

Cooling system

22 Overheating

1 Insufficient coolant in system (Chapter 1).
2 Water pump drivebelt defective or not adjusted properly (Chapter 1).
3 Radiator core blocked or radiator grille dirty and restricted (Chapter 3).
4 Thermostat faulty (Chapter 3).
5 Fan blades broken or cracked (Chapter 3).
6 Radiator cap not maintaining proper pressure. Have cap pressure tested by gas station or repair shop.
7 Ignition timing incorrect (Chapter 1).

23 Overcooling

1 Thermostat faulty (Chapter 3).
2 Inaccurate temperature gauge (Chapter 12)

24 External coolant leakage

1 Deteriorated or damaged hoses or loose clamps. Replace hoses and/or tighten clamps at hose connections (Chapter 1).
2 Water pump seals defective. If this is the case, water will drip from the weep hole in the water pump body (Chapter 1).
3 Leakage from radiator core or header tank. This will require the radiator to be professionally repaired (see Chapter 3 for removal procedures).
4 Engine drain plugs or water jacket core plugs leaking (see Chapter 2).

25 Internal coolant leakage

Note: Internal coolant leaks can usually be detected by examining the oil. Check the dipstick and inside of the rocker arm cover for water deposits and an oil consistency like that of a milkshake.
1 Leaking cylinder head gasket. Have the cooling system pressure tested.
2 Cracked cylinder bore or cylinder head. Dismantle engine and inspect (Chapter 2).

26 Coolant loss

1 Too much coolant in system (Chapter 1).
2 Coolant boiling away due to overheating (see Section 16).
3 Internal or external leakage (see Sections 26 and 27).
4 Faulty radiator cap. Have the cap pressure tested.

27 Poor coolant circulation

1 Inoperative water pump. A quick test is to pinch the top radiator hose closed with your hand while the engine is idling, then let it loose. You should feel the surge of coolant if the pump is working properly (Chapter 1).
2 Restriction in cooling system. Drain, flush and refill the system (Chapter 1). If necessary, remove the radiator (Chapter 3) and have it reverse flushed.
3 Water pump drivebelt defective or not adjusted properly (Chapter 1).
4 Thermostat sticking (Chapter 3).

Clutch

28 Fails to release (pedal pressed to the floor - shift lever does not move freely in and out of Reverse)

1 Clutch hydraulic system low or has air in the system and needs to be bled (Chapter 8).
2 Clutch fork off ball stud. Look under the vehicle, on the left side of transmission.
3 Clutch plate warped or damaged (Chapter 8).

29 Clutch slips (engine speed increases with no increase in vehicle speed)

1 Clutch plate oil soaked or lining worn. Remove clutch (Chapter 8) and inspect.
2 Clutch plate not seated. It may take 30 or 40 normal starts for a new one to seat.
3 Pressure plate worn (Chapter 8).

30 Grabbing (chattering) as clutch is engaged

1 Oil on clutch plate lining. Remove (Chapter 8) and inspect. Correct any leakage source.
2 Worn or loose engine or transmission mounts. These units move slightly when clutch is released. Inspect mounts and bolts.
3 Worn splines on clutch plate hub. Remove clutch components (Chapter 8) and inspect.
4 Warped pressure plate or flywheel. Remove clutch components and inspect.

31 Squeal or rumble with clutch fully engaged (pedal released)

1 Improper adjustment; no free play (Chapter 1).
2 Release bearing binding on transmission bearing retainer. Remove clutch components (Chapter 8) and check bearing. Remove any burrs or nicks, clean and relubricate before reinstallation.
3 Weak linkage return spring. Replace the spring.

32 Squeal or rumble with clutch fully disengaged

(pedal depressed)
1 Worn, defective or broken release bearing (Chapter 8).
2 Worn or broken pressure plate springs (or diaphragm fingers) (Chapter 8).
3 Air in hydraulic line (Chapter 8).

33 Clutch pedal stays on floor when disengaged

1 Bind in linkage or release bearing. Inspect linkage or remove clutch components as necessary.
2 Linkage springs being over extended. Adjust linkage for proper free play. Make sure proper pedal stop (bumper) is installed.
3 Clutch hydraulic cylinder faulty or there is air in the system.

Manual transmission

Note: *All the following references are to Chapter 7, unless noted.*

34 Noisy in Neutral with engine running

1 Input shaft bearing worn.
2 Damaged main drive gear bearing.
3 Worn countershaft bearings.
4 Worn or damaged countershaft end play shims.

35 Noisy in all gears

1 Any of the above causes, and/or:
2 Insufficient lubricant (see checking procedures in Chapter 1).

36 Noisy in one particular gear

1 Worn, damaged or chipped gear teeth for that particular gear.
2 Worn or damaged synchronizer for that particular gear.

37 Slips out of high gear

1 Transmission mounting bolts loose.
2 Shift rods not working freely.
3 Damaged mainshaft pilot bearing.
4 Dirt between transmission case and engine or misalignment of transmission.

38 Difficulty in engaging gears

1 Clutch not releasing completely (see clutch adjustment in Chapter 8).
2 Loose, damaged or out-of-adjustment shift linkage. Make a thorough inspection, replacing parts as necessary.
3 Air in hydraulic system (Chapter 8).

39 Oil leakage

1 Excessive amount of lubricant in transmission (see Chapter 1 for correct checking procedures). Drain lubricant as required.
2 Side cover loose or gasket damaged.
3 Rear oil seal or speedometer oil seal in need of replacement.
4 Clutch hydraulic system leaking (Chapter 8).

Automatic transmission

Note: *Due to the complexity of the automatic transmission, it is difficult for the home mechanic to properly diagnose and service this component. For problems other than the following, the vehicle should be taken to a dealer or reputable mechanic.*

40 General shift mechanism problems

1 Chapter 7 deals with checking and adjusting the shift linkage on automatic transmissions. Common problems which may be attributed to poorly adjusted linkage are:

Engine starting in gears other than Park or Neutral.

Indicator on shifter pointing to a gear other than the one actually being used.

Vehicle moves when in Park.

2 Refer to Chapter 7 to adjust the linkage.

41 Transmission will not downshift with accelerator pedal pressed to the floor

Chapter 7 deals with adjusting the detent cable to enable the transmission to downshift properly.

42 Transmission slips, shifts rough, is noisy or has no drive in forward or reverse gears

1 There are many probable causes for the above problems, but the home mechanic should be concerned with only one possibility - fluid level.
2 Before taking the vehicle to a repair shop, check the level and condition of the fluid as described in Chapter 1. Correct fluid level as necessary or change the fluid and filter if needed. If the problem persists, have a professional diagnose the probable cause.

43 Fluid leakage

1 Automatic transmission fluid is a deep red color. Fluid leaks should not be confused with engine oil, which can easily be blown by air flow to the transmission.
2 To pinpoint a leak, first remove all built-up dirt and grime from around the transmission. Degreasing agents and/or steam cleaning will achieve this. With the underside clean,
drive the vehicle at low speeds so air flow will not blow the leak far from its source. Raise the vehicle and determine where the leak is coming from. Common areas of leakage are:

a) ***Pan:*** *Tighten mounting bolts and/or replace pan gasket as necessary* (see Chapters 1 and 7).
b) ***Filler pipe:*** *Replace the rubber seal where pipe enters transmission case.*
c) *Transmission oil lines: Tighten connectors where lines enter transmission case and/or replace lines.*
d) ***Vent pipe:*** *Transmission overfilled and/or water in fluid* (see checking procedures, Chapter 1).
e) ***Speedometer connector:*** *Replace the O-ring where speedometer cable enters transmission case* (Chapter 7).

Transfer case

44 Transfer case difficult to shift into the desired range

1 Speed may be too great to permit engagement. Stop the vehicle and shift into the desired range.
2 Shift linkage loose, bent or binding. Check the linkage for damage or wear and replace or lubricate as necessary (Chapter 7).
3 Insufficient or incorrect grade or lubricant. Drain and refill the transfer case with the specified lubricant (Chapter 1).
4 Worn or damaged internal components. Disassembly and overhaul of the transfer case may be necessary (Chapter 7).

45 Transfer case noisy in all gears

Insufficient or incorrect grade of lubricant. Drain and refill (Chapter 1).

46 Noisy or jumps out of 4-wheel drive Low range

1 Transfer case not fully engaged. Stop the vehicle, shift into Neutral and then engage 4L.
2 Shift linkage loose, worn or binding. Tighten, repair or lubricate linkage as necessary.
3 Shift fork cracked, inserts worn or fork binding on the rail. Disassemble and repair as necessary (Chapter 7).

47 Lubricant leaks from the vent or output shaft seals

1 Transfer case is overfilled. Drain to the proper level (Chapter 1).
2 Vent is clogged or jammed closed. Clear or replace the vent.
3 Output shaft seal incorrectly installed or damaged. Replace the seal and check contact surfaces for nicks and scoring.

Driveshaft

48 Oil leak at front of driveshaft

Defective transmission rear oil seal. See Chapter 7 for replacement procedures. While this is done, check the splined yoke for burrs or a rough condition which may be damaging the seal. the can be removed with crocus cloth or a fine whetstone.

49 Knock or clunk when the transmission is under initial load (just after transmission is put into gear)

1 Loose or disconnected rear suspension components. Check all mounting bolts, nuts and bushings (Chapter 10).
2 Loose driveshaft bolts. Inspect all bolts and nuts and tighten them to the specified torque.
3 Splined driveshaft in need of lubrication (Chapter 1).
4 Worn or damaged universal joint bearings. Check for wear (Chapter 8).

50 Metallic grating sound consistent with vehicle speed

Pronounced wear in the universal joint bearings. Check as described in Chapter 8.

51 Vibration

Note: *Before assuming that the driveshaft is at fault, make sure the tires are perfectly balanced and perform the following test.*
1 Install a tachometer inside the vehicle to monitor engine speed as the vehicle is driven. Drive the vehicle and note the engine speed at which the vibration (roughness) is most pronounced. Now shift the transmission to a different gear and bring the engine speed to the same point.
2 If the vibration occurs at the same engine speed (rpm) regardless of which gear the transmission is in, the driveshaft is NOT at fault since the driveshaft speed varies.
3 If the vibration decreases or is eliminated when the transmission is in a different gear at the same engine speed, refer to the following probable causes.
4 Bent or dented driveshaft. Inspect and replace as necessary (Chapter 8).
5 Undercoating or built-up dirt, etc. on the driveshaft. Clean the shaft thoroughly and recheck.
6 Worn universal joint bearings. Remove and inspect (Chapter 8).
7 Driveshaft and/or companion flange out of balance. Check for missing weights on the shaft. Remove the driveshaft (Chapter 8) and reinstall 180-degrees from original position, then retest. Have the driveshaft professionally balanced if the problem persists.

Axles

52 Noise

1 Road noise. No corrective procedures available.
2 Tire noise. Inspect the tires and check tire pressures (Chapter 1).
3 Rear wheel bearings loose, worn or damaged (Chapter 10).

53 Vibration

See probable causes under Driveshaft. Proceed under the guidelines listed for the driveshaft. If the problem persists, check the rear wheel bearings by raising the rear of the vehicle and spinning the wheels by hand. Listen for evidence of rough (noisy) bearings. Remove and inspect (Chapter 8).

54 Oil leakage

1 Pinion seal damaged (Chapter 8).
2 Axleshaft oil seals damaged (Chapter 8).
3 Differential inspection cover leaking. Tighten the bolts or replace the gasket as required (Chapter 8).

Brakes

Note: *Before assuming that a brake problem exists, make sure that the tires are in good condition and inflated properly (see Chapter 1), that the front end alignment is correct and that the vehicle is not loaded with weight in an unequal manner.*

55 Vehicle pulls to one side during braking

1 Defective, damaged or oil contaminated brake pads or shoes on one side. Inspect as described in Chapter 9.
2 Excessive wear of brake shoe or pad material or drum/disc on one side. Inspect and correct as necessary.
3 Loose or disconnected front suspension components. Inspect and tighten all bolts to the specified torque (Chapter 10).
4 Defective drum brake or caliper assembly. Remove the drum or caliper and inspect for a stuck piston or other damage (Chapter 9).

56 Noise (high-pitched squeal with the brakes applied)

Disc brake pads worn out. The noise comes from the wear sensor rubbing against the disc (does not apply to all vehicles). Replace the pads with new ones immediately (Chapter 9).

57 Excessive brake pedal travel

1 Partial brake system failure. Inspect the entire system (Chapter 9) and correct as required.
2 Insufficient fluid in the master cylinder. Check (Chapter 1), add fluid and bleed the system if necessary (Chapter 9).
3 Brakes not adjusting properly. Make a series of starts and stops with the vehicle is in Reverse. If this does not correct the situation, remove the drums and inspect the self-adjusters (Chapter 9).

58 Brake pedal feels spongy when depressed

1 Air in the hydraulic lines. Bleed the brake system (Chapter 9).
2 Faulty flexible hoses. Inspect all system hoses and lines. Replace parts as necessary.
3 Master cylinder mounting bolts/nuts loose.
4 Master cylinder defective (Chapter 9).

59 Excessive effort required to stop vehicle

1 Power brake booster not operating properly (Chapter 9).
2 Excessively worn linings or pads. Inspect and replace if necessary (Chapter 9).
3 One or more caliper pistons or wheel cylinders seized or sticking. Inspect and rebuild as required (Chapter 9).
4 Brake linings or pads contaminated with oil or grease. Inspect and replace as required (Chapter 9).
5 New pads or shoes installed and not yet seated. It will take a while for the new material to seat against the drum (or rotor).

60 Pedal travels to the floor with little resistance

Little or no fluid in the master cylinder reservoir caused by leaking wheel cylinder(s), leaking caliper piston(s), loose, damaged or disconnected brake lines. Inspect the entire system and correct as necessary.

61 Brake pedal pulsates during brake application

1 Wheel bearings not adjusted properly or in need of replacement (Chapter 1).
2 Caliper not sliding properly due to improper installation or obstructions. Remove and inspect (Chapter 9).
3 Rotor or drum defective. Remove the rotor or drum (Chapter 9) and check for excessive lateral runout, out of round and parallelism. Have the drum or rotor resurfaced or replace it with a new one.

Suspension and steering systems

62 Vehicle pulls to one side

1 Tire pressures uneven (Chapter 1).

2 Defective tire (Chapter 1).
3 Excessive wear in suspension or steering components (Chapter 10).
4 Front end in need of alignment.
5 Front brakes dragging. Inspect the brakes as described in Chapter 9.

63 Shimmy, shake or vibration

1 Tire or wheel out-of-balance or out-of-round. Have professionally balanced.
2 Loose, worn or out-of-adjustment wheel bearings (Chapters 1 and 8).
3 Shock absorbers and/or suspension components worn or damaged (Chapter 10).

64 Excessive pitching and/or rolling around corners or during braking

1 Defective shock absorbers. Replace as a set (Chapter 10).
2 Broken or weak springs and/or suspension components. Inspect as described in Chapter 10.

65 Excessively stiff steering

1 Lack of fluid in power steering fluid reservoir (Chapter 1).
2 Incorrect tire pressures (Chapter 1).
3 Lack of lubrication at steering joints (Chapter 1).
4 Front end out of alignment.
5 See also section titled Lack of power assistance.

66 Excessive play in steering

1 Loose front wheel bearings (Chapter 1).
2 Excessive wear in suspension or steering components (Chapter 10).
3 Steering gearbox out of adjustment (Chapter 10).

67 Lack of power assistance

1 Steering pump drivebelt faulty or not adjusted properly (Chapter 1).
2 Fluid level low (Chapter 1).
3 Hoses or lines restricted. Inspect and replace parts as necessary.
4 Air in power steering system. Bleed the system (Chapter 10).

68 Excessive tire wear (not specific to one area)

1 Incorrect tire pressures (Chapter 1).
2 Tires out of balance. Have professionally balanced.
3 Wheels damaged. Inspect and replace as necessary.
4 Suspension or steering components excessively worn (Chapter 10).

69 Excessive tire wear on outside edge

1 Inflation pressures incorrect (Chapter 1).
2 Excessive speed in turns.
3 Front end alignment incorrect (excessive toe-in). Have professionally aligned.
4 Suspension arm bent or twisted (Chapter 10).

70 Excessive tire wear on inside edge

1 Inflation pressures incorrect (Chapter 1).
2 Front end alignment incorrect (toe-out). Have professionally aligned.
3 Loose or damaged steering components (Chapter 10).

71 Tire tread worn in one place

1 Tires out of balance.
2 Damaged or buckled wheel. Inspect and replace if necessary.
3 Defective tire (Chapter 1).

Chapter 1
Tune-up and routine maintenance

Contents

	Section
Air filter and PCV filter replacement	24
Automatic transmission fluid and filter change	38
Automatic transmission fluid level check	6
Battery check and maintenance	8
Brake check	22
Carburetor choke check	26
Carburetor/throttle body mounting nut torque check	27
Chassis lubrication	13
CHECK ENGINE light	See Chapter 6
Clutch pedal free play check and adjustment	17
Cooling system check	9
Cooling system servicing (draining, flushing and refilling)	40
Differential oil change	37
Differential oil level check	20
Distributor cap and rotor check and replacement	46
Drivebelt check and adjustment	31
EFE (heat riser) system check	16
Engine oil and filter change	12
Evaporative emissions control system check	42
Exhaust Gas Recirculation (EGR) system check	43
Exhaust system check	15
Fluid level checks	4
Front wheel bearing (2-wheel drive) check, repack and adjustment	39
Fuel filter replacement	25
Fuel system check	23
Idle speed check and adjustment	30
Ignition timing check and adjustment	48
Ignition point replacement	47
Introduction	1
Maintenance schedule	2
Manual transmission oil change	36
Manual transmission oil level check	18
Positive Crankcase Ventilation (PCV) valve check and replacement	41
Power steering fluid level check	7
Seat back latch check	33
Seat belt check	32
Spark plug replacement	44
Spark plug wire check and replacement	45
Starter safety switch check	34
Suspension and steering check	14
Thermostatic air cleaner check	29
Throttle linkage inspection	28
Tire and tire pressure checks	5
Tire rotation	21
Transfer case oil change	35
Transfer case oil level check	19
Tune-up general information	3
Underhood hose check and replacement	10
Wiper blade inspection and replacement	11

Specifications

Recommended lubricants and fluids

Note: *Listed here are the manufacturer recommendations at the time this manual was printed. Manufacturers occasionally upgrade their fluid and lubricant specifications, so check with your local auto parts store for the most current fluid and lubricant recommendations.*

Engine oil type	SH, SH/CC or SH/CD
Engine oil viscosity	See accompanying chart

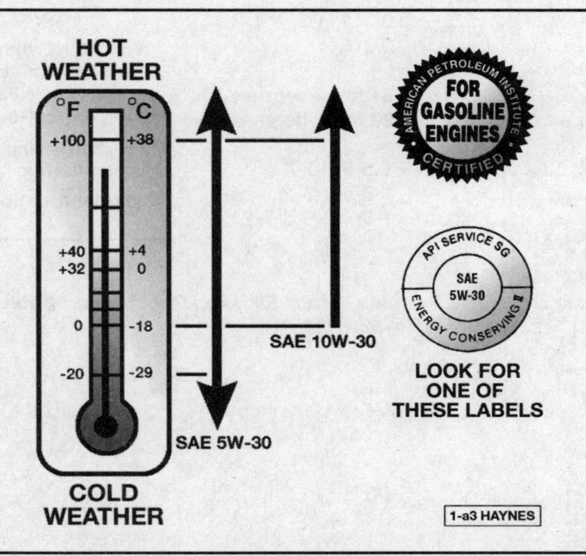

Engine oil viscosity chart

Recommended lubricants and fluids

Automatic transmission	Dexron type ATF
Manual transmission	
all except 4-speed overdrive	SAE 80W GL-5 gear lubricant
4-speed overdrive	Dexron type ATF
Transfer case	
1967 through 1980	
conventional	SAE 80W or SAE 80W-90 GL-5 gear lubricant
full time four wheel drive	SAE 10W-30 or 10W-40 SE or SF engine oil
1981 through 1987	Dexron type ATF
Differential	
US	SAE 80W or SAE 80W-90 GL-5 gear lubricant
Canada	SAE 80W GL-5 gear lubricant
limited slip (all)	Add GM limited-slip additive to the specified lubricant
Chassis grease fittings	GM lubricant 6031 or equivalent NLGI No. 2 chassis grease
Driveshaft splines	GM lubricant 6031 or equivalent NLGI No. 2 chassis grease
Four wheel drive front driveshaft CV joint	GM lubricant 1052497 or equivalent
Engine coolant	Mixture of water and ethylene glycol base antifreeze
Brake fluid	Delco Supreme II or DOT-3 fluid
Clutch fluid	Delco Supreme II or DOT-3 fluid
Power steering fluid	GM power steering fluid or equivalent
Manual steering box lubricant	GM lubricant 4673M or equivalent
Hydro boost fluid	GM power steering fluid or equivalent
Wheel bearing grease	GM lubricant 1051344 or NLGI No. 2 moly-base wheel bearing grease

Capacities

Engine oil (with filter change, approximate)	
Inline 6-cylinder	
230 cu in	5 qts
250 cu in	5 qts
292 cu in	6 qts
V6	5 qts
Small-block V8	5 qts
Big-block V8	6 qts
Cooling system (approximate)	
Inline 6-cylinder	15 qts
V6	11 qts
Small-block V8	16 qts
Big-block V8	24 qts

Ignition system

Ignition timing	Refer to *Vehicle Emission Control Information* label in engine compartment
Spark plug type and gap	See Chapter 5
Ignition point gap	
new	0.019 in
used	0.016 in
Dwell angle	
6-cylinder engine	31 to 34-degrees
V8 engine	29 to 31-degrees
Firing order	
Inline 6-cylinder engine	1-5-3-6-2-4
V6 engine	1-6-5-4-3-2
V8 engine	1-8-4-3-6-5-7-2

0784H
INLINE 6-CYLINDER
Firing order 1-5-3-6-2-4

0785H-X
V6 ENGINE
Firing order 1-6-5-4-3-2

0786H
V8 ENGINE with points-type ignition

0787H
V8 ENGINE with electronic (HEI) ignition

Firing order 1-8-4-3-6-5-7-2

The blackened terminal shown on the distributor cap indicates the Number One spark plug wire position

Cylinder location and distributor rotation

General

Engine idle speed	See *Vehicle Emission Control Information* label in engine compartment
Radiator pressure cap rating	15 psi

Clutch

Clutch pedal free play	3/4 to 1-1/2 inch

Brakes

Brake pad wear limit	1/8 in
Brake shoe wear limit	1/32 in

Torque specifications

	Ft-lbs (unless otherwise indicated)
Differential (axle) fill plug	10 to 20
Spark plugs	22
Oil pan drain plug	20
Wheel lug nuts	
5 lug 1/2 and 7/16-inch	103
6 lug 1/2 and 7/16-inch	88
8 lug 9/16-inch	118
Manual transmission check/fill plug	15 to 25
Manual transmission drain plug	15 to 25
Transfer case check/fill/drain plugs	32
Automatic transmission oil pan bolts	120 in-lbs

1 Introduction

This Chapter is designed to help the home mechanic maintain his or her vehicle for peak performance, economy, safety and long life.

On the following pages you will find a maintenance schedule along with Sections which deal specifically with each item on the schedule. Included are visual checks, adjustments and item replacements.

Servicing your vehicle using the time/mileage maintenance schedule and the sequenced Sections will give you a planned program of maintenance. Keep in mind that it is a full plan, and maintaining only a few items at the specified intervals will not give you the same results.

In many cases the manufacturer will recommend additional owner checks such as warning lamp operation, defroster operation, condition of window glass, etc. We assume these to be obvious, and thus, have not included such items in our maintenance plan. Consult your owner's manual for additional information.

You will find as you service your vehicle that many of the procedures can, and should, be grouped together, due to the nature of the job at hand. Examples of this are as follows:

If the vehicle is raised for a chassis lubrication, for example, it is an ideal time for the following checks: exhaust system, suspension, steering and fuel system.

If the tires and wheels are removed, as during a routine tire rotation, check the brakes and wheel bearings at the same time.

If you must borrow or rent a torque wrench, service the spark plugs and check the carburetor mounting nut torque all in the same day to save time and money.

The first step of the maintenance plan is to prepare yourself before the actual work begins. Read through the appropriate Sections of this Chapter for all work that is to be performed before you begin. Gather together all the necessary parts and tools. If it appears that you could have a problem during a particular job, don't hesitate to seek advice from a mechanic or experienced do-it-yourselfer.

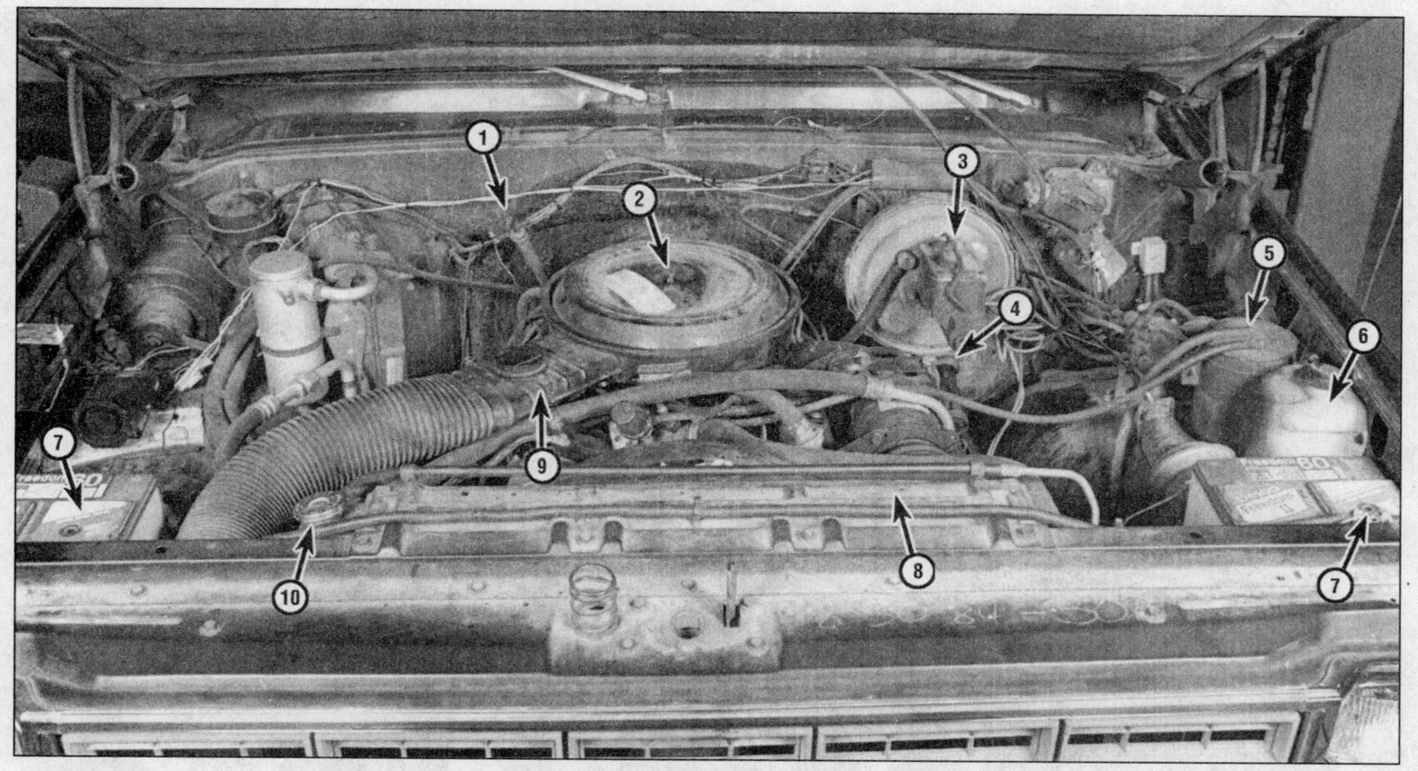

Typical small block V8 underhood components

1 *Automatic transmission fluid dipstick (Section 6)*
2 *Air cleaner (Section 24)*
3 *Brake master cylinder (Section 4)*
4 *Engine oil dipstick (Section 4)*
5 *EVAP system canister (Section 42)*
6 *Windshield washer reservoir (Section 4)*
7 *Battery (Section 8)*
8 *Radiator (Sections 4 and 9)*
9 *Thermostatic air cleaner motor (Section 29)*
10 *Radiator cap (Section 4)*

Typical big block V8 underhood components

1. *Air cleaner (Section 24)*
2. *Brake master cylinder (Section 4)*
3. *Jack*
4. *Alternator (Section 31)*
5. *Upper radiator hose (Section 9)*
6. *Oil fill cap (Section 4)*
7. *Battery (Section 8)*
8. *Radiator cap (Section 4)*
9. *Coolant recovery bottle (Section 4)*
10. *Automatic transmission fluid dipstick (Section 6)*
11. *Engine oil dipstick (Section 4)*

Typical front underside components - 2WD model

1. *Stabilizer bar*
2. *Automatic transmission cooler lines*
3. *Engine oil cooler lines*
4. *Steering gear box*
5. *Ball joint grease fittings (Section 13)*
6. *Starter motor*
7. *Engine oil drain plug (Section 12)*
8. *Oil filter (Section 12)*

Typical front underside components - 4WD model

1 Oil filter (Section 12)
2 Automatic transmission fluid pan (Section 38)
3 Front driveshaft slip yoke grease fitting (Section 13)
4 Front leaf spring (Section 14)
5 Front differential drain plug (Section 37)
6 Lower radiator hose (Sections 9 and 10)
7 Drivebelts (Section 31)
8 Steering damper (Chapter 10)
9 Shock absorber (Section 14)
10 Engine oil drain plug (Section 12)

Typical rear underside components

1 Rear driveshaft universal joint
2 Rear leaf spring (Section 14)
3 Parking brake cable (Section 22)
4 Drum brake assembly (Section 22)
5 Rear shock absorber (Section 14)
6 Exhaust pipe (Section 15)
7 Rear differential cover (Section 37)
8 Rear brake hose (Section 22)

2 Chevrolet and GMC Pick-ups Maintenance schedule

The following recommendations are given with the assumption that the vehicle owner will be doing the maintenance or service work, as opposed to having a dealer service department do the work. The following are factory maintenance recommendations. However, the owner, interested in keeping his or her vehicle in peak condition at all times and with the vehicle's ultimate resale in mind, may want to perform many of these operations more often. Specifically, we would encourage the shortening of oil and filter replacement intervals.

Every 250 miles or weekly, whichever comes first

Check the engine oil level (Section 4)
Check the engine coolant level (Section 4)
Check the windshield washer fluid level (Section 4)
Check the brake fluid and clutch fluid levels (Section 4)
Check the tires and tire pressures (Section 5)

Every 3000 miles or 3 months, whichever comes first

All items listed above plus:
Check the automatic transmission fluid level (Section 6)*
Check the power steering fluid level (Section 7)*
Check and service the battery (Section 8)
Check the cooling system (Section 9)
Inspect and replace, if necessary, all underhood hoses (Section 10)
Inspect and replace, if necessary, the windshield wiper blades (Section 11)

Every 7500 miles or 12 months, whichever comes first

All items listed above plus:
Change the engine oil and oil filter (Section 12)*
Lubricate the chassis components (Section 13)
Inspect the suspension and steering components (Section 14)*
Inspect the exhaust system (Section 15)*
Check the EFE (heat riser) system (Section 16)*
Check and adjust, if necessary, the clutch pedal free play (Section 17)
Check the manual transmission oil level (Section 18)*
Check the transfer case oil level (Section 19)*
Check the differential (axle) oil level (Section 20)*
Rotate the tires (Section 21)
Check the brakes (Section 22)*
Inspect the fuel system (Section 23)
Replace the air filter and PCV filter (Section 24)
Replace the fuel filter (Section 25)
Check the carburetor choke operation (Section 26)
Check the carburetor/throttle body mounting nut torque (Section 27)
Check the throttle linkage (Section 28)
Check the thermostatically controlled air cleaner (Section 29)
Check and adjust, if necessary, the engine idle speed (Section 30)
Check the engine drivebelts (Section 31)
Check the seat belts (Section 32)
Check the seat back latches (Section 33)
Check the starter safety switch (Section 34)

Every 30,000 miles or 24 months, whichever comes first

All items listed above plus:
Change the transfer case fluid (Section 35)*
Change the manual transmission fluid (Section 36)*
Change the differential (axle) fluid (Section 37)*
Change the automatic transmission fluid (Section 38)**
Check and repack the front wheel bearings (2-wheel drive models) (Section 39)*
Service the cooling system (drain, flush and refill) (Section 40)
Inspect and replace, if necessary, the PCV valve (Section 41)
Inspect the evaporative emissions control system (Section 42)
Inspect the EGR system (Section 43)
Replace the spark plugs (Section 44)
Inspect the spark plug wires, distributor cap and rotor (Sections 45 and 46)
Replace the points (Section 47)
Check and adjust, if necessary, the ignition timing (Section 48)

Note: *If your vehicle is operated under severe conditions as described below, perform all maintenance indicated with an asterisk (*) at 7500 mile/12 month intervals.*

Severe conditions include the following:
Operating in dusty areas
Towing a trailer
Idling for extended periods and/or low speed operation
Operating when outside temperatures remain below freezing and when most trips are less than 4 miles in length

** If operated under one or more of the following conditions, change the automatic transmission fluid every 12,000 miles:
In heavy city traffic where the outside temperature regularly reaches 90-degrees F (32-degrees C) or higher
In hilly or mountainous terrain
Frequent trailer pulling

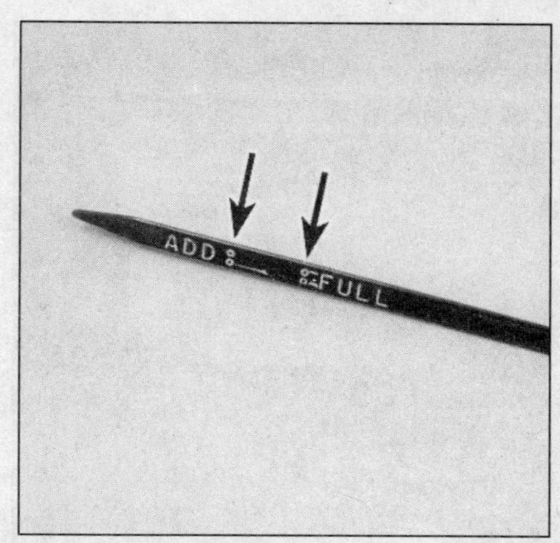

4.4 The engine oil level must be maintained between the marks at all times - it takes one quart of oil to raise the level from the ADD mark to the FULL mark

4.6 Oil is added to the engine through the filler cap in the rocker arm cover

4.8 Typical late model radiator coolant reservoir (arrow)

3 Tune-up general information

The term tune-up is used in this manual to represent a combination of individual operations rather than one specific procedure.

If, from the time the vehicle is new, the routine maintenance schedule is followed closely and frequent checks are made of fluid levels and high wear items, as suggested throughout this manual, the engine will be kept in relatively good running condition and the need for additional work will be minimized.

More likely than not, however, there will be times when the engine is running poorly due to lack of regular maintenance. This is even more likely if a used vehicle, which has not received regular and frequent maintenance checks, is purchased. In such cases an engine tune-up will be needed outside of the regular routine maintenance intervals.

The first step in any tune-up or engine diagnosis to help correct a poor running engine would be a cylinder compression check. A check of the engine compression will give valuable information regarding the overall condition of many internal components and should be used as a basis for tune-up and repair procedures. If, for instance, a compression check indicates serious internal engine wear, a conventional tune-up will not improve engine performance and would be a waste of time and money. Due to its importance, compression checking should be performed by someone who has the proper compression testing gauge and who is knowledgeable with its use. Further information on compression testing can be found in Chapter 2 of this manual.

The following series of operations are those most often needed to bring a generally poor running engine back into a proper state of tune.

Minor tune-up

Clean, inspect and test the battery
Check all engine related fluids
Check and adjust the drivebelts
Replace the spark plugs
Inspect the distributor cap and rotor
Inspect the spark plug and coil wires
Check/clean/adjust the ignition points
Check and adjust the ignition timing
Check the PCV valve
Check and adjust the idle speed
Check the air and PCV filters
Check the cooling system
Check all underhood hoses

Major tune-up

All operations listed under Minor tune-up plus . . .

Check the EGR system
Check the ignition system
Check the charging system
Check the fuel system
Replace the air and PCV filters
Replace the distributor cap and rotor
Replace the ignition points and adjust the dwell (1967 through 1974 models)
Replace the spark plug wires

4 Fluid level checks

Refer to illustrations 4.4, 4.6, 4.8, 4.19 and 4.22

Note: *The following are fluid level checks to be done on a 250 mile or weekly basis. Additional fluid level checks can be found in specific maintenance intervals which follow. Regardless of intervals, be alert to fluid leaks under the vehicle which would indicate a fault to be corrected immediately.*

1 There are a number of components on a vehicle which rely on the use of fluids to perform their job. During normal operation of the vehicle, these fluids are used up and must be replenished before damage occurs. See *Recommended lubricants and fluids* at the beginning of this Chapter for the specific fluid to be used when addition is required. When checking fluid levels it is important to have the vehicle on a level surface.

Engine oil

2 The engine oil level is checked with a dipstick that travels through a tube and into the oil pan to the bottom of the engine.

3 The oil level should be checked before the vehicle has been driven, or about 15 minutes after the engine has been shut off. If the oil is checked immediately after driving the vehicle, some of the oil will remain in the upper engine components, producing an inaccurate reading on the dipstick.

4 Pull the dipstick from the tube and wipe all the oil from the end with a clean rag or paper towel. Insert the clean dipstick all the way back into the oil pan and pull it out again. Observe the oil at the end of the dipstick. Add oil as necessary to keep the level between the ADD mark and the FULL mark on the dipstick **(see illustration)**.

5 Do not overfill the engine by adding too much oil since this may result in oil fouled spark plugs, oil leaks or oil seal failures.

6 Oil is added to the engine after removing a twist off cap located on the rocker arm cover **(see illustration)**. An oil can spout or funnel may help to reduce spills.

7 Checking the oil can be an important preventive maintenance step. If you find the oil level dropping abnormally, it is an indication of oil leakage or internal engine wear which should be corrected. If there are water droplets in the oil, or if it is milky looking, component failure is indicated and the engine should be checked immediately.

Engine coolant

8 Later vehicles covered by this manual are equipped with a pressurized coolant recovery system. A white coolant reservoir located at the front the engine compartment on early models and at the rear of the engine compartment on later models, is connected by a hose to the base of the radiator cap **(see illustration)**. As the engine heats up during operation, coolant is forced from the radiator, through the connecting tube and into the reservoir. As the engine cools, the coolant is automatically drawn back into the radiator to

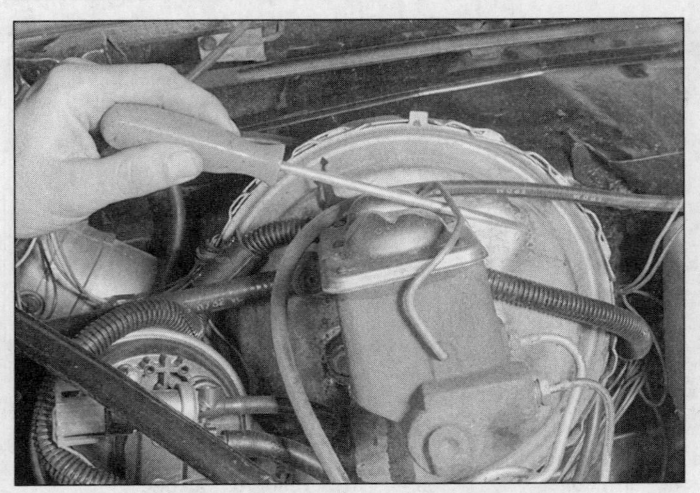

4.19 Early models have a cast iron brake master cylinder with a cover which must be removed to check the fluid level - use a screwdriver to unsnap the retainer

4.22 Later models have a translucent brake fluid reservoir - keep the level above the MIN mark

keep the level correct.

9 On reservoir equipped models coolant level checks are very easy; merely locate the coolant reservoir and observe the level of fluid in the reservoir. With the engine cold, this should be at or slightly above the FULL COLD mark on the reservoir. Some reservoirs also have a FULL HOT mark to check the level when the engine is hot. On a periodic basis, or if the coolant in the reservoir runs completely dry, check the coolant level in the radiator as well (see Step 10).

10 If your particular vehicle is not equipped with a coolant recovery system, the level should be checked by removing the radiator cap. **Warning:** *Under no circumstances should the radiator cap be removed when the system is hot, because escaping steam and scalding liquid could cause serious injury. Wait until the engine has completely cooled, then wrap a thick cloth around the cap and turn it to its first stop. If any steam escapes from the cap, allow the engine to cool further, then remove the cap and check the level in the radiator. It should be just below the bottom of the filler neck.*

11 If only a small amount of coolant is required to bring the system up to the proper level, plain water can be used. However, to maintain the proper antifreeze/water mixture in the system, both should be mixed together to replenish a low level. High quality antifreeze offering protection to -34-degrees F should be mixed with water in the proportion specified on the container. Do not allow antifreeze to come in contact with your skin or painted surfaces of the vehicle. Flush contacted areas immediately with plenty of water.

12 As the coolant level is checked, note the condition of the coolant. It should be relatively transparent. If it is brown or a rust color, the system should be drained, flushed and refilled (Section 40).

13 If the cooling system requires repeated additions to maintain the proper level, have the radiator cap checked for proper sealing ability and check for leaks in the system from cracked hoses, loose hose connections, leaking gaskets, etc.

Windshield washer fluid

14 On later models, fluid for the windshield washer system is located in a plastic reservoir in the engine compartment. Use caution not to confuse the windshield washer fluid reservoir and the coolant overflow reservoir.

15 The reservoir should be kept no more than 2/3 full to allow for expansion should the fluid freeze. The use of an additive such as windshield washer concentrate, available at auto parts stores, will help lower the freezing point of the fluid and will result in better cleaning of the windshield surface. Do not use cooling system antifreeze - it will cause damage to the vehicle's paint.

16 To help prevent icing in cold weather, warm the windshield with the defroster before using the washer.

Battery electrolyte

Warning: *There are certain precautions to be taken when working on or near a battery. Never expose a battery to open flame or sparks which could ignite the hydrogen gas given off by the battery. Wear protective clothing and eye protection to reduce the possibility of the corrosive sulfuric acid solution inside the battery harming you. If the fluid is splashed or spilled, flush the contacted area immediately with plenty of water. Remove all metal jewelry which could contact the positive terminal and another grounded metal source, thus causing a short circuit. Always keep batteries and battery acid out of the reach of children.*

17 Later model vehicles are equipped with Freedom maintenance-free type batteries which are permanently sealed (except for vent holes) and have no filler caps. Water does not have to be added to these batteries at any time. If a maintenance-type battery is installed, the caps on the top of the battery should be removed periodically to check for a low water level. This check is most critical during the warm summer months.

Brake and clutch fluid

18 The brake master cylinder is mounted on the firewall or on the front of the power booster unit in the engine compartment. The clutch cylinder used on later manual transmission-equipped models is mounted adjacent to it.

19 On most models, it will be necessary to remove the cover to check the brake fluid level. Clean the area around the sealing lip and use a long screwdriver to pry the heavy wire retainer off the cover **(see illustration)**.

20 Carefully lift off the cover and check the fluid inside. It should be approximately 1/4-inch below the top edge of the reservoir.

21 If additional fluid is necessary to bring the level up to the proper height, carefully pour the specified brake fluid into the master cylinder. Be careful not to spill any.

22 On some later model reservoirs the fluid inside is readily visible. The level should kept above the MIN marks on the reservoirs **(see illustration)**. If a low level is indicated, be sure to wipe the top of the reservoir cover with a clean rag to prevent contamination of the brake and/or clutch system before removing the cover. Remove the cover by prying up on the tabs.

23 When adding fluid, pour it carefully into the reservoir, taking care not to spill any onto surrounding painted surfaces. Be sure the specified fluid is used, since mixing different types of brake fluid can cause damage to the system. See *Recommended lubricants and fluids* at the front of this Chapter or your owner's manual.

24 At this time the fluid and cylinder can be inspected for contamination. The system should be drained and refilled if rust deposits, dirt particles or water droplets are seen in the fluid.

5.6 Tire pressure gauges are available in a variety of styles - since service station gauges are often inaccurate, keep a gauge in your glove compartment

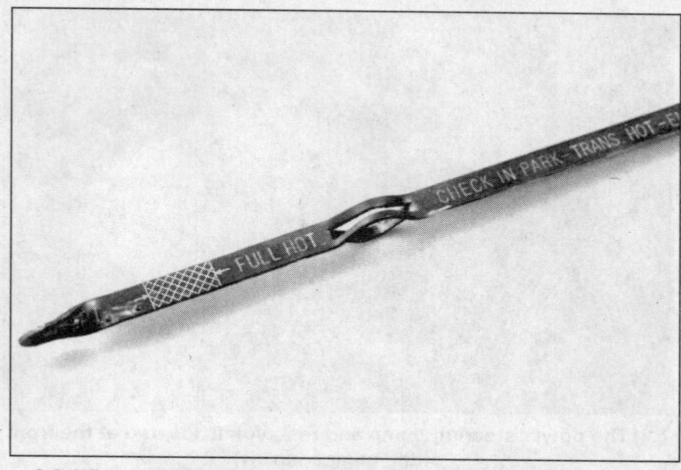

6.6 When checking the automatic transmission fluid level it is important to note the fluid temperature

25 After filling the reservoir to the proper level, make sure the cover is on tight to prevent fluid leakage and contamination.

26 The brake fluid in the master cylinder will drop slightly as the pads or shoes at each wheel wear down during normal operation. If the master cylinder requires repeated replenishing to keep it at the proper level, this is an indication of leakage in the brake system, which should be corrected immediately. Check all brake lines and connections (see Section 22 for more information).

27 If, upon checking the master cylinder fluid level, you discover one or both reservoirs empty or nearly empty, the brake system should be bled (Chapter 9).

Hydro boost pump fluid

28 Some models use a hydraulic pump to provide boost for the braking system. On power steering equipped models, the power steering pump provides the hydraulic pressure for the hydro boost unit. On manual steering models, a separate belt-driven hydro boost pump is used.

29 Both pumps are checked using the power steering fluid level procedure (Section 7). Use only the specified power steering fluid (not brake fluid) when adding fluid to these pumps. See *Recommended lubricants and fluids* at the front of this Chapter or your owner's manual.

5 Tire and tire pressure checks

Refer to illustration 5.6

1 Periodically inspecting the tires may not only prevent you from being stranded with a flat tire, but can also give you clues as to possible problems with the steering and suspension systems before major damage occurs.

2 Proper tire inflation adds miles to the lifespan of the tires, allows the vehicle to achieve maximum miles per gallon of gas and contributes to the overall quality of the ride.

3 When inspecting the tires, first check the wear of the tread. Irregularities in the tread pattern (cupping, flat spots, more wear on one side than the other) are indications of front end alignment and/or balance problems. If any of these conditions are noted, take the vehicle to a repair shop to correct the problem.

4 Check the tread area for cuts and punctures. Many times a nail or tack will embed itself into the tire tread and yet the tire may hold air pressure for a period of time. In most cases, a repair shop or gas station can repair the punctured tire.

5 It is important to check the sidewalls of the tires, both inside and outside. Check for deteriorated rubber, cuts, and punctures. Inspect the inboard side of the tire for signs of brake fluid leakage, indicating that a thorough brake inspection is needed immediately.

6 Incorrect tire pressure cannot be determined merely by looking at the tire. This is especially true for radial tires. A tire pressure gauge must be used **(see illustration)**. If you do not already have a reliable gauge, it is a good idea to purchase one and keep it in the glovebox. Built-in pressure gauges at gas stations are often inaccurate.

7 Always check tire inflation when the tires are cold. Cold, in this case, means the vehicle has not been driven more than one mile after sitting for three hours or more. It is normal for the pressure to increase four to eight pounds when the tires are hot.

8 Unscrew the valve cap protruding from the wheel or hubcap and press the gauge firmly onto the valve. Observe the reading on the gauge and compare the figure to the recommended tire pressure listed on the tire placard. The tire placard is usually attached to the driver's door jamb. The recommended maximum pressure for the tires is usually also stamped on the tire sidewall. However, the information on the tire placard should always be top priority since it applies to tires used on your particular vehicle.

9 Check all tires and add air as necessary to bring them up to the recommended pressure levels. Do not forget the spare tire. Be sure to reinstall the valve caps, which will keep dirt and moisture out of the valve stem mechanism.

6 Automatic transmission fluid level check

Refer to illustration 6.6

1 The level of the automatic transmission fluid should be carefully maintained. Low fluid level can lead to slipping or loss of drive, while overfilling can cause foaming and loss of fluid.

2 With the parking brake set, start the engine, then move the shift lever through all the gear ranges, ending in Park. The fluid level must be checked with the vehicle level and the engine running at idle. **Note:** *Incorrect fluid level readings will result if the vehicle has just been driven at high speeds for an extended period, in hot weather in city traffic, or if it has been pulling a trailer. If any of these conditions apply, wait until the fluid has cooled (about 30 minutes).*

3 With the transmission at normal operating temperature, remove the dipstick from the filler tube. The dipstick is located at the rear of the engine compartment on the passenger's side.

4 Carefully touch the fluid at the end of the dipstick to determine if the fluid is cool, warm or hot. Wipe the fluid from the dipstick with a clean rag and push it back into the filler tube until the cap seats.

5 Pull the dipstick out again and note the fluid level.

6 If the fluid felt cool, the level should be about 1/8 to 3/8-inch above the ADD mark **(see illustration)**. If it felt warm, the level should be between the ADD and FULL HOT marks. If the fluid was hot, the level should be at the FULL HOT mark. If additional fluid is required, add the recommended fluid directly into the tube using a funnel. It takes about

7.2 The power steering pump and reservoir is located at the front of the engine (arrow)

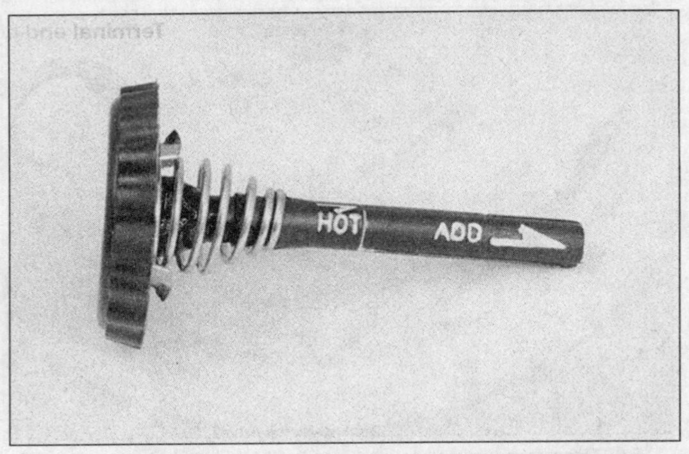

7.6 The marks on the power steering fluid dipstick indicate the safe range with the fluid either HOT or COLD

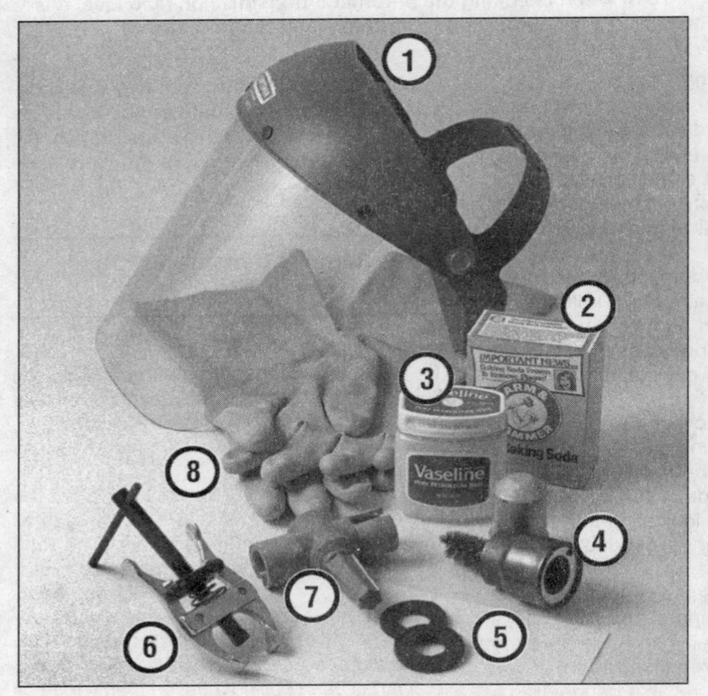

8.1a Tools and materials required for battery maintenance - standard top-terminal battery

1 ***Face shield/safety goggles*** - *When removing corrosion with a brush, the acidic particles can easily fly up into your eyes*
2 ***Baking soda*** - *A solution of baking soda and water can be used to neutralize corrosion*
3 ***Petroleum jelly*** - *A layer of this on the battery posts will help prevent corrosion*
4 ***Battery post/cable cleaner*** - *This wire brush cleaning tool will remove all traces of corrosion from the battery posts and cable clamps*
5 ***Treated felt washers*** - *Placing one of these on each post, directly under the cable clamps, will help prevent corrosion*
6 ***Puller*** - *Sometimes the cable clamps are very difficult to pull off the posts, even after the nut/bolt has been completely loosened. This tool pulls the clamp straight up and off the post without damage*
7 ***Battery post/cable cleaner*** - *Here is another cleaning tool which is a slightly different version of number 4 above, but it does the same thing*
8 ***Rubber gloves*** - *Another safety item to consider when servicing the battery; remember that's acid inside the battery!*

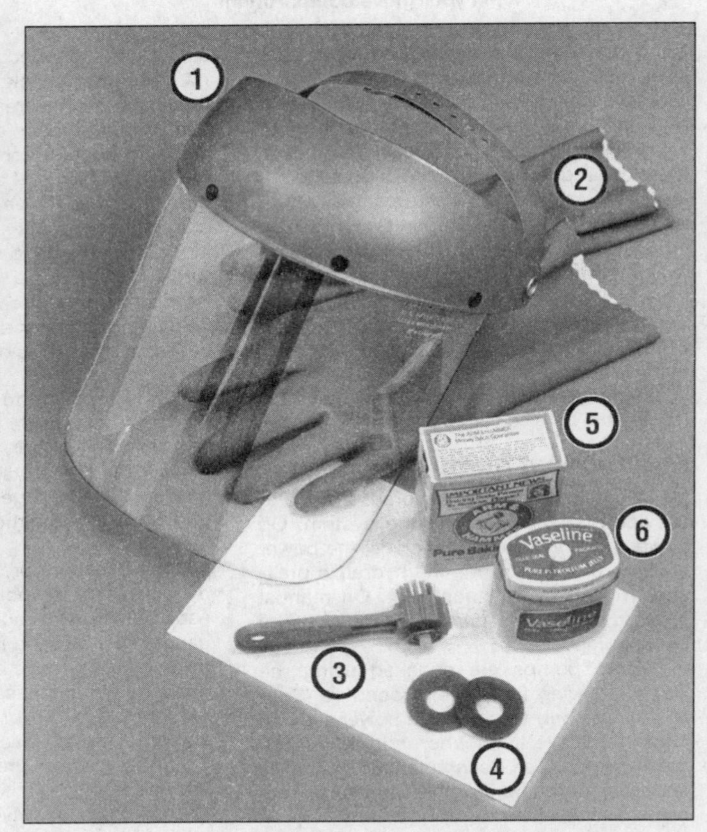

8.1b Tools and materials required for battery maintenance - side-terminal battery

1 ***Face shield/safety goggles*** - *When removing corrosion with a brush, the acidic particles can easily fly up into your eyes*
2 ***Rubber gloves*** - *Another safety item to consider when servicing the battery; remember that's acid inside the battery!*
3 ***Battery terminal/ cable cleaner*** - *This wire brush cleaning tool will remove all traces of corrosion from the battery and cable*
4 ***Treated felt washers*** - *Placing one of these on each terminal, directly under the cable end, will help prevent corrosion (be sure to get the correct type for side terminal batteries)*
5 ***Baking soda*** - *A solution of baking soda and water can be used to neutralize corrosion*
6 ***Petroleum jelly*** - *A layer of this on the battery terminal bolts will help prevent corrosion*

8.4 Remove the cell caps to check the water level in the battery - if the level is low, add distilled water only!

one pint to raise the level from the ADD mark to the FULL HOT mark with a hot transmission, so add the fluid a little at a ti;me and keep checking the level until it is correct.

7 The condition of the fluid should also be checked along with the level. If the fluid at the end of the dipstick is a dark reddish-brown color, or if the fluid has a burned smell, the fluid should be changed. If you are in doubt about the condition of the fluid, purchase some new fluid and compare the two for color and smell.

7 Power steering fluid level check

Refer to illustrations 7.2 and 7.6

1 Unlike manual steering, the power steering system relies on fluid which may, over a period of time, require replenishing.

2 The fluid reservoir for the power steering pump is located on the pump body at the front of the engine **(see illustration)**.

3 For the check, the front wheels should be pointed straight ahead and the engine should be off.

4 Use a clean rag to wipe off the reservoir cap and the area around the cap. This will help prevent any foreign matter from entering the reservoir during the check.

5 Twist off the cap and check the temperature of the fluid at the end of the dipstick with your finger.

6 Wipe off the fluid with a clean rag, reinsert it, then withdraw it and read the fluid level. The level should be at the HOT mark if the fluid was hot to the touch **(see illustration)**. It should be at the COLD mark if the fluid was cool to the touch. At no time should the fluid level drop below the ADD mark.

7 If additional fluid is required, pour the specified type directly into the reservoir, using a funnel to prevent spills.

8 If the reservoir requires frequent fluid additions, all power steering hoses, hose connections and the power steering pump should be carefully checked for leaks.

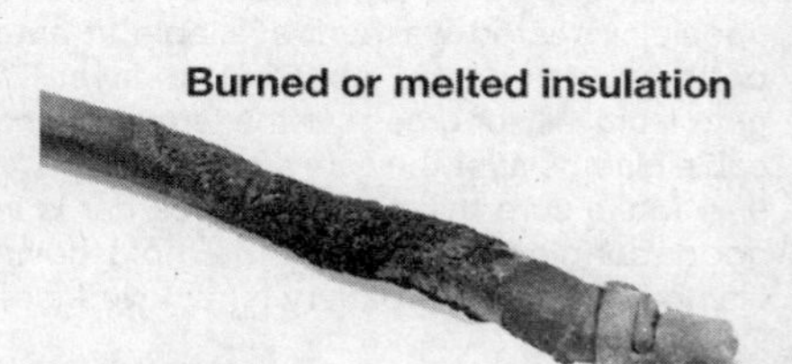

8.7 Typical battery cable problems

8 Battery check and maintenance

Refer to illustrations 8.1a, 8.1b, 8.4, 8.7, 8.8a, 8.8b, 8.8c and 8.8d

Warning: *Certain precautions must be followed when checking and servicing the battery. Hydrogen gas, which is highly flammable, is always present in the battery cells, so keep lighted tobacco and all other open flames and sparks away from the battery. The electrolyte inside the battery is actually dilute sulfuric acid, which will cause injury if splashed on your skin or in your eyes. It will also ruin clothes and painted surfaces. When removing the battery cables, always detach the negative cable first and hook it up last!*

1 Battery maintenance is an important procedure which will help ensure that you are not stranded because of a dead battery. Several tools are required for this procedure **(see illustrations)**.

2 When checking/servicing the battery, always turn the engine and all accessories off.

3 A sealed (sometimes called maintenance-free), side-terminal battery is standard equipment an all later model vehicles. The cell caps cannot be removed, no electrolyte checks are required and water cannot be added to the cells. However, if you have a vehicle with a standard top-terminal battery, or if a standard aftermarket battery has been installed, the following maintenance procedure can be used.

8.8a Battery terminal corrosion usually appears as light, fluffy powder

8.8b Removing the cable from a battery post with a wrench - sometimes special battery pliers are required for this procedure if corrosion has caused deterioration of the nut hex (always remove the ground cable first and hook it up last!)

4 Remove the caps and check the electrolyte level in each of the battery cells **(see illustration)**. It must be above the plates. There's usually a split-ring indicator in each cell to indicate the correct level. If the level is low, add distilled water only, then reinstall the cell caps. **Caution:** *Overfilling the cells may cause electrolyte to spill over during periods of heavy charging, causing corrosion and damage to nearby components.*

5 If the positive terminal and cable clamp on your vehicle's battery is equipped with a rubber protector, make sure that it's not torn or damaged. It should completely cover the terminal.

6 The external condition of the battery should be checked periodically. Look for damage such as a cracked case.

7 Check the tightness of the battery cable clamps to ensure good electrical connections and inspect the entire length of each cable, looking for cracked or abraded insulation and frayed conductors **(see illustration)**.

8 If corrosion (visible as white, fluffy deposits) is evident, remove the cables from

8.8c Regardless of the type of tool used to clean the battery posts, a clean, shiny surface should be the result

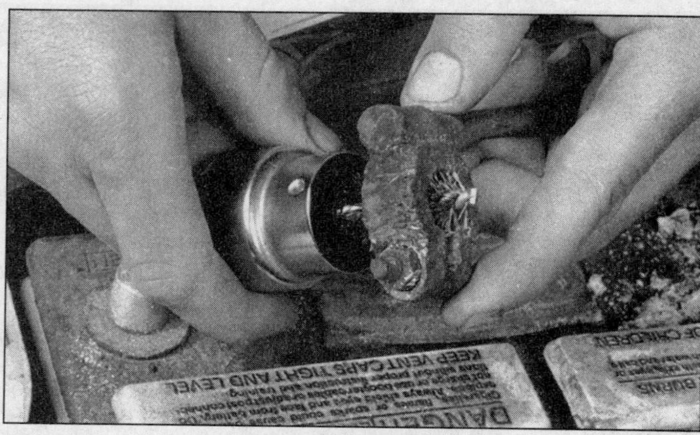

8.8d When cleaning the cable clamps, all corrosion must be removed (the inside of the clamp is tapered to match the taper on the post, so don't remove too much material)

Check for a chafed area that could fail prematurely.

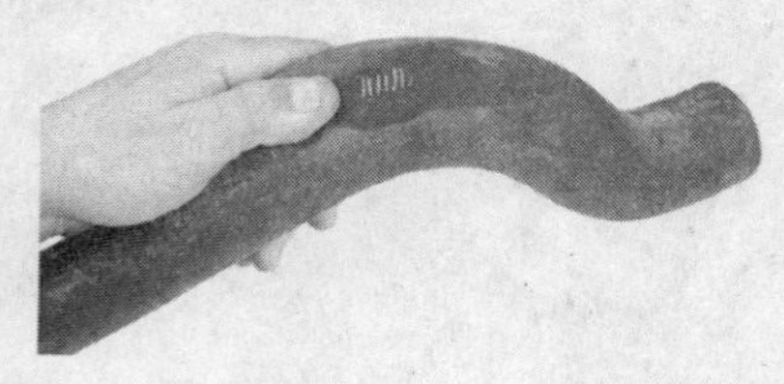

Check for a soft area indicating the hose has deteriorated inside.

Overtightening the clamp on a hardened hose will damage the hose and cause a leak.

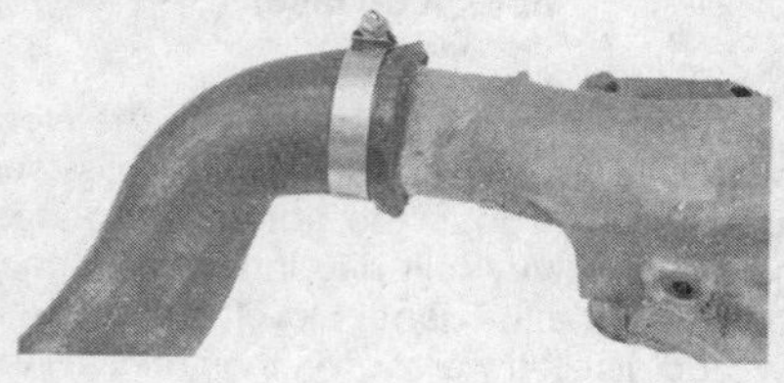

Check each hose for swelling and oil-soaked ends. Cracks and breaks can be located by squeezing the hose.

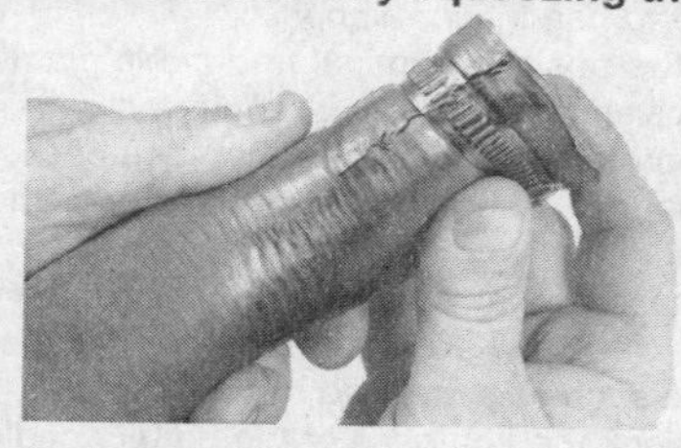

9.4 Cooling system hoses should be carefully inspected to prevent being stranded on the road - regardless of condition, it is a good idea to replace the hoses with new ones every two years

the terminals, clean them with a battery brush and reinstall them **(see illustrations)**. Corrosion can be kept to a minimum by installing specially treated washers available at auto parts stores or by applying a layer of petroleum jelly or grease to the terminals and cable clamps after they are assembled.

9 Make sure that the battery carrier is in good condition and that the hold-down clamp is tight. If the battery is removed (see Chapter 5
for the removal and installation procedure), make sure that no parts remain in the bottom of the carrier when it's reinstalled. When reinstalling the hold-down clamp, don't overtighten the bolt.

10 Corrosion on the carrier, battery case and surrounding areas can be removed with a solution of water and baking soda. Apply the mixture with a small brush, let it work, then rinse it off with plenty of clean water.

11 Any metal parts of the vehicle damaged by corrosion should be coated with a zinc-based primer, then painted.

12 Additional information on the battery, charging and jump starting can be found in the front of this manual and in Chapter 5.

9 Cooling system check

Refer to illustration 9.4

1 Many major engine failures can be attributed to a faulty cooling system. If the vehicle is equipped with an automatic transmission, the cooling system also cools the transmission fluid and thus plays an important role in prolonging transmission life.

2 The cooling system should be checked with the engine cold. Do this before the vehicle is driven for the day or after it has been shut off for at least three hours.

3 Remove the radiator cap by turning it to the left until it reaches a "stop." If you hear a hissing sound (indicating there is still pressure in the system), wait until this stops. Now press down on the cap with the palm of your hand and continue turning to the left until the cap can be removed. Thoroughly clean the cap, inside and out, with clean water. Also clean the filler neck on the radiator. All traces of corrosion should be removed. The coolant inside the radiator should be relatively transparent. If it is rust colored, the system should be drained and refilled (Section 40). If the coolant level is not up to the top, add additional antifreeze/coolant mixture (see Section 4).

4 Carefully check the large upper and lower radiator hoses along with the smaller diameter heater hoses which run from the engine to the firewall. On some models the heater return hose runs directly to the radiator. Inspect each hose along its entire length, replacing any hose which is cracked, swollen or shows signs of deterioration. Cracks may become more apparent if the hose is squeezed **(see illustration)**.

5 Make sure that all hose connections are tight. A leak in the cooling system will usually show up as white or rust colored deposits on the areas adjoining the leak. If wire-type clamps are used at the ends of the hoses, it may be wise to replace them with more secure screwtype clamps.

6 Use compressed air or a soft brush to remove bugs, leaves, etc. from the front of the radiator or air conditioning condenser. Be careful not to damage the delicate cooling fins or cut yourself on them.

7 Every other inspection, or at the first indication of cooling system problems, have the cap and system pressure tested. If you do not have a pressure tester, most gas stations and repair shops will do this for a minimal charge.

10 Underhood hose check and replacement

Warning: *Replacement of air conditioning hoses must be left to a dealer or air conditioning specialist who has the proper equipment to depressurize the system safely. Never remove air conditioning components or hoses until the system has been depressurized.*

General

1 High temperatures under the hood can cause the deterioration of the rubber and plastic hoses used for engine, accessory and emission systems operation. Periodic inspection should be made for cracks, loose clamps, material hardening and leaks.

2 Information specific to the cooling system hoses can be found in Section 9.

3 Some, but not all, hoses use clamps to secure the hoses to fittings. Where clamps are used, check to be sure they haven't lost their tension, allowing the hose to leak. Where clamps are not used, make sure the hose has not expanded and/or hardened where it slips over the fitting, allowing it to leak.

Vacuum hoses

4 It is quite common for vacuum hoses, especially those in the emissions system, to be color coded or identified by colored stripes molded into the hose. Various systems require hoses with different wall thicknesses, collapse resistance and temperature resistance. When replacing hoses, be sure to use the same hose material on the new hose.

5 Often the only effective way to check a hose is to remove it completely from the vehicle. Where more than one hose is removed, be sure to label the hoses and their attaching points to insure proper reattachment.

6 When checking vacuum hoses, be sure to include any plastic T-fittings in the check. Check the fittings for cracks and the hose where it fits over the fitting for enlargement, which could cause leakage.

7 A small piece of vacuum hose (1/4-inch inside diameter) can be used as a stethoscope to detect vacuum leaks. Hold one end of the hose to your ear and probe around vacuum hoses and fittings, listening for the "hissing" sound characteristic of a vacuum leak. **Warning:** *When probing with the vacuum hose stethoscope, be careful not to allow your body or the hose to come into contact with moving engine components such as the drivebelt, cooling fan, etc.*

Fuel hose

Warning: *There are certain precautions which must be taken when inspecting or servicing fuel system components. Work in a well ventilated area and do not allow open flames (cigarettes, appliance pilot lights, etc.) or bare light bulbs near the work area. Mop up any spills immediately and do not store fuel soaked rags where they could ignite. On vehicles equipped with fuel injection, the fuel system is under pressure, so if any fuel lines are to be disconnected, the pressure in the system must be relieved (see Chapter 4 for more information).*

8 Check all rubber fuel lines for deterioration and chafing. Check especially for cracking in areas where the hose bends and just before clamping points, such as where a hose attaches to the fuel filter.

9 High quality fuel line, usually identified by the word *Fluroelastomer* printed on the hose, should be used for fuel line replacement. Under no circumstances should unreinforced vacuum line, clear plastic tubing or water hose be used for fuel line replacement.

10 Spring-type clamps are commonly used on fuel lines. These clamps often lose their tension over a period of time, and can be "sprung" during the removal process. Therefore it is recommended that all spring type clamps be replaced with screw clamps whenever a hose is replaced.

Metal lines

11 Sections of metal line are often used for fuel line between the fuel pump and carburetor or fuel injection unit. Check carefully to be sure the line has not been bent or crimped and that cracks have not started in the line.

12 If a section of metal fuel line must be replaced, only seamless steel tubing should be used, since copper and aluminum tubing do not have the strength necessary to withstand normal engine operating vibration.

13 Check the metal brake lines where they enter the master cylinder and brake proportioning unit (if used) for cracks in the lines or loose fittings. Any sign of brake fluid leakage calls for an immediate thorough inspection of the brake system.

11 Wiper blade inspection and replacement

1 The windshield wiper and blade assembly should be inspected periodically for damage, loose components and cracked or worn blade elements.

2 Road film can build up on the wiper blades and affect their efficiency, so they should be washed regularly with a mild detergent solution.

3 The action of the wiping mechanism can loosen the bolts, nuts and fasteners, so they should be checked and tightened, as necessary, at the same time the wiper blades are checked.

4 If the wiper blade elements (sometimes called "inserts") are cracked, worn or warped, they should be replaced with new ones.

5 Pull the wiper blade/arm assembly away from the glass.

6 Depress the blade-to-arm connector and slide the blade assembly off the wiper arm and over the retaining stud.

7 Pinch the tabs at the end and then slide the element out of the blade assembly.

8 Compare the new element with the old for length, design, etc.

9 Slide the new element into place. It will automatically lock at the correct location.

10 Reinstall the blade assembly onto the arm, wet the windshield and test for proper operation.

12 Engine oil and filter change

Refer to illustrations 12.3, 12.9 and 12.14

1 Frequent oil changes may be the best form of preventive maintenance available to the home mechanic. When engine oil ages, it becomes diluted and contaminated, which leads to premature engine wear.

2 Although some sources recommend oil filter changes every other oil change, we feel that the minimal cost of an oil filter and the relative ease with which it is installed dictate that a new filter be used whenever the oil is changed.

3 Gather together all necessary tools and materials before beginning the procedure **(see illustration)**.

4 You should have plenty of clean rags and newspapers handy to mop up any spills. Access to the underside of the vehicle is greatly improved if the vehicle can be lifted on a hoist, driven onto ramps or supported

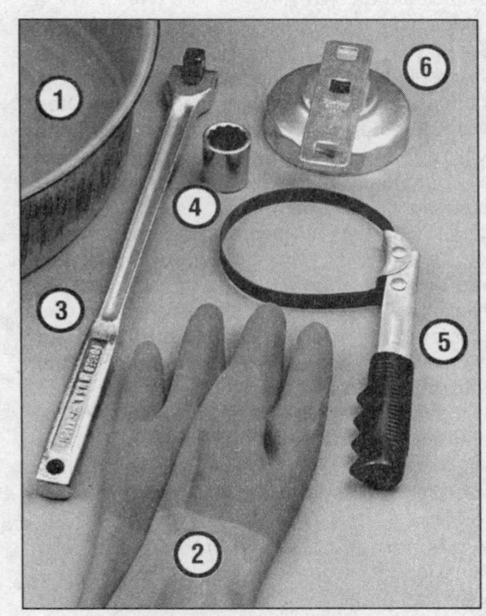

12.3 These tools are required when changing the engine oil and filter

1 ***Drain pan*** *- It should be fairly shallow in depth, but wide in order to prevent spills*
2 ***Rubber gloves*** *- When removing the drain plug and filter it is inevitable that you will get oil on your hands (the gloves will prevent burns)*
3 ***Breaker bar*** *- Sometimes the oil drain plug is pretty tight and a long breaker bar is needed to loosen it*
4 ***Socket*** *- To be used with the breaker bar or a ratchet (must be the correct size to fit the drain plug)*
5 ***Filter wrench*** *- This is a metal band-type wrench, which requires clearance around the filter to be effective*
6 ***Filter wrench*** *- This type fits on the bottom of the filter and can be turned with a ratchet or breaker bar (different size wrenches are available for different types of filters)*

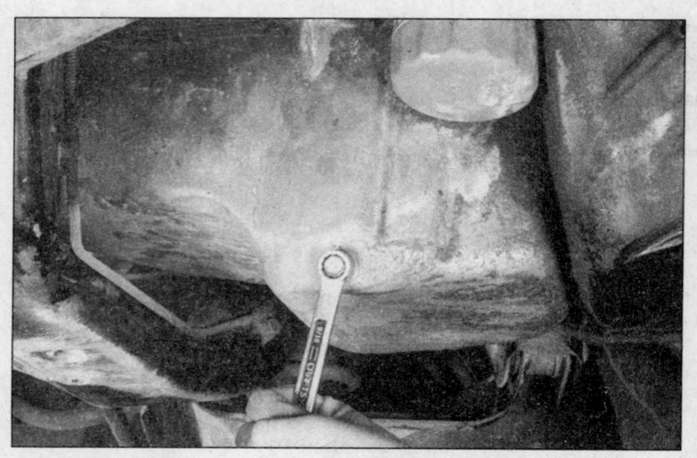

12.9 The oil drain plug is located at the bottom of the pan and should be removed using either a socket or a box-end wrench - DO NOT use an open-end wrench, as the corners on the bolt can easily be rounded off

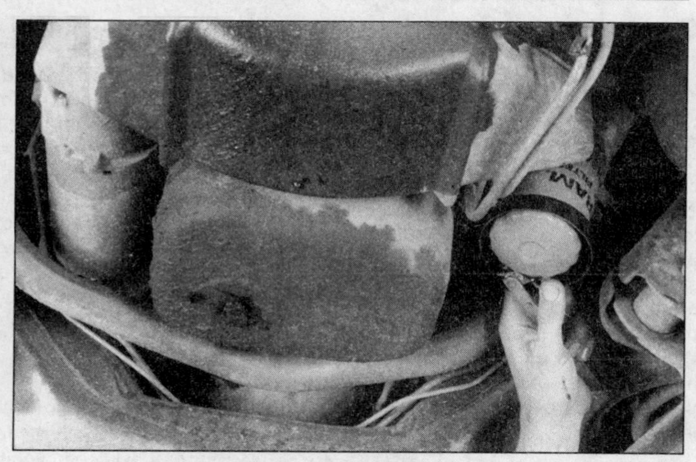

12.14 A strap-type oil filter wrench is used to loosen the oil filter (if access makes removal difficult, other types of filter wrenches are available)

by jackstands. **Warning:** *Do not work under a vehicle which is supported only by a jack.*

5 If this is your first oil change, get under the vehicle and familiarize yourself with the locations of the oil drain plug and the oil filter. The engine and exhaust components will be warm during the actual work, so figure out any potential problems before the engine and accessories are hot.

6 Warm the engine to normal operating temperature. If new oil or any tools are needed, use this warm-up time to gather everything necessary for the job. The correct type of oil for your application can be found in *Recommended lubricants and fluids* at the beginning of this Chapter.

7 With the engine oil warm (warm engine oil will drain better and more built-up sludge will be removed with the oil), raise and support the vehicle. Make sure it is safely supported!

8 Move all necessary tools, rags and newspapers under the vehicle. Position the drain pan under the drain plug. Keep in mind that the oil will initially flow from the pan with some force, so place the pan accordingly.

9 Being careful not to touch any of the hot exhaust components, use the wrench to remove the drain plug near the bottom of the oil pan **(see illustration)**. Depending on how hot the oil has become, you may want to wear gloves while unscrewing the plug the final few turns.

10 Allow the old oil to drain into the pan. It may be necessary to move the pan farther under the engine as the oil flow slows to a trickle.

11 After all the oil has drained, wipe off the drain plug with a clean rag. Small metal particles may cling to the plug which would immediately contaminate the new oil.

12 Clean the area around the drain plug opening and reinstall the plug. Tighten the plug securely with the wrench. If a torque wrench is available, use it to tighten the plug.

13 Move the drain pan into position under the oil filter.

14 Use the filter wrench to loosen the oil filter **(see illustration)**. Chain or metal band filter wrenches may distort the filter canister, but this is of no concern as the filter will be discarded anyway.

15 Completely unscrew the old filter. Be careful, as it is full of oil. Empty the oil inside the filter into the drain pan.

16 Compare the old filter with the new one to make sure they are the same type.

17 Use a clean rag to remove all oil, dirt and sludge from the area where the oil filter mounts to the engine. Check the old filter to make sure the rubber gasket is not stuck to the engine mounting surface. If the gasket is stuck to the engine (use a flashlight if necessary), remove it.

18 Apply a light coat of oil around the full circumference of the rubber gasket of the new oil filter.

19 Attach the new filter to the engine, following the tightening directions printed on the filter canister or packing box. Most filter manufacturers recommend against using a filter wrench due to the possibility of overtightening and damage to the seal.

20 Remove all tools, rags, etc. from under the vehicle, being careful not to spill the oil in the drain pan, then lower the vehicle.

21 Move to the engine compartment and locate the oil filler cap.

22 If an oil can spout is used, push the spout into the top of the oil can and pour the fresh oil through the filler opening. A funnel may also be used.

23 Pour four quarts of fresh oil into the engine. Wait a few minutes to allow the oil to drain into the pan, then check the level on the oil dipstick (see Section 4 if necessary). If the oil level is above the ADD mark, start the engine and allow the new oil to circulate.

24 Run the engine for only about a minute and then shut it off. Immediately look under the vehicle and check for leaks at the oil pan drain plug and around the oil filter. If either is leaking, tighten with a bit more force.

25 With the new oil circulated and the filter now completely full, recheck the level on the dipstick and add more oil as necessary.

26 During the first few trips after an oil change, make it a point to check frequently for leaks and proper oil level.

27 The old oil drained from the engine cannot be reused in its present state and should be disposed of. Oil reclamation centers, auto repair shops and gas stations will normally accept the oil, which can be refined and used again. After the oil has cooled it can be drained into a suitable container (capped plastic jugs, topped bottles, milk cartons, etc.) for transport to one of these disposal sites.

13 Chassis lubrication

Refer to illustrations 13.1, 13.6, 13.9 and 13.10

1 Refer to *Recommended lubricants and fluids* at the front of this Chapter to obtain the necessary grease, etc. You will also need a grease gun **(see illustration)**. Occasionally plugs will be installed rather than grease fittings. If so, grease fittings will have to be purchased and installed.

2 Look under the vehicle and see if grease fittings or plugs are installed. If there are plugs, remove them and buy grease fittings, which will thread into the component. A dealer or auto parts store will be able to supply the correct fittings. Straight, as well as angled, fittings are available.

3 For easier access under the vehicle, raise it with a jack and place jackstands under the frame. Make sure it is safely supported by the stands. If the wheels are to be removed at this interval for tire rotation or brake inspection, loosen the lug nuts slightly while the vehicle is still on the ground.

4 Before beginning, force a little grease out of the nozzle to remove any dirt from the end of the gun. Wipe the nozzle clean with a rag.

5 With the grease gun and plenty of clean

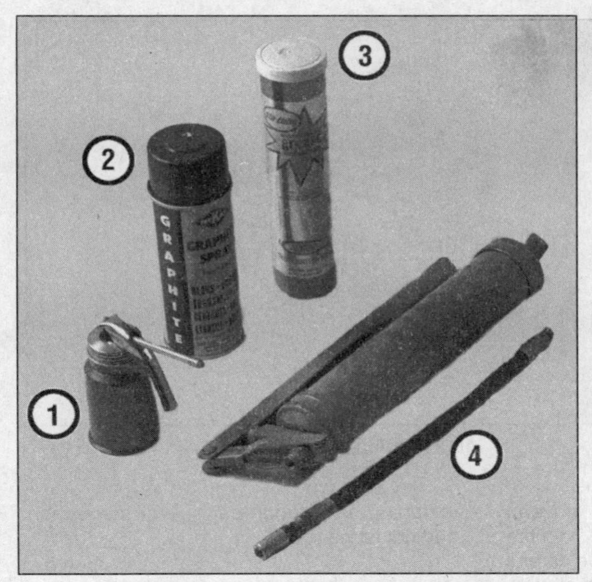

13.1 Materials required for chassis and body lubrication

1 ***Engine oil*** *- Light engine oil in a can like this can be used for door and hood hinges*
2 ***Graphite spray*** *- Used to lubricate lock cylinders*
3 ***Grease*** *- Grease, in a variety of types and weights, is available for use in a grease gun. Check the Specifications for your requirements*
4 ***Grease gun*** *- A common grease gun, shown here with a detachable hose and nozzle, is needed for chassis lubrication. After use, clean it thoroughly!*

13.6 After wiping the grease fitting clean, push the nozzle firmly into place and pump the grease into the component - usually about two pumps of the gun will be sufficient

13.9 In addition to the conventional universal joints at each end of the driveshaft, 4-wheel drive models have a CV (constant velocity) joint, some of which require a special "needle" nozzle on the grease gun

rags, crawl under the vehicle and begin lubricating the components.

6 Wipe the balljoint grease fitting nipple clean and push the nozzle firmly over it **(see illustration)**. Squeeze the trigger on the grease gun to force grease into the component. The balljoints should be lubricated until the rubber seal is firm to the touch. Do not pump too much grease into the fittings as it could rupture the seal. For all other suspension and steering components, continue pumping grease into the fitting until it oozes out of the joint between the two components. If it escapes around the grease gun nozzle, the nipple is clogged or the nozzle is not completely seated on the fitting. Resecure the gun nozzle to the fitting and try again. If necessary, replace the fitting with a new one.

7 Wipe the excess grease from the components and the grease fitting. Repeat the procedure for the remaining fittings.

8 If equipped with a manual transmission, lubricate the shift linkage with a little multipurpose grease. Later 4-speed transmissions also have a grease fitting so the grease gun can be used.

9 On 4-wheel drive models, lubricate the front driveshaft Constant Velocity (CV) joint, located at the transfer case end, using a special "needle" nozzle on the grease gun **(see illustration)**. Lubricate the transfer case shift mechanism contact surfaces with clean engine oil.

10 If equipped, lubricate the driveshaft slip joints by pumping grease into the fitting until it can be seen coming out of the vent hole **(see illustration)**. On Dana model driveshafts, unscrew the grease cap and slide it back. Cover the vent hole with your finger and continue pumping grease until it can be seen leaving at the slip yoke seal. Install the grease cap securely.

11 On manual transmission-equipped models, lubricate the clutch linkage pivot points with clean engine oil. Lubricate the pushrod-to-fork contact points with chassis grease. Some models also have a grease fitting on the clutch cross-shaft) and at the clutch ball pivot on the bellhousing.

12 While you are under the vehicle, clean and lubricate the parking brake cable along with the cable guides and levers. This can be done by smearing some of the chassis grease onto the cable and its related parts with your fingers.

13 The steering gear seldom requires the addition of lubricant, but if there is obvious leakage of grease at the seals, remove the plug or cover and check the lubricant level. If the level is low, add the specified lubricant.

14 Lubricate the clutch cross-shaft (located in the passenger compartment, under the dash) at the grease fitting.

15 Open the hood and smear a little chassis grease on the hood latch mechanism. Have an assistant pull the hood release lever

13.10 Lubricate the driveshaft slip joint by pumping grease into the fitting until the grease can be seen coming out of the vent hole (arrow)

from inside the vehicle as you lubricate the cable at the latch.

16 Lubricate all the hinges (door, hood, etc.) with engine oil to keep them in proper working order.

17 The key lock cylinders can be lubricated with spray-on graphite or silicone lubricant, which is available at auto parts stores.

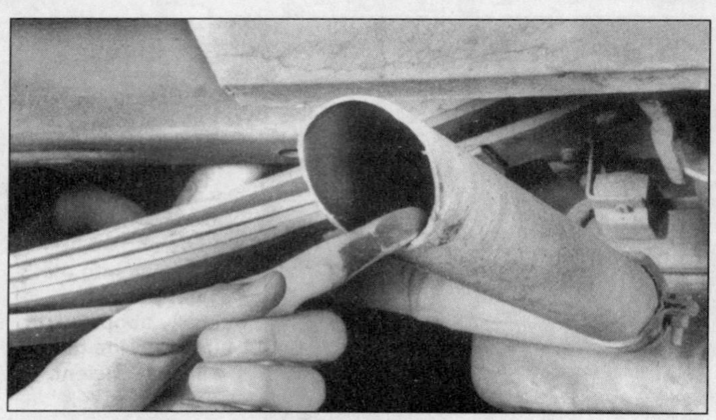

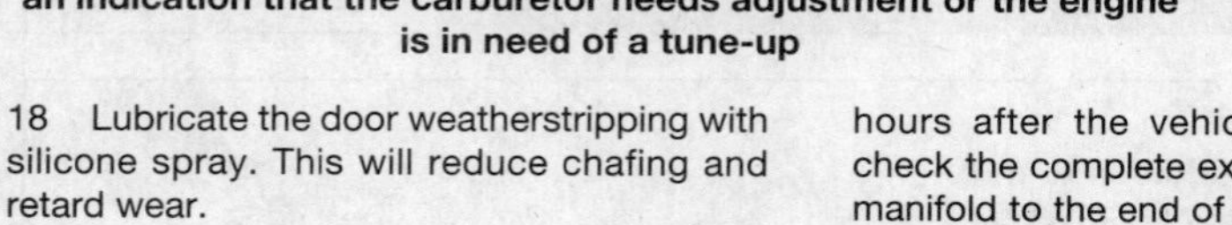

15.5 Black, sooty deposits at the end of the exhaust pipe may be an indication that the carburetor needs adjustment or the engine is in need of a tune-up

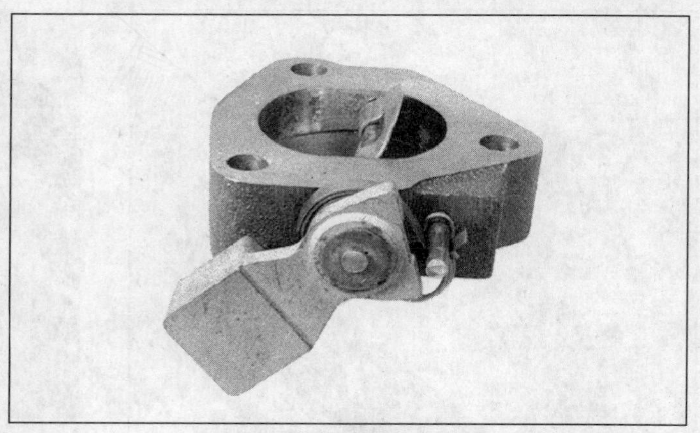

16.2 Typical heat riser

18 Lubricate the door weatherstripping with silicone spray. This will reduce chafing and retard wear.

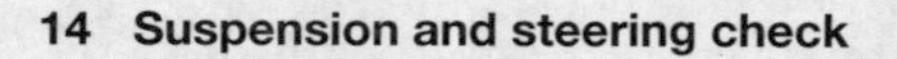

14 Suspension and steering check

1 Raise the front of the vehicle periodically and visually check the suspension and steering components for wear.

2 Indications of a fault in these systems are excessive play in the steering wheel before the front wheels react, excessive sway around corners, body movement over rough roads or binding at some point as the steering wheel is turned.

3 Raise the front end of the vehicle and support it securely on jackstands placed under the frame rails. Because of the work to be done, make sure the vehicle cannot fall from the stands.

4 Check the wheel bearings. Do this by spinning the front wheels. Listen for any abnormal noises and watch to make sure the wheel spins true (does not wobble). Grabbing the top and bottom of the tire, pull in and out on the tire, noticing any movement which would indicate a loose wheel bearing assembly. If the bearings are suspect, refer to Section 39 and Chapter 8 for more information.

5 From under the vehicle check for loose bolts, broken or disconnected parts and deteriorated rubber bushings on all suspension and steering components. Look for grease or fluid leaking from the steering assembly. Check the power steering hoses and connections for leaks.

6 Have an assistant turn the steering wheel from side-to-side and check the steering components for free movement, chafing and binding. If the steering does not react with the movement of the steering wheel, try to determine where the slack is located.

15 Exhaust system check

Refer to illustration 15.5

1 With the engine cold (at least three hours after the vehicle has been driven), check the complete exhaust system from the manifold to the end of the tailpipe. Be careful around the catalytic converter, which may be hot even after three hours. The inspection should be done on a hoist where unrestricted access is available.

2 Check the pipes and connections for signs of leakage and/or corrosion indicating a potential failure. Make sure that all brackets and hangers are in good condition and tight.

3 Inspect the underside of the body for holes, corrosion, open seams, etc. which may allow exhaust gases to enter the passenger compartment. Seal all body openings with silicone or body putty.

4 Rattles and other noises can often be traced to the exhaust system, especially the hangers, mounts and heat shields. Try to move the pipes, mufflers and catalytic converter. If the components can come in contact with the body or suspension parts, secure the exhaust system with new brackets and hangers.

5 This is an ideal time to check the running condition of the engine by inspecting the very end of the tail pipe. The exhaust deposits here are an indication of engine state-of-tune. If the pipe is black and sooty **(see illustration)**, or bright white deposits are found here, the engine is in need of a tune-up including a thorough carburetor inspection and adjustment.

16 EFE (heat riser) system check

Refer to illustrations 16.2 and 16.5

1 The heat riser and the Early Fuel Evaporation (EFE) system both perform the same job, but each functions in a slightly different manner.

2 The heat riser is a valve inside the exhaust pipe, near the junction between the exhaust manifold and pipe. It can be identified by an external weight and spring **(see illustration)**.

3 With the engine and exhaust pipe cold, try moving the weight by hand. It should move freely.

4 Again with the engine cold, start the engine and watch the heat riser. Upon starting, the weight should move to the closed position. As the engine warms to normal operating temperature, the weight should move the valve to the open position, allowing a free flow of exhaust through the tailpipe. Since it could take several minutes for the system to heat up, you could mark the cold weight position, drive the vehicle, and then check the weight.

16.5 Typical EFE system component layout

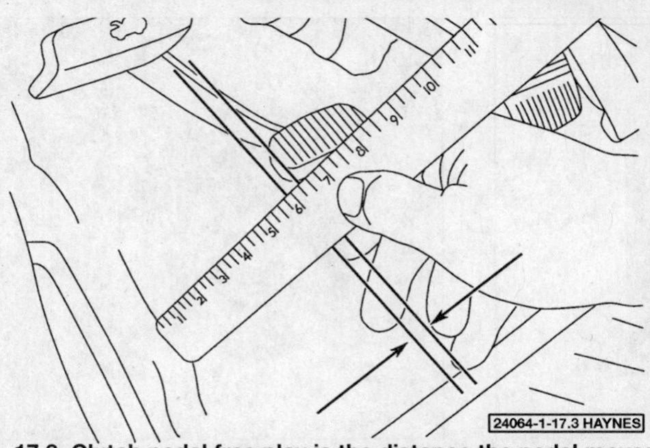

17.3 Clutch pedal free play is the distance the pedal moves before resistance is felt

18.1 The manual transmission has two plugs; the upper one for checking and filling and the lower one for draining

5 The EFE system also blocks off exhaust flow when the engine is cold. However, this system uses more precise temperature sensors and vacuum to open and close the exhaust pipe valve **(see illustration)**.

6 Locate the EFE actuator, which is bolted to a bracket on the right side of the engine. It will have an actuating rod attached to it which will lead down to the valve inside the pipe. In some cases the entire mechanism, including actuator, will be located at the exhaust pipe-to-manifold junction.

7 With the engine cold, have an assistant start the engine as you watch the actuating rod. It should immediately move to close off the valve as the engine warms. This process may take some time, so you might want to mark the position of the rod when the valve is closed, drive the vehicle to reach normal operating temperature, then open the hood and check that the rod has moved to the open position.

8 Further information on the EFE system can be found in Chapter 6.

17 Clutch pedal free play check and adjustment

Refer to illustration 17.3

Check

1 On 1967 through 1984 manual transmission models, it is important to have the clutch free play at the proper point. Free play is the distance between the clutch pedal when it is all the way up and the point at which the clutch starts to disengage.

2 Slowly depress the clutch pedal until you can feel resistance. Do this a number of times until you can pinpoint exactly where the resistance is felt.

3 Now measure the distance the pedal travels before the resistance is felt **(see illustration)**.

4 If the distance is not as specified, the clutch pedal free play should be adjusted.

Adjustment

Early models

5 Alter the effective length of the pushrod by loosening and then turning the locknuts as necessary.

Later models

6 Disconnect the clutch fork return spring.

7 Rotate the clutch lever and shaft assembly until the clutch pedal is firmly against the rubber bumper on the brake pedal bracket.

8 Push the outer end of the clutch fork to the rear until the release bearing lightly contacts the pressure plate fingers or bracket.

9 Loosen the locknut, and adjust the fork rod length so that the swivel slips freely into the gauge hole. With the swivel located in the gauge hole, turn the fork rod until all play is removed from the clutch system.

10 Remove the swivel from the gauge hole and insert it into the lower hole on the lever.

11 Tighten the locknut, taking care not to change the rod length, and reconnect the return spring.

12 Recheck the free play.

18 Manual transmission oil level check

Refer to illustration 18.1

1 Manual transmissions do not have a dipstick. The lubricant level is checked by removing the plug in the side of the transmission case **(see illustration)**.

2 If the oil level is not at the bottom of the plug opening, use a syringe to squeeze the appropriate lubricant into the opening until it just starts to run out of the hole.

3 Install the plug and tighten it securely. Drive the vehicle a short distance, then check for leaks.

19 Transfer case oil level check

Refer to illustration 19.1

1 The transfer case oil is checked by removing the upper plug located in the side of the case **(see illustration)**.

2 After removing the plug, reach inside the hole. The oil level should be just at the bottom of the hole. If not, add the appropriate lubricant through the opening.

20 Differential oil level check

1 Remove the oil check/fill plug from the side of the differential or the rear cover.

2 The oil level should be at the bottom of the plug opening. If not, use a syringe to add the proper lubricant until it just starts to run out of the opening. On some models a tag is located in the area of the plug which gives information regarding lubricant type, particularly on models equipped with a limited slip differential.

21 Tire rotation

Refer to illustration 21.2

1 The tires should be rotated at the specified intervals and whenever uneven wear is noticed.

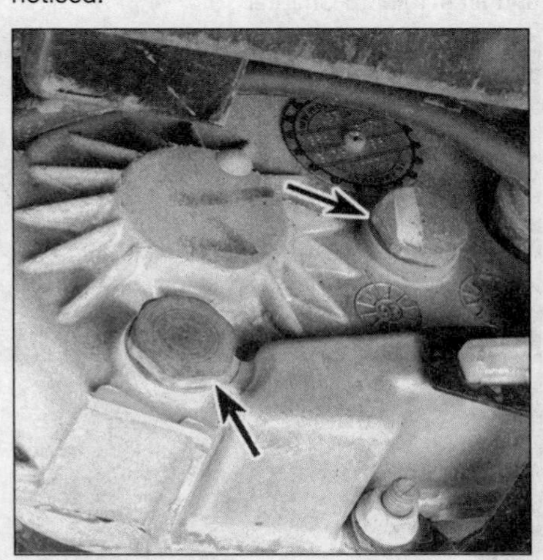

19.1 The transfer case oil check and drain plugs (arrows)

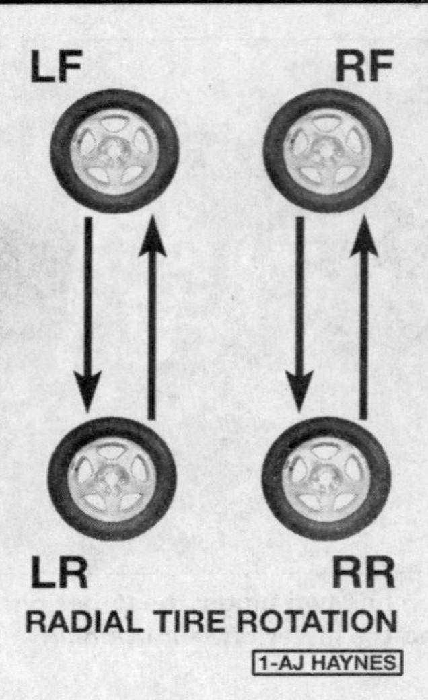

21.2 Tire rotation diagram

2 Refer to the **accompanying illustration** for the preferred tire rotation pattern. On vehicles with dual rear wheels, the tire with the largest diameter or the least wear should be mounted on the outside.

3 Refer to the information in *Jacking and towing* at the front of this manual for the proper procedures to follow when raising the vehicle and changing a tire. If the brakes are to be checked, do not apply the parking brake as stated. Make sure the tires are blocked to prevent the vehicle from rolling as it is raised.

4 Preferably, the entire vehicle should be raised at the same time. This can be done on a hoist or by jacking up each corner and then lowering the vehicle onto jackstands placed under the frame rails. Always use four jackstands and make sure the vehicle is safely supported.

5 After rotation, check and adjust the tire pressures as necessary and be sure to check the lug nut tightness.

6 For further information on the wheels and tires, refer to Chapter 10.

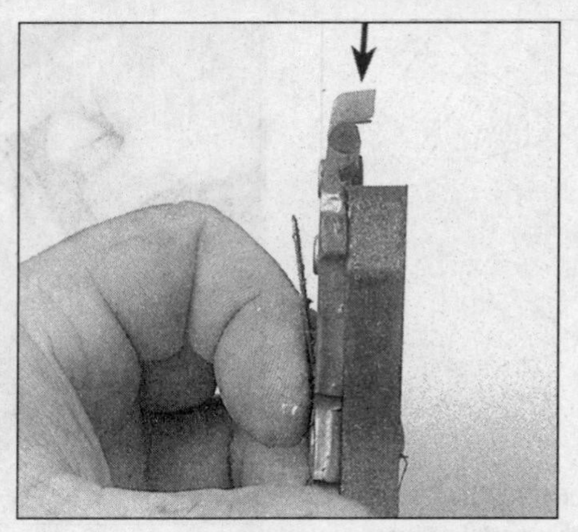

22.4 Front disc brake wear indicators make a screeching sound to alert you to the need for pad replacement

22 Brake check

Refer to illustrations 22.4, 22.6, 22.11, 22.12, 22.14 and 22.16

Note: *For detailed photographs of the brake system, refer to Chapter 9.*

1 In addition to the specified intervals, the brakes should be inspected every time the wheels are removed or whenever a defect is suspected.

2 To check the brakes, raise the vehicle and place it securely on jackstands. Remove the wheels (see Jacking and towing at the front of the manual, if necessary).

Disc brakes

3 Disc brakes are used on the front wheels of most models. Extensive rotor damage can occur if the pads are not replaced when needed.

4 Most later models are equipped with a wear sensor attached to the inner pad. This is a small, bent piece of metal which is visible from the inboard side of the brake caliper. When the pad wears to the specified limit, the metal sensor rubs against the rotor and makes a screeching sound **(see illustration)**.

22.6 Check the pad lining thickness by looking through the opening in the caliper

5 The disc brake calipers, which contain the pads, are visible with the wheels removed. There is an outer pad and an inner pad in each caliper. All pads should be inspected.

6 The caliper has a "window" to inspect the pads. Check the thickness of the lining by looking into the caliper at each end and down through the inspection window at the top of the housing **(see illustration)**. If the wear sensor is very close to the rotor or the pad material has worn to about 1/8-inch or less, the pads require replacement.

7 If you are unsure about the exact thickness of the remaining lining material, remove the pads for further inspection or replacement (refer to Chapter 9).

8 Before installing the wheels, check for leakage and/or damage (cracking, splitting, etc.) around the brake hose connections. Replace the hose or fittings as necessary, referring to Chapter 9.

9 Check the condition of the rotor. Look for scoring, gouging and burned spots. If these conditions exist, the hub/rotor assembly should be removed for servicing (Chapter 9).

Drum brakes

10 On front drum brakes, remove the

22.11 Use a hammer and chisel to remove the plug in the face of the drum

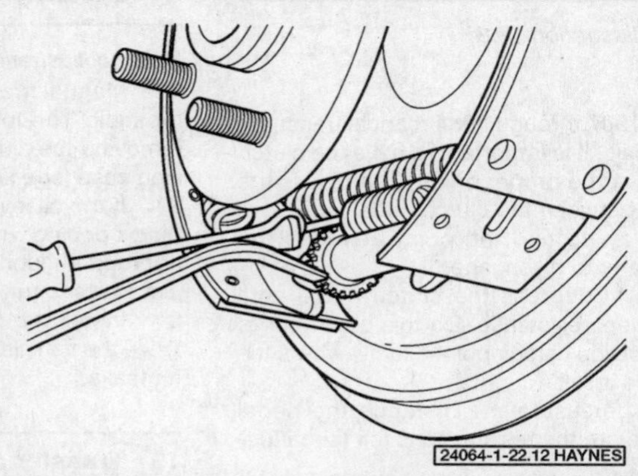

22.12 Hold the adjuster lever out of the way with a screwdriver while rotating the star wheel to back off the adjuster

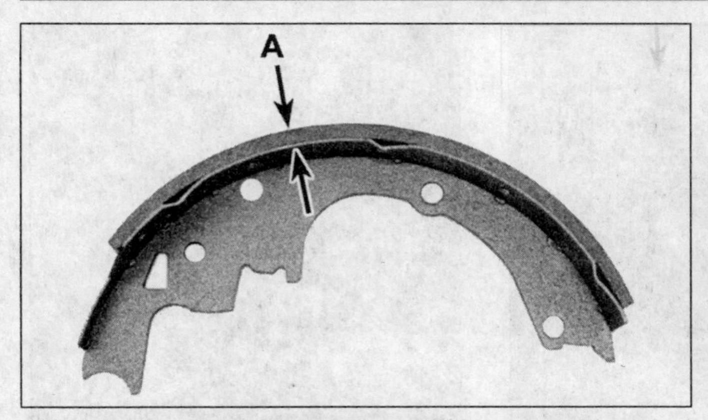

22.14 The brake shoe lining thickness (A) is measured from the outer surface of the lining to the metal shoe

22.16 Leakage often occurs from the wheel cylinder located at the top of the brake shoes

hub/drum (see Section 39 [2WD] or Chapter 8 [4WD]).

11 On rear brakes, remove the drum by pulling it off the axle and brake assembly. If this proves difficult, make sure the parking brake is released, then squirt penetrating oil around the center hub areas. Allow the oil to soak in and try again to pull the drum off. If the drum still cannot be pulled off, the brake shoes will have to be adjusted. This is done by first removing the plug from the drum or backing plate with a hammer and chisel **(see illustration)**.

12 With the plug removed, rotate the drum until the opening lines up with the adjuster wheel. Push or pull the lever off the star wheel and then use a small screwdriver to turn the star wheel, which will move the linings away from the drum **(see illustration)**.

13 With the drum removed, do not touch any brake dust. **Warning:** *Brake system dust may contain asbestos, which is harmful to your health. Never blow it out with compressed air and do not inhale any of it. Wash the assembly with the brake system cleaner.*

14 Note the thickness of the lining material on both the front and rear brake shoes. If the material has worn away to within 1/32-inch of the recessed rivets or metal backing, the shoes should be replaced **(see illustration)**. The shoes should also be replaced if they are cracked, glazed (shiny surface) or wet with brake fluid.

15 Check that all the brake assembly springs are connected and in good condition.

16 Check the brake components for any signs of fluid leakage. With your finger, carefully pry back the rubber cups on the wheel cylinders located at the top of the brake shoes **(see illustration)**. Any leakage is an indication that the wheel cylinders should be overhauled immediately (Chapter 9). Also check brake hoses and connections for signs of leakage.

17 Wipe the inside of the drum with a clean rag and brake cleaner or denatured alcohol. Again, be careful not to breath the dangerous asbestos dust.

18 Check the inside of the drum for cracks, score marks, deep scratches and hard spots, which will appear as small discolorations. If these imperfections cannot be removed with fine emery cloth, the drum must be taken to a machine shop equipped to turn the drums.

19 If after the inspection process all parts are in good working condition, reinstall the brake drum (using a metal or rubber plug if the knockout was removed).

20 Install the wheels and lower the vehicle.

Parking brake

21 The parking brake operates from a hand lever or foot pedal and locks the rear brake system. The easiest, and perhaps most obvious method of periodically checking the operation of the parking brake assembly is to park the vehicle on a steep hill with the parking brake set and the transmission in Neutral. If the parking brake cannot prevent the vehicle from rolling, it is in need of adjustment (see Chapter 9).

Brake pedal travel

22 The brakes should be periodically checked for pedal travel, which is the distance the brake pedal moves toward the floor from a fully released position. The brakes must be cold while performing this test.

23 Using a ruler, measure the distance from the floor to the brake pedal.

24 Pump the brakes at least three times. On power brake models do this without starting the engine. Press firmly on the brake pedal and measure the distance between the floor and the pedal.

25 The distance the pedal travels should not exceed 3-1/2 inches (manual brakes) or 4-1/2 inches (power brakes).

23 Fuel system check

Warning: *There are certain precautions to take when inspecting or servicing the fuel system components. Work in a well ventilated area and do not allow open flames (cigarettes, appliance pilot lights, etc.) in the work area. Mop up spills immediately and do not store fuel soaked rags where they could ignite. On fuel injection equipped models the fuel system is under pressure and no component should be disconnected without first relieving the pressure (see Chapter 4).*

1 On some models the main fuel tank is located behind the seat, inside the passenger cab. On other models the fuel tank may be located under one or both sides of the bed liner, just behind the cab.

2 The fuel system is most easily checked with the vehicle raised on a hoist so the components underneath the vehicle are readily visible and accessible.

3 If the smell of gasoline is noticed while driving or after the vehicle has been in the sun, the system should be thoroughly inspected immediately.

4 Remove the gas filler cap and check for damage, corrosion and an unbroken sealing imprint on the gasket. Replace the cap with a new one if necessary.

5 With the vehicle raised, inspect the gas tank and filler neck for punctures, cracks and other damage. The connection between the filler neck and the tank is especially critical. Sometimes a rubber filler neck will leak due to loose clamps or deteriorated rubber, problems a home mechanic can usually rectify. **Warning:** *Do not, under any circumstances, try to repair a fuel tank yourself (except rubber components) unless you have had considerable experience. A welding torch or any open flame can easily cause the fuel vapors to explode if the proper precautions are not taken.*

6 Carefully check all rubber hoses and metal lines leading away from the fuel tank. Check for loose connections, deteriorated hoses, crimped lines and other damage. Follow the lines to the front of the vehicle, carefully inspecting them all the way. Repair or replace damaged sections as necessary.

7 If a fuel odor is still evident after the inspection, refer to Section 42.

24 Air filter and PCV filter replacement

Refer to illustrations 24.7 and 24.13

1 At the specified intervals, the air filter and (on later models) PCV filter should be

24.7 When replacing the air filter, be sure to wipe the inside of the housing clean

24.13 The PCV filter on most models is located in the air cleaner housing

25.7 Two wrenches are required to loosen the fuel inlet nuts

replaced with new ones. A thorough program of preventive maintenance would call for the two filters to be inspected between changes. The engine air cleaner also supplies filtered air to the PCV system.

Oil bath air cleaner

2 Release the clamp screw at the base of the reservoir and lift the cleaner assembly off the carburetor.

3 Remove the wing nut and take off the cover and element.

4 Release the clamp screw and remove the air intake horn from the carburetor. Loosen the stud wing nut to allow removal of the reservoir.

5 Drain the oil from the reservoir and clean all components with solvent.

6 Reassemble and install the air cleaner components. Fill the reservoir with SAE 50 engine oil when operating in above freezing temperatures, or SAE 20 below freezing.

Paper element air cleaner

7 The filter is located on top of the carburetor or Throttle Body Injection (TBI) unit and is replaced by unscrewing the wing nut from the top of the filter housing and lifting off the cover **(see illustration)**.

8 While the top plate is off, be careful not to drop anything down into the carburetor, TBI or air cleaner assembly.

9 Lift the air filter element out of the housing and wipe out the inside of the air cleaner housing with a clean rag.

10 If so equipped, remove the polywrap band from the paper element and discard the element. If the band is in good condition, rinse it in kerosene or solvent and squeeze it dry. Dip the band in clean engine oil and gently squeeze out the excess. Install the band on a new paper element and reassemble.

11 Place the new filter into the air cleaner housing. Make sure it seats properly in the bottom of the housing.

12 The PCV filter is also located inside the air cleaner housing. Remove the top plate and air filter as previously described, then locate the PCV filter on the inside of the housing.

13 Remove the old filter **(see illustration)**.

14 Install the new PCV filter and the new air filter.

15 Install the top plate and any hoses which were disconnected.

25 Fuel filter replacement

Refer to illustrations 25.7, 25.9 and 25.16

Warning: *Gasoline is extremely flammable, so extra precautions must be taken when working on any part of the fuel system. Do not smoke or allow open flames or bare light bulbs near the work area. Also, don't work in a garage if a natural gas-type appliance with a pilot light is present.*

Early models

1 Unscrew the filter bowl for access to the disposable element. Replace the element and install the bowl, making sure the sealing ring is in good condition.

Later models

Carburetor equipped

2 On these models the fuel filter is located inside the fuel inlet nut at the carburetor. It is made of either pleated paper or porous bronze and cannot be cleaned or reused.

3 The job should be done with the engine cold (after sitting at least three hours). The necessary tools include open-end wrenches to fit the fuel line nuts. Flare nut wrenches (which wrap around the nut) should be used if available. In addition, you have to obtain the replacement filter (make sure it is for your specific vehicle and engine) and some clean rags.

4 Remove the air cleaner assembly. If vacuum hoses must be disconnected, be sure to note their positions and/or tag them to ensure that they are reinstalled correctly.

5 Follow the fuel line from the fuel pump to the point where it enters the carburetor. In most cases the fuel line will be metal all the way from the fuel pump to the carburetor.

6 Place some rags under the fuel inlet fittings to catch spilled fuel as the fittings are disconnected.

7 With the proper size wrench, hold the fuel inlet nut immediately next to the carburetor body. Now loosen the fitting at the end of the metal fuel line. Make sure the fuel inlet nut next to the carburetor is held securely while the fuel line is disconnected **(see illustration)**.

8 After the fuel line is disconnected, move it aside for better access to the inlet nut. Do not crimp the fuel line.

9 Unscrew the fuel inlet nut, which was previously held steady. As this fitting is drawn away from the carburetor body, be careful not to lose the thin washer-type gasket on

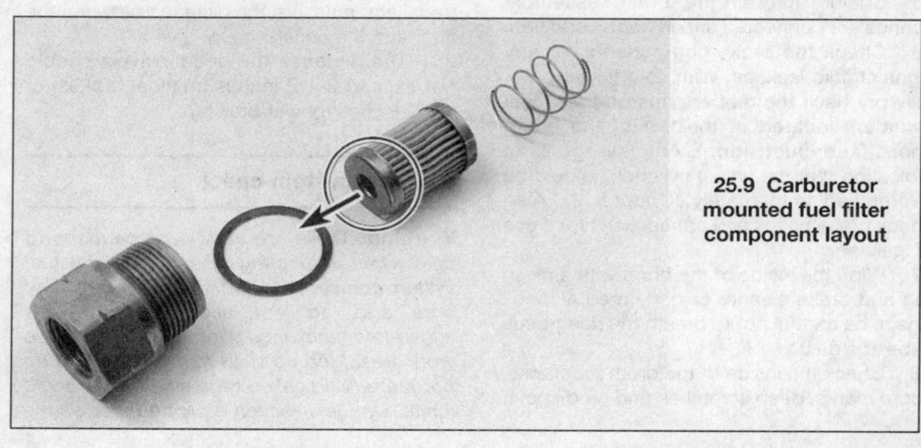

25.9 Carburetor mounted fuel filter component layout

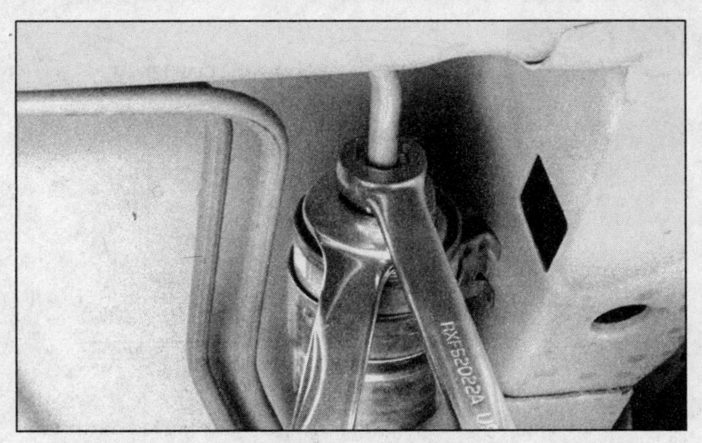

25.16 Two wrenches are required to loosen the fuel injection filter lines

26.3 The carburetor choke plate is visible after removing the air cleaner top plate

the nut or the spring, located behind the fuel filter. Also pay close attention to how the filter is installed **(see illustration)**.

10 Compare the old filter with the new one to make sure they are the same length and design.

11 Reinstall the spring in the carburetor body.

12 Place the filter in position (a gasket is usually supplied with the new filter) and tighten the nut. Make sure it is not cross-threaded. Tighten it securely, but be careful not to overtighten as the threads can strip easily, causing fuel leaks. Reconnect the fuel line to the fuel inlet nut, again using caution to avoid cross-threading the nut. Use a back-up wrench on the fuel inlet nut while tightening the fuel line fitting.

13 Start the engine and check carefully for leaks. If the fuel line fitting leaks, disconnect it and check for stripped or damaged threads. If the fuel line fitting has stripped threads, remove the entire line and have a repair shop install a new fitting. If the threads look all right, purchase some thread sealing tape and wrap the threads with it. Inlet nut repair kits are available at most auto parts stores to overcome leaking at the fuel inlet nut.

Fuel injected models

14 Fuel injected engines employ an in-line fuel filter. The filter is located on the left side frame rail.

15 With the engine cold, place a container, newspapers or rags under the fuel filter.

16 Use wrenches to disconnect the fuel lines and detach the filter from the frame, noting its direction of installation **(see illustration)**.

17 Install the new filter by reversing the removal procedure. Tighten the fittings securely, but do not cross-thread them.

26 Carburetor choke check

Refer to illustration 26.3

1 The choke operates when the engine is cold, and thus this check can only be performed before the vehicle has been started for the day.

2 Open the hood and remove the top plate of the air cleaner assembly. It is usually held in place by a wing nut at the center. If any vacuum hoses must be disconnected, make sure you tag the hoses for reinstallation in their original positions. Place the top plate and wing nut aside, out of the way of moving engine components.

3 Look at the center of the air cleaner housing. You will notice a flat plate at the carburetor opening **(see illustration)**.

4 Have an assistant press the accelerator pedal to the floor. The plate should close completely. Start the engine while you watch the plate at the carburetor. Do not position your face directly over the carburetor, as the engine could backfire, causing serious burns. When the engine starts, the choke plate should open slightly.

5 Allow the engine to continue running at an idle speed. As the engine warms up to operating temperature, the plate should slowly open, allowing more cold air to enter through the top of the carburetor.

6 After a few minutes, the choke plate should be fully open to the vertical position. Blip the throttle to make sure the fast idle cam disengages.

7 You will notice that the engine speed corresponds with the plate opening. With the plate fully closed, the engine should run at a fast idle speed. As the plate opens and the throttle is moved to disengage the fast idle cam, the engine speed will decrease.

8 Refer to Chapter 4 for specific information on adjusting and servicing the choke components.

27 Carburetor/throttle body mounting nut torque check

1 The carburetor or TBI unit is attached to the top of the intake manifold by two to four bolts or nuts. These fasteners can sometimes work loose from vibration and temperature changes during normal engine operation and cause a vacuum leak.

2 If you suspect that a vacuum leak exists at the bottom of the carburetor or throttle body, obtain a length of hose about the diameter of fuel hose. Start the engine and place one end of the hose next to your ear as you probe around the base with the other end. You will hear a hissing sound if a leak exists (be careful of hot or moving engine components).

3 Remove the air cleaner assembly, tagging each hose to be disconnected with a piece of numbered tape to make reassembly easier.

4 Locate the mounting nuts or bolts at the base of the carburetor or throttle body. Decide what special tools or adapters will be necessary, if any, to tighten the fasteners.

5 Tighten the nuts or bolts securely and evenly. Do not overtighten them, as the threads could strip.

6 If, after the nuts or bolts are properly tightened, a vacuum leak still exists, the carburetor or throttle body must be removed and a new gasket installed. See Chapter 4 for more information.

7 After tightening the fasteners, reinstall the air cleaner and return all hoses to their original positions.

28 Throttle linkage inspection

1 Inspect the throttle linkage for damaged or missing parts and for binding and interference when the accelerator pedal is operated.

2 Lubricate the various linkage pivot points with engine oil.

29 Thermostatic air cleaner check

Refer to illustrations 29.5 and 29.6

1 Later model engines are equipped with a thermostatically controlled air cleaner which draws air to the carburetor from different locations, depending upon engine temperature.

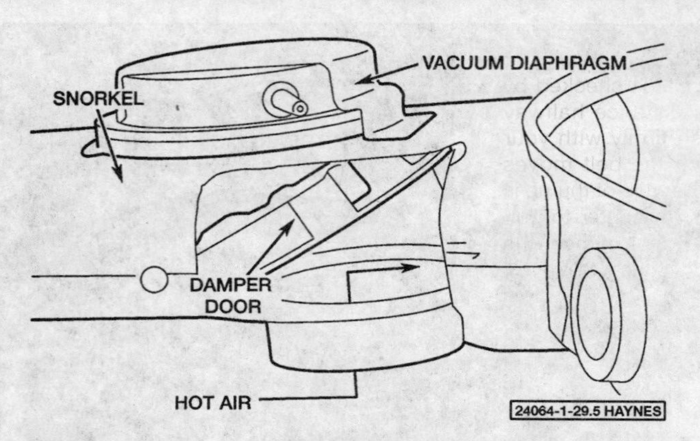

29.5 When the engine is cold, the damper door closes off the snorkel passage, allowing air warmed by the exhaust manifold to enter the carburetor

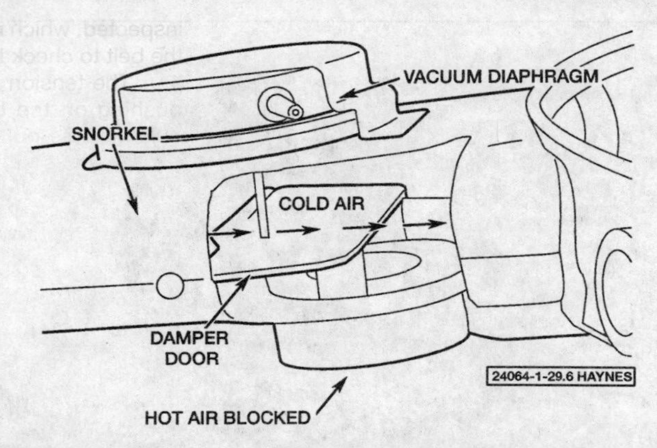

29.6 As the engine warms up, the damper door moves down to close off the heat stove passage and open the snorkel passage so outside air can enter the carburetor

2 This is a visual check. If access is limited, a small mirror may have to be used.

3 Open the hood and locate the damper door inside the air cleaner assembly. It is inside the long snorkel of the metal air cleaner housing.

4 If there is a flexible air duct attached to the end of the snorkel, leading to an area behind the grille, disconnect it at the snorkel. This will enable you to look through the end of the snorkel and see the damper inside.

5 The check should be done when the engine is cold. Start the engine and look through the snorkel at the damper, which should move to a closed position. With the damper closed, air cannot enter through the end of the snorkel, but instead enters the air cleaner through the flexible duct attached to the exhaust manifold and the heat stove passage **(see illustration)**.

6 As the engine warms up to operating temperature, the damper should open to allow air through the snorkel end **(see illustration)**. Depending on outside temperature, this may take 10 to 15 minutes. To speed up this check you can reconnect the snorkel air duct, drive the vehicle, then check to see if the damper is completely open.

7 If the thermo-controlled air cleaner is not operating properly see Chapter 6 for more information.

30 Idle speed check and adjustment

Refer to illustrations 30.4a and 30.4b

1 Engine idle speed is the speed at which the engine operates when no accelerator pedal pressure is applied. On fuel injected models this speed is governed by the ECM, while on carbureted models the idle speed can be adjusted. The idle speed is critical to the performance of the engine itself, as well as many engine sub-systems.

2 A hand-held tachometer must be used when adjusting idle speed to get an accurate reading. The exact hook-up for these meters varies with the manufacturer, so follow the particular directions included with the instrument.

3 Since the manufacturer used many different carburetors over the time period covered by this manual, and each varies somewhat when setting idle speed, it would be impractical to cover all types in this Section. Refer to Chapter 4 for information on specific carburetors. Most later models have a label located in the engine compartment with instructions for setting idle speed.

4 For most applications, the idle speed is set by turning an adjustment screw located on the side of the carburetor. This screw changes the amount the throttle plate is held open by the throttle linkage. The screw may be on the linkage itself or may be part of a device such as an idle stop solenoid **(see illustrations)**. Refer to the tune-up label or Chapter 4.

5 Once you have found the idle screw, experiment with different length screwdrivers until the adjustment can be easily made without coming into contact with hot or moving engine components.

6 Follow the instructions on the tune-up decal or in Chapter 4, which will probably include disconnecting certain vacuum or electrical connections. To plug a vacuum hose after disconnecting, insert a suitable

30.4a The idle speed screw (arrow) is usually found on the throttle linkage side of a carburetor, but can be easily confused with other carburetor adjustment screws - if you are unsure about this procedure, seek advice

30.4b On some models, the idle speed is controlled by an electric solenoid - the adjustment is made at the end of the solenoid

size metal rod into the opening, or thoroughly wrap the open end with tape to prevent any vacuum loss through the hose.

7 If the air cleaner is removed, the vacuum hose to the snorkel should be plugged.

8 Make sure the parking brake is firmly set and the wheels blocked to prevent the vehicle from rolling. This is especially true if the transmission is to be in Drive. An assistant inside the vehicle pressing on the brake pedal is the safest method.

9 For all applications, the engine must be completely warmed-up to operating temperature, which will automatically render the choke fast idle inoperative.

31 Drivebelt check and adjustment

Standard V-belts

Refer to illustrations 31.1, 31.3 and 31.4

1 The drivebelts, or V-belts as they are often called, are located at the front of the engine and play an important role in the overall operation of the vehicle and its components. Due to their function and material make-up, the belts are prone to failure after a period of time and should be inspected and adjusted periodically to prevent major engine damage.

2 The number of belts used on a particular vehicle depends on the accessories installed. Drivebelts are used to turn the alternator, power steering pump, water pump and air-conditioning compressor. Depending on the pulley arrangement, more than one of these components may be driven by a single belt.

3 With the engine off, open the hood and locate the various belts at the front of the engine. Using your fingers (and a flashlight, if necessary), move along the belts checking for cracks and separation of the belt plies. Also check for fraying and glazing, which gives the belt a shiny appearance **(see illustration)**. Both sides of each belt should be inspected, which means you will have to twist the belt to check the underside.

4 The tension of each belt is checked by pushing on the belt at a distance halfway between the pulleys. Push firmly with your thumb and see how much the belt moves (deflects) **(see illustration)**. A rule of thumb is that if the distance from pulley center-to-pulley center is between 7 and 11 inches, the belt should deflect 1/4-inch. If the belt travels between pulleys spaced 12 to 16 inches apart, the belt should deflect 1/2-inch.

5 If it is necessary to adjust the belt tension, either to make the belt tighter or looser, it is done by moving the belt-driven accessory on the bracket.

6 For each component there will be an adjusting bolt and a pivot bolt. Both bolts must be loosened slightly to enable you to move the component.

7 After the two bolts have been loosened, move the component away from the engine to tighten the belt or toward the engine to loosen the belt. Hold the accessory in position and check the belt tension. If it is correct, tighten the two bolts until just snug, then recheck the tension. If the tension is all right, tighten the bolts.

8 It will often be necessary to use some sort of pry bar to move the accessory while the belt is adjusted. If this must be done to gain the proper leverage, be very careful not to damage the component being moved or the part being pried against.

Serpentine drivebelt (some later models)

Refer to illustrations 31.10, 31.13 and 31.15

9 A single serpentine drivebelt is located at the front of the engine and plays an important role in the overall operation of the engine accessories. Due to its function and material makeup, the belt is prone to failure after a period of time and should be inspected periodically.

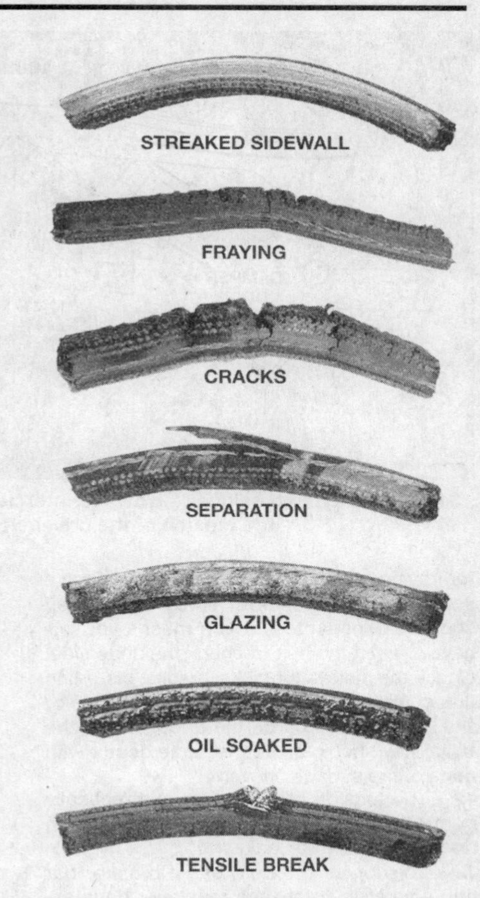

31.3 Here are some of the more common problems associated with drivebelts (check the belts very carefully to prevent an untimely breakdown)

10 With the engine off, locate the drive belt at the front of the engine. Using your fingers (and a flashlight if necessary), move along the belt, checking for cracks and separations of the belt plies **(see illustration)**. Also check

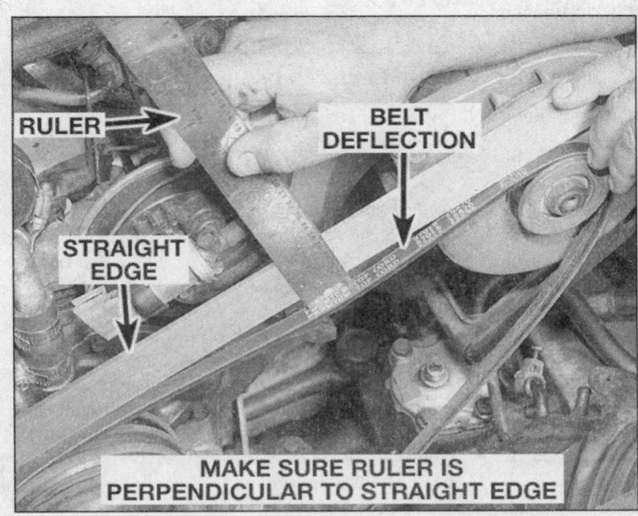

31.4 Drivebelt tension can be checked with a straightedge and a ruler

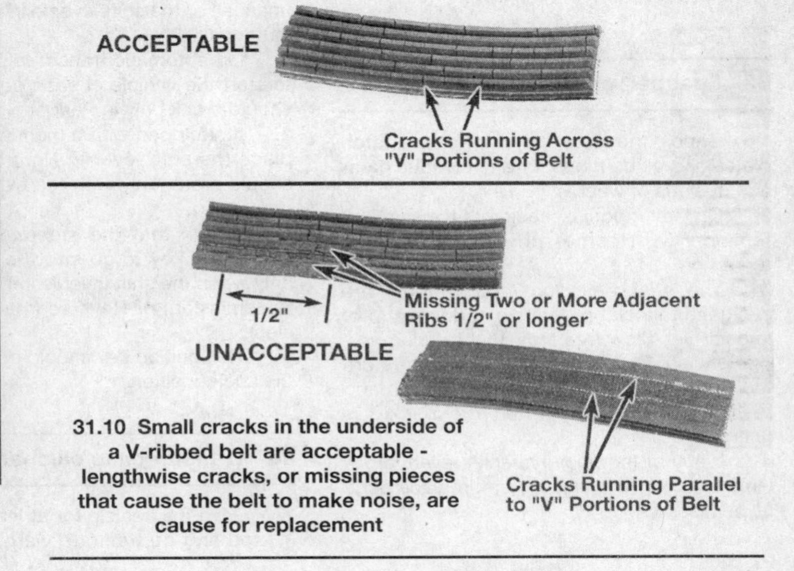

31.10 Small cracks in the underside of a V-ribbed belt are acceptable - lengthwise cracks, or missing pieces that cause the belt to make noise, are cause for replacement

31.13 Rotate the tensioner (arrow) counterclockwise to release the tension of the drivebelt

31.15 Typical drivebelt routing sticker

for fraying and glazing, which gives the belt a shiny appearance. Both sides of the belt should be inspected, which means you will have to twist the belt to check the underside. Check the pulleys for nicks, cracks, distortion and corrosion.

11 Check the ribs on the underside of the belt. They should all be the same depth, with none of the surface uneven.

12 The tension of the belt is automatically controlled by the tensioner, so the tension does not need to be adjusted.

13 To replace the belt, use a breaker bar and socket to rotate the tensioner counterclockwise **(see illustration)**. This will release the tension so the belt can be removed. When the belt is out of the way, release the tensioner slowly so you don't damage it.

14 Take the old belt with you when you purchase a new one to make a direct comparison for length, width and design.

15 When installing the new belt, make sure it is routed correctly (refer to the label in the engine compartment **(see illustration)**. Also, the belt must completely engage the grooves in the pulleys.

32 Seat belt check

1 Check the seat belts, buckles, latch plates and guide loops for any obvious damage or signs of wear.

2 On later models, check that the seat belt reminder light comes on when the key is turned on.

3 The seat belts on later models are designed to lock up during a sudden stop or impact, yet allow free movement during normal driving. Check that the retractors return the belt against your chest while driving and rewind the belt fully when the buckle is unlatched.

4 If any of the above checks reveal problems with the seat belt system, replace parts as necessary.

33 Seat back latch check

1 It is important to periodically check the seatback latch mechanism to prevent the seatback from pivoting forward during a sudden stop or an accident.

2 Grasping the top of the seat, attempt to tilt the seatback forward. It should tilt only when the latch mechanism is released.

3 When returned to the upright position, the seatback should latch securely.

34 Starter safety switch check

Warning: *During the following checks there is a chance that the vehicle could lunge forward, possibly causing damage or injuries. Allow plenty of room around the vehicle, firmly apply the parking brake and hold down the regular brake pedal during the checks.*

1 Later models are equipped with a starter safety switch which prevents the engine from starting unless the clutch pedal is depressed (manual) or the shift lever is in Neutral or Park (automatic).

2 On automatic transmission vehicles, try to start the vehicle in each gear. The engine should crank only in Park or Neutral.

3 If equipped with a manual transmission, place the shift lever in Neutral. The engine should crank only with the clutch pedal fully depressed.

4 Check that the steering column lock allows the key to go into the Lock position only when the shift lever is in Park (automatic transmission) or Reverse (manual transmission).

5 The ignition key should come out only in the Lock position.

35 Transfer case oil change

1 Drive the vehicle for at least 15 minutes in stop and go traffic to warm the oil in the case. Perform this warm-up procedure in 4-wheel drive. The manual locking hubs should be in the Lock position if the vehicle is so equipped. Use all gears including Reverse to ensure that the lubricant is sufficiently warm to drain completely.

2 Raise the vehicle and support it securely on jackstands.

3 Remove the filler plug from the case **(see illustration 19.1)**.

4 Remove the drain plug from the lower part of the case and allow the old oil to drain completely.

5 Carefully clean and install the drain plug after the case is completely drained. Tighten the plug to the specified torque.

6 Fill the case with the specified lubricant until it is level with the lower edge of the filler hole.

7 Install the filler plug and tighten it securely.

8 Drive the vehicle for a short distance and recheck the oil level. In some instances a small amount of additional lubricant will have to be added.

36 Manual transmission oil change

1 Raise the vehicle and support it securely on jackstands.

2 Move a drain pan, rags, newspapers and wrenches under the transmission.

3 Remove the transmission drain plug at the bottom of the case and allow the oil to drain into the pan **(see illustration 18.1)**.

4 After the oil has drained completely, reinstall the plug and tighten it securely.

5 Remove the fill plug in the side of the transmission case. Using a hand pump, syringe or funnel, fill the transmission with the correct amount of the specified lubricant. Reinstall the fill plug and tighten it securely.

6 Lower the vehicle.

7 Drive the vehicle for a short distance then check the drain and fill plugs for any sign of leakage.

38.7 After allowing some of the fluid to drain, completely remove the remaining bolts and lower the pan - be careful, there is still plenty of fluid in the pan!

38.10 The filter is held in place by bolts or screws

37 Differential oil change

1 Some differentials can be drained by removing the drain plug, while on others it is necessary to remove the cover plate on the differential housing. As an alternative, a hand suction pump can be used to remove the differential lubricant through the filler hole. If there is no drain plug and a suction pump is not available, be sure to obtain a new gasket at the same time the gear lubricant is purchased.
2 Move a drain pan, rags, newspapers and wrenches under the vehicle.
3 Remove the fill plug.
4 If equipped with a drain plug, remove the plug and allow the differential oil to drain completely. After the oil has drained, install the plug and tighten it securely.
5 If a suction pump is being used, insert the flexible suction hose. Work the hose down to the bottom of the differential housing and pump the oil out.
6 If the differential is being drained by removing the cover plate, remove the bolts on the lower half of the plate. Loosen the bolts on the upper half and use them to keep the cover loosely attached. Allow the oil to drain into the pan, then completely remove the cover.
7 Using a lint-free rag, clean the inside of the cover and the accessible areas of the differential housing. As this is done, check for chipped gears and metal particles in the lubricant, indicating that the differential should be more thoroughly inspected and/or repaired.
8 Thoroughly clean the gasket mating surfaces of the differential housing and the cover plate. Use a gasket scraper or putty knife to remove all traces of the old gasket.
9 Apply a thin layer of RTV-type gasket sealant to the cover flange and then press a new gasket into position on the cover. Make sure the bolt holes align properly.
10 Place the cover on the differential housing and install the bolts. Tighten the bolts securely.
11 On all models, use a hand pump, syringe or funnel to fill the differential housing with the specified lubricant until it is level with the bottom of the plug hole.
12 Install the filler plug and tighten it securely.

38 Automatic transmission fluid and filter change

Refer to illustrations 38.7, 38.10 and 38.12

1 At the specified time intervals, the transmission fluid should be drained and replaced. Since the fluid will remain hot long after driving, perform this procedure only after sufficient cooling.
2 Before beginning work, purchase the specified transmission fluid (see *Recommended lubricants and fluids* at the front of this Chapter) and filter.
3 Other tools necessary for this job include jackstands to support the vehicle in a raised position, a drain pan capable of holding at least 8 pints, newspapers and clean rags.
4 Raise the vehicle and support it securely on jackstands. **Note:** *Later models may be equipped with a transmission fluid drain plug. The transmission fluid can be drained before the pan is removed on these models (like changing the engine oil). Those models without a transmission drain plug must use the following procedure.*
5 With a drain pan in place, remove the front and side pan bolts.
6 Loosen the rear pan bolts approximately four turns.
7 Carefully pry the transmission pan loose with a screwdriver, allowing the fluid to drain **(see illustration)**.
8 Remove the remaining bolts, pan and gasket. Carefully clean the gasket surface of the transmission to remove all traces of the old gasket and sealant.

38.12 With the new filter in place, the pan is ready for installation (remember to tighten each of the bolts a little at a time to prevent warping)

9 Drain the fluid from the transmission pan, clean it with solvent and dry it with compressed air.
10 Remove the filter from the mount inside the transmission **(see illustration)**.
11 Install a new filter screen and gasket or O-ring.
12 Make sure the gasket surface on the transmission pan is clean, then install a new gasket. Put the pan in place against the transmission and, working around the pan, tighten each bolt a little at a time until the final torque figure is reached **(see illustration)**.
13 Lower the vehicle and add the specified amount of automatic transmission fluid through the filler tube (Section 6).
14 With the selector in Park and the parking brake set, run the engine at a fast idle, but do not race the engine.
15 Move the gear selector through each range and back to Park. Check the fluid level.
16 Check under the vehicle for leaks during the first few trips.

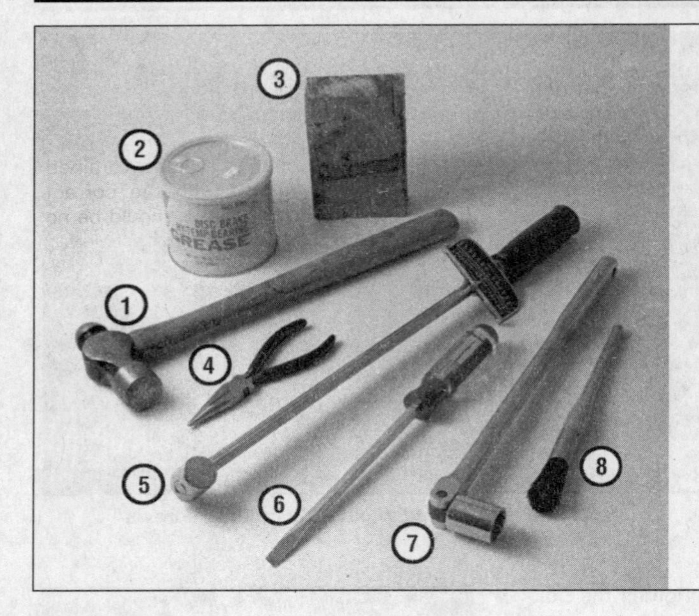

39.1 Tools and materials needed for front wheel bearing maintenance

1 ***Hammer*** *- A common hammer will do just fine*
2 ***Grease*** *- High-temperature grease which is formulated specially for front wheel bearings should be used*
3 ***Wood block*** *- If you have a scrap piece of 2x4, it can be used to drive the new seal into the hub*
4 ***Needle-nose pliers*** *- Used to straighten and remove the cotter pin in the spindle*
5 ***Torque wrench*** *- This is very important in this procedure; if the bearing is too tight, the wheel won't turn freely - if it is too loose, the wheel will "wobble' on the spindle. Either way, it could mean extensive damage*
6 ***Screwdriver*** *- Used to remove the seal from the hub (a long screwdriver would be preferred)*
7 ***Socket/breaker bar*** *- Needed to loosen the nut on the spindle if it is extremely tight*
8 ***Brush*** *- Together with some clean solvent, this will be used to remove old grease from the hub and spindle*

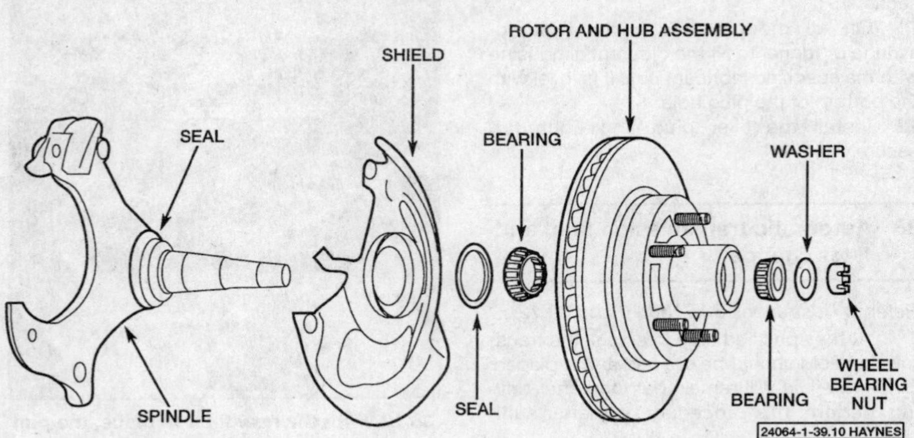

39.10 Front wheel bearing component layout for vehicles with disc brakes - drum brake equipped vehicles use the same bearing assembly, except that the bearings are mounted in the drum and the backing plate contains the brake shoe assembly

39.11 Use a screwdriver to pry the grease seal from the back side of the hub

39 Front wheel bearings (2-wheel drive) check, repack and adjustment

Refer to illustrations 39.1, 39.10, 39.11 and 39.15

Note: *See Chapter 8 for the 4-wheel drive front wheel bearing check, repack and adjustment procedure.*

1 In most cases the front wheel bearings will not need servicing until the brake pads are changed. However, the bearings should be checked whenever the front wheels are raised for any reason. Several items, including a torque wrench and special grease, are required for this procedure **(see illustration)**.

2 With the vehicle securely supported on jackstands, spin each wheel and check for noise, rolling resistance and free play.

3 Grasp the top of each tire with one hand and the bottom with the other. Move the wheel in-and-out on the spindle. If there is any noticeable movement the bearings should be checked and then repacked with grease or replaced if necessary.

4 Remove the wheel.

5 On disc brake equipped models fabricate a wood block (1-1/16 inch by 1/16-inch by 2-inches long) which can be slid between the brake pads to keep them separated. Remove the brake caliper (Chapter 9) and hang it out of the way on a piece of wire.

6 Pry the dust cap out of the hub using a screwdriver or hammer and chisel.

7 Use needle-nose pliers or a screwdriver to straighten the bent ends of the cotter pin and then pull the cotter pin out of the locking nut. Discard the cotter pin and use a new one during reassembly.

8 Remove the spindle nut and washer from the end of the spindle.

9 Pull the hub assembly out slightly and then push it back into its original position. This should force the outer bearing off the spindle enough so that it can be removed. On models equipped with drum brakes at the front it may be necessary to retract the brake shoes slightly to release the drum. If the drum will not slide off the brake shoes, refer to Chapter 9 for the brake drum removal procedure.

10 Pull the hub assembly off the spindle **(see illustration)**.

11 On the rear side of the hub, use a screwdriver to pry out the seal **(see illustration)**. As this is done, note the direction in which the seal is installed.

12 Again noting the direction in which it is installed, remove the inner bearing from the hub.

13 Use solvent to remove all traces of the old grease from the bearings, hub and spindle. A small brush may prove helpful; however make sure no bristles from the brush embed themselves inside the bearing rollers. Allow the parts to air dry.

14 Carefully inspect the bearings for

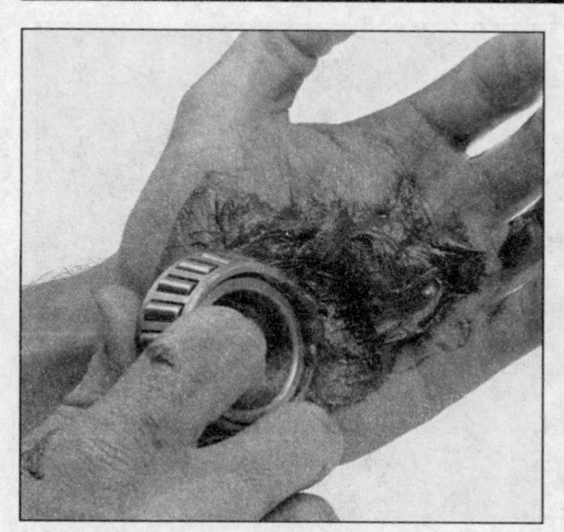

39.15 Work the grease completely into the rollers

cracks, heat discoloration, worn rollers, etc. Check the bearing races inside the hub for cracks, scoring and uneven surfaces. If the bearing races are defective, the hubs should be taken to a machine shop with the facilities to remove the old races and press new ones in. Note that the bearings and races come as matched sets, and old bearings should never be installed on new races.

15 Use high-temperature front wheel bearing grease to pack the bearings. Work the grease completely into the bearings, forcing it between the rollers, cone and cage from the back side **(see illustration)**.

16 Apply a thin coat of grease to the spindle at the outer bearing seat, inner bearing seat, shoulder and seal seat.

17 Put a small quantity of grease inboard of each bearing race inside the hub. Using your finger, form a dam at these points to provide extra grease availability and to keep thinned grease from flowing out of the bearing.

18 Place the grease-packed inner bearing into the rear of the hub and put a little more grease outboard of the bearing.

19 Place a new seal over the inner bearing and tap the seal evenly into place with a hammer and block of wood until it is flush with the hub.

20 Carefully place the hub assembly onto the spindle and push the grease-packed outer bearing into position.

21 Install the washer and spindle nut. Tighten the nut only slightly (no more than 12 ft-lbs of torque).

22 Spin the hub in a forward direction to seat the bearings and remove any grease or burrs which could cause excessive bearing play later.

23 Check to see that the tightness of the spindle nut is still approximately 12 ft-lbs.

24 Loosen the spindle nut until it is just loose, no more.

25 Using your hand (not a wrench of any kind), tighten the nut until it is snug. Install a new cotter pin through the hole in the spindle and spindle nut. If the nut slots do not line up, loosen the nut slightly until they do. From the hand-tight position, the nut should not be loosened more than one-half flat to install the cotter pin.

26 Bend the ends of the cotter pin until they are flat against the nut. Cut off any extra length which could interfere with the dust cap.

27 Install the dust cap, tapping it into place with a hammer.

28 If the vehicle is equipped with front disc brakes, place the brake caliper near the rotor and carefully remove the wood spacer. Install the caliper (Chapter 9).

29 If drum brakes are used at the front, adjust the brakes as described in Chapter 9.

30 Install the tire/wheel assembly on the hub and tighten the lug nuts.

31 Grasp the top and bottom of the tire and check the bearings in the manner described earlier in this Section.

32 Lower the vehicle.

40 Cooling system servicing (draining, flushing and refilling)

Warning: *Make sure the engine is completely cool before beginning this procedure.*

1 Periodically, the cooling system should be drained, flushed and refilled to replenish the antifreeze mixture and prevent formation of rust and corrosion, which can impair the performance of the cooling system and cause engine damage.

2 At the same time the cooling system is serviced, all hoses and the radiator cap should be inspected and replaced if defective (see Section 9).

3 Since antifreeze is a corrosive and poisonous solution, be careful not to spill any of the coolant mixture on the vehicle's paint or your skin. If this happens, rinse immediately with plenty of clean water. Consult your local authorities about the dumping of antifreeze before draining the cooling system. In many areas, reclamation centers have been set up to collect automobile oil and drained antifreeze/water mixtures, rather than allowing them to be added to the sewage system.

4 With the engine cold, remove the radiator cap.

5 Move a large container under the radiator to catch the coolant as it is drained.

6 Drain the radiator by opening the fitting at the bottom. If the drain has excessive corrosion and cannot be turned easily, or if the radiator is not equipped with a drain, disconnect the lower radiator hose to allow the coolant to drain. Be careful that none of the solution is splashed on your skin or into your eyes - wear safety glasses!.

7 If equipped, disconnect the hose from the coolant reservoir and remove the reservoir. Flush it out with clean water.

8 Place a garden hose in the radiator filler neck and flush the system until the water runs clear at all drain points.

9 In severe cases of contamination or clogging of the radiator, remove it (see Chapter 3) and reverse flush it. This involves inserting the hose in the bottom radiator outlet to allow the water to run against the normal flow, draining through the top. A radiator repair shop should be consulted if further cleaning or repair is necessary.

10 When the coolant is regularly drained and the system refilled with the correct antifreeze/water mixture, there should be no need to use chemical cleaners or descalers.

11 To refill the system, reconnect the radiator hoses and install the reservoir and the overflow hose.

12 Fill the radiator with the proper mixture of antifreeze and water (see Section 4) to the base of the filler neck and then add more coolant to the reservoir until it reaches the lower mark.

13 With the radiator cap still removed, start the engine and run until normal operating temperature is reached. With the engine idling, add additional coolant to the radiator to bring up to the proper level. Install the radiator and reservoir caps.

14 Keep a close watch on the coolant level and the cooling system hoses during the first few miles of driving. Tighten the hose clamps and/or add more coolant as necessary.

41 Positive Crankcase Ventilation (PCV) valve check and replacement

Refer to illustration 41.2

1 The PCV valve is usually located in the rocker arm cover.

2 With the engine idling at normal operating temperature, pull the valve (with hose attached) from the rubber grommet in the cover **(see illustration)**.

3 Place your finger over the end of the valve. If there is no vacuum at the valve, check for a plugged hose, manifold port, or the valve itself. Replace any plugged or deteriorated hoses.

4 Turn off the engine and shake the PCV valve, listening for a rattle. If the valve does not rattle, replace it with a new one.

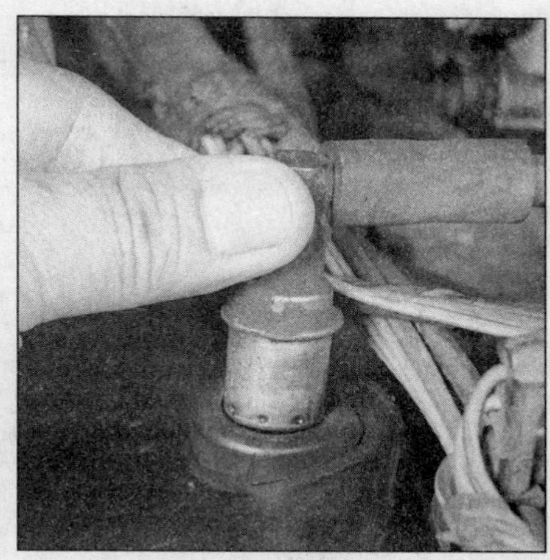

41.2 Most PCV valves are mounted on the valve cover, in a rubber grommet

42.2a Typical charcoal canister

42.2b On vehicles with two batteries, the charcoal canister is located under the battery tray

5 To replace the valve, pull it from the end of the hose, noting its installed position and direction.

6 When purchasing a replacement PCV valve, make sure it is for your particular vehicle and engine size. Compare the old valve with the new one to make sure they are the same.

7 Push the valve into the end of the hose until it is seated.

8 Inspect the rubber grommet for damage and replace it with a new one if necessary.

9 Push the PCV valve and hose securely into position.

42 Evaporative emissions control system check

Refer to illustration 42.2a and 42.2b

1 The function of the evaporative emissions control system is to draw fuel vapors from the gas tank and fuel system, store them in a charcoal canister and route them to the intake manifold during normal engine operation.

2 The most common symptom of a fault in the evaporative emissions system is a strong fuel odor in the engine compartment. If a fuel odor is detected, inspect the charcoal canister, located in the engine compartment **(see illustrations)**. Check the canister and all hoses for damage and deterioration.

3 The evaporative emissions control system is explained in more detail in Chapter 6.

43 Exhaust Gas Recirculation (EGR) system check

Refer to illustration 43.2

1 The EGR valve is usually located on the intake manifold, adjacent to the carburetor. Most of the time when a problem develops in this emissions system, it is due to a stuck or corroded EGR valve.

2 With the engine cold to prevent burns, reach under the EGR valve and push on the diaphragm. Using moderate pressure, you should be able to press the diaphragm up-and-down within the housing **(see illustration)**.

3 If the diaphragm does not move or moves only with much effort, replace the EGR valve with a new one. If in doubt about the condition of the valve, compare the free movement of your EGR valve with a new valve.

4 Refer to Chapter 6 for more information on the EGR system.

43.2 The diaphragm, located under the EGR valve, should be checked for free movement

44 Spark plug replacement

Refer to illustrations 44.1, 44.4a, 44.4b, 44.5 and 44.9

1 Before beginning, obtain the necessary tools, which will include a special spark plug wrench or socket and a gap measuring tool **(see illustration)**.

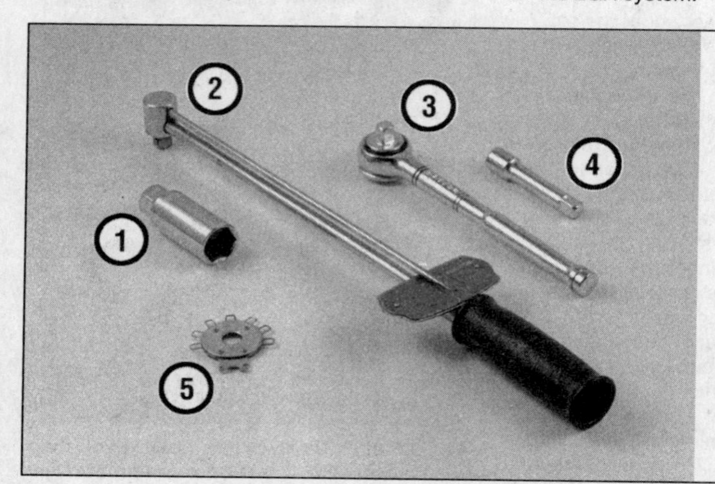

44.1 Tools required for changing spark plugs

1 ***Spark plug socket*** *- This will have special padding inside to protect the spark plug porcelain insulator*

2 ***Torque wrench*** *- Although not mandatory, use of this tool is the best way to ensure that the plugs are tightened properly*

3 ***Ratchet*** *- Standard hand tool to fit the plug socket*

4 ***Extension*** *- Depending on model and accessories, you may need special extensions and universal joints to reach one or more of the plugs*

5 ***Spark plug gap gauge*** *- This gauge for checking the gap comes in a variety of styles. Make sure the gap for your engine is included*

44.4a Spark plug manufacturers recommend using a wire type gauge when checking the gap - if the wire does not slide between the electrodes with a slight drag, adjustment is required

44.4b To change the gap, bend the side electrode only, as indicated by the arrows, and be very careful not to crack or chip the porcelain insulator surrounding the center electrode

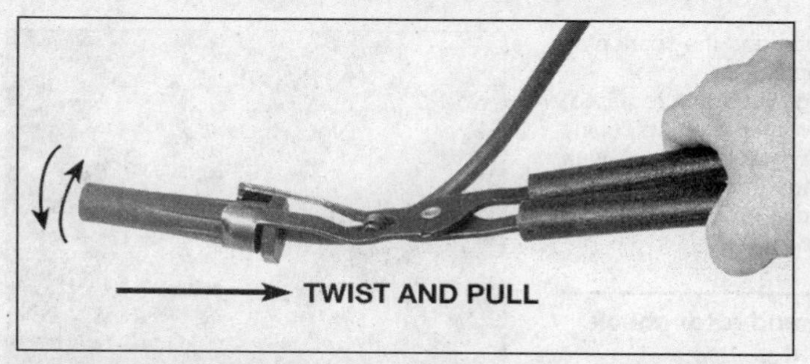

44.5 When removing a spark plug wire from a spark plug it is important to pull on the end of the boot (as shown on the left) and not on the wire itself (as shown at the right) - a slight twisting motion will also help

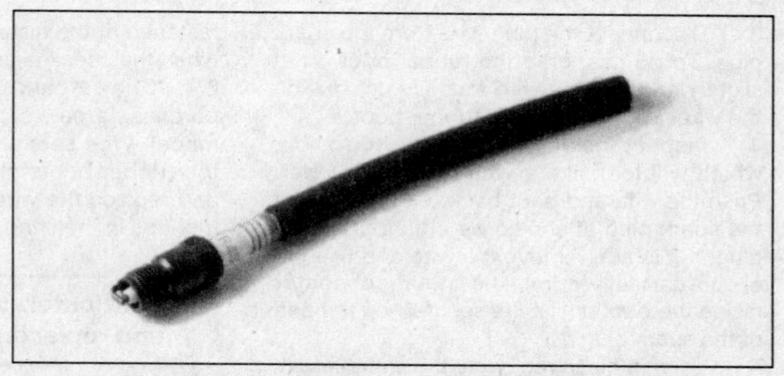

44.9 A length of snug-fitting rubber hose will save time and prevent damaged threads when installing the spark plugs

2 The best procedure to follow when replacing the spark plugs is to purchase the new spark plugs beforehand, adjust them to the proper gap, and then replace each plug one at a time. When buying the new spark plugs it is important to obtain the correct plugs for your specific engine. This information can be found on the tune-up or Vehicle Emissions Control Information label located under the hood or in the owner's manual. If differences exist between these sources, purchase the spark plug type specified on the label, because the information was printed for your specific engine.

3 With the new spark plugs on hand, allow the engine to cool completely before attempting plug removal. During this time, each of the new spark plugs can be inspected for defects and the gaps can be checked.

4 The gap is checked by inserting the proper thickness gauge between the electrodes at the tip of the plug **(see illustration)**. The gap between the electrodes should be the same as that given in the Specifications or on the Emissions Control label. The wire should just touch each of the electrodes. If the gap is incorrect, use the notched adjuster on the feeler gauge body to bend the curved side electrode slightly until the proper gap is achieved **(see illustration)**. If the side electrode is not exactly over the center electrode, use the notched adjuster to align the two. Check for cracks in the porcelain insulator, indicating the spark plug should not be used.

5 With the engine cool, remove the spark plug wire from one spark plug. Do this by grabbing the boot at the end of the wire, not the wire itself **(see illustration)**. Sometimes it is necessary to use a twisting motion while the boot and plug wire are pulled free.

6 If compressed air is available, use it to blow any dirt or foreign material away from the spark plug area. A common bicycle pump will also work. The idea here is to eliminate the possibility of debris falling into the cylinder as the spark plug is removed.

7 Place the spark plug socket over the plug and remove it from the engine by turning in a counterclockwise direction.

8 Compare the spark plug with those shown on the inside back cover of this manual to get an indication of the overall running condition of the engine.

9 Thread the new plug into the engine until you can no longer turn it with your fingers, then tighten it with the socket. Where there might be difficulty in inserting the spark plugs into the spark plug holes, or the possibility of cross-threading them into the head, a short piece of rubber tubing can be attached to the end of the spark plug **(see illustration)**. The flexible tubing will act as a universal joint to help align the plug with the plug hole, and should the plug begin to cross-thread, the hose will slip on the spark plug, preventing thread damage. If one is available, use a torque wrench to tighten the plug to ensure that it is seated correctly. The correct torque figure is included in the Specifications at the front of this Chapter.

10 Before pushing the spark plug wire onto the end of the plug, inspect the wire following the procedures outlined in Section 45.

11 Attach the plug wire to the new spark plug, again using a twisting motion on the boot until it is firmly seated on the spark plug. Make sure the wire is routed away from the exhaust manifold.

12 Follow the above procedure for the remaining spark plugs, replacing them one at a time to prevent mixing up the spark plug wires.

45 Spark plug wire check and replacement

1 The spark plug wires should be checked at the recommended intervals and whenever new spark plugs are installed in the engine.

2 The wires should be inspected one at a time to prevent mixing up the order, which is essential for proper engine operation.

46.3 After the screws or latches have been released, the cap can be lifted away for access to the rotor

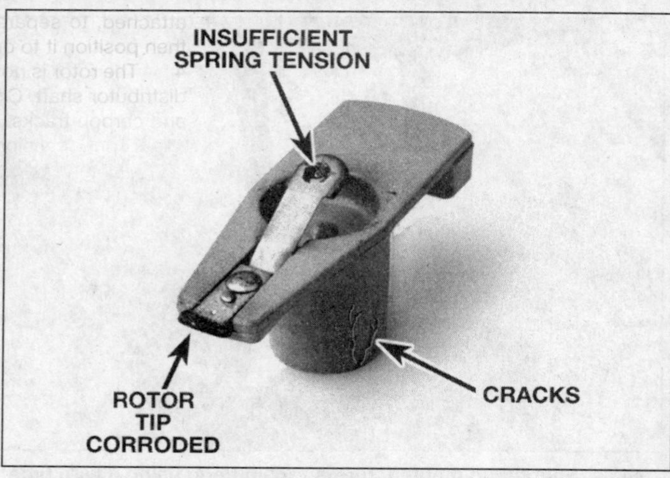

46.4 The ignition rotor should be checked for wear and corrosion as indicated here (if in doubt about its condition, buy a new one)

3 Disconnect the plug wire from the spark plug. To do this, grab the rubber boot, twist slightly and pull the wire free. Do not pull on the wire itself, only on the rubber boot.

4 Inspect inside the boot for corrosion, which will look like a white crusty powder. Push the wire and boot back onto the end of the spark plug. It should be a tight fit on the plug. If it is not, remove the wire and use pliers to carefully crimp the metal connector inside the boot until it fits securely on the end of the spark plug.

5 Using a clean rag, wipe the entire length of the wire to remove any built-up dirt and grease. Once the wire is clean, check for burns, cracks and other damage. Do not bend the wire excessively or pull the wire since the conductor inside might break.

6 Disconnect the wire from the distributor cap. A retaining ring at the top of the distributor may have to be removed to free the wires. Again, pull only on the rubber boot. Check for corrosion and a tight fit in the same manner as the spark plug end. Replace the wire into the distributor cap.

7 Check the remaining spark plug wires one at a time, making sure they are securely fastened at the distributor and the spark plug when the check is complete.

8 If new spark plug wires are required, purchase a new set for your specific engine model. Wire sets are available pre-cut, with the rubber boots already installed. Remove and replace the wires one at a time to avoid mix-ups in the firing order.

46 Distributor cap and rotor check and replacement

Refer to illustrations 46.3, 46.4, 46.5, 46.7 and 46.13

1 It's common practice to install a new distributor cap and rotor whenever new spark plug wires are installed. Although the pointless electronic distributor used on later models requires much less maintenance than a conventional distributor, periodic inspections should be performed when the plug wires are checked. **Note:** *On later models with electronic ignition, if the distributor cap must be replaced, the ignition coil will have to be removed from the cap and installed in the new one.*

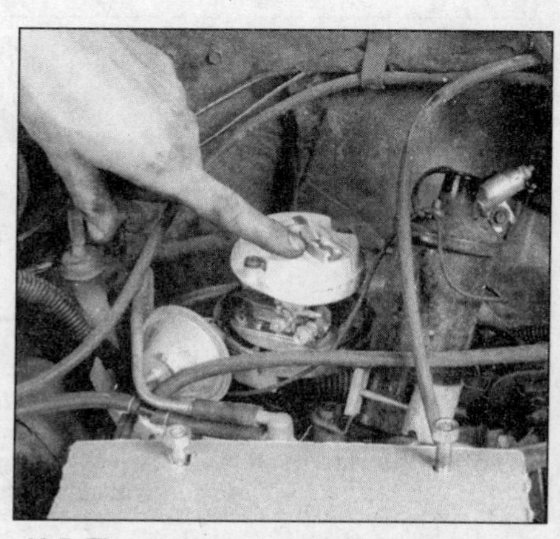

46.5 The rotor is attached by two screws on V8 engines and models with electronic ignition

Check

2 To gain access to the distributor cap it may be necessary to remove either the air cleaner assembly or the top plate.

3 Loosen the distributor cap mounting screws (note that the screws have a shoulder

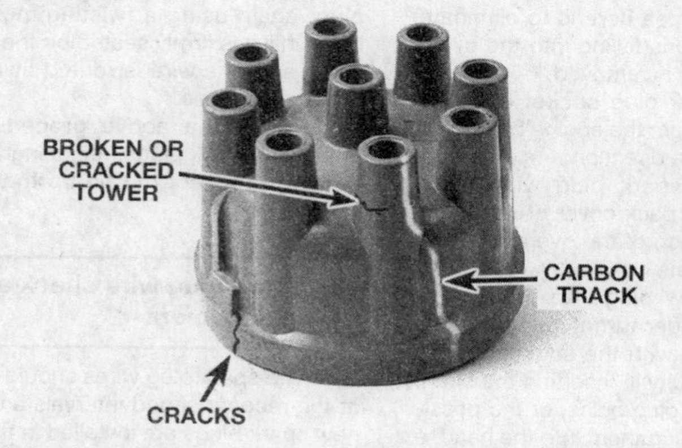

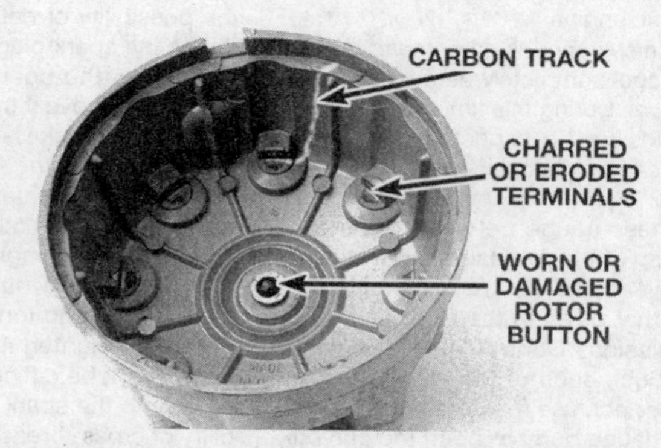

46.7 Shown here are some of the common defects to look for when inspecting the distributor cap (if in doubt about its condition, install a new one)

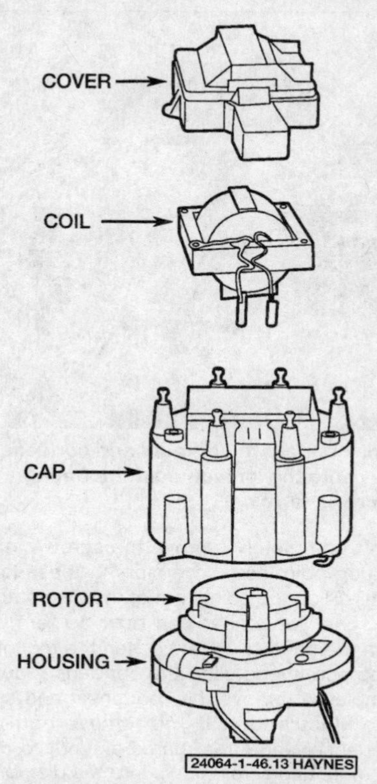

46.13 Typical later model distributor cap mounting details (the ignition coil is mounted in the cap)

so they don't come completely out). On many models, the cap is held in place with latches that look like screws - to release them, push down with a screwdriver and turn them about 1/2-turn. Pull up on the cap, with the wires attached, to separate it from the distributor, then position it to one side **(see illustration)**.

4 The rotor is now visible on the end of the distributor shaft. Check it carefully for cracks and carbon tracks. Make sure the center terminal spring tension is adequate and look for corrosion and wear on the rotor tip **(see illustration)**. If in doubt as to its condition, replace it with a new one.

5 If replacement is required, detach the rotor from the shaft and install a new one. On some models, the rotor is press fit on the shaft and can be pulled off. On other models, particularly with V8 engines and electronic ignition, the rotor is attached to the distributor shaft with two screws **(see illustration)**.

6 The rotor is indexed to the shaft so it can only be installed one way. Press fit rotors have an internal key that must line up with a slot in the end of the shaft. Rotors held in place with screws have one square and one round peg on the underside that must fit into holes with the same shape.

7 Check the distributor cap for carbon tracks, cracks and other damage. Closely examine the terminals on the inside of the cap for excessive corrosion and damage **(see illustration)**. Slight deposits are normal. Again, if in doubt as to the condition of the cap, replace it with a new one.

Replacement

Conventional distributor

8 On models with a separately mounted ignition coil, simply separate the cap from the distributor and transfer the spark plug wires, one at a time, to the new cap. Be very careful not to mix up the wires!

9 Reattach the cap to the distributor, then tighten the screws or reposition the latches to hold it in place.

Coil-in-cap distributor

10 Use your thumbs to push the spark plug wire retainer latches away from the coil cover.

11 Lift the retainer ring away from the distributor cap with the spark plug wires attached to the ring. It may be necessary to work the wires off the distributor cap towers so they remain with the ring.

12 Disconnect the battery/tachometer/coil electrical connector from the distributor cap.

13 Remove the two coil cover screws and lift off the coil cover **(see illustration)**.

14 There are three small spade connectors on wires extending from the coil into the electrical connector hood at the side of the distributor cap. Note which terminals the wires are attached to, then use a small screwdriver to push them free.

15 Remove the four coil mounting screws and lift the coil out of the cap.

16 When installing the coil in the new cap, be sure to install the rubber arc seal in the cap.

17 Install the coil screws, the wires in the connector hood, and the coil cover.

18 Install the cap on the distributor.

19 Plug in the coil electrical connector to the distributor cap.

20 Install the spark plug wire retaining ring on the distributor cap.

47 Ignition point replacement

All models

Refer to illustrations 47.1 and 47.2

1 The ignition points must be replaced at regular intervals on vehicles not equipped with electronic ignition. Occasionally the rubbing block will wear enough to require adjust-

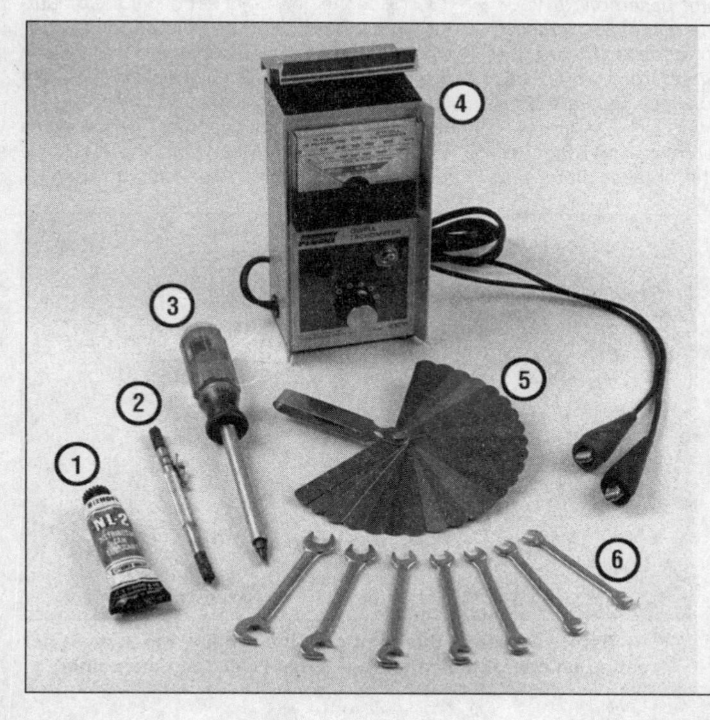

47.1 Tools and materials needed for ignition point replacement and dwell angle adjustment

1 ***Distributor cam lube*** *- Sometimes this special lubricant comes with the new points; however, its a good idea to buy a tube and have it on hand*
2 ***Screw starter*** *- This tool has special claws which hold the screw securely as it is started, which helps prevent accidental dropping of the screw*
3 ***Magnetic screwdriver*** *- Serves the same purpose as 2 above. If you do not have one of these special screwdrivers, you risk dropping the point mounting screws down into the distributor body*
4 ***Dwell meter*** *- A dwell meter is the only accurate way to determine the point setting (gap). Connect the meter according to the instructions supplied with it*
5 ***Blade-type feeler gauges*** *- These are required to set the initial point gap (space between the points when they are open)*
6 ***Ignition wrenches*** *- These special wrenches are made to work within the tight confines of the distributor. Specifically, they are needed to loosen the nut/bolt which secures the leads to the points*

47.2 Although it is possible to restore ignition points that are pitted (as shown here), burned and corroded, they should be replaced instead

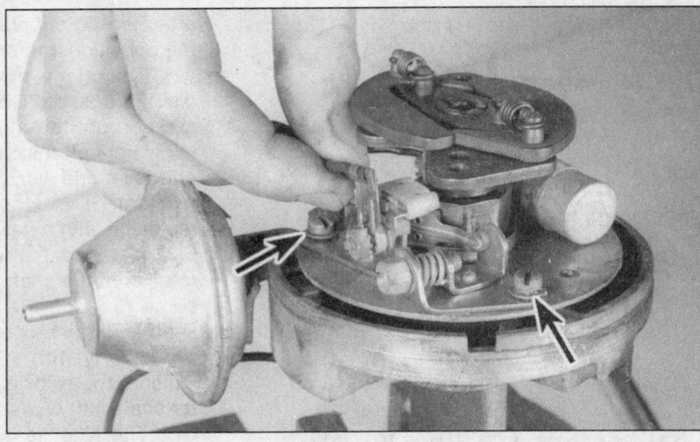

47.7 Loosen the nut and disconnect the primary and condenser wires from the points - note the ignition point mounting screws (arrows)

ment of the points. It's also possible to clean and dress them with a fine file, but replacement is recommended since they are relatively inaccessible and very inexpensive. Several special tools are required for this procedure **(see illustration)**.

2 After removing the distributor cap and rotor (Section 46), the ignition points are plainly visible. They can be examined by gently prying them open to reveal the condition of the contact surfaces **(see illustration)**. If they're rough, pitted, covered with oil or burned, they should be replaced, along with the condenser. **Caution:** *This procedure requires the removal of small screws which can easily fall down into the distributor. To retrieve them, the distributor would have to be removed and disassembled. Use a magnetic or spring-loaded screwdriver and be extra careful.*

V8 engines

Refer to illustrations 47.7, 47.9, 47.16, 47.17 and 47.29

Point replacement

3 If not already done, remove the distributor cap by positioning a screwdriver in the slotted head of each latch. Press down on the latch and rotate it 1/2-turn to release the cap from the distributor body (see Section 46).

4 Position the cap (with the spark plug wires still attached) out of the way. Use a length of wire to hold it out of the way if necessary.

5 Remove the rotor, which is held in place with two screws.

6 If equipped with a radio frequency interference shield (RFI), remove the mounting screws and the two-piece shield to gain access to the ignition points.

7 Note how they are routed, then disconnect the primary and condenser wire leads from the points **(see illustration)**. The wires may be attached with a small nut (which should be loosened, but not removed) a small screw or by a spring loaded terminal. **Note:** *Some models are equipped with ignition points which include the condenser as an integral part of the point assembly. If your vehicle has this type, the condenser removal procedure below will not apply and there will be only one wire to detach from the points, rather than the two used with a separate condenser assembly.*

8 Loosen the two screws which secure the ignition points to the breaker plate, but don't completely remove the screws (most ignition point sets have slots at these locations). Slide the points out of the distributor.

9 The condenser can now be removed from the breaker plate. Loosen the mounting strap screw and slide the condenser out or completely remove the condenser and strap **(see illustration)**. If you remove both the condenser and strap, be careful not to drop the mounting screw down into the distributor body.

10 Before installing the new points and condenser, clean the breaker plate and the cam on the distributor shaft to remove all dirt, dust and oil.

11 Apply a small amount of distributor cam lube (usually supplied with the new points, but also available separately) to the cam lobes.

12 Position the new condenser and tighten the mounting strap screw securely.

13 Slide the new point set under the mounting screw heads and make sure the protrusions on the breaker plate fit into the holes in the point base (to properly position the point set), then tighten the screws securely.

14 Attach the primary and condenser wires to the new points. Make sure the wires are routed so they don't interfere with breaker

47.9 The condenser is attached to the breaker plate by a single screw

47.16 Before adjusting the point gap, the rubbing block must be resting on one of the cam lobes (which will open the points)

47.17 With the points open, insert a 0.019-inch thick feeler gauge and turn the adjustment screw with an Allen wrench

47.29 With the door open, a 1/8-inch Allen wrench can be inserted into the adjustment screw socket and turned to adjust the point dwell

plate or advance weight movement.

15 Although the gap between the contact points (dwell angle) will be adjusted later, make the initial adjustment now, which will allow the engine to be started.

16 Make sure that the point rubbing block is resting on one of the high points of the cam **(see illustration)**. If it isn't, turn the ignition switch to Start in short bursts to reposition the cam. You can also turn the crankshaft with a breaker bar and socket attached to the large bolt that holds the vibration damper in place.

17 With the rubbing block on a cam high point (points open), insert a 0.019-inch feeler gauge between the contact surfaces and use an Allen wrench to turn the adjustment screw until the point gap is equal to the thickness of the feeler gauge **(see illustration)**. The gap is correct when a slight amount of drag is felt as the feeler gauge is withdrawn.

18 If equipped, install the RFI shield.

19 Before installing the rotor, check it as described in Section 46.

20 Install the rotor. The rotor is indexed with a square peg underneath on one side and a round peg on the other side, so it will fit on the advance mechanism only one way. Tighten the rotor mounting screws securely.

21 Before installing the distributor cap, inspect it as described in Section 46.

22 Install the distributor cap and lock the latches under the distributor body by depressing and turning them with a screwdriver.

23 Start the engine and check the dwell angle and ignition timing.

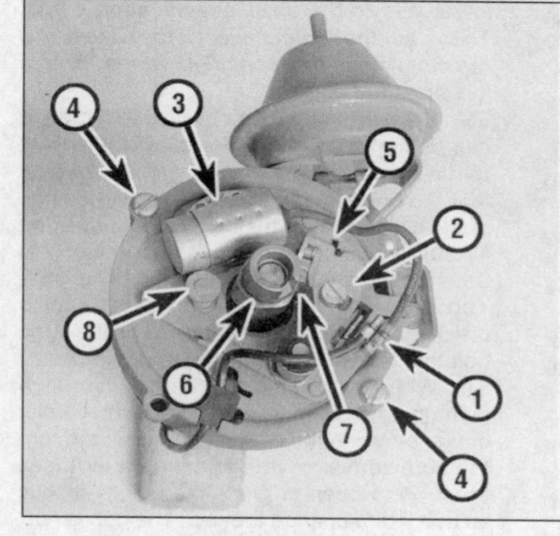

47.39 Distributor components - 6-cylinder engine

1 *Quick disconnect terminal*
2 *Point assembly mounting screw*
3 *Condenser mounting screw*
4 *Breaker plate mounting screws*
5 *Adjustment slot*
6 *Point cam*
7 *Point rubbing block*
8 *Point cam lubricator*

Dwell angle adjustment

24 Whenever new ignition points are installed or the original points are cleaned, the dwell angle must be checked and adjusted.

25 Precise adjustment of the dwell angle requires an instrument called a dwell meter. Combination tach/dwell meters are commonly available at reasonable cost from auto parts stores. An approximate setting can be obtained if a meter isn't available.

26 If a dwell meter is available, hook it up following the manufacturer's instructions.

27 Start the engine and allow it to run at idle until normal operating temperature is reached (the engine must be warm to obtain an accurate reading). Turn off the engine.

28 Raise the metal door in the distributor cap. Hold it in the open position with tape if necessary.

29 Just inside the opening is the ignition point adjustment screw. Insert a 1/8-inch Allen wrench into the adjustment screw socket **(see illustration)**.

30 Start the engine and turn the adjustment screw as required to obtain the specified dwell reading on the meter. Dwell angle specifications can be found at the beginning of this Chapter and on the tune-up decal in the engine compartment. If there's a discrepancy between the two, assume the tune-up decal is correct. **Note:** *When adjusting the dwell, aim for the lower end of the dwell specification range. Then, as the points wear, the dwell will remain within the specified range over a longer period of time.*

31 Remove the Allen wrench and close the door on the distributor. Turn off the engine and disconnect the dwell meter, then check the ignition timing (see Section 48).

32 If a dwell meter isn't available, use the following procedure to obtain an approximate dwell setting.

33 Start the engine and allow it to idle until normal operating temperature is reached.

34 Raise the metal door in the distributor cap. Hold it in the open position with tape if necessary.

35 Just inside the opening is the ignition point adjustment screw. Insert a 1/8-inch Allen wrench into the adjustment screw socket **(see illustration 47.29)**.

36 Turn the Allen wrench clockwise until the engine begins to misfire, then turn the screw 1/2-turn counterclockwise.

37 Remove the Allen wrench and close the door. As soon as possible have the dwell angle checked and/or adjusted with a dwell meter to ensure optimum performance.

6-cylinder engine

Refer to illustrations 47.39 and 47.48

Point replacement

38 Loosen the mounting screws, then detach the distributor cap, with the wires attached, and position it out of the way. Remove the rotor and the dust cover (if equipped).

39 Note how they are routed, then disconnect the primary circuit and condenser wires from the contact point assembly quick dis-

connect terminal **(see illustration)**.

40 Remove the point assembly mounting screw and detach the points from the breaker plate.

41 Remove the screw and detach the condenser from the breaker plate.

42 Before installing the new points and condenser, clean the breaker plate and the cam on the distributor shaft to remove all dirt, dust and oil.

43 Apply a small amount of distributor cam lube (usually supplied with the new points, but also available separately) to the cam lobes. If the cam lubricator is in good shape, several drops of oil can be applied to it instead of using distributor cam lube.

44 Position the new condenser and tighten the mounting screw securely.

45 Install the new points, but don't tighten the mounting screw completely. Make sure the peg on the point assembly fits into the hole in the breaker plate before installing the screw.

46 Reconnect the primary circuit and condenser wires.

Dwell adjustment

47 If a dwell meter is available, hook it up by following the manufacturer's instructions.

48 Make sure the point assembly mounting screw is snug, but not tight, then insert a screwdriver into the adjustment slot **(see illustration)**.

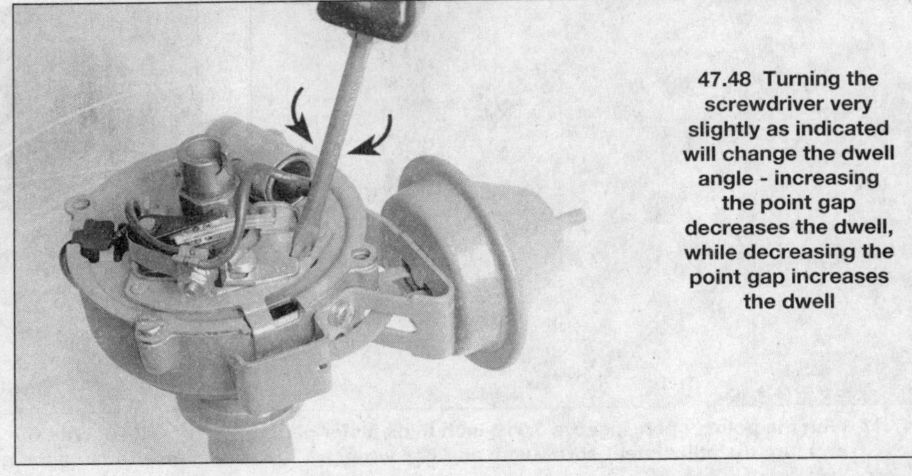

47.48 Turning the screwdriver very slightly as indicated will change the dwell angle - increasing the point gap decreases the dwell, while decreasing the point gap increases the dwell

49 Have an assistant crank the engine over with the starter while you note the dwell reading on the meter. If the dwell is incorrect, turn the screwdriver as required to bring it into the specified range, then tighten the point assembly mounting screw and recheck the dwell. Dwell angle specifications can be found at the beginning of this Chapter and on the tune-up decal in the engine compartment. If there's a discrepancy between the two, assume the tune-up decal is correct.

Note: *When adjusting the dwell, aim for the lower end of the dwell specification range. Then, as the points wear, the dwell will remain within the specified range over a longer period of time.*

50 If a dwell meter isn't available, the dwell can be set close enough to allow the engine to run by adjusting the point gap. Make sure that the point rubbing block is resting on one of the high points of the cam **(see illustration 47.39)**. If it isn't, turn the ignition switch to Start in short bursts to reposition the cam. You can also turn the crankshaft with a breaker bar and socket attached to the large bolt that holds the vibration damper in place.

51 With the rubbing block on a cam high point (points open), insert a 0.019-inch feeler gauge between the contact surfaces and turn the screwdriver in the adjustment slot (see Step 49) to open or close the points slightly as needed to change the gap. The gap is correct when a slight amount of drag is felt as the feeler gauge is withdrawn. Tighten the point mounting screw, then recheck the gap.

52 Check the distributor cap and rotor as described in Section 46, then install them, along with the shield (if equipped).

53 Start the engine and allow it to reach normal operating temperature, then check the dwell again. **Note:** *If you don't have a dwell meter, have the dwell checked/adjusted with a meter as soon as possible to ensure optimum performance.*

54 If the dwell isn't as specified, remove the distributor cap, readjust the dwell, reinstall the cap and check it again.

48 Ignition timing check and adjustment

Refer to illustrations 48.2 and 48.6

Note: *It is imperative that the procedures included on the Tune-up or Vehicle Emissions Control Information (VECI) label be followed when adjusting the ignition timing. The label will include all information concerning prelim-*

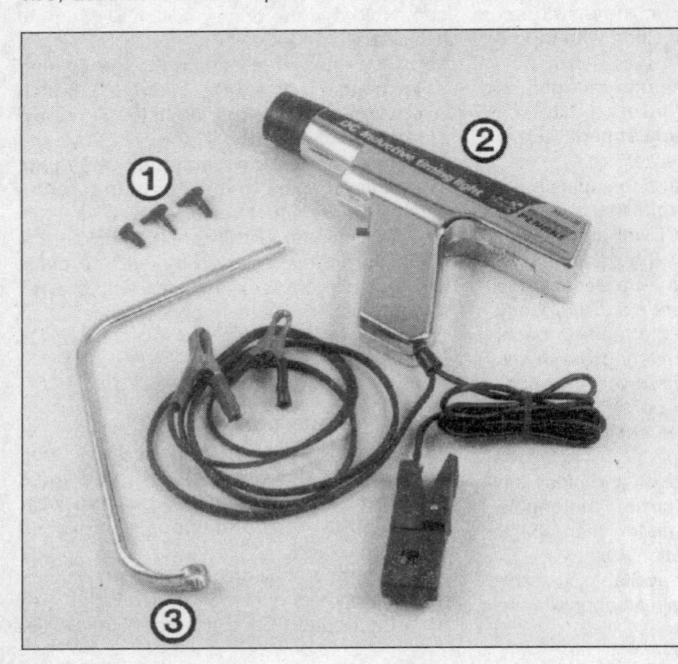

48.2 Tools needed to check and adjust the ignition timing

1 ***Vacuum plugs*** *- Vacuum hoses will, in most cases, have to be disconnected and plugged. Molded plugs in various shapes and sizes are available for this*

2 ***Inductive pick-up timing light*** *- Flashes a bright concentrated beam of light when the number one spark plug fires. Connect the leads according to the instructions supplied with the light*

3 ***Distributor wrench*** *- On some models, the hold-down bolt for the distributor is difficult to reach and turn with conventional wrenches or sockets. A special wrench like this must be used*

48.6 The ignition timing marks are located on the front of the engine

inary steps to be performed before adjusting the timing, as well as the timing specifications.

1 At the specified intervals, whenever the ignition points have been replaced, the distributor removed or a change made in the fuel type, the ignition timing should be checked and adjusted.

2 Locate the Tune-up or VECI label under the hood and read through and perform all preliminary instructions concerning ignition timing. Some special tools will be needed for this procedure **(see illustration)**.

3 Before attempting to check the timing, make sure the ignition point dwell angle is correct (Section 47), and the idle speed is as specified (Section 30).

4 If specified on the tune-up label, disconnect the vacuum hose from the distributor and plug the open end of the hose with a rubber plug, rod or bolt of the proper size. Also, disconnect the 'set timing' connector if specified on the VECI label.

5 Connect a timing light in accordance with the manufacturer's instructions. Generally, the light will be connected to power and ground sources and to the number 1 spark plug wire. The number 1 spark plug is the first spark plug on 6-cylinder engines or the first one on the left cylinder head (driver's side) on V6 and V8 engines.

6 Locate the numbered timing tag on the front cover of the engine **(see illustration)**. It is located just behind the lower crankshaft pulley. Clean it off with solvent if necessary to read the printing and small grooves.

7 Use chalk or paint to mark the groove in the crankshaft pulley.

8 Put a mark on the timing tab in accordance with the number of degrees called for on the VECI label or the tune-up label in the engine compartment. Each peak or notch on the timing tab represents 2-degrees. The word Before or the letter A indicates advance and the letter 0 indicates Top Dead Center (TDC). As an example, if your vehicle specifications call for 8-degrees BTDC (Before Top Dead Center), you will make a mark on the timing tab 4 notches before the 0.

9 Check that the wiring for the timing light is clear of all moving engine components, then start the engine and warm it up to normal operating temperature.

10 Aim the flashing timing light at the timing mark by the crankshaft pulley, again being careful not to come in contact with moving parts. The marks should appear to be stationary. If the marks are in alignment, the timing is correct.

11 If the notch is not lining up with the correct mark, loosen the distributor hold-down bolt and rotate the distributor until the notch is lined up with the correct timing mark.

12 Retighten the hold-down bolt and recheck the timing.

13 Turn off the engine and disconnect the timing light. Reconnect the vacuum advance hose if removed, and any other components which were disconnected.

Notes

Chapter 2 Part A
V8 and V6 engines

Contents

	Section
Camshaft, bearings and lifters - removal, inspection and installation	11
Crankshaft oil seals - replacement	14
Cylinder compression check	See Chapter 2C
Cylinder heads - removal and installation	8
Engine mounts - check and replacement	15
Engine oil and filter change	See Chapter 1
Exhaust manifolds - removal and installation	7
General information	1
Intake manifold - removal and installation	6
Oil pan - removal and installation	12
Oil pump - removal and installation	13
Repair operations possible with the engine in the vehicle	2
Rocker arms and pushrods - removal, inspection and installation	4
Rocker arm covers - removal and installation	3
Spark plug replacement	See Chapter 1
Timing cover, chain and sprockets - removal and installation	10
Top Dead Center (TDC) for number 1 piston - locating	9
Valve springs, retainers and seals - replacement in vehicle	5
Water pump - removal and installation	See Chapter 3

Specifications

General

Cylinder numbers (front-to-rear)	
V8	
left (driver's) side	1-3-5-7
right side	2-4-6-8
V6	
left (driver's) side	1-3-5
right side	2-4-6
Firing order	
V8	1-8-4-3-6-5-7-2
V6	1-6-5-4-3-2
Bore and stroke	
V8 engines	
283	3.875 x 3.00 in
305	3.736 x 3.48 in
307	3.875 x 3.25 in
327	4.000 x 3.25 in
350	4.000 x 3.48 in
396/402	4.125 x 3.76 in
454	4.250 x 4.00 in
V6 engine	4.000 x 3.48 in

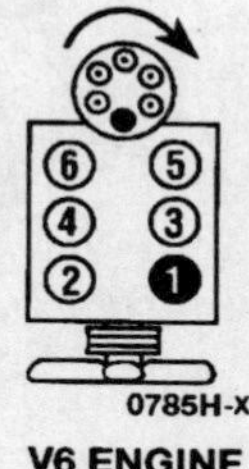

V6 ENGINE
Firing order
1-6-5-4-3-2

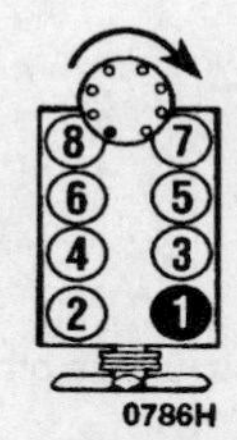

V8 ENGINE with points-type ignition

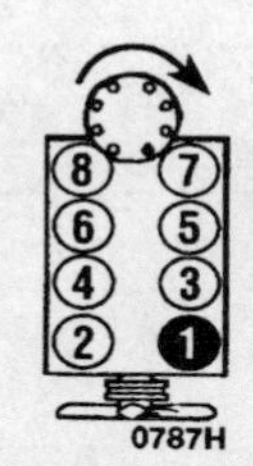

V8 ENGINE with electronic (HEI) ignition

Firing order
1-8-4-3-6-5-7-2

The blackened terminal shown on the distributor cap indicates the Number One spark plug wire position

Cylinder location and distributor rotation

Camshaft

Bearing journal diameter	
Small block V8	
400	1.9482 to 1.9492 in
All others	1.8682 to 1.8692 in
Big block V8	1.9482 to 1.9492 in
V6	1.8682 to 1.8692 in
Lobe lift	
Intake lobe	
262 V6	0.3570 in
283 V8	0.2658 in
305 V8	
to 1986	0.2484 in
1987	0.2336 in
307 V8	0.2600 in
327	0.3980 in
350 V8	
to 1986	0.2600 in
1987	0.2565 in
396 V8	0.2714 in
402 V8	0.2365 in
454 V8	0.2343 in
Exhaust lobe	
262 V6	0.3900 in
283 V8	0.2658 in
305 V8	
to 1986	0.2667 in
1987	0.2565 in
307 V8	0.2733 in
327	0.3980 in
350 V8	
to 1986	0.2733 in
1987	0.2690 in
396 V8	0.2824 in
402 V8	0.2411 in
454 V8	0.2530 in

Torque specifications

	Ft-lbs (unless otherwise noted)
Rocker arm cover	
nuts	
small block V8	65 in-lbs
big block V8	50 in-lbs
bolts	50 in-lbs
1988 through 1991 (5.7L V8 engines)	100 in-lbs
V6 through-bolts	88 in-lbs
Intake manifold bolts	
small block V8 and V6	30
big block V8	30
Exhaust manifold bolts	
small block V8 cast iron	
inside port bolts	30
outside port bolts	30
small block stainless steel	26
V6	
center bolts	26
outer bolts	20
big block V8	
stainless steel	40
cast iron	20
Cylinder head bolts	**Ft-lbs** (unless otherwise noted)
small block V8 and V6	65
big block V8	80
Timing cover bolts	
V6	92 in-lbs
small block V8	
to 1985	80 in-lbs
1986 on	100 in-lbs
big block V8	80 in-lbs

	Ft-lbs (unless otherwise noted)
Camshaft sprocket bolts	
small block V8 and V6	20
big block V8	20
Torsional (vibration) damper bolt	
small block V8 and V6	
to 1985	60
1986 on	70
big block V8	85
Oil pan	
1986 and later small block V8 and V6	
bolts	100 in-lbs
nuts	200 in-lbs
to 1985 small block V8	
1/4-20	80 in-lbs
5/16-18	165 in-lbs
big block oil pan-to-front cover bolts	
to 1985	55 in-lbs
1986 on	70 in-lbs
big block oil pan-to-block bolts	145 in-lbs
Oil pump bolt	65
Rear main bearing cap bolts	
V6	75
small block V8	80
big block V8	110
Rear oil seal housing bolts	135 in-lbs
Flywheel-to-crankshaft bolts	
V6	75
small block V8	
1985 and earlier	60
1986 and later	75
big block V8	65

1 General information

This Part of Chapter 2 is devoted to in-vehicle repair procedures for V8 and V6 engines. All information concerning engine removal and installation and engine block and cylinder head overhaul can be found in Part C of this Chapter.

Since the repair procedures included in this Part are based on the assumption that the engine is still installed in the vehicle, if they are being used during a complete engine overhaul (with the engine already out of the vehicle and on a stand) many of the steps included here will not apply.

The Specifications included in this Part of Chapter 2 apply only to the procedures found here. The specifications necessary for rebuilding the block and cylinder heads are included in Part C.

The V8 engines used in Chevrolet and GMC trucks vary in size from 283 cubic inches to 454 cubic inches. The 283, 305, 307, 327, 350 and 400 cubic inch engines are collectively known as *Small Block* engines. The 396, 402 and 454 cubic inch engines are known as *Big Block* or *Mark IV* engines. Note that the 396 engine, the 402 engine, and the occasionally found Big Block based 400 engine are all the same except for external decals, with an actual displacement of 401.9 cubic inches.

The V6 engine, used from 1986 on, displaces 262 cubic inches (4.3 liters), and in design is very similar to the 350 cubic inch small block V8 engine. Almost all removal, inspection, installation and replacement procedures for the V6 engine are the same as those for the small block V8, with the exceptions noted within this Chapter.

2 Repair operations possible with the engine in the vehicle

Many major repair operations can be accomplished without removing the engine from the vehicle.

Clean the engine compartment and the exterior of the engine with some type of pressure washer before any work is done. A clean engine will make the job easier and will help keep dirt out of the internal areas of the engine.

Depending on the components involved, it may be a good idea to remove the hood to improve access to the engine as repairs are performed (refer to Chapter 11 if necessary).

If oil or coolant leaks develop, indicating a need for gasket or seal replacement, the repairs can generally be made with the engine in the vehicle. The oil pan gasket, the cylinder head gaskets, intake and exhaust manifold gaskets, timing cover gaskets and the crankshaft oil seals are accessible with the engine in place.

Exterior engine components, such as the water pump, the starter motor, the alternator, the distributor and the carburetor or fuel injection unit, as well as the intake and exhaust manifolds, can be removed for repair with the engine in place.

Since the cylinder heads can be removed without pulling the engine, valve component servicing can also be accomplished with the engine in the vehicle.

Replacement of, repairs to or inspection of the timing chain and sprockets and the oil pump are all possible with the engine in place.

In extreme cases caused by a lack of necessary equipment, repair or replacement of piston rings, pistons, connecting rods and rod bearings is possible with the engine in the vehicle. However, this practice is not recommended because of the cleaning and preparation work that must be done to the components involved.

3 Rocker arm covers - removal and installation

Refer to illustrations 3.5a and 3.5b

1 Disconnect the negative cable from the battery, then remove the air cleaner assembly and the heat stove duct.

3.5a Most small block and big block rocker arm covers (small block shown) are held down by studs or bolts around the edges

3.5b The rocker arm covers on V6 engines (and some later V8 engines) are held down by three bolts which pass through the center of the cover

Removal

Right side

2 Detach the hose from the AIR system exhaust check valve (if equipped).

3 Refer to Chapter 3 if necessary and remove the A/C compressor (if equipped) from the bracket and position it out of the way. Do not loosen or disconnect the compressor lines.

4 On V6 models, remove the oil filler tube, PCV valve and the choke wires.

5 Remove the four rocker arm cover mounting nuts or bolts on V8 engines or the three through-bolts on V6 engines **(see illustrations)**. If equipped, slip the spark plug wire clip brackets and washers off the lower rocker arm cover studs and position the brackets/wires out of the way.

6 Detach the spark plug wires from the plugs, then lay the spark plug wire harness up over the distributor and remove the rocker arm cover. **Note:** *If the cover is stuck to the head, bump the front of the cover with a block of wood and a hammer to release it. If it still will not come loose, try to slip a flexible putty knife between the head and cover to break the seal. Do not pry at the cover-to-head joint or damage to the sealing surface and cover flange will result and oil leaks will develop.*

Left side

7 On V8 engines pull the PCV valve out of the cover and detach the hose from the intake manifold fitting.

8 Detach the vacuum brake line from the manifold fitting.

9 Refer to Chapter 5 if necessary and detach the alternator. Lay it aside, out of the way, without disconnecting the wires.

10 On V6 models remove the accelerator and TVS cable bracket from the intake manifold.

11 Refer to Steps 5 and 6 above for the remainder of the procedure.

Installation

12 The mating surfaces of each cylinder head and rocker arm cover must be perfectly clean when the covers are installed. Use a gasket scraper to remove all traces of sealant or old gasket, then wipe the mating surfaces with a cloth saturated with lacquer thinner or acetone. If there is sealant or oil on the mating surfaces when the cover is installed, oil leaks may develop.

13 If studs are used to mount the rocker covers, clean the threads with a die to remove any corrosion and restore damaged threads. On models that utilize bolts, make sure the threaded holes in the head are clean. Run a tap into them to remove corrosion and restore damaged threads.

14 Mate the new gaskets to the covers before the covers are installed. Apply a thin coat of RTV sealant to the cover flange, then position the gasket inside the cover lip and allow the sealant to set up so the gasket adheres to the cover (if the sealant is not allowed to set, the gasket may fall out of the cover as it is installed on the engine).

15 Carefully position the cover on the head and install the nuts or bolts. Don't forget to slip the spark plug wire clip brackets and washers over the studs before the nuts are threaded on.

16 Tighten the nuts or bolts in three or four steps to the specified torque.

17 The remaining installation steps are the reverse of removal.

18 Start the engine and check carefully for oil leaks as the engine warms up.

4 Rocker arms and pushrods - removal, inspection and installation

Refer to illustration 4.13

Removal

1 Refer to Section 3 and detach the rocker arm covers from the cylinder heads.

2 Beginning at the front of one cylinder head, loosen and remove the rocker arm stud nuts. Store them separately in marked containers to ensure that they will be reinstalled in their original locations. **Note:** *If the pushrods are the only items being removed, loosen each nut just enough to allow the rocker arms to be rotated to the side so the pushrods can be lifted out.*

3 Lift off the rocker arms and pivot balls and store them in the marked containers with the nuts (they must be reinstalled in their original locations).

4 Remove the pushrods and store them in order to make sure they will not get mixed up during installation.

Inspection

5 Check each rocker arm for wear, cracks and other damage, especially where the pushrods and valve stems contact the rocker arm faces.

6 Make sure the hole at the pushrod end of each rocker arm is open.

7 Check each rocker arm pivot area for wear, cracks and galling. If the rocker arms are worn or damaged, replace them with new ones and use new pivot balls as well.

8 Inspect the pushrods for cracks and excessive wear at the ends. Roll each pushrod across a piece of plate glass to see if it is bent (if it wobbles, it is bent).

Installation

9 Lubricate the lower end of each pushrod with clean engine oil or moly-base grease and install them in their original locations. Make sure each pushrod seats completely in the lifter socket.

10 Apply moly-base grease to the ends of the valve stems and the upper ends of the pushrods before positioning the rocker arms over the studs.

11 Set the rocker arms in place, then install the pivot balls and nuts. Apply moly-base grease to the pivot balls to prevent damage

4.13 Rotate each pushrod as the rocker arm nut is tightened to determine the point at which all play is removed, then tighten each nut an additional 3/4-turn

5.8 Once the spring is depressed, the keepers can be removed with a small magnet or a pair of needle-nose pliers

5.9a The flat O-ring type valve stem seal fits into a groove just below the keeper groove on the valve stem

to the mating surfaces before engine oil pressure builds up. Be sure to install each nut with the flat side against the pivot ball.

Adjustment

12 Refer to Section 9 and bring the number one piston to top dead center on the compression stroke.

13 Tighten the rocker arm nuts (number one cylinder only) until all play is removed at the pushrods. This can be determined by rotating each pushrod between your thumb and index finger as the nut is tightened **(see illustration)**. At the point where a slight drag is just felt as you spin the pushrod, all lash has been removed.

14 Tighten each nut an additional 3/4-turn to center the lifters. Valve adjustment for cylinder number one is now complete.

V6 engine

15 If your vehicle is equipped with a V6 engine, you can also adjust the number two and three intake valves and number five and six exhaust valves at this time. When these valves have been adjusted, turn the crankshaft one full revolution (360-degrees) and adjust the number four, five and six intake valves and the number two, three and four exhaust valves.

V8 engine

16 If the vehicle is equipped with a V8 engine, you can also adjust the number two, five and seven intake valves and the number three, four and eight exhaust valves at this time. After adjusting these valves, turn the crankshaft one complete revolution (360-degrees) and adjust the number three, four, six and eight intake valves and the number two, five, six and seven exhaust valves.

All engines

17 Refer to Section 3 and install the rocker arm covers. Start the engine, listen for unusual valve train noises and check for oil leaks at the rocker arm cover joints.

5 Valve springs, retainers and seals - replacement in vehicle

Refer to illustrations 5.8, 5.9a, 5.9b and 5.18

Note: *Broken valve springs and defective valve stem seals can be replaced without removing the cylinder head. Two special tools and a compressed air source are normally required to perform this operation, so read through this Section carefully and rent or buy the tools before beginning the job. If compressed air is not available, a length of nylon rope can be used to keep the valves from falling into the cylinder during this procedure.*

1 Refer to Section 3 and remove the rocker arm cover from the affected cylinder head. If all of the valve stem seals are being replaced, remove both rocker arm covers.

2 Remove the spark plug from the cylinder which has the defective component. If all of the valve stem seals are being replaced, all of the spark plugs should be removed.

3 Turn the crankshaft until the piston in the affected cylinder is at top dead center on the compression stroke (refer to Section 9 for instructions). If you are replacing all of the valve stem seals, begin with cylinder number one and work on the valves for one cylinder at a time. Move from cylinder-to-cylinder following the firing order sequence (1-8-4-3-6-5-7-2 for V8 engines, 1-6-5-4-3-2 for V6 engines).

4 Thread an adapter into the spark plug hole and connect an air hose from a compressed air source to it. Most auto parts stores can supply the air hose adapter. **Note:** *Many cylinder compression gauges utilize a screw-in fitting that may work with your air hose quick-disconnect fitting.*

5 Remove the nut, pivot ball and rocker arm for the valve with the defective part and pull out the pushrod. If all of the valve stem seals are being replaced, all of the rocker arms and pushrods should be removed (refer to Section 4).

6 Apply compressed air to the cylinder. The valves should be held in place by the air pressure. If the valve faces or seats are in poor condition, leaks may prevent the air pressure from retaining the valves- refer to the alternative procedure below.

7 If you do not have access to compressed air, an alternative method can be used. Position the piston at a point approximately 45-degrees before TDC on the compression stroke, then feed a long piece of nylon rope through the spark plug hole until it fills the combustion chamber. Be sure to leave the end of the rope hanging out of the engine so it can be removed easily. Use a large breaker bar and socket to rotate the crankshaft in the normal direction of rotation until slight resistance is felt as the piston comes up against the rope in the combustion chamber.

8 Stuff shop rags into the cylinder head holes above and below the valves to prevent parts and tools from falling into the engine, then use a valve spring compressor to compress the spring/damper assembly. Remove the keepers with a pair of small needle-nose pliers or a magnet **(see illustration)**. **Note:** *A couple of different types of tools are available for compressing the valve springs with the head in place. One type grips the lower spring coils and presses on the retainer as the knob is turned, while the other type utilizes the rocker arm stud and nut for leverage. Both types work very well, although the lever type is usually less expensive.*

9 Remove the spring retainer or rotator, oil shield and valve spring assembly, then remove the valve stem O-ring seal. Three different types of valve stem oil seals are used on these engines, depending on year and engine size. The most common is a small O-ring which simply fits around the valve stem just above the guide boss. A second type is a flat O-ring which fits into a groove in the valve stem just below the valve stem keeper groove **(see illustration)**. On other applica-

5.9b Some engines use an umbrella-type oil seal which fits over the valve guide boss

5.18 Apply a small amount of grease to the valve keepers to hold them in place on the valve stem until the spring compressor is released

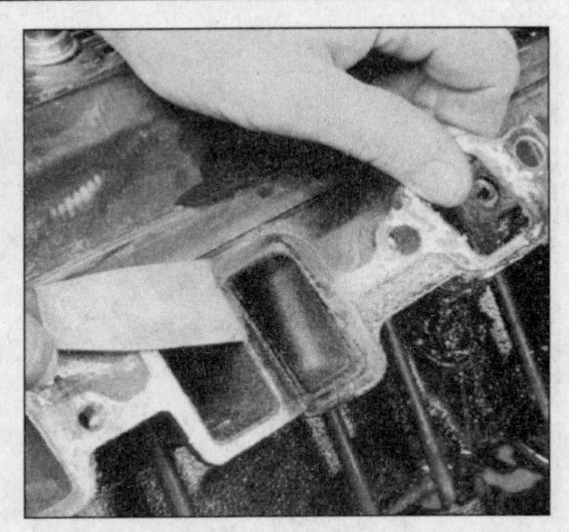

6.9 After covering the lifter valley, use a gasket scraper to remove all traces of sealant and old gasket material from the head and manifold mating surfaces

tions, an umbrella type seal which extends down over the valve guide boss is used over the valve stem **(see illustration)**. In most cases the umbrella type seal is used in conjunction with the flat O-ring type seal. The O-ring type seals will most likely be hardened and will probably break when removed, so plan on installing a new one each time the original is removed. **Note:** *If air pressure fails to hold the valve in the closed position during this operation, the valve face or seat is probably damaged. If so, the cylinder head will have to be removed for additional repair operations.*

10 Wrap a rubber band or tape around the top of the valve stem so the valve will not fall into the combustion chamber, then release the air pressure. **Note:** *If a rope was used instead of air pressure, turn the crankshaft slightly in the direction opposite normal rotation.*

11 Inspect the valve stem for damage. Rotate the valve in the guide and check the end for eccentric movement, which would indicate that the valve is bent.

12 Move the valve up-and-down in the guide and make sure it doesn't bind. If the valve stem binds, either the valve is bent or the guide is damaged. In either case, the head will have to be removed for repair.

13 Inspect the rocker arm studs for wear. Worn studs on most small block engines can only be replaced by an automotive machine shop, since they must be pressed into place a precise depth. On some big block engines, however, and some high performance versions of the small block (not normally found in trucks), the studs are threaded into the head and can be replaced if worn. In addition, in some applications of screw-in studs, a guide plate is installed between the stud and head to aid in locating pushrod position in relation to the rocker arm. Be sure to replace the guide plate if the studs are removed and reinstalled, and use gasket sealer on the studs when threading them into the head.

14 Reapply air pressure to the cylinder to retain the valve in the closed position, then remove the tape or rubber band from the valve stem. If a rope was used instead of air pressure, rotate the crankshaft in the normal direction of rotation until slight resistance is felt.

15 Lubricate the valve stem with engine oil and install a new oil seal of the type originally used on the engine (see Step 9).

16 Install the spring/damper assembly and shield in position over the valve.

17 Install the valve spring retainer or rotator and compress the valve spring assembly.

18 Position the keepers in the upper groove. Apply a small dab of grease to the inside of each keeper to hold it in place if necessary **(see illustration)**. Remove the pressure from the spring tool and make sure the keepers are seated.

19 Disconnect the air hose and remove the adapter from the spark plug hole. If a rope was used in place of air pressure, pull it out of the cylinder.

20 Refer to Section 4 and install the rocker arms and pushrods.

21 Install the spark plugs and hook up the wires.

22 Refer to Section 3 and install the rocker arm covers.

23 Start and run the engine, then check for oil leaks and unusual sounds coming from the rocker arm cover area.

6 Intake manifold - removal and installation

Refer to illustrations 6.9, 6.10a, 6.10b, 6.12, 6.15a, 6.15b, 6.5c and 6.15d

Warning: *The engine must be completely cool before performing this procedure.*

Removal

1 Disconnect the negative cable from the battery, then refer to Chapter 1 and drain the cooling system.

2 Refer to Chapter 4 and remove the air cleaner assembly.

3 While in Chapter 4, refer to the appropriate Sections and remove the carburetor and choke components or the fuel injection unit as necessary to expose the intake manifold mounting bolts.

4 Refer to Chapter 3 and detach the upper radiator hose from the thermostat housing cover and remove the alternator brace (if not already done). The thermostat housing cover may have to be detached to provide room for removal of the left-front manifold bolt.

5 Remove the rocker arm covers (Section 3).

6 Refer to Chapter 5 and remove the distributor.

7 If there is a heater hose fitting in the intake manifold, remove the hose.

8 Loosen the manifold mounting bolts in 1/4-turn increments until they can be removed by hand. The manifold will probably be stuck to the cylinder heads and force may be required to break the gasket seal. A large pry bar can be positioned under the cast-in lug near the thermostat housing to pry up the front of the manifold. **Caution:** *Do not pry between the block and manifold or the heads and manifold or damage to the gasket sealing surfaces will result and vacuum leaks could develop.*

Installation

Note: *The mating surfaces of the cylinder heads, block and manifold must be perfectly clean when the manifold is installed. Gasket removal solvents in aerosol cans are available at most auto parts stores and may be helpful when removing old gasket material that is stuck to the heads and manifold. Be sure to follow the directions printed on the container.*

9 Use a gasket scraper to remove all traces of sealant and old gasket material, then wipe the mating surfaces with a cloth saturated with lacquer thinner or acetone. If there is old sealant or oil on the mating surfaces when the manifold is installed, oil or vacuum leaks may develop. When working on the heads and block, cover the lifter valley with shop rags to keep debris out of the engine **(see illustration)**. Use a vacuum

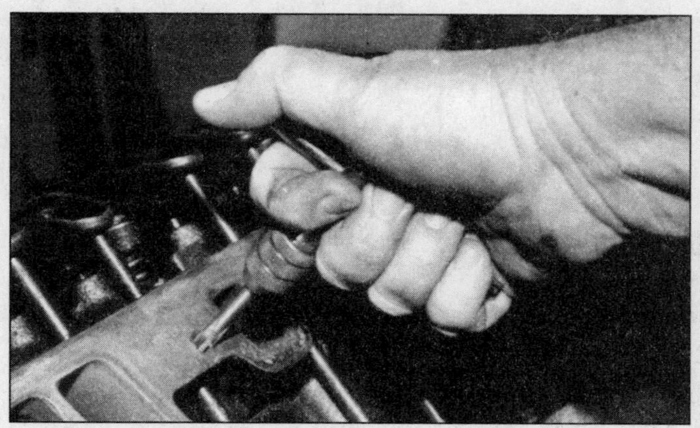

6.10a The bolt hole threads must be clean and dry to ensure accurate torque readings when the manifold mounting bolts are installed

6.10b Clean the bolt holes with compressed air, but be careful - wear safety goggles!

6.12 Make sure the intake manifold gaskets are installed right side up or all the passages and bolt holes may not line up properly!

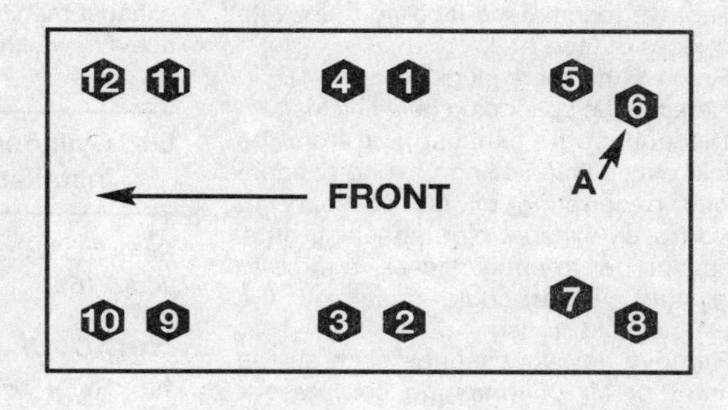

6.15a Intake manifold bolt tightening sequence for small block V8 engines

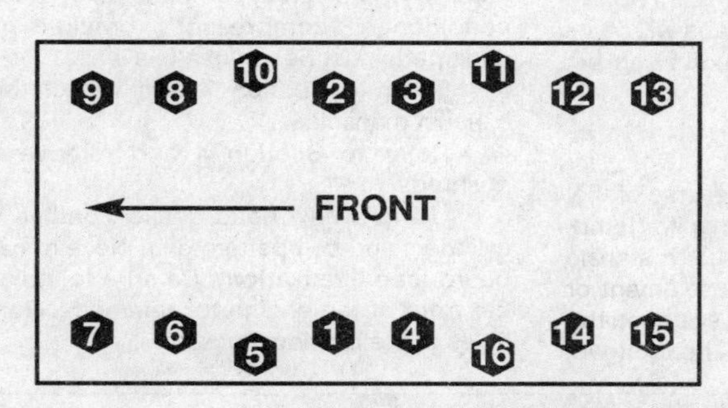

6.15b Intake manifold bolt tightening sequence for big block V8 engines (except 1991)

cleaner to remove any gasket material that falls into the intake ports in the heads.

10 Use a tap of the correct size to chase the threads in the bolt holes, then use compressed air (if available) to remove the debris from the holes **(see illustrations)**. **Warning:** *Wear safety glasses or a face shield to protect your eyes when using compressed air.*

11 Apply a thin coat of RTV sealant around the coolant passage holes on the cylinder head side of the new intake manifold gaskets (there is normally one hole at each end).

12 Position the gaskets on the cylinder heads. Make sure all intake port openings, coolant passage holes and bolt holes are aligned correctly and that the This side up is visible **(see illustration)**.

13 Install the front and rear end seals on the block. Note that most seals have either rubber spikes which fit into matching holes in the block, or rubber tabs which fit over the edge of the block to locate the seals. On some early models there is a tab on the rear seal which may have to be cut off for proper fit. Refer to the instructions with the gasket set for further information.

14 Carefully set the manifold in place. Do not disturb the gaskets and do not move the manifold fore-and-aft after it contacts the front and rear seals.

15 Apply a non-hardening sealant to the manifold bolt threads, then install the bolts. While the sealant is still wet, tighten the bolts to the specified torque following the recommended sequence **(see illustrations)**. Work up to the final torque in three steps. Note that the V6 manifold uses different sequences for initial and final tightening **(see illustration)**.

16 The remaining installation steps are the reverse of removal. Start the engine and check carefully for oil, vacuum and coolant leaks at the intake manifold joints.

7 Exhaust manifolds - removal and installation

Removal

1 Disconnect the negative cable from the battery.

2 Remove the carburetor heat stove pipe between the exhaust manifold and air cleaner snorkle (if equipped).

3 Disconnect the electrical connector to the oxygen sensor (if equipped).

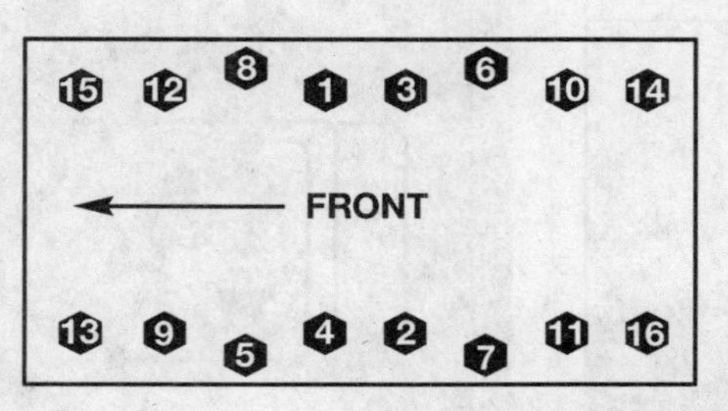

6.15c Intake manifold bolt tightening sequence for big block V8 engines (1991 models only)

4 Remove the AIR hose at the check valve and the AIR pipe bracket from the manifold stud (if equipped).

5 Disconnect the spark plug wires from the spark plugs (refer to Chapter 1 if necessary). If there is any danger of mixing the plug wires up, we recommend labeling them with small pieces of tape.

6 Remove the spark plugs (Chapter 1).

7 Remove the spark plug heat shields.

8 Disconnect the exhaust pipe from the manifold outlet. Often a short period of soaking with penetrating oil is necessary to remove frozen exhaust pipe attaching nuts. Use caution not to apply excessive force to frozen nuts, which could shear off the exhaust manifold studs.

9 Remove any accessories such as the alternator or air conditioning compressor bolted to the exhaust manifold, along with any mounting brackets.

10 Remove the two front and two rear manifold mounting bolts first, then the two center bolts to separate the manifold from the head. Some models use tabbed washers under the manifold bolts to keep the bolts from vibrating loose. On these models the tabs will have to be straightened before the bolts can be removed.

Installation

11 Installation is basically the reverse of the removal procedure. Clean the manifold and head gasket surfaces of old gasket material, then install new gaskets. Do not use any gasket cement or sealer on exhaust system gaskets. Be sure to transfer the heat stove assembly if a new manifold is being installed.

12 Install all the manifold bolts and tighten them to the specified torque. Work from the center to the ends and approach the final torque in three steps.

13 Apply anti-seize compound to the exhaust manifold-to-exhaust pipe studs, and use a new exhaust "doughnut" gasket.

8 Cylinder heads - removal and installation

Refer to illustrations 8.5, 8.12, 8.16a, 8.16b and 8.16c

Removal

1 Refer to Section 3 and remove the rocker arm covers.

2 Refer to Section 6 and remove the intake manifold. Note that the cooling system must be drained to prevent coolant from getting into internal areas of the engine when the manifold and heads are removed. The alternator, power steering pump, air pump and air conditioner compressor brackets (if equipped) must be removed as well.

3 Refer to Section 7 and detach both exhaust manifolds.

4 Refer to Section 4 and remove the pushrods.

5 Using a new head gasket, outline the cylinders and bolt pattern on a piece of cardboard **(see illustration)**. Be sure to indicate the front of the engine for reference. Punch holes at the bolt locations.

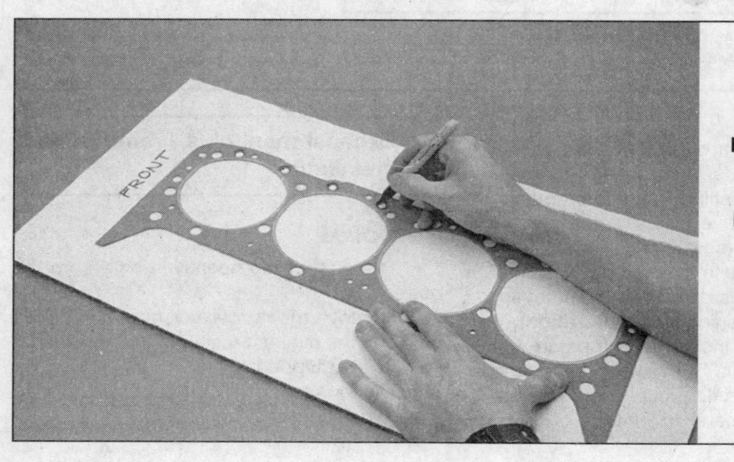

8.5 To avoid mixing up the head bolts, use a new gasket to transfer the bolt hole pattern to a piece of cardboard, then punch holes to accept the bolts

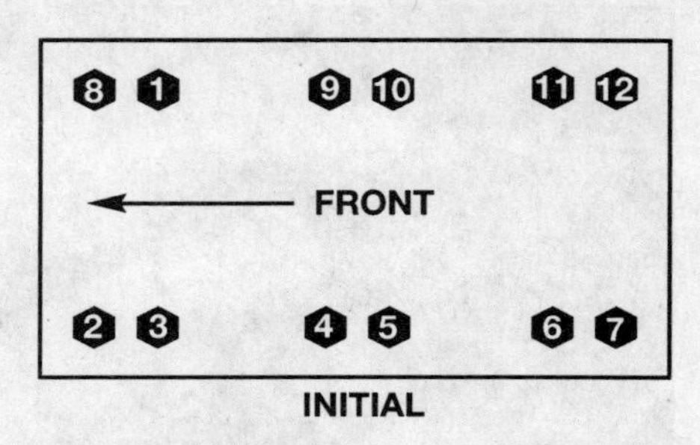

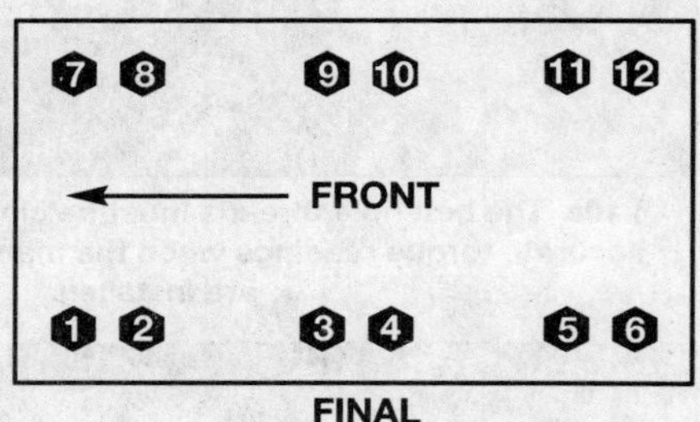

6.15d Intake manifold bolt tightening sequence for V6 engines

6 Loosen the head bolts in 1/4-turn increments until they can be removed by hand. Work from bolt-to-bolt in a pattern that is the reverse of the tightening sequence. **Note:** *Don't overlook the row of bolts on the lower edge of each head, near the spark plug holes. Store the bolts in the cardboard holder as they are removed. This will ensure that the bolts are reinstalled in their original holes.*

7 Lift the heads off the engine. If resistance is felt, do not pry between the head and block as damage to the mating surfaces will result. To dislodge the head, place a block of wood against the end of it and strike the wood block with a hammer. Store the heads on blocks of wood to prevent damage to the gasket sealing surfaces.

8 Cylinder head disassembly and inspection procedures are covered in detail in Chapter 2, Part C.

Installation

9 The mating surfaces of the cylinder heads and block must be perfectly clean when the heads are installed.

10 Use a gasket scraper to remove all traces of carbon and old gasket material, then wipe the mating surfaces with a cloth saturated with lacquer thinner or acetone. If there is oil on the mating surfaces when the heads are installed, the gaskets may not seal correctly and leaks may develop. When working on the block, cover the lifter valley with shop rags to keep debris out of the engine. Use a vacuum cleaner to remove any debris that falls into the cylinders.

11 Check the block and head mating surfaces for nicks, deep scratches and other damage. If damage is slight, it can be removed with emery cloth. If it is excessive, machining may be the only alternative.

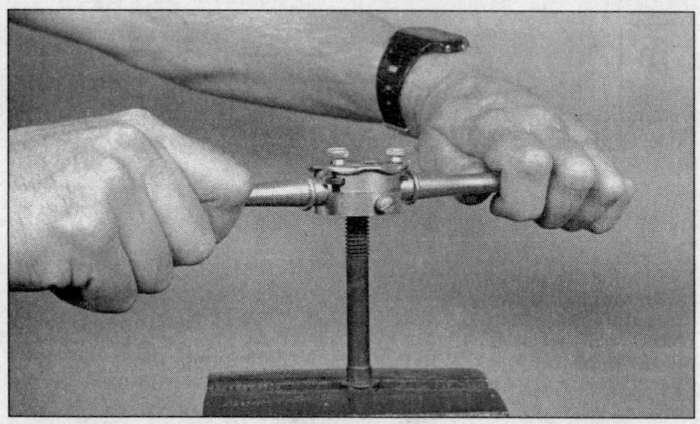

8.12 A die should be used to remove sealant and corrosion from the head bolt threads prior to installation

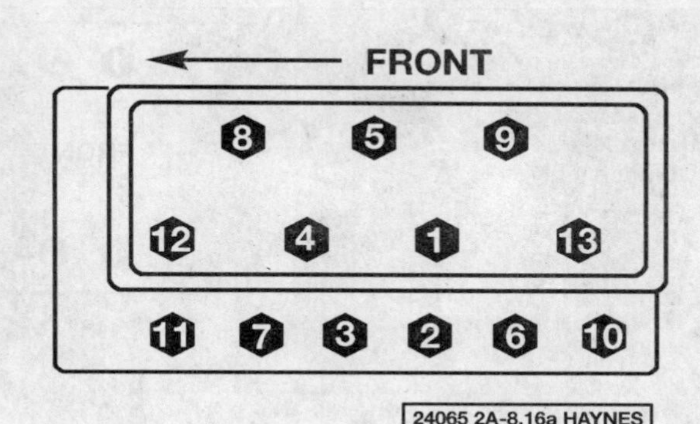

8.16a Cylinder head bolt tightening sequence for V6 engines

12 Use a tap of the correct size to chase the threads in the head bolt holes in the block. Mount each bolt in a vise and run a die down the threads to remove corrosion and restore the threads **(see illustration)**. Dirt, corrosion, sealant and damaged threads will affect torque readings.

13 Position the new gaskets over the dowel pins in the block. **Note:** *If a steel gasket is used (shim-type gasket), apply a thin, even coat of sealant such as K&W Copper Coat to both sides prior to installation. Steel gaskets must be installed with the raised bead UP. Composition gaskets must be installed dry; do not use sealant.*

14 Carefully position the heads on the block without disturbing the gaskets.

15 Before installing the head bolts, coat the threads with a non-hardening sealant such as Permatex No. 2.

16 Install the bolts in their original locations and tighten them finger tight. Following the recommended sequence, tighten the bolts in several steps to the specified torque **(see illustrations)**.

17 The remaining installation steps are the reverse of removal.

9 Top Dead Center (TDC) for number 1 piston - locating

Refer to illustration 9.6

1 Top Dead Center (TDC) is the highest point in the cylinder that each piston reaches as it travels up-and-down when the crankshaft turns. Each piston reaches TDC on the compression stroke and again on the exhaust stroke, but TDC generally refers to piston position on the compression stroke. The timing marks on the vibration damper installed on the front of the crankshaft are referenced to the number one piston at TDC on the compression stroke.

2 Positioning the pistons at TDC is an essential part of many procedures such as rocker arm removal, valve adjustment and distributor removal.

3 In order to bring any piston to TDC, the crankshaft must be turned using one of the methods outlined below. When looking at the front of the engine, normal crankshaft rotation is clockwise. **Warning:** *Before beginning this procedure, be sure to place the transmission in Neutral and disable the ignition system by removing the coil wire from the distributor cap and grounding it (remote coil systems) or disconnecting the BAT wire to the coil-in-cap type distributor.*

a) The preferred method is to turn the crankshaft with a large socket and breaker bar attached to the vibration damper bolt that is threaded into the front of the crankshaft on most engines.

b) A remote starter switch, which may save some time, can also be used. Attach the switch leads to the S (switch) and B (battery) terminals on the starter solenoid. Once the piston is close to TDC, use a socket and breaker bar as described in the previous paragraph.

c) If an assistant is available to turn the ignition switch to the Start position in short bursts, you can get the piston close to TDC without a remote starter switch. Use a socket and breaker bar as described in Paragraph a) to complete the procedure.

4 Scribe or paint a small mark on the distributor body directly below the number one spark plug wire terminal in the distributor cap or make a mark on the intake manifold

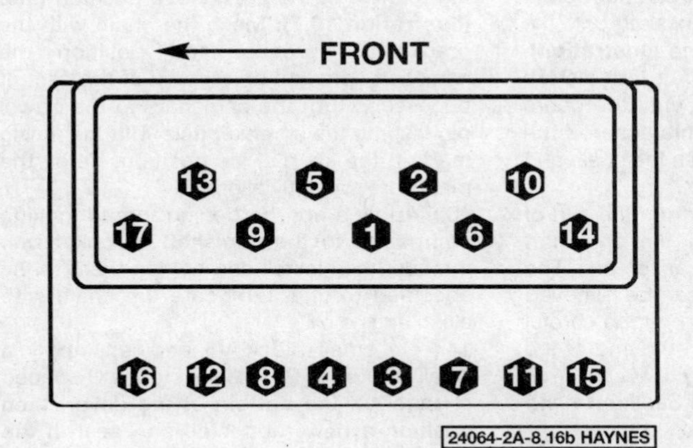

8.16b Cylinder head bolt tightening sequence for small block V8 engines

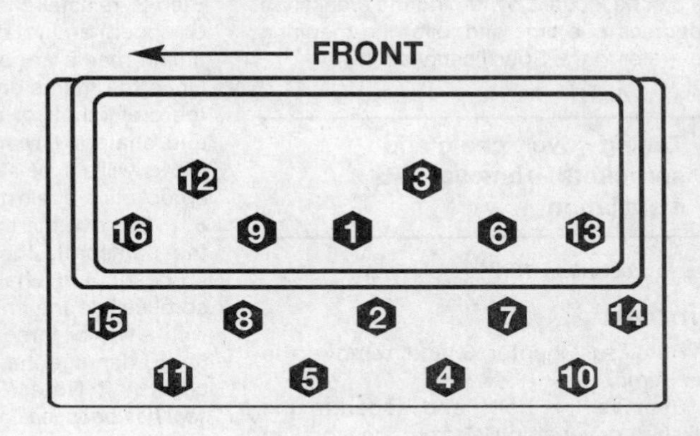

8.16c Cylinder head bolt tightening sequence for big block V8 engines

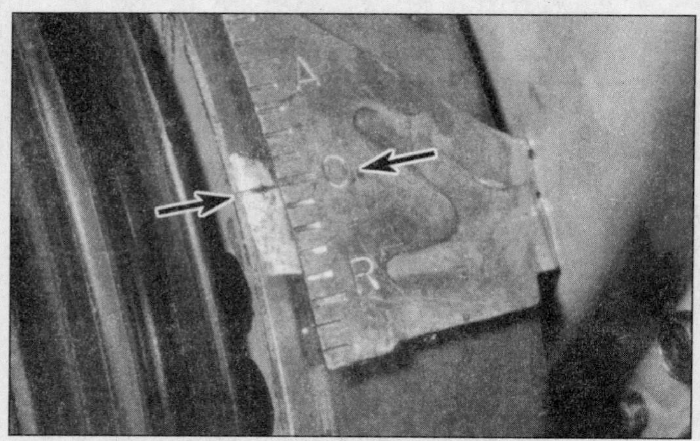

9.6 Turn the crankshaft until the line on the vibration damper is directly opposite the zero mark on the timing plate as shown here

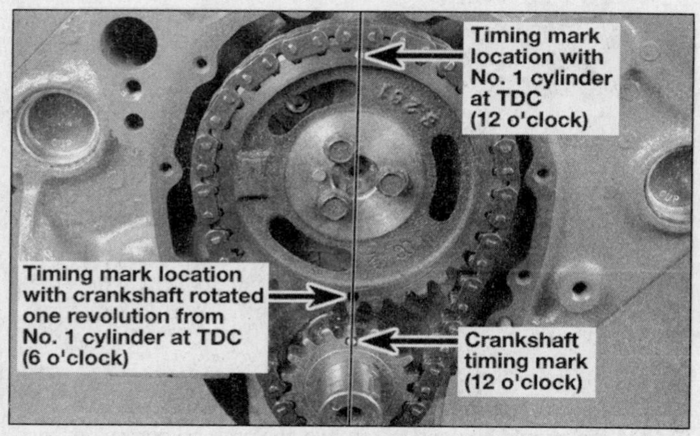

10.7 Align the timing marks as shown - the cam gear mark should be in the six o'clock position and the crankshaft gear mark should be in the 12 o'clock position

directly across from the number one spark plug wire terminal in the distributor cap.

5 Remove the distributor cap as described in Chapter 1.

6 Turn the crankshaft (see Step 3 above) until the line on the vibration damper is aligned with the zero mark on the timing plate **(see illustration)**. The timing plate and vibration damper are located low on the front of the engine, behind the pulley that turns the drivebelts.

7 The rotor should now be pointing directly at the mark on the distributor base or intake manifold. If it isn't, the piston is at TDC on the exhaust stroke.

8 To get the piston to TDC on the compression stroke, turn the crankshaft one complete turn (360-degrees) clockwise. The rotor should now be pointing at the mark. When the rotor is pointing at the number one spark plug wire terminal in the distributor cap (which is indicated by the mark on the distributor body or intake manifold) and the timing marks are aligned, the number one piston is at TDC on the compression stroke.

9 After the number one piston has been positioned at TDC on the compression stroke, TDC for any of the remaining cylinders can be located by turning the crankshaft 90-degrees at a time and following the firing order (refer to the Specifications).

10 Timing cover, chain and sprockets - removal and installation

Refer to illustration 10.7

Removal

1 Refer to Chapter 3 and remove the water pump.

2 Remove the bolts and separate the crankshaft drivebelt pulley from the vibration damper.

3 Most, but not all V8 and V6 engines, use a large bolt threaded into the nose of the crankshaft to secure the vibration damper in position. If your engine has a bolt, remove it from the front of the crankshaft, then use a puller to detach the vibration damper. **Caution:** *Do not use a puller with jaws that grip the outer edge of the damper. The puller must be the type that utilizes bolts to apply force to the damper hub only.*

4 On most small block V8 engines the timing cover cannot be removed with the oil pan in place. The pan bolts will have to be loosened and the pan lowered approximately 1/4-inch for the timing cover to be removed. If the pan has been in place for an extended period of time, it is likely the pan gasket will break when the pan is lowered. In this case the pan should be removed and a new gasket installed.

5 Remove the bolts and separate the timing chain cover from the block. It may be stuck. If so, use a putty knife to break the gasket seal. The cover is easily distorted, so do not attempt to pry it off.

6 On big block engines, pull the cover forward far enough to insert a knife between the cover and the block, cut the forward portion of the oil pan gasket on each side, then remove the cover.

7 Following the procedure in Section 9, Step 3, reinstall the vibration damper bolt (if equipped) and rotate the crankshaft until the timing marks are aligned **(see illustration)**. Once the marks are aligned, do not disturb the crankshaft or camshaft until the gears and chain are reinstalled (this assures the marks will still be aligned when the chain and sprockets are reinstalled).

8 Remove the three bolts from the end of the camshaft, then detach the camshaft sprocket and chain as an assembly. The sprocket on the crankshaft can be removed with a two or three jaw puller, but be careful not to damage the threads in the end of the crankshaft. **Note:** *If the timing chain cover oil seal has been leaking, refer to Section 14 and install a new one.*

Installation

9 Use a gasket scraper to remove all traces of old gasket material and sealant from the cover and engine block. Stuff a shop rag into the opening at the front of the oil pan to keep debris out of the engine. Wipe the cover and block sealing surfaces with a cloth saturated with lacquer thinner or acetone.

10 Check the cover flange for distortion, particularly around the bolt holes. If necessary, place the cover on a block of wood and use a hammer to flatten and restore the gasket surface.

11 If new parts are being installed, be sure to align the keyway in the crankshaft sprocket with the Woodruff key in the end of the crankshaft. **Note:** *Timing chains must be replaced as a set with the camshaft and crankshaft gears. Never put a new chain on old gears.* Align the sprocket with the Woodruff key and press the sprocket onto the crankshaft with the vibration damper bolt, a large socket and some washers, or tap it gently into place until it is completely seated. **Caution:** *If resistance is encountered, do not hammer the sprocket onto the crankshaft. It may eventually move onto the shaft, but it may be cracked in the process and fail later, causing extensive engine damage.*

12 Loop the new chain over the camshaft sprocket, then turn the sprocket until the timing mark is in the six o'clock position **(see illustration 10.7)**. Mesh the chain with the crankshaft sprocket and position the camshaft sprocket on the end of the cam. If necessary, turn the camshaft so the dowel pin fits into the sprocket hole with the timing mark in the six o'clock position. Verify the marks are correctly aligned.

13 Apply a non-hardening thread locking compound to the camshaft sprocket bolt threads, then install and tighten them to the specified torque. Lubricate the chain with clean engine oil.

14 On small block V8 engines, apply a small amount of RTV sealant to the U-shaped channel on the bottom of the cover, then position a new rubber oil pan seal in the channel. The sealant should hold it in place as the cover is installed.

15 On big block V8 engines, cut the tabs from a new front oil pan seal and use gasket

11.3 When checking the camshaft lobe lift, the dial indicator plunger must be positioned directly above and in line with the pushrod

11.11 The lifters on an engine that has accumulated many miles may have to be removed with a special tool (arrow) - store the lifters in an organized manner to make sure they are reinstalled in their original locations

sealer to hold the seal in the bottom of the front cover. Apply a 1/8-inch bead of RTV-type gasket sealer to the junction of the oil pan and front face of the block on each side.

16 The V6 engine uses a one-piece oil pan gasket, which should be checked for cracks and deformation before installing the timing cover. If the gasket has deteriorated, it must be replaced before reinstalling the timing cover.

17 Apply a thin layer of RTV sealant to both sides of the new gasket, then position it on the engine block. The dowel pins and sealant will hold it in place.

18 Install the timing chain cover on the block, tightening the bolts finger tight.

19 If the oil pan was removed on a small block V8 engine, reinstall it (Section 12). If it was only loosened, tighten the oil pan bolts, bringing the oil pan up against the lower seal in the timing chain cover.

20 Tighten the timing chain cover bolts to the specified torque.

21 Lubricate the oil seal contact surface of the vibration damper hub with moly-base grease or clean engine oil, then install the damper on the end of the crankshaft. The keyway in the damper must be aligned with the Woodruff key in the crankshaft nose. If the damper cannot be seated by hand, slip a large washer over the bolt, install the bolt and tighten it to push the damper into place. Remove the large washer and tighten the bolt to the specified torque.

22 The remaining installation steps are the reverse of removal.

11 Camshaft, bearings and lifters - removal, inspection and installation

Refer to illustrations 11.3, 11.11, 11.15, 11.18a, 11.18b, 11.20 and 11.22

Camshaft lobe lift check

1 In order to determine the extent of cam lobe wear, the lobe lift should be checked prior to camshaft removal. Position the number one piston at TDC on the compression stroke (see Section 9).

2 Refer to Section 3 and remove the rocker arm covers. Refer to Section 4 and loosen the rocker arm nuts, then pivot the rocker arms to the side for access to the pushrods.

3 Beginning with the number one cylinder, mount a dial indicator on the engine and position the plunger against the top of the first pushrod. The plunger should be directly in line with the pushrod **(see illustration)**.

4 Zero the dial indicator, then very slowly turn the crankshaft in the normal direction of rotation until the indicator needle stops and begins to move in the opposite direction. The point at which it stops indicates maximum cam lobe lift.

5 Record this figure for future reference, then reposition the piston at TDC on the compression stroke.

6 Move the dial indicator to the other number one cylinder pushrod and repeat the check. Be sure to record the results for each valve.

7 Repeat the check for the remaining valves. Since each piston must be at TDC on the compression stroke for this procedure, work from cylinder-to-cylinder following the firing order sequence.

8 After the check is complete, compare the results to the Specifications. If camshaft lobe lift is less than specified, cam lobe wear has occurred and a new camshaft should be installed.

Removal

9 Refer to the appropriate Sections and remove the intake manifold, the rocker arms, the pushrods and the timing chain and camshaft sprocket. Remove the radiator (see Chapter 3) and the mechanical fuel pump and rod (see Chapter 4). **Note:** *If the vehicle is equipped with air conditioning it may be necessary to remove the air conditioning condenser to remove the camshaft. If the condenser must be removed, the system must first be depressurized by a dealer service department or air conditioning shop. Do not disconnect any air conditioning lines until the system has been properly depressurized.*

10 There are several ways to extract the lifters from the bores. A special tool designed to grip and remove lifters is manufactured by many tool companies and is widely available, but it may not be required in every case. On newer engines without a lot of varnish buildup, the lifters can often be removed with a small magnet or even with your fingers. A machinist's scribe with a bent end can be used to pull the lifters out by positioning the point under the retainer ring inside the top of each lifter. **Caution:** *Do not use pliers to remove the lifters unless you intend to replace them with new ones (along with the camshaft). The pliers will damage the precision machined and hardened lifters, rendering them useless.*

11 Before removing the lifters, arrange to store them in a clearly labeled box to ensure that they are reinstalled in their original locations. Remove the lifters and store them where they will not get dirty **(see illustration)**. Do not attempt to withdraw the camshaft with the lifters in place.

12 Thread a 6-inch long 5/16 - 18 bolt into one of the camshaft sprocket bolt holes to use as a handle when removing the camshaft from the block.

13 Carefully pull the camshaft out. Support the cam near the block so the lobes do not nick or gouge the bearings as it is withdrawn.

Inspection

Camshaft and bearings

14 After the camshaft has been removed from the engine, cleaned with solvent and dried, inspect the bearing journals for uneven wear, pitting and evidence of seizure. If the journals are damaged, the bearing inserts in the block are probably damaged as well. Both the camshaft and bearings will have to be replaced. Replacement of the camshaft bearings requires special tools and techniques which place it beyond the scope of the home mechanic. The block will have to be

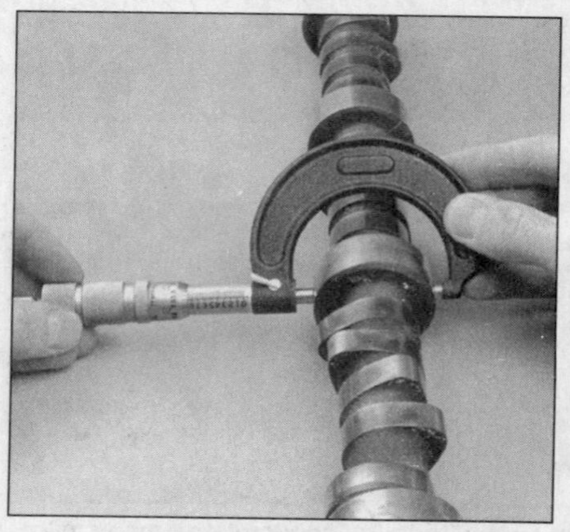

11.15 The camshaft bearing journal diameter is checked to pinpoint excessive wear and out-of-round conditions

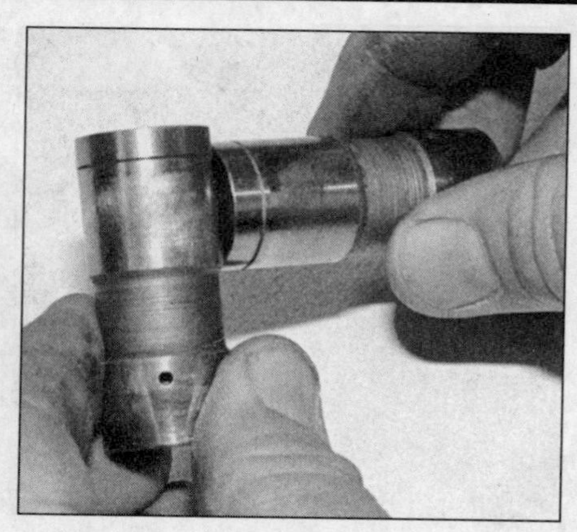

11.18a The foot of each lifter must be slightly convex - the side of another lifter can be used as a straightedge to check it; if it appears flat, it's worn and must not be reused

11.18b If the foot of any lifter is worn concave, scratched or galled, both the entire lifter set and the camshaft will have to be replaced with new parts

11.20 Be sure to apply camshaft assembly lube to the cam lobes and bearing journals before installing the camshaft

11.22 After the camshaft is in place, turn it until the dowel pin (arrow) is in the 9 o'clock position as shown here (for small block engines only)

removed from the vehicle and taken to an automotive machine shop for this procedure.

15 Measure the bearing journals with a micrometer to determine if they are excessively worn or out-of-round **(see illustration)**.

16 Check the camshaft lobes for heat discoloration, score marks, chipped areas, pitting and uneven wear. If the lobes are in good condition and if the lobe lift measurements are as specified, the camshaft can be reused.

Hydraulic lifters

17 Clean the lifters with solvent and dry them thoroughly without mixing them up.

18 Check each lifter wall, pushrod seat and foot for scuffing, score marks and uneven wear. Each lifter foot (the surface that rides on the cam lobe) must be slightly convex, although this can be difficult to determine by eye. If the base of the lifter is concave **(see illustrations)**, the lifters and camshaft must be replaced. If the lifter walls are damaged or worn (which is not very likely), inspect the lifter bores in the engine block as well. If the pushrod seats are worn, check the pushrod ends.

19 If new lifters are being installed, a new camshaft must also be installed. If a new camshaft is installed, then use new lifters as well. Never install used lifters unless the original camshaft is used and the lifters can be installed in their original locations.

Installation

20 Lubricate the camshaft bearing journals and cam lobes with camshaft assembly lube **(see illustration)**.

21 Slide the camshaft into the engine. Support the cam near the block and be careful not to scrape or nick the bearings.

22 Turn the camshaft until the dowel pin is in the 9 o'clock position (for small block engines only) **(see illustration)**.

23 Refer to Section 10 and install the timing chain and sprockets.

24 Lubricate the lifters with clean engine oil and install them in the block. If the original lifters are being reinstalled, be sure to return them to their original locations. If a new camshaft was installed, be sure to install new lifters as well.

25 The remaining installation steps are the reverse of removal.

26 Before starting and running the engine, change the oil and install a new oil filter (see Chapter 1).

12 Oil pan - removal and installation

Removal

1 Disconnect the negative cable from the battery.

2 Remove the distributor cap to prevent breakage against the firewall as the engine is lifted.

3 Unbolt the radiator shroud and move it back over the fan (Chapter 3). You may wish to insert a piece of heavy cardboard between the fan and radiator to protect the radiator fins from the fan when the engine is raised.

4 Raise the vehicle and support it securely on jackstands.

5 Drain the engine oil (Chapter 1).

6 If equipped with single exhaust, unbolt the crossover pipe at the exhaust manifolds.

7 If equipped with strut rods, remove the

13.3 Make sure the nylon sleeve is in place between the oil pump and oil pump driveshaft

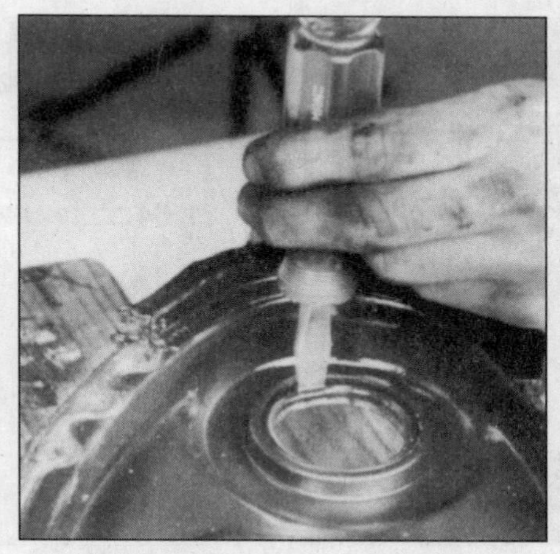

14.2 While supporting the cover near the seal bore, drive the old seal out from the inside with a hammer and punch or screwdriver

14.4 Clean the bore, then apply a small amount of oil to the outer edge of the seal and drive it squarely into the opening with a large socket or a piece of pipe and a hammer - do not damage the seal in the process!

strut rods at the flywheel cover and the strut rod brackets at the front engine mounts.

8 If equipped with an automatic transmission, remove the converter inspection cover.

9 If equipped with a manual transmission, remove the starter (Chapter 5) and flywheel inspection cover.

10 If equipped with a V6 engine, remove the starter (Chapter 5).

11 On models with big block engines, remove the oil filter and disconnect the oil pressure gauge line from the side of the block.

12 Turn the crankshaft until the timing mark on the on the vibration damper is pointed straight down.

13 Remove the motor mount through bolts.

14 Use an engine hoist or a floor jack and block of wood under the oil pan to lift the engine approximately three inches. **Caution:** *On most engines the oil pump pickup is very close to the bottom of the oil pan, and it can be damaged easily if concentrated pressure from a jack is applied to the pan. Only use a jack with a large piece of wood to spread the load over a wide area. When lifting the engine check to make sure the distributor isn't hitting the firewall and the fan isn't hitting the radiator.*

15 Place blocks of wood between the crossmember and the engine block in the area of the motor mounts to hold the engine in the raised position, then remove the floor jack or engine hoist.

16 Remove the oil pan bolts. Note that some later models use studs and nuts in some positions, and that earlier models use larger bolts at the front and rear of each pan rail.

17 Remove the pan by tilting the back downward and working it free of the crankshaft throws, oil pump pickup and the front crossmember.

Installation

18 Thoroughly clean the mounting surfaces of the oil pan and engine block of old gasket material and sealer.

19 Apply a thin layer of RTV-type sealer to the oil pan and install a new oil pan gasket. On some later models a one-piece oil pan gasket is used which does not require the installation of front and rear pan oil seals.

20 On earlier models, install a new seal in the front cover and in the rear main bearing cap.

21 Lift the pan into position, being careful not to disturb the gasket, and install the bolts/nuts finger tight.

22 Starting at the ends and alternating from side-to-side towards the center, tighten the bolts to the specified torque.

23 The remainder of the installation procedure is the reverse of removal. Fill the pan with new oil, start the engine and check for leaks before placing the vehicle back in service.

13 Oil pump - removal and installation

Refer to illustration 13.3

1 Remove the oil pan as described in Section 12.

2 While supporting the oil pump, remove the pump-to-rear main bearing cap bolt.

3 Lower the pump and remove it along with the pump driveshaft. Note that on most models a hard nylon sleeve is used to align the oil pump driveshaft and the oil pump shaft. Make sure this sleeve is in place on the oil pump driveshaft **(see illustration)**. If it is not there, check the oil pan for the pieces of the sleeve, clean them out of the pan, then get a new sleeve for the oil pump driveshaft.

4 If a new oil pump is installed, make sure the pump driveshaft is mated with the shaft inside the pump.

5 Position the pump on the engine and make sure the slot in the upper end of the driveshaft is aligned with the tang on the lower end of the distributor shaft. The distributor drives the oil pump, so it is absolutely essential that the components mate properly.

6 Install the mounting bolt and tighten it to the specified torque.

7 Install the oil pan.

14 Crankshaft oil seals - replacement

Refer to illustrations 14.2, 14.4, 14.9, 14.11, 14.12a, 14.12b, 14.13, 14.14 and 14.19

Front seal

1 Remove the timing chain cover as described in Section 10.

2 Use a punch or screwdriver and hammer to drive the seal out of the cover from the backside. Support the cover as close to the seal bore as possible **(see illustration)**. Be careful not to distort the cover or scratch the wall of the seal bore. If the engine has accumulated a lot of miles, apply penetrating oil to the seal-to-cover joint on each side and allow it to soak in before attempting to drive the seal out.

3 Clean the bore to remove any old seal material and corrosion. Support the cover on blocks of wood and position the new seal in the bore with the open end of the seal facing in. A small amount of oil applied to the outer edge of the new seal will make installation easier.

4 Drive the seal into the bore with a large socket and hammer until it is completely seated **(see illustration)**. Select a socket that is the same outside diameter as the seal (a section of pipe can be used if a socket is not available).

5 Reinstall the timing chain cover.

Rear seal

Small Block V8 engines (before 1986) and all Big Block engines

6 The rear main seal can be replaced with

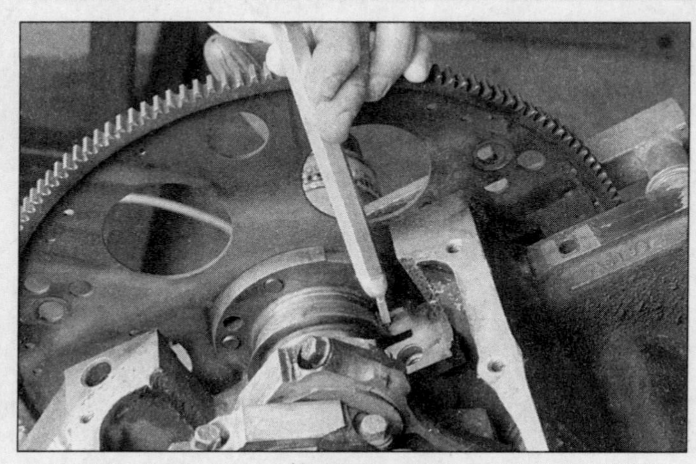

14.9 Tap the seal end with a brass punch or wood dowel and hammer, until it can be gripped with a pair of pliers and pulled out

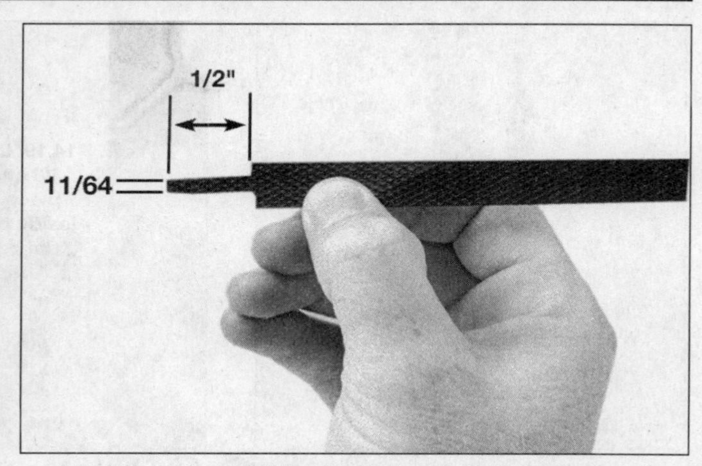

14.11 If the new seal did not include an installation tool, make one from a piece of brass, plastic or shim stock

the engine in the vehicle. Refer to the appropriate Sections and remove the oil pan and oil pump.

7 Remove the bolts and detach the rear main bearing cap from the engine.

8 The seal section in the bearing cap can be pried out with a screwdriver.

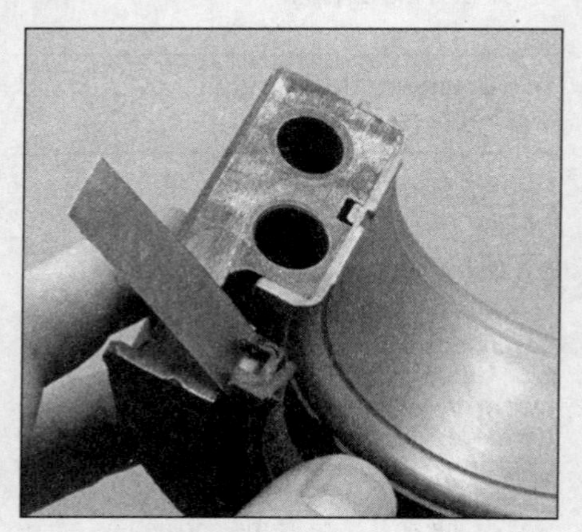

14.12a Using the tool like a "shoehorn", attach the seal section to the bearing cap . . .

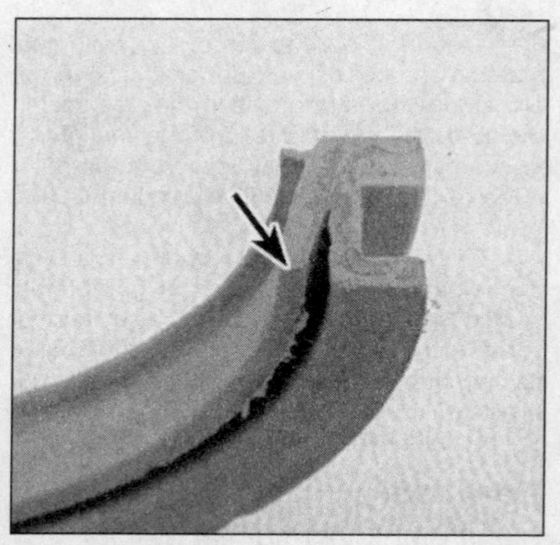

14.12b . . . with the oil seal lip pointing toward the front of the engine

9 To remove the seal section in the block, tap on one end with a hammer and brass punch or wood dowel until the other end protrudes far enough to grip it with a pair of pliers and pull it out **(see illustration)**. Be very careful not to nick or scratch the crankshaft journal or seal surface as this is done.

10 Inspect the bearing cap and engine block mating surfaces, as well as the cap seal grooves, for nicks, burrs and scratches. Remove any defects with a fine file or deburring tool.

11 A small seal installation tool is usually included when a new seal is purchased. If you didn't receive one, they can also be purchased separately at most auto parts stores or you can make one from an old feeler gauge or a piece of brass shim stock **(see illustration)**.

12 Using the tool, install one seal section in the cap with the lip facing the front of the engine (if the seal has two lips, the one with the helix must face the front) **(see illustrations)**. The ends should be flush with the mating surface of the cap. Make sure it is completely seated.

14.13 Position the tool to protect the backside of the seal as it passes over the sharp edge of the ridge

13 Position the narrow end of the tool so that it will protect the backside of the seal as it passes over the sharp edge of the ridge in the block **(see illustration)**.

14 Lubricate the seal lips and the groove in the backside with molybase grease or clean engine oil - do not get any lubricant on the seal ends. Insert the seal into the block, over the tool **(see illustration)**. **Caution:** *Make sure that the lip points toward the front of the engine when the seal is installed.*

15 Push the seal into place, using the tool like a "shoehorn". Turning the crankshaft may help to draw the seal into place. When both ends of the seal are flush with the block surface, remove the tool.

16 Lubricate the cap seal lips with molybase grease or clean engine oil.

17 Carefully position the bearing cap on the block, install the bolts and tighten them to 10-to-12 ft-lbs only. Tap the crankshaft forward and backward with a lead or brass hammer to line up the main bearing and crankshaft thrust surfaces, then tighten the rear bearing cap bolts to the specified torque.

18 Install the oil pump and oil pan.

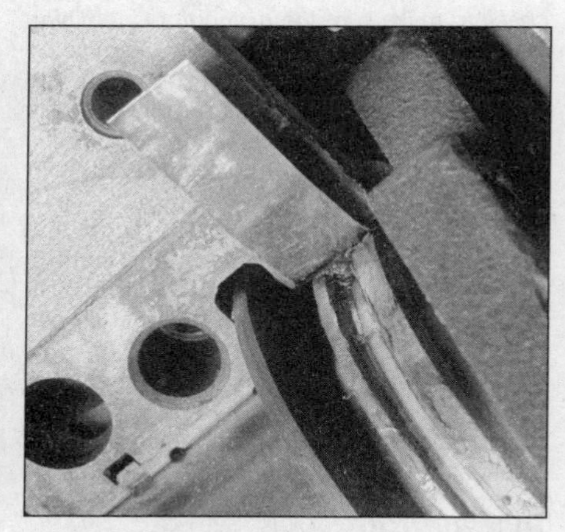

14.14 Make sure the seal lip faces the front of the engine and hold the tool in place to protect the seal as it is installed- note that the seal straddles the ridge

14.19 Later model V6 and V8 engines use a one-piece rear oil seal held inside a housing bolted to the back of the block - notches are provided in the housing to pry the old oil seal out

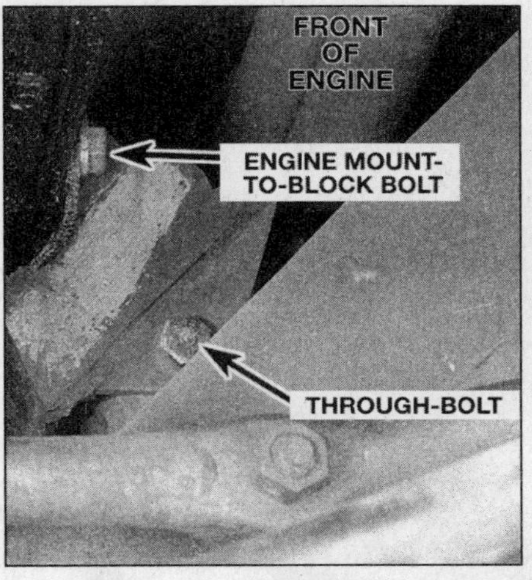

15.1a Typical 2WD engine mount details

Small Block V8 engines (1986 on) and all V6 engines

19 Later model small block and V6 engines use a one-piece rear main oil seal which is installed in a bolt-on housing **(see illustration)**. Replacing this seal requires that the transmission, clutch assembly and flywheel (manual transmission) or torque converter and driveplate (automatic transmission) be removed. Refer to Chapter 7 for the transmission removal procedures.

20 Remove the oil pan (Section 12).

21 Although the seal can be removed by prying it out of the retainer by inserting a screwdriver in the notches provided, installation with the retainer still mounted on the block is most easily accomplished with the use of a special tool which attaches to the threaded holes in the crankshaft flange and then presses the new seal into place.

22 If the special installation tool is not available, remove the screws securing the retainer to the block and detach the retainer and gasket. Whenever the retainer is removed from the block a new seal and gasket must be installed.

23 Insert a screwdriver blade into the notches in the seal housing and pry out the old seal. Be sure to note how far it is recessed into the housing bore before removal so the new seal can be installed to the same depth.

24 Clean the housing thoroughly then apply a thin coat of engine oil to the new seal. Set the seal squarely into the recess in the housing, then, using two pieces of wood, one each side of the housing, use a large vise to press the seal into place.

25 Carefully slide the seal over the crankshaft and bolt the seal housing to the block. Be sure to use a new gasket, but do not use any gasket sealer.

26 The remainder of installation is the reverse of the removal procedure.

15 Engine mounts - check and replacement

Refer to illustrations 15.1a and 15.1b

1 Engine mounts seldom require attention, but broken or deteriorated mounts should be replaced immediately or the added strain placed on the driveline components may cause damage **(see illustrations)**.

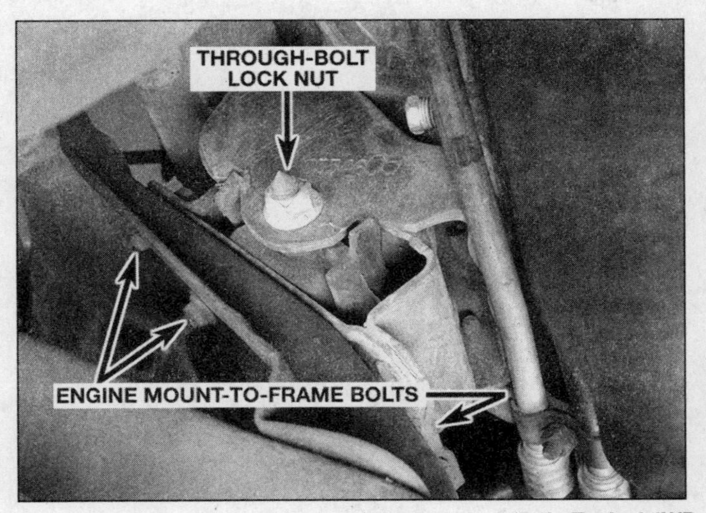

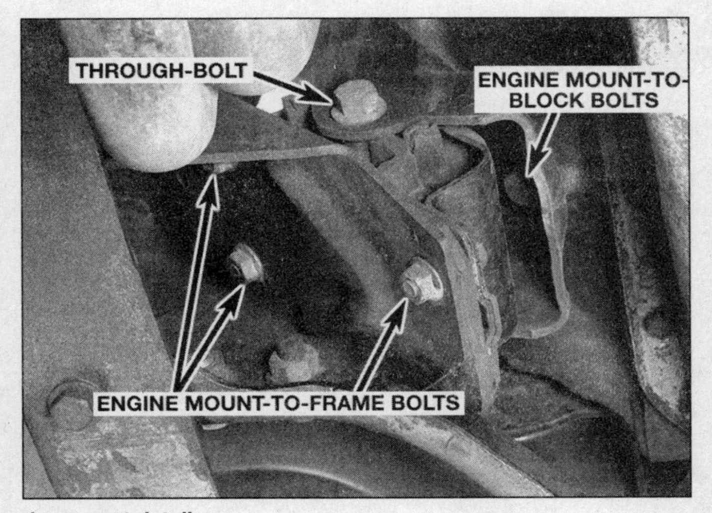

15.1b Typical 4WD engine mount details

Check

2 During the check, the engine must be raised slightly to remove the weight from the mounts. Refer to Chapter 1 and remove the distributor cap before raising the engine.

3 Raise the vehicle and support it securely on jackstands, then position the jack under the engine oil pan. Place a large block of wood between the jack head and the oil pan, then carefully raise the engine just enough to take the weight off the mounts.

4 Check the mounts to see if the rubber is cracked, hardened or separated from the metal plates. Sometimes the rubber will split right down the center. Rubber preservative or WD-40 should be applied to the mounts to slow deterioration.

5 Check for relative movement between the mount plates and the engine or frame (use a large screwdriver or pry bar to attempt to move the mounts). If movement is noted, lower the engine and tighten the mount fasteners.

Replacement

6 Disconnect the negative cable from the battery, then raise the vehicle and support it securely on jackstands.

7 Remove the nut and withdraw the mount through-bolt from the frame bracket.

8 Raise the engine slightly, then remove the mount-to-frame bolts and detach the mount.

9 Installation is the reverse of removal. Use thread locking compound on the mount bolts and be sure to tighten them securely.

Chapter 2 Part B
Six-cylinder inline engines

Contents

	Section
Crankshaft oil seals - replacement	14
Cylinder compression check	See Chapter 2C
Cylinder head - removal and installation	10
Engine mounts - check and replacement	15
Engine oil and filter change	See Chapter 1
Exhaust manifold (integrated head) - removal and installation	9
General information	1
Hydraulic lifters - removal and installation	5
Intake and exhaust manifold (non-integrated head) - removal and installation	8
Oil pan and oil pump - removal and installation	13
Pushrod cover - removal and installation	4
Repair operations possible with the engine in the vehicle	2
Rocker arm cover - removal and installation	3
Rocker arms and pushrods - removal, inspection and installation	6
Spark Plug replacement	See Chapter 1
Timing cover, gears and camshaft - removal, inspection and installation	12
Top Dead Center (TDC) for number 1 piston - locating	11
Valve springs, retainers and seals - replacement in vehicle	7
Water pump - removal and installation	See Chapter 3

Specifications

General

Firing order	1-5-3-6-2-4
Bore	3.875 in
Stroke	
230	3.250 in
250	3.530 in
292	4.120 in

INLINE 6-CYLINDER
Firing order
1-5-3-6-2-4

The blackened terminal shown on the distributor cap indicates the Number One spark plug wire position

Cylinder location and distributor rotation- inline 6-cylinder engine

Camshaft

Bearing journal diameter	1.8677 to 1.8697 in
Out-of-round limit	0.020 in
End play	0.003 to 0.008 in
Lobe lift	
230	0.1896 in
250	0.2217 in
292	0.2315 in

Torque specifications

	Ft-lbs (unless otherwise noted)
Rocker arm cover bolts	45 in-lbs
Pushrod cover bolts	50 in-lbs
Intake manifold	
Outer bolts	20
Inner bolts	30
Exhaust manifold (non-integrated head)	
To 1982	30
1983 on	20
Exhaust manifold (integrated head)	
Outer bolts	18 to 25
Top and bottom center bolts	40 to 45
Center inner bolts	35 to 40
Exhaust manifold-to-intake manifold bolts/nuts	45
Cylinder head bolts	
1985 and earlier	95
1985 on	
left front	85
all others	95
Timing cover bolts	80 in-lbs
Camshaft thrust plate bolts	80 in-lbs
Torsional (vibration) damper bolt	60
Oil pan bolts	
To front cover	45 in-lbs
1/4-20	80 in-lbs
5/16-18	165 in-lbs
Oil pump bolt	115 in-lbs
Rear main bearing cap bolt	65
Flywheel-to-crankshaft bolts	
230 and 250	60
292	110

1 General information

This Part of Chapter 2 is devoted to in-vehicle repair procedures for 6-cylinder inline engines. All information concerning engine removal and installation and engine block and cylinder head overhaul can be found in Part C of this Chapter.

Since the repair procedures included in this Part are based on the assumption that the engine is still installed in the vehicle, if they are being used during a complete engine overhaul (with the engine already out of the vehicle and on a stand) many of the steps included here will not apply.

The specifications included in this Part of Chapter 2 apply only to the procedures found here. The specifications necessary for rebuilding the block and cylinder heads are included in Part C.

The 6-cylinder inline engines used in Chevrolet and GMC trucks come in three sizes: 230 cubic inches, 250 cubic inches and 292 cubic inches. Mechanically the engines are basically the same, with the exception that some engines use a head with a cast-in intake manifold, rather than a separate bolt-on manifold.

2 Repair operations possible with the engine in the vehicle

Many major repair operations can be accomplished without removing the engine from the vehicle.

Clean the engine compartment and the exterior of the engine with some type of pressure washer before any work is done. A clean engine will make the job easier and will help keep dirt out of the internal areas of the engine.

Depending on the components involved, it may be a good idea to remove the hood to improve access to the engine as repairs are performed (refer to Chapter 11 if necessary).

If oil or coolant leaks develop, indicating a need for gasket or seal replacement, the repairs can generally be made with the engine in the vehicle. The oil pan gasket, the cylinder head gasket, intake and exhaust manifold gaskets, timing cover gaskets and the crankshaft oil seals are accessible with the engine in place.

Exterior engine components, such as the water pump, the starter motor, the alternator, the distributor and the carburetor, as well as the intake and exhaust manifolds, can be removed for repair with the engine in place.

Since the cylinder head can be removed without pulling the engine, valve component servicing can also be accomplished with the engine in the vehicle.

Replacement of, repairs to or inspection of the timing gears and the oil pump are all possible with the engine in place.

In extreme cases caused by a lack of necessary equipment, repair or replacement of piston rings, pistons, connecting rods and rod bearings is possible with the engine in the vehicle. However, this practice is not recommended because of the cleaning and preparation work that must be done to the components involved.

3 Rocker arm cover - removal and installation

Removal

1 Remove the PCV hose or valve (depending on how equipped) from the rocker arm cover (refer to Chapter 6 if necessary).

2 Remove the air cleaner assembly.

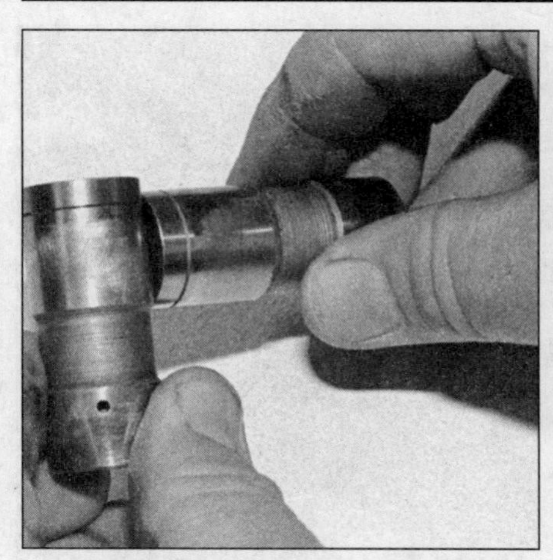

5.5a The foot of each lifter must be slightly convex - the side of another lifter can be used as a straightedge to check it; if it appears flat, it's worn and must not be reused

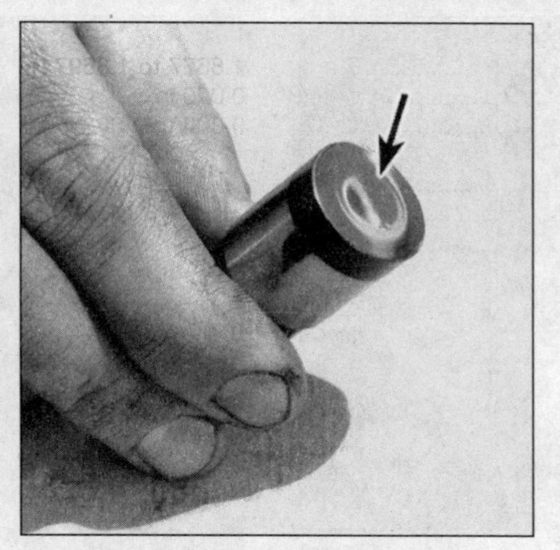

5.5b If the bottom (foot) of any lifter (arrow) is worn concave, scratched or galled, replace the entire set and the camshaft with new parts

3 Remove the wiring, fuel and vacuum lines from the rocker arm clips.

4 Note which rocker arm bolts are used to retain clips, then remove the bolts, reinforcements and clips.

5 Remove the rocker arm cover. If the cover is stuck to the head, do not try to pry it off with a screwdriver or knife. Instead use a rubber mallet to bump the end of the cover or place a block of wood against the end of the cover and hit it with a hammer. If the cover will still not release, carefully insert a putty knife or gasket scraper under the sealing flange and cut the gasket loose from the head. Use caution not to distort the sealing flange of the rocker arm cover.

6 Carefully clean all traces of old gasket and sealer from the sealing flange of the rocker arm cover and the head. Use caution to prevent debris from falling into the oil drain-back holes in the head.

Installation

7 Check the rocker arm cover sealing flange for straightness and straighten or replace as necessary.

8 Using a small amount of RTV sealant in the rocker arm cover to hold it in place, install a new gasket. Do not use sealant on the gasket-to-head surface.

9 The remainder of installation is the reverse of removal.

4 Pushrod cover - removal and installation

1 Disconnect the negative cable at the battery.

2 If the rear cover or both covers are to be removed, remove the engine oil dipstick and the dipstick tube. If only the front cover is to be removed the dipstick can remain in place.

3 Remove the pushrod cover bolts and carefully pry the pushrod covers from the side of the block, using a thin-blade knife if necessary to release them from the block. Use caution not to distort the cover sealing flanges.

4 Scrape all old gasket and sealer from the pushrod covers and the block.

5 Using a small amount of RTV sealer to hold them in place, install new gaskets on the pushrod covers.

6 Installation is the reverse of the removal procedure.

5 Hydraulic lifters - removal and installation

Refer to illustrations 5.5a and 5.5b

1 Remove the rocker arm cover (Section 3), pushrods (Section 6) and pushrod covers (Section 4).

2 There are several ways to extract the lifters from the bores. A special tool designed to grip and remove lifters is manufactured by many tool companies and is widely available, but it may not be required in every case. On newer engines without a lot of varnish buildup, the lifters can often be removed with a small magnet or even with your fingers. A machinist's scribe with a bent end can be used to pull the lifters out by positioning the point under the retainer ring inside the top of each lifter. **Caution:** *Do not use pliers to remove the lifters unless you intend to replace them with new ones (along with the camshaft). The pliers will damage the precision machined and hardened lifters, rendering them useless.*

3 Before removing the lifters, arrange to store them in a clearly labeled box to ensure that they are reinstalled in their original locations. Remove the lifters and store them where they will not get dirty.

4 Clean the lifters with solvent and dry them thoroughly without mixing them up.

5 Check each lifter wall, pushrod seat and foot for scuffing, score marks and uneven wear. Each lifter foot (the surface that rides on the cam lobe) must be slightly convex, although this can be difficult to determine by eye. If the base of the lifter is concave **(see illustrations)**, the lifters and camshaft must be replaced. If the lifter walls are damaged or worn (which is not very likely), inspect the lifter bores in the engine block as well. If the pushrod seats are worn, check the pushrod ends.

6 If new lifters are being installed, a new camshaft must also be installed. If a new camshaft is installed, then use new lifters as well. Never install used lifters unless the original camshaft is used and the lifters can be installed in their original locations.

7 Coat each lifter foot with engine assembly lube or moly-base grease before installation.

8 The remainder of installation is the reverse of removal.

6 Rocker arms and pushrods - removal, inspection and installation

Removal

1 Refer to Section 3 and detach the rocker arm cover from the cylinder head.

2 Beginning at the front of the head, loosen and remove the rocker arm stud nuts. Store them separately in marked containers to ensure that they will be reinstalled in their original locations. **Note:** *If the pushrods are the only items being removed, loosen each nut just enough to allow the rocker arms to be rotated to the side so the pushrods can be lifted out.*

3 Lift off the rocker arms and pivot balls and store them in the marked containers with the nuts (they must be reinstalled in their original locations).

4 Remove the pushrods and store them separately to make sure they will not get mixed up during installation.

Inspection

5 Check each rocker arm for wear, cracks and other damage, especially where the pushrods and valve stems contact the rocker arm faces.

6 Make sure the hole at the pushrod end of each rocker arm is open.

7 Check each rocker arm pivot area for wear, cracks and galling. If the rocker arms are worn or damaged, replace them with new ones and use new pivot balls as well.

8 Inspect the pushrods for cracks and excessive wear at the ends. Roll each pushrod across a piece of plate glass to see if it is bent (if it wobbles, it is bent).

Installation

9 Lubricate the lower end of each pushrod with clean engine oil or moly-base grease

6.16 Rotate each pushrod as the rocker arm nut is tightened to determine the point at which all play is removed, then tighten each nut an additional 3/4-turn

7.8 A tool which attaches to the rocker arm stud can be used to compress the valve springs

and install them in their original locations. Make sure each pushrod seats completely in the lifter socket.

10 Apply moly-base grease to the ends of the valve stems and the upper ends of the pushrods before positioning the rocker arms over the studs.

11 Set the rocker arms in place, then install the pivot balls and nuts. Apply moly-base grease to the pivot balls to prevent damage to the mating surfaces before engine oil pressure builds up. Be sure to install each nut with the flat side against the pivot ball.

Valve adjustment

Refer to illustration 6.16

12 Using chalk or paint mark the distributor housing under the number one and number six spark plug wire terminals.

13 Remove the distributor cap.

14 Turn the engine over until the rotor is pointed at the mark on the distributor housing for the number one plug wire terminal.

15 The intake valves for the number one, two and four cylinders and the exhaust valves for the number one, three and five cylinders can now be adjusted.

16 To adjust a valve, tighten the rocker arm nut until all play is removed at the pushrod **(see illustration)**. This can be determined by rotating the pushrod between your thumb and index finger as the nut is tightened. At the point where a slight drag is just felt as you spin the pushrod, all lash has been removed.

17 Tighten each nut an additional 3/4-turn to center the lifters.

18 Turn the engine over until the rotor is pointing at the mark made on the distributor body under the number six plug wire.

19 The intake valves for the number three, five and six cylinders and the exhaust valves for the number two, four and six cylinders can now be adjusted as described in Steps 16 and 17.

20 Install the distributor cap and the rocker arm cover.

7 Valve springs, retainers and seals - replacement in vehicle

Refer to illustrations 7.8, 7.9, 7.16 and 7.17

Note: *Broken valve springs and defective valve stem seals can be replaced without removing the cylinder head. Two special tools and a compressed air source are normally required to perform this operation, so read through this Section carefully and rent or buy the tools before beginning the job. If compressed air is not available, a length of nylon rope can be used to keep the valves from falling into the cylinder during this procedure.*

1 Refer to Section 3 and remove the rocker arm cover.

2 Remove the spark plug from the cylinder which has the defective component. If all of the valve stem seals are being replaced, all of the spark plugs should be removed.

3 Turn the crankshaft until the piston in the affected cylinder is at top dead center on the compression stroke (refer to Section 11 for instructions). If you are replacing all of the valve stem seals, begin with cylinder number one and work on the valves for one cylinder at a time. Move from cylinder-to-cylinder following the firing order sequence (1-5-3-6-2-4).

4 Thread an adapter into the spark plug hole and connect an air hose from a compressed air source to it. Most auto parts stores can supply the air hose adapter. **Note:** *Many cylinder compression gauges utilize a screw-in fitting that may work with your air hose quick-disconnect fitting.*

5 Remove the nut, pivot ball and rocker arm for the valve with the defective part and pull out the pushrod. If all of the valve stem seals are being replaced, all of the rocker arms and pushrods should be removed (refer to Section 6).

6 Apply compressed air to the cylinder. The valves should be held in place by the air pressure. If the valve faces or seats are in poor condition, leaks may prevent the air pressure from retaining the valves - refer to the alternative procedure below.

7 If you do not have access to compressed air, an alternative method can be used. Position the piston at a point approximately 45-degrees before TDC on the compression stroke, then feed a long piece of nylon rope through the spark plug hole until it fills the combustion chamber. Be sure to leave the end of the rope hanging out of the engine so it can be removed easily. Use a large breaker bar and socket to rotate the crankshaft in the normal direction of rotation until slight resistance is felt.

8 Stuff shop rags into the cylinder head holes around the valves to prevent parts and tools from falling into the engine, then use a valve spring compressor to compress the spring/damper assembly **(see illustration)**. Remove the keepers with small needle-nose pliers or a magnet. **Note:** *A couple of different types of tools are available for compressing the valve springs with the head in place. One type grips the lower spring coils and presses on the retainer as the knob is turned, while the other type utilizes the rocker arm stud and nut for leverage. Both types work very well, although the lever type is usually less expensive.*

9 Remove the spring retainer or rotator, oil shield and valve spring assembly, then remove the valve stem O-ring seal and the umbrella-type guide seal (usually installed on the intake valves only) **(see illustration)**. The O-ring seal will most likely be hardened and will probably break when removed, so plan on installing a new one each time the original is removed. **Note:** *If air pressure fails to hold the valve in the closed position during this operation, the valve face or seat is probably damaged. If so, the cylinder head will have to be removed for additional repair operations.*

10 Wrap a rubber band or tape around the top of the valve stem so the valve will not fall into the combustion chamber, then release the air pressure. **Note:** *If a rope was used instead of air pressure, turn the crankshaft slightly in the direction opposite normal rotation.*

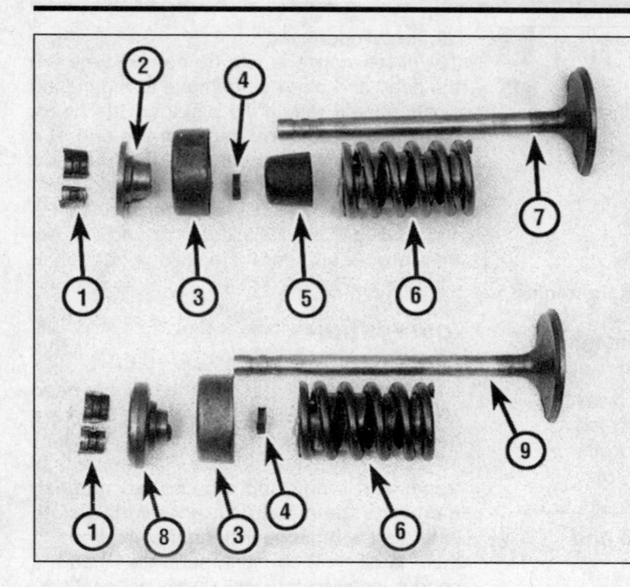

7.9 Valve spring components - exploded view

1 *Keepers*
2 *Retainer*
3 *Oil shield*
4 *O-ring oil seal*
5 *Umbrella seal*
6 *Spring and damper*
7 *Intake valve*
8 *Retainer/rotator*
9 *Exhaust valve*

7.16 Make sure the O-ring seal under the retainer is seated in the groove and not twisted before installing the keepers

7.17 Apply a small dab of grease to each keeper as shown here before installation - it will hold them in place on the valve stem as the spring is released

11 Inspect the valve stem for damage. Rotate the valve in the guide and check the end for eccentric movement, which would indicate that the valve is bent.

12 Move the valve up-and-down in the guide and make sure it doesn't bind. If the valve stem binds, either the valve is bent or the guide is damaged. In either case, the head will have to be removed for repair.

13 Reapply air pressure to the cylinder to retain the valve in the closed position, then remove the tape or rubber band from the valve stem. If a rope was used instead of air pressure, rotate the crankshaft in the normal direction of rotation until slight resistance is felt.

14 Lubricate the valve stem with engine oil and install a new umbrella-type guide seal on the intake valve.

15 Install the spring/damper assembly and shield in position over the valve.

16 Install the valve spring retainer or rotator. Compress the valve spring assembly and carefully install the new O-ring seal in the lower groove of the valve stem. Make sure the seal is not twisted - it must lie perfectly flat in the groove **(see illustration)**.

17 Position the keepers in the upper groove. Apply a small dab of grease to the inside of each keeper to hold it in place if necessary **(see illustration)**. Remove the pressure from the spring tool and make sure the keepers are seated.

18 Disconnect the air hose and remove the adapter from the spark plug hole. If a rope was used in place of air pressure, pull it out of the cylinder.

19 Refer to Section 6 and install the rocker arms and pushrods.

20 Install the spark plugs and hook up the wire(s).

21 Refer to Section 3 and install the rocker arm cover.

22 Start and run the engine, then check for oil leaks and unusual sounds coming from the rocker arm cover area.

8 Intake and exhaust manifold (non-integrated head) - removal and installation

Note: *Some engines have a cylinder head with an integral intake manifold (integrated head), while others use a separate bolt-on intake manifold (non-integrated head). The procedure in this Section is for removal of the intake and exhaust manifolds for the non-integrated head. For removal of the exhaust manifold from a head with an integrated intake manifold, refer to Section 9.*

Removal

1 Disconnect the negative cable at the battery.

2 Remove the air cleaner assembly.

3 Disconnect the throttle cable or throttle rod (depending on application) and remove the throttle return spring.

4 Disconnect the fuel and vacuum lines at the carburetor. If there is any chance the lines may get mixed up, use tape to label them so they can be reinstalled in their original locations.

5 Remove the PCV hose from the rocker arm cover.

6 Disconnect the vapor hose at the EVAP canister (if equipped).

7 Disconnect the exhaust pipe from the exhaust manifold. Applying penetrating oil, then waiting a few minutes for it to work may make removal of frozen exhaust pipe nuts easier. Use caution not to apply excessive force or you might shear off the studs.

8 Remove the manifold assembly bolts and clamps and separate the manifold assembly from the head.

9 Remove the one bolt and two nuts from the center of the assembly to separate the intake and exhaust manifolds.

Installation

10 Clean all old gasket material and sealant from the mating surfaces of both manifolds and the head.

11 Lay a straightedge along the exhaust manifold mating surface and use a feeler gauge to check for straightness. If a gap of 0.030-inch or more exists at any point, the manifold is warped and will not seal. Replace the exhaust manifold with a new one.

12 Install a new gasket and assemble the intake and exhaust manifolds by replacing the one bolt and two nuts at the center of the manifold. Tighten them only finger tight at this time.

13 Put a new gasket over the manifold end studs on the head, hold the manifold assembly in place and install the bolts, clamps and washers.

14 Tighten all the manifold assembly-to-cylinder head bolts to the specified torque.

15 Tighten the intake-to-exhaust manifold bolt and two nuts to the specified torque.

16 The remainder of installation is the reverse of removal.

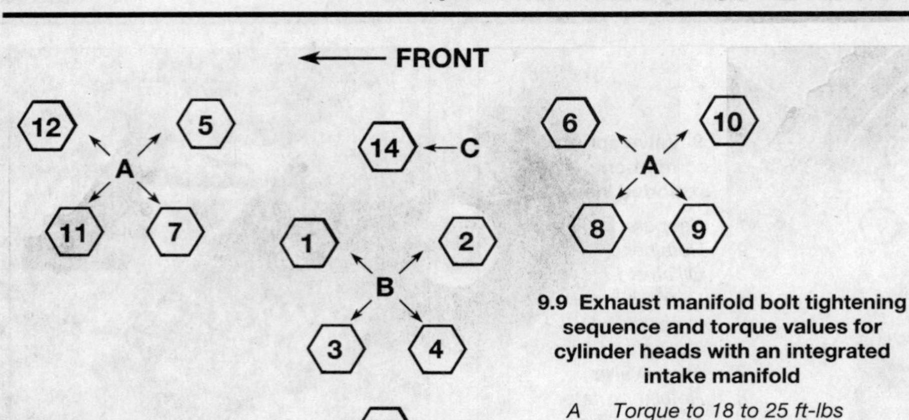

9.9 Exhaust manifold bolt tightening sequence and torque values for cylinder heads with an integrated intake manifold

A Torque to 18 to 25 ft-lbs
B Torque to 35 to 40 ft-lbs
C Torque to 40 to 45 ft-lbs

9 Exhaust manifold (integrated head) - removal and installation

Refer to illustration 9.9

1 Disconnect the negative cable at the battery.
2 Remove the air cleaner assembly.
3 Remove the power steering pump and AIR pump brackets, if equipped.
4 Disconnect the four pulse air valve pipe fittings from the cylinder head and remove the check valve pipes from the plenum grommets.
5 Raise the vehicle and support it securely on jackstands.
6 Disconnect the exhaust pipe from the exhaust manifold and the converter bracket from the transmission mount.
7 Lower the vehicle and remove the exhaust manifold bolts. Detach the exhaust manifold.
8 Clean the gasket surface of the exhaust manifold and check the manifold for cracks.
9 Installation is the reverse of the removal procedure. Tighten the exhaust manifold bolts to the specified torque in the order shown **(see illustration)**.

10 Cylinder head - removal and installation

Removal

1 Disconnect the negative cable at the battery.
2 Remove the air cleaner assembly.
3 Remove the intake and exhaust manifolds (non-integrated head) or carburetor and exhaust manifold (integrated head), referring to Sections 8 or 9 as necessary.
4 Remove the rocker arm cover (Section 3).
5 Remove the rocker arms and pushrods (Section 6).
6 Drain the cooling system (Chapter 1).
7 Remove the fuel and vacuum lines from the retaining clips and disconnect the wire from the temperature sending unit.
8 Disconnect the air injection hose from the check valve (if so equipped).
9 Disconnect the upper radiator hose from the thermostat housing.
10 Remove the battery ground strap from the head.
11 Remove the cylinder head bolts, then lift off the cylinder head.
12 If resistance is felt, do not pry between the head and block as damage to the mating surfaces will result. To dislodge the head, place a block of wood against the end of it and strike the wood block with a hammer. Store the head on blocks of wood to prevent damage to the gasket sealing surface.
13 Cylinder head disassembly and inspection procedures are covered in detail in Chapter 2, Part C.

Installation

Refer to illustrations 10.17 and 10.21

14 The mating surface of the cylinder head and block must be perfectly clean when the head is installed.
15 Use a gasket scraper to remove all traces of carbon and old gasket material, then wipe the mating surfaces with a cloth saturated with lacquer thinner or acetone. If there is oil on the mating surfaces when the head is installed the gasket may not seal correctly and leaks may develop.
16 Check the block and head mating surfaces for nicks, deep scratches and other damage. If damage is slight, it can be removed with a large file. If it is excessive, machining may be the only alternative.
17 Use a tap of the correct size to chase the threads in the head bolt holes in the block. Mount each bolt in a vise and run a die down the threads to remove corrosion and restore the threads **(see illustration)**. Dirt, corrosion, sealant and damaged threads will affect torque readings.
18 Position the new gasket over the dowel pins in the block. **Note:** *If a steel gasket is used (shim-type gasket), apply a thin, even coat of sealant such as KW Copper Coat to both sides prior to installation. Steel gaskets must be installed with the raised bead UP. Composition gaskets must be installed dry; do not use sealant.*
19 Carefully position the head on the block without disturbing the gasket.
20 Before installing the head bolts, coat the

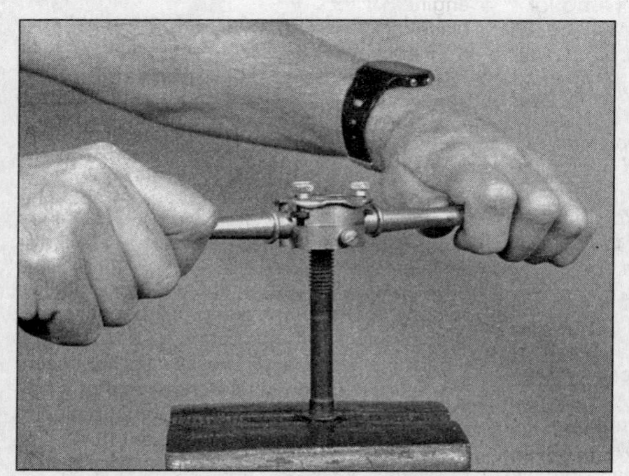

10.17 A die should be used to remove sealant and corrosion from the head bolt threads prior to installation

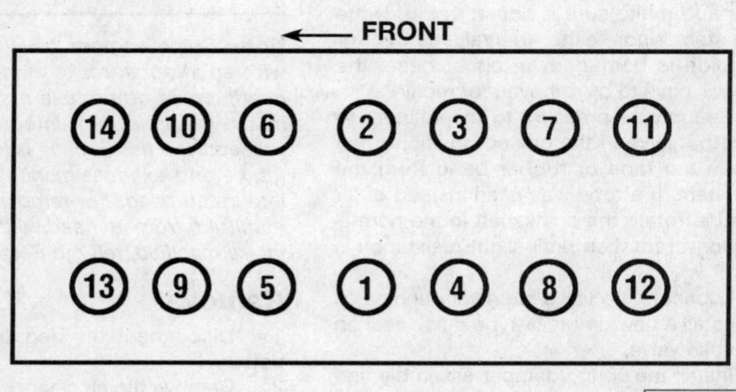

10.21 Cylinder head bolt tightening sequence

12.3 Use the recommended puller to remove the vibration damper - if a puller that applies force to the outer edge is used, the damper will be damaged

12.9 Cam and crankshaft gear timing mark alignment

threads with a non-hardening sealant such as Permatex No. 2.

21 Install the bolts and tighten them finger tight. Following the recommended sequence **(see illustration)**, tighten the bolts in several steps to the specified torque.

22 The remaining installation steps are the reverse of removal.

11 Top Dead Center (TDC) for number 1 piston - locating

1 Top Dead Center (TDC) is the highest point in the cylinder that each piston reaches as it travels up and down when the crankshaft turns. Each piston reaches TDC on the compression stroke and again on the exhaust stroke, but TDC generally refers to piston position on the compression stroke. The timing marks on the vibration damper installed on the front of the crankshaft are referenced to the number one piston at TDC on the compression stroke.

2 Positioning the pistons at TDC is an essential part of many procedures such as rocker arm removal, valve adjustment, timing gear replacement and distributor removal.

3 In order to bring any piston to TDC, the crankshaft must be turned using one of the methods outlined below. When looking at the front of the engine, normal crankshaft rotation is clockwise. **Warning:** *Before beginning this procedure, be sure to place the transmission in Neutral and disable the ignition system by removing the coil wire from the distributor cap and grounding it against the block.*

a) *The preferred method is to turn the crankshaft with a large socket and breaker bar attached to the vibration damper bolt threaded into the front of the crankshaft.*

b) *A remote starter switch, which may save some time, can also be used. Attach the switch leads to the S (switch) and B (battery) terminals on the starter solenoid. Once the piston is close to TDC, use a socket and breaker bar as described in the previous paragraph.*

c) *If an assistant is available to turn the ignition switch to the Start position in short bursts, you can get the piston close to TDC without a remote starter switch. Use a socket and breaker bar as described in Paragraph a) to complete the procedure.*

4 Scribe or paint a small mark on the distributor body directly below the number one spark plug wire terminal.

5 Remove the distributor cap as described in Chapter 1.

6 Turn the crankshaft (see Step 3 above) until the line on the vibration damper is aligned with the zero mark on the timing plate. The timing plate and vibration damper are located low on the front of the engine, near the pulley that turns the drivebelts.

7 The rotor should now be pointing directly at the mark on the distributor body. If it isn't, the piston is at TDC on the exhaust stroke.

8 To get the piston to TDC on the compression stroke, turn the crankshaft one complete turn (360°) clockwise. The rotor should now be pointing at the mark.

9 After the number one piston has been positioned at TDC on the compression stroke, TDC for any of the remaining cylinders is located by turning the engine over in the normal direction of rotation until the rotor is pointing to the spark plug wire terminal in the distributor cap (you will have to install and remove the distributor cap as you are turning the engine over) for the piston you wish to bring to top dead center.

12 Timing cover, gears and camshaft - removal, inspection and installation

Removal

Refer to illustrations 12.3, 12.9 and 12.11

Note: *Removal of the camshaft with the engine in the vehicle requires the removal of the radiator and (if equipped) the air conditioning condenser. If the air conditioning condenser has to be removed have the system discharged by an dealer service department or air conditioning shop before disconnecting any lines. Before removing the camshaft, check the lobe lift with a dial indicator* (see Chapter 2, Part A).

1 Drain the cooling system (Chapter 1), remove the drivebelts (Chapter 1) and the radiator (Chapter 3).

2 Remove the drivebelt pulley from the front of the crankshaft.

3 Remove the bolt from the front of the crankshaft, then use a puller to detach the vibration damper **(see illustration). Caution:** *Do not use a puller with jaws that grip the outer edge of the damper. The puller must be the type shown in the illustration that utilizes bolts to apply force to the damper hub only.*

4 Remove the two bolts holding the oil pan to the bottom of the timing cover.

5 Using a sharp, thin-blade knife, cut the pan gasket flush with the front edge of the block on both sides. On some late model engines a bead of RTV sealant is used in place of a gasket.

6 Remove the bolts and detach the timing cover.

7 Remove the rocker arm cover (Section 3), the pushrod covers (Section 4), the rocker arms and pushrods (Section 6) and the hydraulic lifters (Section 5).

8 Remove the fuel pump (Chapter 4).

9 Turn the crankshaft until the marks on the timing gears are aligned **(see illustration)**.

10 Mount a dial indicator on the front face of the block with the plunger resting against the camshaft gear. Push the camshaft as far back in the block as it will go, zero the dial indicator, then use a screwdriver to lightly pry the camshaft forward. Check the dial indicator for the end play and compare the measurement to the Specifications.

11 Reach in through the holes in the camshaft gear and remove the thrust plate

12.11 Reach through the holes in the camshaft gear to remove the thrust plate bolts

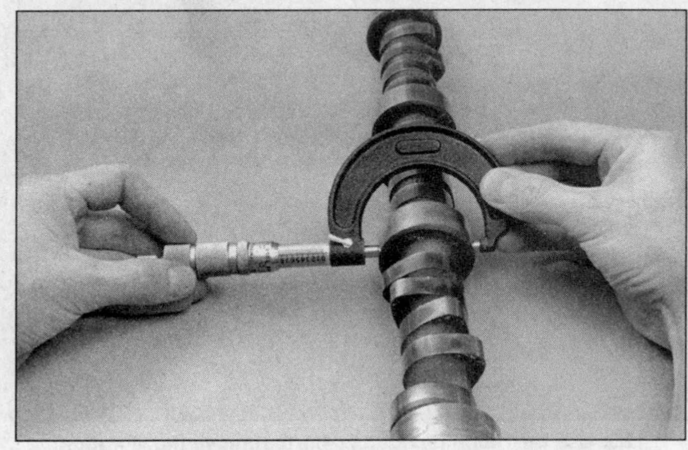

12.14 The camshaft bearing journal diameter is checked to pinpoint excessive wear and out-of-round conditions

bolts **(see illustration)**.

12 Carefully pull the camshaft out. Support the cam near the block so the lobes do not nick or gouge the bearings as it is withdrawn.

Inspection

Camshaft and bearings

Refer to illustration 12.14

13 After the camshaft has been removed from the engine, cleaned with solvent and dried, inspect the bearing journals for uneven wear, pitting and evidence of seizure. If the journals are damaged, the bearing inserts in the block are probably damaged as well. Both the camshaft and bearings will have to be replaced.

14 Measure the bearing journals with a micrometer **(see illustration)** to determine if they are excessively worn or out-of-round.

Timing gears

15 Check the gears for cracked, galled, excessively worn or missing teeth. Make sure the keyway in the bore is not deformed. If wear is noted, or the teeth are damaged, replace both gears with new parts.

16 The crankshaft gear can be removed from the crankshaft with a claw-type gear puller. The camshaft gear must be pressed from the camshaft with a special adapter. If the camshaft, gear or the thrust plate must be replaced, take the camshaft to a dealer service department or an automotive machine shop to have the gear/thrust plate removed and new components installed.

Bearing replacement

17 Camshaft bearing replacement requires special tools and expertise that places it outside the scope of the home mechanic. Take the block to an automotive machine shop to ensure that the job is done correctly.

Installation

Refer to illustration 12.19

18 If it was removed, reinstall the crankshaft timing gear.

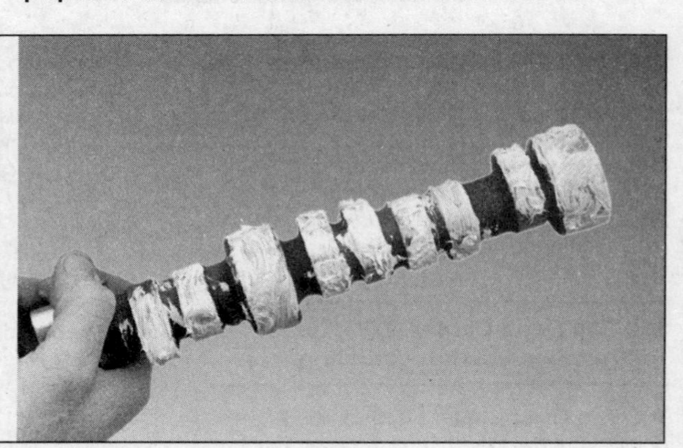

12.19 Coat both the cam lobes and the journals with camshaft assembly lube before installing the camshaft

19 Thoroughly coat the camshaft lobes with camshaft assembly lube **(see illustration)** and install the camshaft in the block, using caution not to damage the camshaft bearings during installation.

20 Just before the camshaft bottoms in the block, turn it so the timing marks on the camshaft and crankshaft gears are in alignment.

21 Seat the camshaft in the block.

22 Install the thrust plate bolts and tighten them to the specified torque.

23 Using RTV sealant to hold it in place, install a new gasket on the timing cover.

24 If the engine has an oil pan gasket, cut the tabs from a new oil pan front seal and install it on the front cover, pressing the tips into the holes provided in the cover.

25 If the engine had RTV sealant instead of a pan gasket, carefully clean all traces of old sealant from the pan and front cover, then apply a 3/16-inch bead of RTV sealant to the cover-to-pan sealing surface.

26 Apply a 1/8-inch bead of RTV sealant to the junction where the oil pan and block meet on each side.

27 Install the front cover, tightening the bolts finger tight only.

28 Apply moly-base grease to the outside of the hub, then install the vibration damper, using a piece of pipe and a hammer to press it through the seal until it seats against the timing gear.

29 Tighten the front cover bolts and the two pan bolts to the specified torque.

30 The remainder of installation is the reverse of the removal procedure.

13 Oil pan and oil pump - removal and installation

1 Disconnect the negative cable from the battery.

2 Raise the vehicle and support it securely on jackstands.

3 Drain the engine oil (Chapter 1).

4 Remove the starter (Chapter 5).

5 Remove the flywheel or torque converter inspection cover.

6 Remove the engine mount through bolts.

7 Using a floor jack under the pan with a block of wood to spread the load, lift the engine until the engine portions of the engine mounts are clear of the frame portions.

8 Reinstall the engine mount through bolts and set the engine down with the through bolts resting on top of the frame portion of the mounts, holding the engine in a raised position.

9 Remove the oil pan bolts and detach the oil pan.

10 Remove the two flange mounting bolts

14.9 Use the protector tool (arrow) when pushing the seal into place (if this tool wasn't furnished with the new seal, refer to Chapter 2A, illustration 14.11)

14.10 Use a small punch and a hammer to start the old seal out of the block, then pull it free with pliers

and the pickup pipe bolt then remove the oil pump and screen as an assembly.

11 Installation is the reverse of the removal procedure. Be sure to install a new gasket on the oil pan and tighten all bolts to the specified torque.

14 Crankshaft oil seals - replacement

Front seal

1 Remove the front cover (Section 12).

2 Pry the old seal out of the cover with a screwdriver, using care not to scratch or distort the front cover.

3 Lightly coat the outside of the new seal with RTV sealant, then place it in the cover, making sure it is seated straight.

4 Use a socket slightly smaller in diameter than the outside diameter of the seal and a hammer to tap the seal into the front cover. Use caution when tapping the seal into place to make sure it does not become cocked while being driven into the cover.

5 Replace the front cover on the engine (Section 12).

Rear seal

Refer to illustrations 14.9, 14.10 and 14.13

6 Remove the oil pan and oil pump (Section 13).

7 Remove the rear main bearing cap, taking care not to disturb or damage the main bearing.

8 Use a small screwdriver to remove the seal half from the main bearing cap.

9 Press a new seal half into the cap, making sure the ends fit flush with the cap-to-block mating surface **(see illustration)**.

10 Use a small punch to tap one end of the block half of the seal until the other side protrudes approximately 1/4-inch from the block **(see illustration)**. Use caution not to scratch the crankshaft with the punch when tapping the seal around.

11 Grip the exposed end of the old seal with a pair of pliers and pull it out of the block.

12 Liberally coat the new seal half with engine oil then start it into the block. Turning the crankshaft while forcing the seal into the slot will aid in moving it around the crankshaft.

13 Apply a small amount of RTV sealant to the ends of the seal and the adjoining areas of the block **(see illustration)**, being careful not to get any on the crankshaft or bearing, and reinstall the rear main cap, tightening the bolts to the specified torque.

14 The remainder of the installation is the reverse of the removal procedure.

14.13 Apply a thin film of RTV sealant to the ends of the seal and the adjacent areas on the block (arrows) before installing the rear main bearing cap

15 Engine mounts - check and replacement

1 Engine mounts seldom require attention, but broken or deteriorated mounts should be replaced immediately or the added strain placed on the driveline components may cause damage.

Check

2 During the check, the engine must be raised slightly to remove the weight from the mounts.

3 Raise the vehicle and support it securely on jackstands, then position the jack under the engine oil pan. Place a large block of wood between the jack head and the oil pan, then carefully raise the engine just enough to take the weight off the mounts.

4 Check the mounts to see if the rubber is cracked, hardened or separated from the metal plates. Sometimes the rubber will split right down the center. Rubber preservative or WD-40 should be applied to the mounts to slow deterioration.

5 Check for relative movement between the mount plates and the engine or frame (use a large screwdriver or pry bar to attempt to move the mounts). If movement is noted, lower the engine and tighten the mount fasteners.

Replacement

6 Disconnect the negative cable from the battery, then raise the vehicle and support it securely on jackstands.

7 Remove the nut and withdraw the mount through bolt from the frame bracket.

8 Raise the engine slightly, then remove the mount-to-frame bolts and detach the mount.

9 Installation is the reverse of removal. Use thread locking compound on the mount bolts and be sure to tighten them securely.

Notes

Chapter 2 Part C
General engine overhaul procedures

Contents

	Section
Crankshaft - inspection	18
Crankshaft - installation and main bearing oil clearance check	22
Crankshaft - removal	13
Cylinder compression check	4
Cylinder head - cleaning and inspection	9
Cylinder head - disassembly	8
Cylinder head - reassembly	11
Cylinder honing	16
Engine block - cleaning	14
Engine block - inspection	15
Engine overhaul - disassembly sequence	7
Engine overhaul - general information	3
Engine overhaul - reassembly sequence	20
Engine rebuilding alternatives	5
Engine - removal and installation	6
Engine removal - methods and precautions	2
General information	1
Initial start-up and break-in after overhaul	25
Main and connecting rod bearings - inspection	19
Piston/connecting rod assembly - inspection	17
Piston/connecting rod assembly - installation and rod bearing oil clearance check	23
Piston/connecting rod assembly - removal	12
Piston rings - installation	21
Pre-oiling engine after overhaul	24
Valves - servicing	10

Specifications

General

Compression pressure
- Inline 6-cylinder engine: 130 psi
- V6 and V8 engines: 150 psi

Maximum variation between cylinders: 20 psi

Block

Bore diameter
- 262 V6: 3.9995 to 4.0025 in
- 283 and 307 V8: 3.8745 to 3.8775 in
- 305 V8: 3.7350 to 3.7385 in
- 327 and 350 V8: 3.9995 to 4.0025 in
- 396, 400 and 402 V8: 4.1246 to 4.1274 in
- 454 V8: 4.2495 to 4.2525 in (4.2500 to 4.2507 for 1991 models)
- 230, 250 and 292 inline 6-cylinder: 3.8750 to 3.8780 in

Taper limit: 0.001 in

Out-of-round limit: 0.002 in

Heads and valve train

Item	Specification
Warpage limit	0.003 in per 6 in
Valve seat angle	46 degrees
Valve seat width	
Intake	1/32 to 1/16 in
Exhaust	1/16 to 3/32 in
Valve face angle	45 degrees
Minimum valve margin	1/32 in
Valve stem-to-valve guide clearance	
Intake valves	0.0010 to 0.0027 in
Exhaust valves	
230, 250 and 292 inline 6-cylinder	0.0015 to 0.0032 in
283 and 307 V8	0.0012 to 0.0029 in
305, 327 and 350 V8	0.0010 to 0.0027 in
396, 402 and 454 V8	0.0012 to 0.0029 in
262 V6	0.0010 to 0.0027 in
Valve spring free length	
Inline 6-cylinder	2.08 in
Small block V8 and V6	2.03 in
Big block V8	2.12 in
Valve spring installed height	
Inline 6-cylinder	1-21/32 in
Small block V8 and V6	1-23/32 in
Big block V8	1-51/64 in

Crankshaft

Item	Specification
Main journal diameter	
262 V6	
1	2.4484 to 2.4493 in
2 and 3	2.4481 to 2.4490 in
4	2.4479 to 2.4488 in
283 V8 and 1967 327 V8	
1	2.2987 to 2.2997 in
2, 3 and 4	2.2983 to 2.2993 in
5	2.2978 to 2.2998 in
305, 307, 327 (1968) and 350 V8	
1	2.4484 to 2.4493 in
2, 3 and 4	2.4481 to 2.4490 in
5	2.4479 to 2.4488 in
396 V8	
1 and 2	2.7487 to 2.7496 in
3 and 4	2.7481 to 2.7490 in
5	2.7478 to 2.7488 in
402 V8	
1, 2, 3 and 4	2.7481 to 2.7490 in
5	2.7473 to 2.7483 in
454 V8	
1, 2, 3 and 4 (and 5 on 1991 models)	2.7481 to 2.7490 in
5	2.7476 to 2.7486 in
230 inline 6-cylinder (all)	2.2983 to 2.2993 in
250 and 292 inline 6-cylinder (all)	2.2979 to 2.2994 in
Main journal taper limit	0.001 in
Main journal out-of-round limit	0.001 in
Main bearing oil clearance	
262 V6	
1	0.0008 to 0.0020 in
2 and 3	0.0011 to 0.0023 in
4	0.0017 to 0.0032 in
283 and 327 V8	
1, 2, 3 and 4	0.0008 to 0.0024 in
5	0.0010 to 0.0026 in
305 V8	
1	0.0008 to 0.0020 in
2, 3 and 4	0.0011 to 0.0023 in
5	0.0017 to 0.0032 in
307 V8	
1	0.0003 to 0.0015 in
2, 3 and 4	0.0006 to 0.0018 in
5	0.0008 to 0.0023 in

350 V8	
1	0.0008 to 0.0020 in
2, 3 and 4	0.0011 to 0.0023 in
5	0.0017 to 0.0032 in
396 V8	
1	0.0007 to 0.0019 in
2, 3 and 4	0.0013 to 0.0025 in
5	0.0024 to 0.0040 in
402 V8	
1	0.0003 to 0.0016 in
2, 3 and 4	0.0006 to 0.0018 in
5	0.0008 to 0.0023 in
454 V8	
1, 2, 3 and 4	0.0013 to 0.0025 in (0.0017 to 0.0030 for 1991 models)
5	0.0024 to 0.0040 in
230 inline 6-cylinder (all)	0.0003 to 0.0029 in
250 and 292 inline 6-cylinder	
1 through 6	0.0010 to 0.0024 in
7	0.0016 to 0.0035 in
Rod journal diameter	
262 V6	2.2487 to 2.2497 in
283 V8 and 1967 327 V8	1.999 to 2.000 in
305 and 350 V8	2.0988 to 2.0998 in
307 V8, 1968 327 V8	2.099 to 2.100 in
396 V8	2.1985 to 2.1995 in
402 V8	2.0990 to 2.1000 in
454 V8	2.1990 to 2.2000 in
230 and 250 inline 6-cylinder	1.999 to 2.000 in
292 inline 6-cylinder	2.099 to 2.100 in
Rod journal taper limit	0.001 in
Rod journal out-of-round limit	0.001 in
Rod bearing oil clearance	
262 V6	0.0013 to 0.0035 in
283, 307 and 327 V8	0.0007 to 0.0028 in
305 and 350 V8	0.0013 to 0.0035 in
396, 402 and 454 V8	0.0009 to 0.0025 in
230 inline 6-cylinder	0.0007 to 0.0027 in
250 and 292 inline 6-cylinder	0.0010 to 0.0026 in
Rod end play	
262 V6	0.006 to 0.014 in
Small block V8	0.008 to 0.014 in
Big block V8	0.015 to 0.025 in
230 inline 6-cylinder	0.0085 to 0.0135 in
250 and 292 inline 6-cylinder	0.006 to 0.017 in
Crankshaft end play	
262 V6	0.002 to 0.006 in
283 and 327 V8	0.003 to 0.011 in
305, 307 and 350 V8	0.002 to 0.006 in
396, 402 and 454 V8	0.006 to 0.010 in
All inline 6-cylinder	0.002 to 0.006 in

Pistons and rings

Piston-to-bore clearance	
262 V6	
Standard	0.0007 to 0.0017 in
Service limit	0.0027 in
283, 307 and 327 V8	
Standard	0.0005 to 0.0011 in
Service limit	0.0025 in
305 and 350 V8	
Standard	0.0007 to 0.0017 in
Service limit	0.0027 in
396 V8	
Standard	0.0018 to 0.0026 in
Service limit	0.0045 in
402 V8	
Standard	0.0014 to 0.0020 in
Service limit	0.0034 in

Pistons and rings (continued)

Piston-to-bore clearance	
454 V8	
Standard	0.0030 to 0.0040 in
Service limit	0.0050 in
230 inline 6-cylinder	
Standard	0.0005 to 0.0014 in
Service limit	0.0025 in
250 inline 6-cylinder	
Standard	0.0010 to 0.0020 in
Service limit	0.0030 in
292 inline 6-cylinder	
Standard	0.0026 to 0.0036 in
Service limit	0.0045 in
Piston ring side clearance	
262 V6	
Compression (both)	0.0012 to 0.0032 in
Oil control	0.002 to 0.007 in
283 V8	
Top compression	0.0007 to 0.0027 in
2nd compression	0.0012 to 0.0032 in
Oil control	0.000 to 0.005 in
305, 307 and 350 V8	
Compression (both)	0.0012 to 0.0032 in
Oil control	0.002 to 0.007 in
396 V8	
Top compression	0.0018 to 0.0032 in
2nd compression	0.0010 to 0.0030 in
Oil control	0.002 to 0.0035 in
402 V8	
Compression (both)	0.0012 to 0.0032 in
Oil control	0.0050 to 0.0065 in
454 V8	
Compression (both)	0.0017 to 0.0032 in (0.0012 to 0.0029 for 1991 models)
Oil control	0.0050 to 0.0065 in
230 and 250 inline 6-cylinder	
Compression (both)	0.0012 to 0.0027 in
Oil control	0.000 to 0.005 in
292 inline 6-cylinder	
Compression (both)	0.0020 to 0.0040 in
Oil control	0.0050 to 0.0055 in
Piston ring end gap (all)	
Top compression	0.010 to 0.020 in (0.010 to 0.018 for 1991 454 models)
2nd compression	0.010 to 0.025 in (0.016 to 0.024 for 1991 454 models)
Oil control	0.015 to 0.055 in (0.010 to 0.030 for 1991 454 models)

Torque specifications

	Ft-lbs (unless otherwise indicated)
Connecting rod cap nuts	
230 and 250 inline 6-cylinder	35
292 inline 6-cylinder	40
262 V6	45
283 and 327 V8	35
305, 307 and 350 V8	45
396, 402 and 454 V8	50
Main bearing cap bolts	
All inline six-cylinder	65
262 V6	75
Small block V8	
2-bolt main caps	80
4-bolt main caps (to 1976)	
Inner	70
Outer	65
4-bolt main caps (1977 on)	
Inner	80
Outer	70
Big block V8 (all)	110

1 General information

Included in this portion of Chapter 2 are the general overhaul procedures for cylinder head and internal engine components. The information ranges from advice concerning preparation for an overhaul and the purchase of replacement parts to detailed, step-by-step procedures covering removal and installation of internal engine components and the inspection of parts.

The following Sections have been written based on the assumption that the engine has been removed from the vehicle. For information concerning in-vehicle engine repair, as well as removal and installation of the external components necessary for the overhaul, see Parts A and B of this Chapter and Section 7 of this Part.

The Specifications included here in Part C are only those necessary for the inspection and overhaul procedures which follow. Refer to Parts A and B for additional Specifications.

2 Engine removal - methods and precautions

If you have decided that an engine must be removed for overhaul or major repair work, several preliminary steps should be taken.

Locating a suitable work area is extremely important. A shop is, of course, the most desirable place to work. Adequate workspace, along with storage space for the vehicle, will be needed. If a shop or garage is not available, at the very least a flat, level, clean work surface made of concrete or asphalt is required.

Cleaning the engine compartment and engine before beginning the removal procedure will help keep tools clean and organized.

An engine hoist or A-frame will be needed. Make sure that the equipment is rated in excess of the combined weight of the engine and its accessories. Safety is of primary importance, considering the potential hazards involved in lifting the engine out of the vehicle.

If a novice is removing the engine, a helper should be available. Advice and aid from someone more experienced would also be helpful. There are many instances when one person cannot simultaneously perform all of the operations required when lifting the engine out of the vehicle.

Plan the operation ahead of time. Arrange for or obtain all of the tools and equipment you will need prior to beginning the job. Some of the equipment necessary to perform engine removal and installation safely and with relative ease are (in addition to an engine hoist) a heavy duty floor jack, complete sets of wrenches and sockets as described in the front of this manual, wooden blocks and plenty of rags and cleaning solvent for mopping up spilled oil, coolant and gasoline. If the hoist is to be rented, make sure that you arrange for it in advance and perform beforehand all of the operations possible without it. This will save you money and time.

Plan for the vehicle to be out of use for a considerable amount of time. A machine shop will be required to perform some of the work which the do-it-yourselfer cannot accomplish due to a lack of special equipment. These shops often have a busy schedule, so it would be wise to consult them before removing the engine in order to accurately estimate the amount of time required to rebuild or repair components that may need work.

Always use extreme caution when removing and installing the engine. Serious injury can result from careless actions. Plan ahead, take your time and a job of this nature, although major, can be accomplished successfully.

3 Engine overhaul - general information

Refer to illustration 3.4

It is not always easy to determine when, or if, an engine should be completely overhauled, as a number of factors must be considered.

High mileage is not necessarily an indication that an overhaul is needed, while low mileage does not preclude the need for an overhaul. Frequency of servicing is probably the most important consideration. An engine that has had regular and frequent oil and filter changes, as well as other required maintenance, will most likely give many thousands of miles of reliable service. Conversely, a neglected engine may require an overhaul very early in its life.

Excessive oil consumption is an indication that piston rings and/or valve guides are in need of attention. Make sure that oil leaks are not responsible before deciding that the rings and/or guides are bad. Perform a compression check (see Section 4) or have a cylinder leakdown test performed by an experienced tune-up mechanic to determine the extent of the work required.

If the engine is making obvious knocking or rumbling noises, the connecting rod and/or main bearings are probably at fault. If your vehicle is equipped with an oil pressure warning light instead of an oil pressure gauge, check the pressure with a gauge temporarily installed in place of the oil pressure sending unit and compare it to the Specifications **(see illustration)**. If the pressure is extremely low, the bearings and/or oil pump are probably worn out.

Loss of power, rough running, excessive valve train noise and high fuel consumption rates may also point to the need for an overhaul, especially if they are all present at the same time. If a complete tune-up does not remedy the situation, major mechanical work is the only solution.

An engine overhaul involves restoring the internal parts to the specifications of a new engine. During an overhaul, the piston rings are replaced and the cylinder walls are reconditioned (rebored and/or honed). If a rebore is done, new pistons are required. The main bearings, connecting rod bearings and camshaft bearings are generally replaced with new ones and, if necessary, the crankshaft may be reground to restore the journals. Generally, the valves are serviced as well, since they are usually in less-than-perfect condition at this point. While the engine is being overhauled, other components, such as the distributor, starter and alternator, can be rebuilt as well. The end result should be a like new engine that will give many trouble free miles. **Note:** *Critical cooling system components such as the hoses, the drivebelts, the thermostat and the water pump MUST be replaced with new parts when an engine is overhauled. The radiator should be checked carefully to ensure that it isn't clogged or leaking. If in doubt, replace it with a new one. Also, we do not recommend overhauling the oil pump - always install a new one when an engine is rebuilt.*

Before beginning the engine overhaul, read through the entire procedure to familiarize yourself with the scope and requirements of the job. Overhauling an engine is not difficult, but it is time consuming. Plan on the vehicle being tied up for a minimum of two weeks, especially if parts must be taken to an automotive machine shop for repair or reconditioning. Check on availability of parts and make sure that any necessary special tools and equipment are obtained in advance. Most work can be done with typical hand tools, although a number of precision measuring tools are required for inspecting parts to determine if they must be replaced. Often an automotive machine shop will handle the

3.4 Check the oil pressure by removing the sending unit and attaching an oil pressure gauge. On V6 and small block V8 engines, the sending unit is located near the distributor; on inline six-cylinder engines it's located on the side of the cylinder block, above the starter; on big block V8 engines it's located at the left (driver's side) rear of the cylinder block

inspection of parts and offer advice concerning reconditioning and replacement. **Note:** *Always wait until the engine has been completely disassembled and all components, especially the engine block, have been inspected before deciding what service and repair operations must be performed by an automotive machine shop. Since the block's condition will be the major factor to consider when determining whether to overhaul the original engine or buy a rebuilt one, never purchase parts or have machine work done on other components until the block has been thoroughly inspected. As a general rule, time is the primary cost of an overhaul, so it does not pay to install worn or substandard parts.*

As a final note, to ensure maximum life and minimum trouble from a rebuilt engine, everything must be assembled with care in a spotlessly clean environment.

4 Cylinder compression check

Refer to illustration 4.4

1 A compression check will tell you what mechanical condition the upper end (pistons, rings, valves, head gaskets) of your engine is in. Specifically, it can tell you if the compression is down due to leakage caused by worn piston rings, defective valves and seats or a blown head gasket. **Note:** *The engine must be at normal operating temperature for this check and the battery must be fully charged.*

2 Begin by cleaning the area around the spark plugs before you remove them (compressed air works best for this). This will prevent dirt from getting into the cylinders as the compression check is being done. Remove all of the spark plugs from the engine.

3 Block the throttle wide open and disconnect the wire from the BAT terminal on the distributor cap (coil-in-cap models). On remote coil models, remove the coil wire from the distributor cap and ground the wire to the engine block.

4.4 A compression gauge with a threaded fitting for the spark plug hole is preferred over the type that requires hand pressure to maintain the seal

4 With the compression gauge in the number one spark plug hole, crank the engine over at least four compression strokes and watch the gauge **(see illustration)**. The compression should build up quickly in a healthy engine. Low compression on the first stroke, followed by gradually increasing pressure on successive strokes, indicates worn piston rings. A low compression reading on the first stroke, which does not build up during successive strokes, indicates leaking valves or a blown head gasket (a cracked head could also be the cause). Record the highest gauge reading obtained.

5 Repeat the procedure for the remaining cylinders and compare the results to the Specifications.

6 Add some engine oil (about three squirts from a plunger-type oil can) to each cylinder, through the spark plug hole, and repeat the test.

7 If the compression increases after the oil is added, the piston rings are definitely worn. If the compression does not increase significantly, the leakage is occurring at the valves or head gasket. Leakage past the valves may be caused by burned valve seats and/or faces or warped, cracked or bent valves.

8 If two adjacent cylinders have equally low compression, there is a strong possibility that the head gasket between them is blown. The appearance of coolant in the combustion chambers or the crankcase would verify this condition.

9 If the compression is unusually high, the combustion chambers are probably coated with carbon deposits. If that is the case, the cylinder head(s) should be removed and decarbonized.

10 If compression is way down or varies greatly between cylinders, it would be a good idea to have a leak-down test performed by an automotive repair shop. This test will pinpoint exactly where the leakage is occurring and how severe it is.

5 Engine rebuilding alternatives

The do-it-yourselfer is faced with a number of options when performing an engine overhaul. The decision to replace the engine block, piston/connecting rod assemblies and crankshaft depends on a number of factors, with the number one consideration being the condition of the block. Other considerations are cost, access to machine shop facilities, parts availability, time required to complete the project and the extent of prior mechanical experience on the part of the do-it-yourselfer.

Some of the rebuilding alternatives include:

Individual parts - If the inspection procedures reveal that the engine block and most engine components are in reusable condition, purchasing individual parts may be the most economical alternative. The block, crankshaft and piston/connecting rod assemblies should all be inspected carefully. Even if the block shows little wear, the cylinder bores should be surface honed.

Crankshaft kit - This rebuild package consists of a reground crankshaft and a matched set of pistons and connecting rods. The pistons will already be installed on the connecting rods. Piston rings and the necessary bearings will be included in the kit. These kits are commonly available for standard cylinder bores, as well as for engine blocks which have been bored to a regular oversize.

Short block - A short block consists of an engine block with a crankshaft and piston/connecting rod assemblies already installed. All new bearings are incorporated and all clearances will be correct. The existing camshaft, valve train components, cylinder head(s) and external parts can be bolted to the short block with little or no machine shop work necessary.

Long block - A long block consists of a short block plus an oil pump, oil pan, cylinder head(s), rocker arm cover(s), camshaft and valve train components, timing gears or timing sprockets and chain and timing cover. All components are installed with new bearings, seals and gaskets incorporated throughout. The installation of manifolds and external parts is all that is necessary.

Give careful thought to which alternative is best for you and discuss the situation with local automotive machine shops, auto parts dealers or parts store countermen before ordering or purchasing replacement parts.

6 Engine - removal and installation

Refer to illustrations 6.7, 6.12, 6.13 and 6.84

Note: *Although it is not absolutely necessary, removal of the hood (see Chapter 11) is recommended to facilitate engine removal and installation.*

Inline six-cylinder engine

1 Disconnect the battery cables at the battery (negative cable first, then positive).

2 Remove the air cleaner assembly.

3 Drain the cooling system (Chapter 1).

4 Disconnect the throttle cable from the throttle linkage on the carburetor (Chapter 4).

5 If equipped with an automatic transmission, disconnect the kickdown cable from the throttle linkage.

6 Remove the engine drivebelts (Chapter 1).

7 Label and disconnect all wiring connectors from the engine **(see illustration)**.

8 Remove the heater hoses from the engine and, on models where the heater return hose goes direct to the radiator, from the radiator side tank.

9 Remove the automatic transmission oil cooler lines to the radiator. Plug them to prevent fluid loss and contamination.

10 Remove the radiator hoses and the radiator (Chapter 3).

11 Remove the fan and water pump pulley.

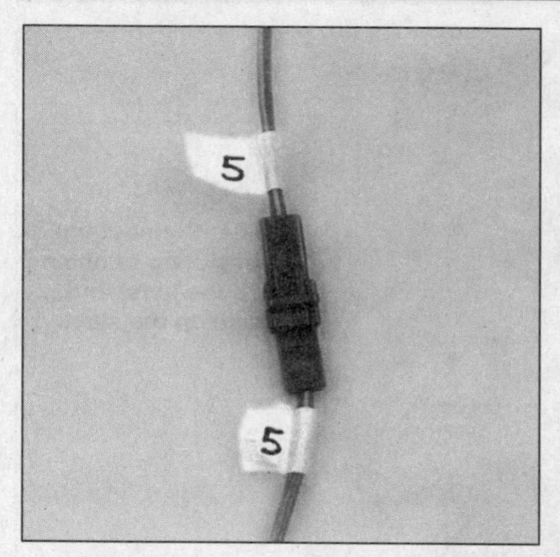

6.7 Label both ends of each wire before unplugging the connector

12 Remove the air conditioning compressor (if used) without disconnecting the lines and support it out of the way **(see illustration)**.
13 Remove the power steering pump (if used) without disconnecting the hoses and support it out of the way **(see illustration)**.
14 Disconnect the alternator wires, remove the alternator and hang it out of the way (Chapter 5).
15 Disconnect the fuel line from the fuel pump (Chapter 4) and plug the line to prevent fuel loss.
16 Label and disconnect all vacuum hoses.
17 Raise the vehicle and support it securely on jackstands.
18 Drain the engine oil (Chapter 1).
19 Disconnect the wires to the starter solenoid and remove the starter (Chapter 5).
20 Remove the flywheel or torque converter inspection cover.
21 Disconnect the exhaust pipe from the exhaust manifold.
22 Remove the engine mount through-bolts.
23 If equipped with an automatic transmission, remove the torque converter-to-driveplate bolts.
24 If equipped with four-wheel drive, remove the strut rods at the motor mounts.
25 Support the transmission with a floor jack and remove the bellhousing-to-engine bolts.
26 Attach a hoist to the engine and lift it slightly **(see illustration 6.84)**. Check to make sure everything has been disconnected, then remove the engine from the vehicle.
27 Installation is the reverse of the removal procedure.

V6 engine

28 Disconnect the negative cable from the battery.
29 Drain the cooling system (Chapter 1).
30 Remove the air cleaner assembly.
31 Remove the drivebelts (Chapter 1).
32 Remove the fan and water pump pulley.
33 If equipped with an automatic transmission, remove the oil cooler lines from the radiator and plug the lines to prevent fluid loss and contamination.
34 Remove the upper and lower radiator hoses.
35 Remove the radiator and shroud (Chapter 3).
36 Remove the heater hoses at the engine.
37 Remove the accelerator cable and kickdown linkage (if used) from the carburetor or throttle body.
38 Remove the air conditioning compressor (if used) without disconnecting the lines and hang it out of the way **(see illustration 6.12)**.
39 Remove the power steering pump (if used) without disconnecting the hoses and hang it out of the way **(see illustration 6.13)**.
40 Disconnect the wires from the alternator then remove the alternator and hang it out of the way (Chapter 5).
41 Label and disconnect all wires from the engine **(see illustration 6.7)**.
42 Disconnect the fuel line at the fuel pump (Chapter 4) and plug the line to prevent fuel loss.
43 Label and remove all vacuum lines from the intake manifold.
44 Raise the vehicle and support it securely on jackstands.
45 Drain the engine oil (Chapter 1).
46 Disconnect the exhaust pipes from the exhaust manifolds.
47 On four-wheel drive models with an automatic transmission, disconnect the strut rods at the engine mounts.
48 Remove the flywheel or torque converter inspection cover.
49 Remove the wiring harness from the clips along the oil pan rail.
50 Disconnect the wires from the starter solenoid and remove the starter (Chapter 5).
51 Disconnect the fuel gauge wire.
52 If equipped with an automatic transmission, remove the torque converter-to-driveplate bolts.
53 Support the transmission with a floor jack.
54 Attach an engine hoist to the engine and take the weight off the engine mounts **(see illustration 6.84)**.
55 Remove the bellhousing-to-engine bolts.
56 Remove the engine mount-to-frame bolts.
57 Lift the engine slightly, check to make sure everything is disconnected, then lift the engine out of the vehicle.
58 Installation is the reverse of the removal procedure.

V8 engines

59 Disconnect the negative cable at the battery.
60 Drain the cooling system (Chapter 1).
61 Remove the air cleaner assembly.
62 Remove the distributor (Chapter 5).
63 Remove the drivebelts (Chapter 1).
64 Remove the fan and water pump pulley.
65 If equipped with an automatic transmission, remove the oil cooler lines from the radiator. Plug the lines to prevent fluid loss and contamination.
66 Remove the upper and lower radiator

6.12 Unbolt the air conditioning compressor from the engine and tie it out of the way with a wire or rope

6.13 Use a wire or rope to tie the power steering pump out of the way - make sure to keep the pump in an upright position

hoses and the heater return hose if it attaches to the radiator.

67 Remove the radiator and shroud (Chapter 3).

68 Remove the heater hoses from the engine.

69 Remove the accelerator, cruise control and kickdown linkage (as equipped) from the carburetor.

70 Remove the air conditioning compressor (if used) without disconnecting the lines and hang it out of the way **(see illustration 6.12)**.

71 Remove the power steering pump (if used) without disconnecting the hoses and hang it out of the way **(see illustration 6.13)**.

72 Disconnect the wires from the alternator, then remove it and hang it out of the way (Chapter 5).

73 Label and disconnect all wires from the engine **(see illustration 6.7)**.

74 Remove the fuel line at the fuel pump (Chapter 4) and plug the line to prevent fuel loss.

75 Label and disconnect the vacuum lines from the intake manifold.

76 Raise the vehicle and support it securely on jackstands.

77 Drain the engine oil (Chapter 1).

78 Disconnect the exhaust pipes from the exhaust manifolds.

79 If equipped with an automatic transmission and four-wheel drive, disconnect the strut rods at the engine mounts.

80 Remove the flywheel or torque converter inspection cover.

81 Disconnect the wires from the starter solenoid and remove the starter (Chapter 5).

82 If equipped with an automatic transmission, remove the torque converter-to-driveplate bolts.

83 Support the transmission with a floor jack.

84 Attach an engine hoist to the engine and take the weight off the engine mounts **(see illustration)**.

85 Remove the bellhousing-to-engine bolts.

86 Remove the engine mount throughbolts.

87 Lift the engine slightly, check to make sure everything is disconnected, then lift the engine out of the vehicle.

88 Installation is the reverse of the removal procedure.

6.84 Connect the lifting sling or chain to the hoist and take up the slack

7 Engine overhaul - disassembly sequence

1 It is much easier to disassemble and work on the engine if it is mounted on a portable engine stand. These stands can often be rented quite cheaply from an equipment rental yard. Before the engine is mounted on a stand, the flywheel/driveplate should be removed from the engine.

2 If a stand is not available, it is possible to disassemble the engine with it blocked up on a sturdy workbench or on the floor. Be extra careful not to tip or drop the engine when working without a stand.

3 If you are going to obtain a rebuilt engine, all external components must come off first, to be transferred to the replacement engine, just as they will if you are doing a complete engine overhaul yourself. These include:

Alternator and brackets
Emissions control components
Distributor, spark plug wires and spark plugs
Thermostat and housing cover
Water pump
Carburetor or EFI components
Intake/exhaust manifolds
Oil filter
Engine mounts
Clutch and flywheel/driveplate

Note: *When removing the external components from the engine, pay close attention to details that may be helpful or important during installation. Note the installed position of gaskets, seals, spacers, pins, washers, bolts and other small items.*

4 If you are obtaining a short block, which consists of the engine block, crankshaft, pistons and connecting rods all assembled, then the cylinder head(s), oil pan and oil pump will have to be removed as well. See *Engine rebuilding alternatives* for additional information regarding the different possibilities to be considered.

5 If you are planning a complete overhaul, the engine must be disassembled and the internal components removed in the following order:

Rocker arm cover(s)
Pushrod cover (6-cylinder only)
Intake and exhaust manifolds
Rocker arms and pushrods
Valve lifters
Cylinder head(s)
Timing chain cover
Timing chain and sprockets (V8 and V6 engines)
Timing gears (inline 6-cylinder engine)
Camshaft
Oil pan
Oil pump
Piston/connecting rod assemblies
Crankshaft and main bearings

6 Critical cooling system components such as the hoses, the drivebelts, the thermostat and the water pump MUST be replaced with new parts when an engine is overhauled. Also, we do not recommend overhauling the oil pump - always install a new one when an engine is rebuilt.

7 Before beginning the disassembly and overhaul procedures, make sure the following items are available:

Common hand tools
Small cardboard boxes or plastic bags for storing parts
Gasket scraper
Ridge reamer
Vibration damper puller
Micrometers
Telescoping gauges
Dial indicator set
Valve spring compressor
Cylinder surfacing hone
Piston ring groove cleaning tool
Electric drill motor
Tap and die set
Wire brushes
Oil gallery brushes
Cleaning solvent

8 Cylinder head - disassembly

Refer to illustrations 8.2, 8.3a and 8.3b

Note: *New and rebuilt cylinder heads are commonly available for most engines at dealerships and auto parts stores. Due to the fact that some specialized tools are necessary for the disassembly and inspection procedures, and replacement parts may not be readily available, it may be more practical and economical for the home mechanic to purchase replacement heads rather than taking the time to disassemble, inspect and recondition the originals.*

Caution: *On big block (396, 402 and 454) engines, two head designs are generally avail-*

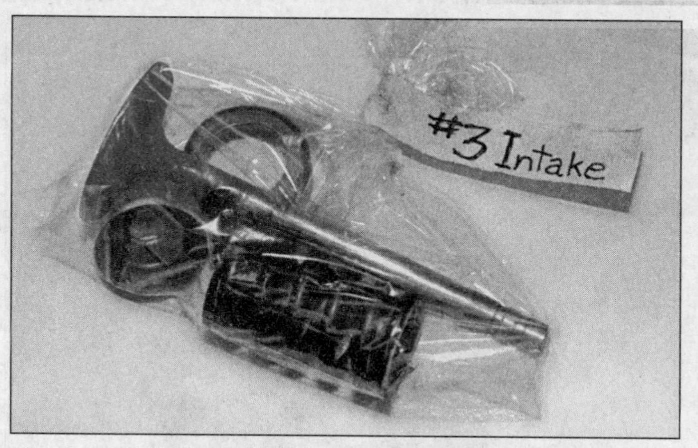

8.2 A small plastic bag with an appropriate label can be used to store the valve and spring components so they can be reinstalled in the correct guide

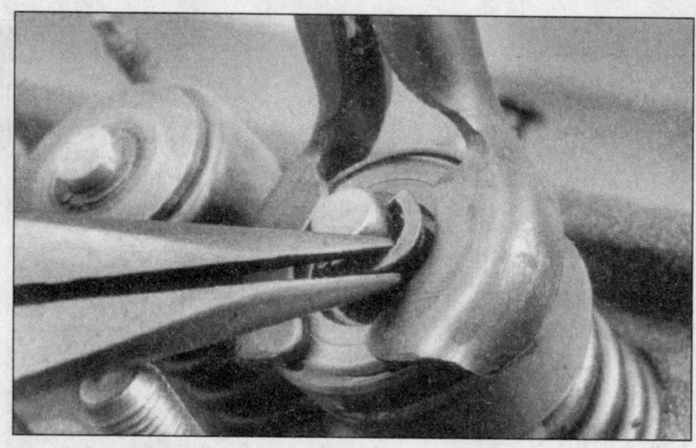

8.3a Compress the valve spring and remove the keepers

able - "open chamber" and "closed chamber." While these heads will interchange as far as bolt patterns, accessory mounting, etc. is concerned, the different combustion chamber designs require different piston dome shapes. If you are purchasing a set of reconditioned cylinder heads, make sure the new heads match your old ones in combustion chamber design.

1 Cylinder head disassembly involves removal of the intake and exhaust valves and related components. If they are still in place, remove the rocker arm nuts, pivot balls and rocker arms from the cylinder head studs. Label the parts or store them separately so they can be reinstalled in their original locations.

2 Before the valves are removed, arrange to label and store them, along with their related components, so they can be kept separate and reinstalled in the same valve guides they are removed from **(see illustration).**

3 Compress the springs on the first valve with a spring compressor and remove the keepers **(see illustration)**. Carefully release the valve spring compressor and remove the retainer and (if used) rotators, the shield, the springs and the spring seat or shims (if used). Remove the oil seal(s) from the valve stem and the umbrella-type seal from over the guide boss (if used), then pull the valve from the head. If the valve binds in the guide (won't pull through), push it back into the head and deburr the area around the keeper groove with a fine file or whetstone **(see illustration).**

4 Repeat the procedure for the remaining valves. Remember to keep all the parts for each valve together so they can be reinstalled in the same locations.

5 Once the valves and related components have been removed and stored in an organized manner, the head should be thoroughly cleaned and inspected. If a complete engine overhaul is being done, finish the engine disassembly procedures before beginning the cylinder head cleaning and inspection process.

8.3b If the valve stem will not pull easily through the guide, use a fine file to deburr the top of the valve stem

9 Cylinder head - cleaning and inspection

Refer to illustrations 9.12, 9.14, 9.16, 9.17 and 9.18

1 Thorough cleaning of the cylinder heads and related valve train components, followed by a detailed inspection, will enable you to decide how much valve service work must be done during the engine overhaul.

Cleaning

2 Scrape away all traces of old gasket material and sealing compound from the head gasket, intake manifold and exhaust manifold sealing surfaces. Be very careful not to gouge the cylinder head. Special gasket removal solvents, which soften gaskets and make removal much easier, are available at auto parts stores.

3 Remove any built up scale from the coolant passages.

4 Run a stiff wire brush through the various holes to remove any deposits that may have formed in them.

5 Run an appropriate size tap into each of the threaded holes to remove any corrosion and thread sealant that may be present. If compressed air is available, use it to clear the holes of debris produced by this operation.

6 Clean the rocker arm pivot stud threads with a wire brush.

7 Clean the cylinder head with solvent and dry it thoroughly. Compressed air will speed the drying process and ensure that all holes and recessed areas are clean. **Note:** *Decarbonizing chemicals are available and may prove very useful when cleaning cylinder heads and valve train components. They are very caustic and should be used with caution. Be sure to follow the instructions on the container.*

8 Clean the rocker arms, pivot balls, nuts and pushrods with solvent and dry them thoroughly (don't mix them up during the cleaning process). Compressed air will speed the drying process and can be used to clean out the oil passages.

9 Clean all the valve springs, shields, keepers and retainers (or rotators) with solvent and dry them thoroughly. Do the components from one valve at a time to avoid mixing up the parts.

10 Scrape off any heavy deposits that may have formed on the valves, then use a motorized wire brush to remove deposits from the valve heads and stems. Again, make sure the valves do not get mixed up.

9.12 Check the cylinder head gasket surface for warpage by trying to slip a feeler gauge under the straightedge (see the Specifications for the maximum warpage allowed and use a feeler gauge of that thickness)

9.14 A dial indicator can be used to determine the valve stem-to-guide clearance (move the valve stem as indicated by the arrows)

Inspection

Cylinder head

11 Inspect the head very carefully for cracks, evidence of coolant leakage and other damage. If cracks are found, a new cylinder head should be obtained.

12 Using a straightedge and feeler gauge, check the head gasket mating surface for warpage **(see illustration)**. If the warpage exceeds the specified limit, it can be resurfaced at an automotive machine shop.

13 Examine the valve seats in each of the combustion chambers. If they are pitted, cracked or burned, the head will require valve service that is beyond the scope of the home mechanic.

14 Check the valve stem-to-guide clearance by measuring the lateral movement of the valve stem with a dial indicator attached securely to the head **(see illustration)**. The valve must be in the guide and approximately 1/16-inch off the seat. The total valve stem movement indicated by the gauge needle must be divided by two to obtain the actual clearance. After this is done, if there is still some doubt regarding the condition of the valve guides they should be checked by an automotive machine shop (the cost should be minimal).

Valves

15 Carefully inspect each valve face for uneven wear, deformation, cracks, pits and burned spots. Check the valve stem for scuffing and galling and the neck for cracks. Rotate the valve and check for any obvious indication that it is bent. Look for pits and excessive wear on the end of the stem. The presence of any of these conditions indicates the need for valve service by an automotive machine shop.

16 Measure the margin width on each valve. Any valve with a margin narrower than 1/32-inch will have to be replaced with a new one **(see illustration)**.

Valve components

17 Check each valve spring for wear (on the ends) and pits. Measure the free length and compare it to the Specifications **(see illustration)**. Any springs that are shorter than specified have sagged and should not be reused. The tension of all springs should be checked with a special fixture before deciding that they are suitable for use in a rebuilt engine (take the springs to an automotive machine shop for this check).

18 Stand each spring on a flat surface and check it for squareness **(see illustration)**. If any of the springs are distorted or sagged, replace all of them with new parts.

19 Check the spring retainers (or rotators) and keepers for obvious wear and cracks. Any questionable parts should be replaced with new ones, as extensive damage will occur if they fail during engine operation.

Rocker arm components

20 Check the rocker arm faces (the areas that contact the pushrod ends and valve stems) for pits, wear, galling, score marks and rough spots. Check the rocker arm pivot contact areas and pivot balls as well. Look for cracks in each rocker arm and nut.

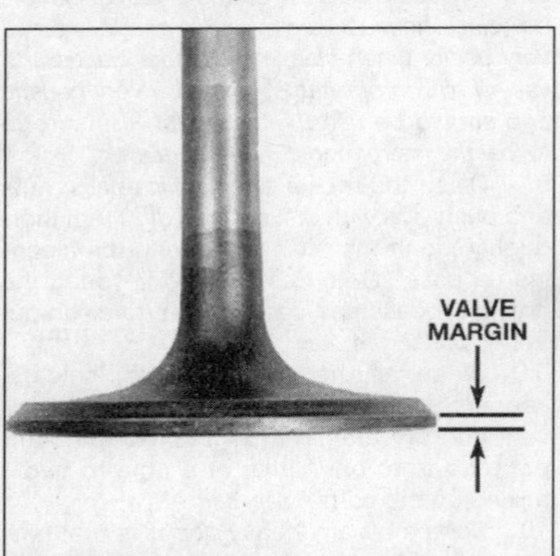

9.16 The margin width on each valve must be as specified (if no margin exists the valve must be replaced)

9.17 Measure the free length of each valve spring and compare it to the Specifications

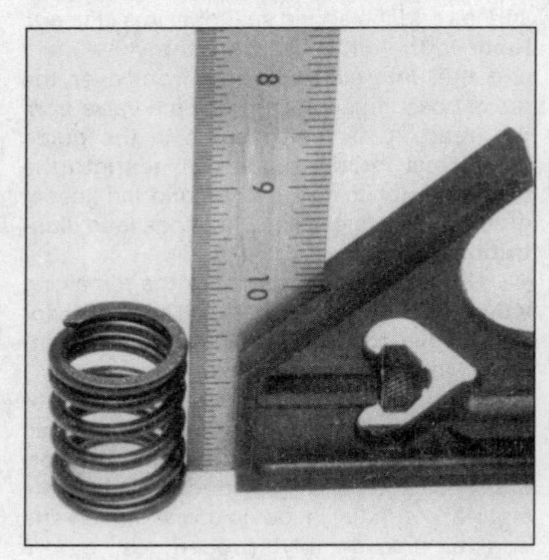

9.18 Check each valve spring for squareness

21 Inspect the pushrod ends for scuffing and excessive wear. Roll each pushrod on a flat surface, such as a piece of plate glass, to determine if it is bent.

22 Check the rocker arm studs in the cylinder heads for damaged threads and secure installation. The press-in rocker arm studs used in most small block applications cannot be replaced by the home mechanic due to the need for precision reaming equipment and a press suitable for the job. If any studs are damaged, the head should be taken to an automotive machine shop for stud replacement.

23 Some small block engines and most big block engines use screwed in rather than pressed in rocker arm studs. Also, most engines with screw-in studs use a guide plate, attached to the head by the studs, to maintain pushrod-to-rocker arm alignment. If an engine equipped with screw-in studs is found to have worn, bent or otherwise damaged studs, the studs can be removed individually and replaced. Be sure to replace the guide plates under the studs and apply RTV sealant to the stud threads.

24 Any damaged or excessively worn parts must be replaced with new ones.

25 If the inspection process indicates that the valve components are in generally poor condition and worn beyond the limits specified, which is usually the case in an engine that is being overhauled, reassemble the valves in the cylinder head and refer to Section 10 for valve servicing recommendations.

26 If the inspection turns up no excessively worn parts, and if the valve faces and seats are in good condition, the valve train components can be reinstalled in the cylinder head without major servicing. Refer to the appropriate Section for the cylinder head reassembly procedure.

10 Valves - servicing

1 Because of the complex nature of the job and the special tools and equipment needed, servicing of the valves, the valve seats and the valve guides, commonly known as a valve job, is best left to a professional.

2 The home mechanic can remove and disassemble the head(s), do the initial cleaning and inspection, then reassemble and deliver the head(s) to a dealer service department or an automotive machine shop for the actual valve servicing.

3 The dealer service department, or automotive machine shop, will remove the valves and springs, recondition or replace the valves and valve seats, recondition the valve guides, check and replace the valve springs, spring retainers or rotators and keepers (as necessary), replace the valve seals with new ones, reassemble the valve components and make sure the installed spring height is correct. The cylinder head gasket surface will also be resurfaced if it is warped.

4 After the valve job has been performed by a professional, the head will be in like new condition. When the head is returned, be sure to clean it again before installation on the engine to remove any metal particles and abrasive grit that may still be present from the valve service or head resurfacing operations. Use compressed air, if available, to blow out all the oil holes and passages.

11 Cylinder head - reassembly

1 Regardless of whether or not the head(s) were sent to an automotive machine shop for valve servicing, make sure they are clean before beginning reassembly.

2 If the head(s) were sent out for valve servicing, the valves and related components will already be in place. Begin the reassembly procedure with Step 8.

3 Beginning at one end of the head, lubricate and install the first valve. Apply moly-base grease or clean engine oil to the valve stem.

4 Three different types of valve stem oil seals are used on these engines, depending on year, engine size and horsepower rating. The most common is a small O-ring which simply fits around the valve stem just above the guide boss. A second type is a flat O-ring which fits into a groove in the valve stem just below the valve stem keeper groove. On most high horsepower applications, an umbrella type seal which extends down over the valve guide boss is used over the valve stem. In most cases the umbrella type seal is used in conjunction with the flat O-ring type seal. If round O-ring or umbrella type seals are used on your head install them at this time.

5 Drop the spring seat or shim(s) over the valve guide and set the valve springs, shield and retainer (or rotator) in place.

6 Compress the springs with a valve spring compressor and if the flat O-ring type seal is used on your head carefully install it in the lower groove of the valve stem. Make sure the seal is not twisted - it must lie perfectly flat in the groove. Position the keepers in the upper groove, then slowly release the compressor and make sure the keepers seat properly. Apply a small dab of grease to each keeper to hold it in place if necessary.

7 Repeat the procedure for the remaining valves. Be sure to return the components to their original locations - do not mix them up!

8 Check the installed valve spring height with a ruler graduated in 1/32-inch increments or a dial caliper. If the heads were sent out for service work, the installed height should be correct (but don't automatically assume that it is). The measurement is taken from the top of each spring seat or shim(s) to the top of the oil shield (or the bottom of the retainer/rotator, the two points are the same). If the height is greater than specified, shims can be added under the springs to correct it. **Caution:** *Do not, under any circumstances, shim the springs to the point where the installed height is less than specified.*

9 Apply moly-base grease to the rocker arm faces and the pivot balls, then install the rocker arms and pivots on the cylinder head studs. Thread the nuts on three or four turns only.

12 Piston/connecting rod assembly - removal

Refer to illustrations 12.1, 12.3 and 12.5

Note: *Prior to removing the piston/connecting rod assemblies, remove the cylinder head(s), the oil pan and the oil pump by referring to the appropriate Sections in Chapter 2, Parts A or B.*

1 Completely remove the ridge at the top of each cylinder with a ridge reaming tool **(see illustration)**. Follow the manufacturer's instructions provided with the tool. Failure to remove the ridge before attempting to remove the piston/connecting rod assemblies will result in piston breakage.

2 After the cylinder ridges have been removed, turn the engine upside-down so the crankshaft is facing up.

3 Before the connecting rods are removed, check the end play with feeler

12.1 Use a ridge reamer to completely remove the ridge from the top of each cylinder before attempting to remove the pistons

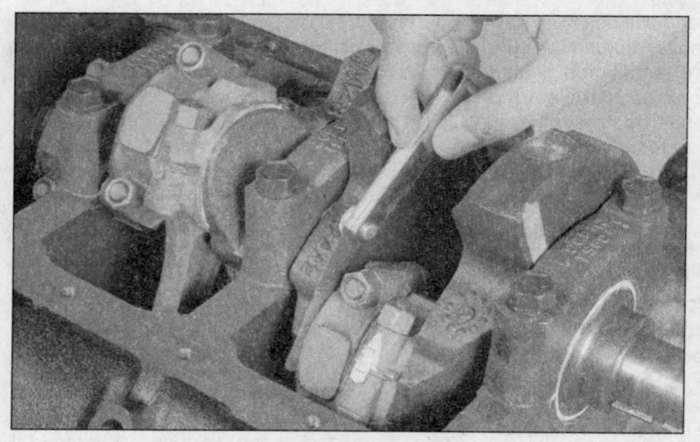

12.3 Use a feeler gauge to check the connecting rod end play

12.5 To prevent damage to the crankshaft journals and cylinder walls, slip sections of hose over the rod bolts before removing the pistons

13.2 Checking crankshaft end play with a dial indicator

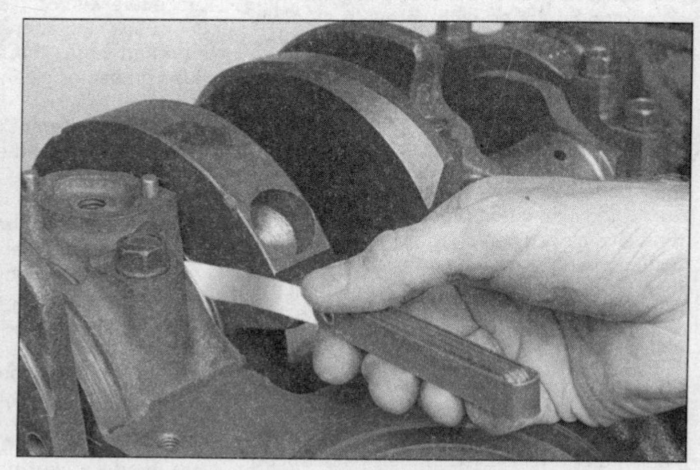

13.3 Checking crankshaft end play by inserting a feeler gauge between the crankshaft and the face of the thrust bearing

gauges. Slide them between the first connecting rod and the crankshaft throw until the play is removed **(see illustration)**. The end play is equal to the thickness of the feeler gauge(s). If the end play exceeds the service limit, new connecting rods will be required. If new rods (or a new crankshaft) are installed, the end play may fall under the specified minimum. If it does, the rods will have to be machined to restore it - consult an automotive machine shop for advice if necessary. Repeat the procedure for the remaining connecting rods.

4 Check the connecting rods and caps for identification marks. If they are not plainly marked, use a small center punch to make the appropriate number of indentations on each rod and cap.

5 Loosen each of the connecting rod cap nuts 1/2-turn at a time until they can be removed by hand. Remove the number one connecting rod cap and bearing insert. Do not drop the bearing insert out of the cap. Slip a short length of plastic or rubber hose over each connecting rod cap bolt to protect the crankshaft journal and cylinder wall when the piston is removed **(see illustration)**. Push the connecting rod/piston assembly out through the top of the engine. Use a wooden hammer handle to push on the upper bearing insert in the connecting rod. If resistance is felt, double-check to make sure that all of the ridge was removed from the cylinder.

6 Repeat the procedure for the remaining cylinders. After removal, reassemble the connecting rod caps and bearing inserts in their respective connecting rods and install the cap nuts finger tight. Leaving the old bearing inserts in place until reassembly will help prevent the connecting rod bearing surfaces from being accidentally nicked or gouged.

13 Crankshaft - removal

Refer to illustrations 13.2, 13.3, 13.4a and 13.4b

Note: *The crankshaft can be removed only after the engine has been removed from the vehicle. It is assumed that the flywheel or driveplate, vibration damper, timing chain or gears, oil pan, oil pump and piston/connecting rod assemblies have already been removed.*

1 Before the crankshaft is removed, check the end play. Mount a dial indicator with the stem in line with the crankshaft and just touching one of the crank throws.

2 Push the crankshaft all the way to the rear and zero the dial indicator. Next, pry the crankshaft to the front as far as possible and check the reading on the dial indicator **(see illustration)**. The distance that it moves is the end play. If it is greater than specified, check the crankshaft thrust surfaces for wear. If no wear is evident, new main bearings should correct the end play.

3 If a dial indicator is not available, feeler gauges can be used. Gently pry or push the crankshaft all the way to the front of the engine. Slip feeler gauges between the crankshaft and the front face of the thrust main bearing to determine the clearance **(see illustration)**.

4 Check the main bearing caps to see if they are marked to indicate their locations. They should be numbered consecutively from the front of the engine to the rear. If they aren't, mark them with number stamping dies or a center punch **(see illustration)**. Main bearing caps generally have a cast-in arrow **(see illustration)**, which points to the front of the engine. Loosen each of the main bearing cap bolts 1/4-turn at a time each, until they

13.4a Use a center punch or number stamping dies to mark the main bearing caps to ensure that they are reinstalled in their original locations on the block (make the punch marks near one of the bolt heads)

13.4b The arrow on the main bearing cap indicates the front of the engine

can be removed by hand.

5 Gently tap the caps with a soft-face hammer, then separate them from the engine block. If necessary, use the bolts as levers to remove the caps. Try not to drop the bearing inserts if they come out with the caps.

6 Carefully lift the crankshaft out of the engine. It is a good idea to have an assistant available, since the crankshaft is quite heavy. With the bearing inserts in place in the engine block and main bearing caps, return the caps to their respective locations on the engine block and tighten the bolts finger tight.

14 Engine block - cleaning

Refer to illustrations 14.1a, 14.1b, 14.8 and 14.10

1 Remove the soft plugs from the engine block. To do this, knock the plugs into the block, using a hammer and punch **(see illustration)**, then grasp them with large pliers and pull them back through the holes **(see illustration)**.

2 Using a gasket scraper, remove all traces of gasket material from the engine block. Be very careful not to nick or gouge the gasket sealing surfaces.

3 Remove the main bearing caps and separate the bearing inserts from the caps and the engine block. Tag the bearings, indicating which cylinder they were removed from and whether they were in the cap or the block, then set them aside.

4 Using a 1/4-inch drive breaker bar or ratchet, remove all of the threaded oil gallery plugs from the rear of the block. Discard the plugs and use new ones when the engine is reassembled.

5 If the engine is extremely dirty it should be taken to an automotive machine shop to be steam cleaned or hot tanked.

6 After the block is returned, clean all oil holes and oil galleries one more time. Brushes specifically designed for this purpose are available at most auto parts stores. Flush the passages with warm water until the water runs clear, dry the block thoroughly and wipe all machined surfaces with a light, rust preventive oil. If you have access to compressed air, use it to speed the drying process and to blow out all the oil holes and galleries.

7 If the block is not extremely dirty or sludged up, you can do an adequate cleaning job with warm soapy water and a stiff brush. Take plenty of time and do a thorough job. Regardless of the cleaning method used, be sure to clean all oil holes and galleries very thoroughly, dry the block completely and coat all machined surfaces with light oil.

8 The threaded holes in the block must be clean to ensure accurate torque readings during reassembly. Run the proper size tap into each of the holes to remove any rust, corrosion, thread sealant or sludge and to

14.1a Use a hammer and punch to drive the soft plugs into the block

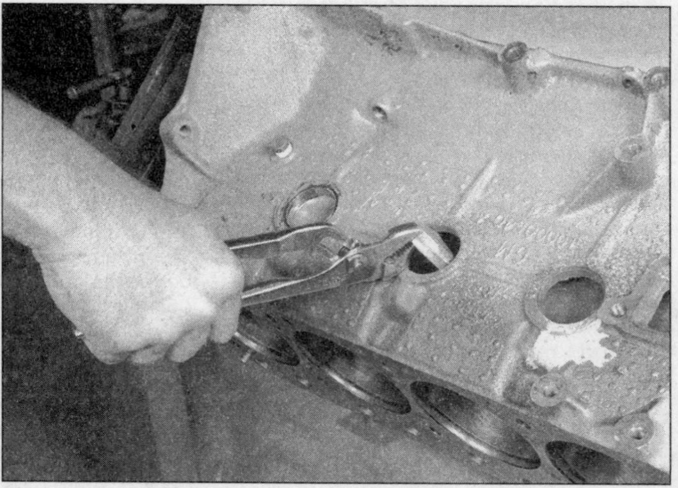

14.1b Grasp each soft plug with a pair of pliers and pull it out of the block

14.8 All bolt holes in the block - particularly the main bearing cap and head bolt holes - should be cleaned and restored with a tap (be sure to remove debris from the holes after this is done)

14.10 A large socket on an extension can be used to drive the new soft plugs into the bores

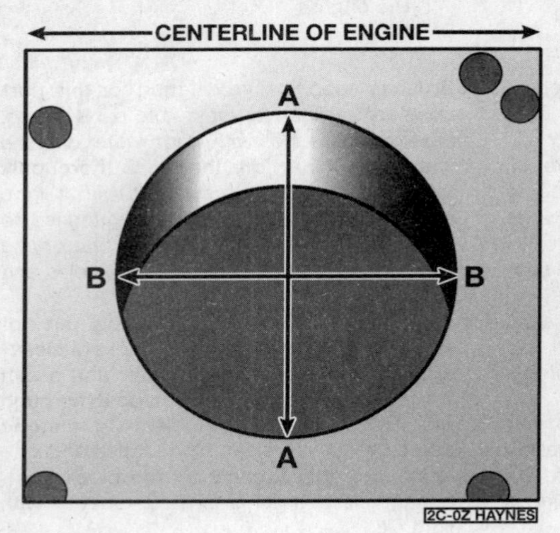

15.4a Measure the diameter of each cylinder at a right angle to engine centerline (A), and parallel to the engine centerline (B) - out-of-round is the difference between A and B; taper is the difference between A and B at the top of the cylinder and A and B at the bottom of the cylinder

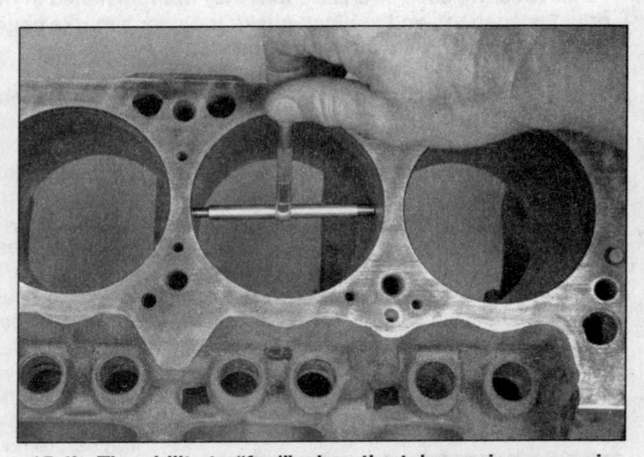

15.4b The ability to "feel" when the telescoping gauge is at the correct point will be developed over time, so work slowly and repeat the check until you are satisfied the bore measurement is accurate

restore any damaged threads **(see illustration)**. If possible, use compressed air to clear the holes of debris produced by this operation. Now is a good time to clean the threads on the head bolts and the main bearing cap bolts as well.

9 Reinstall the main bearing caps and tighten the bolts finger tight.

10 After coating the sealing surfaces of the new soft plugs with a hard setting sealant (such as Permatex No. 1), install them in the engine block. Make sure they are driven in straight and seated properly or leakage could result. Special tools are available for this purpose, but equally good results can be obtained using a large socket, with an outside diameter that will just slip into the soft plug, and a hammer **(see illustration)**.

11 Apply non-hardening sealant (such as Permatex Number 2 or Teflon tape) to the new oil gallery plugs and thread them into the holes at the rear of the block. Make sure they are tightened securely.

12 If the engine is not going to be reassembled right away, cover it with a large plastic trash bag to keep it clean.

15 Engine block - inspection

Refer to illustrations 15.4a, 15.4b and 15.4c

1 Before the block is inspected, it should be cleaned as described in Section 14. Double-check to make sure that the ridge at the top of each cylinder has been completely removed.

2 Visually check the block for cracks, rust and corrosion. Look for stripped threads in the threaded holes. It is also a good idea to have the block checked for hidden cracks by an automotive machine shop that has the special equipment to do this type of work. If defects are found, have the block repaired, if possible, or replaced.

3 Check the cylinder bores for scuffing and scoring.

4 Measure the diameter of each cylinder at the top (just under the ridge area), center and bottom of the cylinder bore, parallel to the crankshaft axis **(see illustrations)**. Next, measure each cylinder's diameter at the same three locations across the crankshaft axis. Compare the results to the Specifications. If the cylinder walls are badly scuffed or scored, or if they are out-of-round or tapered beyond the limits given in the Specifications, have the engine block rebored and honed at an automotive machine shop. If a rebore is

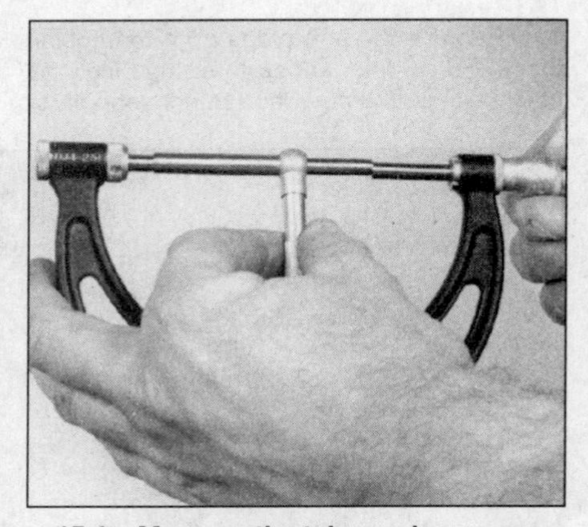

15.4c Measure the telescoping gauge with a micrometer to determine the bore size

16.3a A "bottle brush" hone will produce better results if you have never done cylinder honing before

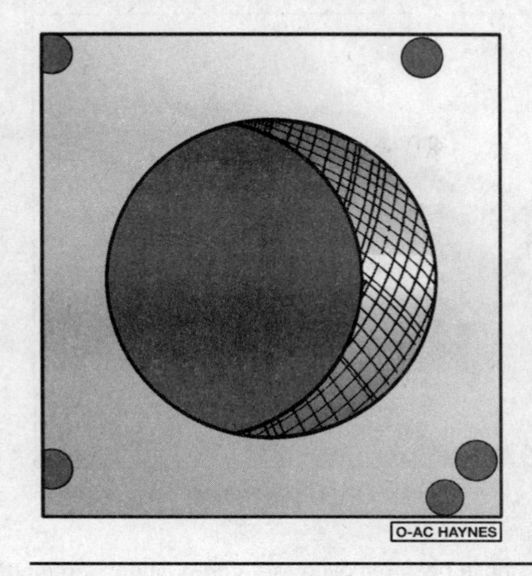

16.3b The hone should leave a crosshatch pattern with the lines intersecting at approximately a 60-degree angle

done, oversize pistons and rings will be required.

5 If the cylinders are in reasonably good condition and not worn to the outside of the limits, and if the piston-to-cylinder clearances can be maintained properly, then they do not have to be rebored. Honing is all that is necessary (Section 16).

16 Cylinder honing

Refer to illustrations 16.3a and 16.3b

1 Prior to engine reassembly, the cylinder bores must be honed so the new piston rings will seat correctly and provide the best possible combustion chamber seal. **Note:** *If you do not have the tools or do not want to tackle the honing operation, most automotive machine shops will do it for a reasonable fee.*

2 Before honing the cylinders, install the main bearing caps and tighten the bolts to the specified torque.

3 Two types of cylinder hones are commonly available - the flex hone or "bottle brush" type and the more traditional surfacing hone with spring-loaded stones. Both will do the job, but for the less experienced mechanic the "bottle brush" hone will probably be easier to use. You will also need plenty of light oil or honing oil, some rags and an electric drill motor. Proceed as follows:

a) *Mount the hone in the drill motor, compress the stones and slip it into the first cylinder* **(see illustration)**.

b) *Lubricate the cylinder with plenty of oil, turn on the drill and move the hone up-and-down in the cylinder at a pace which will produce a fine crosshatch pattern on the cylinder walls. Ideally, the crosshatch lines should intersect at approximately a 60° angle* **(see illustration)**. *Be sure to use plenty of lubricant and do not take off any more material than is absolutely necessary to produce the desired finish.* **Note:** *Piston ring manufacturers may specify a smaller crosshatch angle than the traditional 60° - read and follow any instructions printed on the piston ring packages.*

c) *Do not withdraw the hone from the cylinder while it is running. Instead, shut off the drill and continue moving the hone up-and-down in the cylinder until it comes to a complete stop, then compress the stones and withdraw the hone. If you are using a "bottle brush" type hone, stop the drill motor, then turn the chuck in the normal direction of rotation while withdrawing the hone from the cylinder.*

d) *Wipe the oil out of the cylinder and repeat the procedure for the remaining cylinders.*

4 After the honing job is complete, chamfer the top edges of the cylinder bores with a small file so the rings will not catch when the pistons are installed. Be very careful not to nick the cylinder walls with the end of the file.

5 The entire engine block must be washed again very thoroughly with warm, soapy water to remove all traces of the abrasive grit produced during the honing operation. **Note:** *The bores can be considered clean when a white cloth - dampened with clean engine oil - used to wipe down the bores does not pick up any more honing residue, which will show up as gray areas on the cloth. Be sure to run a brush through all oil holes and galleries and flush them with running water.*

6 After rinsing, dry the block and apply a coat of light rust preventive oil to all machined surfaces. Wrap the block in a plastic trash bag to keep it clean and set it aside until reassembly.

17 Piston/connecting rod assembly - inspection

Refer to illustrations 17.4a, 17.4b, 17.10 and 17.11

1 Before the inspection process can be carried out, the piston/connecting rod assemblies must be cleaned and the original piston rings removed from the pistons. **Note:** *Always use new piston rings when the engine is reassembled.*

2 Using a piston ring installation tool, carefully remove the rings from the pistons. Be careful not to nick or gouge the pistons in the process.

3 Scrape all traces of carbon from the top (known as the crown) of the piston. A hand-held wire brush or a piece of fine emery cloth can be used once the majority of the deposits have been scraped away. Do not, under any circumstances, use a wire brush mounted in a drill motor to remove deposits from the pistons. The piston material is soft and will be eroded away by the wire brush.

4 Use a piston ring groove cleaning tool to remove carbon deposits from the ring grooves **(see illustration)**. If a tool is not available, a piece broken off the old ring will do the job **(see illustration)**. Be very careful to remove only the carbon deposits - don't remove any metal and do not nick or scratch the sides of the ring grooves.

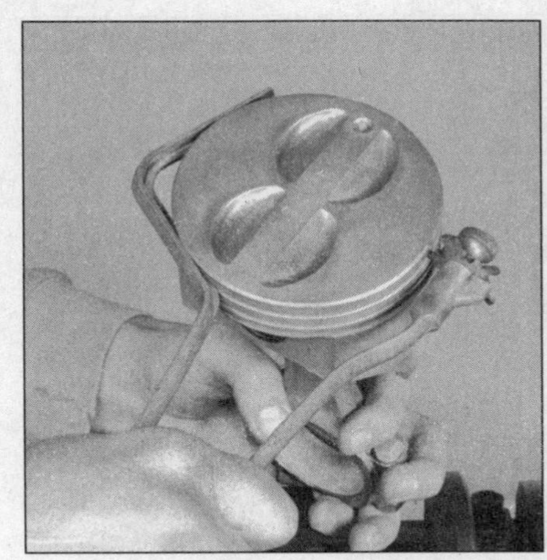

17.4a The piston ring grooves can be cleaned with a special tool, as shown here . . .

17.4b . . . or a section of a broken ring

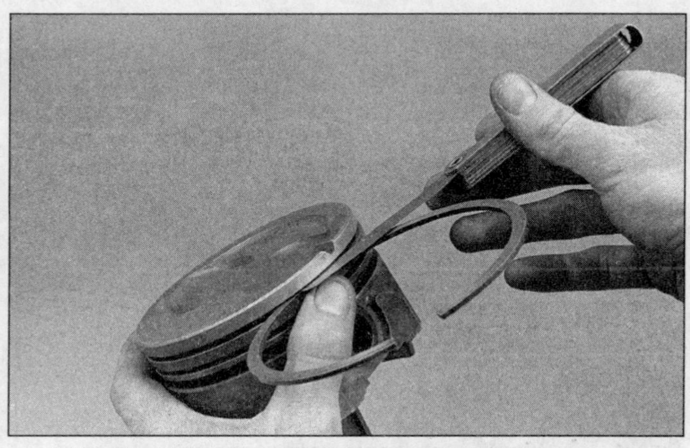

17.10 Check the ring side clearance with a feeler gauge at several points around the groove

5 Once the deposits have been removed, clean the piston/rod assemblies with solvent and dry them with compressed air (if available). Make sure that the oil return holes in the back sides of the ring grooves are clear.

6 If the pistons are not damaged or worn excessively, and if the engine block is not rebored, new pistons will not be necessary. Normal piston wear appears as even vertical wear on the piston thrust surfaces and slight looseness of the top ring in its groove. New piston rings, on the other hand, should always be used when an engine is rebuilt.

7 Carefully inspect each piston for cracks around the skirt, at the pin bosses and at the ring lands.

8 Look for scoring and scuffing on the thrust faces of the skirt, holes in the piston crown and burned areas at the edge of the crown. If the skirt is scored or scuffed, the engine may have been suffering from overheating and/or abnormal combustion, which caused excessively high operating temperatures. The cooling and lubrication systems should be checked thoroughly. A hole in the piston crown is an indication that abnormal combustion (preignition) was occurring. Burned areas at the edge of the piston crown are usually evidence of spark knock (detonation). If any of the above problems exist, the causes must be corrected or the damage will occur again.

9 Corrosion of the piston, in the form of small pits, indicates that coolant is leaking into the combustion chamber and/or the crankcase. Again, the cause must be corrected or the problem may persist in the rebuilt engine.

10 Measure the piston ring side clearance by laying a new piston ring in each ring groove and slipping a feeler gauge in beside it **(see illustration)**. Check the clearance at three or four locations around each groove. Be sure to use the correct ring for each groove; they are different. If the side clearance is greater than specified, new pistons will have to be used.

11 Check the piston-to-bore clearance by measuring the bore (see Section 15) and the piston diameter. Make sure that the pistons and bores are correctly matched. Measure the piston across the skirt, at a 90° angle to and in line with the piston pin **(see illustration)**. Subtract the piston diameter from the bore diameter to obtain the clearance. If it is greater than specified, the block will have to be rebored and new pistons and rings installed.

12 Check the piston-to-rod clearance by twisting the piston and rod in opposite directions. Any noticeable play indicates that there is excessive wear, which must be corrected. The piston/connecting rod assemblies should be taken to an automotive machine shop to have the pistons and rods rebored and new pins installed.

13 If the pistons must be removed from the connecting rods for any reason, they should be taken to an automotive machine shop. While they are there have the connecting rods checked for bend and twist, since automotive machine shops have special equipment for this purpose. **Note:** *Unless new pistons and/or connecting rods must be installed, do not disassemble the pistons and connecting rods.*

14 Check the connecting rods for cracks and other damage. Temporarily remove the rod caps, lift out the old bearing inserts, wipe the rod and cap bearing surfaces clean and inspect them for nicks, gouges and scratches. After checking the rods, replace the old bearings, slip the caps into place and tighten the nuts finger tight.

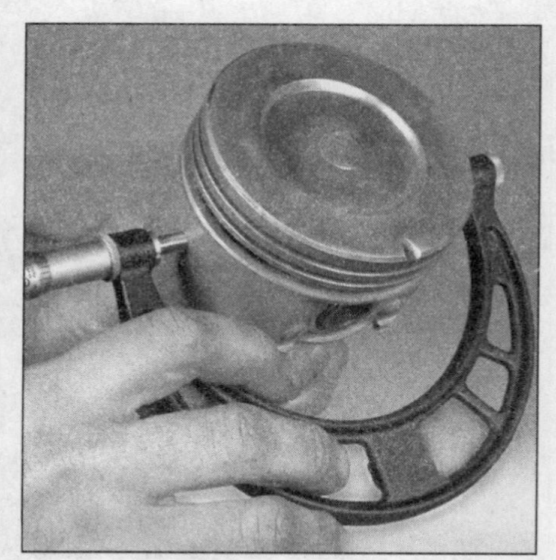

17.11 Measure the piston diameter at a 90-degree angle to the piston pin and in line with it

18 Crankshaft - inspection

Refer to illustration 18.2

1 Clean the crankshaft with solvent and dry it with compressed air (if available). Be sure to clean the oil holes with a stiff brush

18.2 Measure the diameter of each crankshaft journal at several points to detect taper and out-of-round conditions

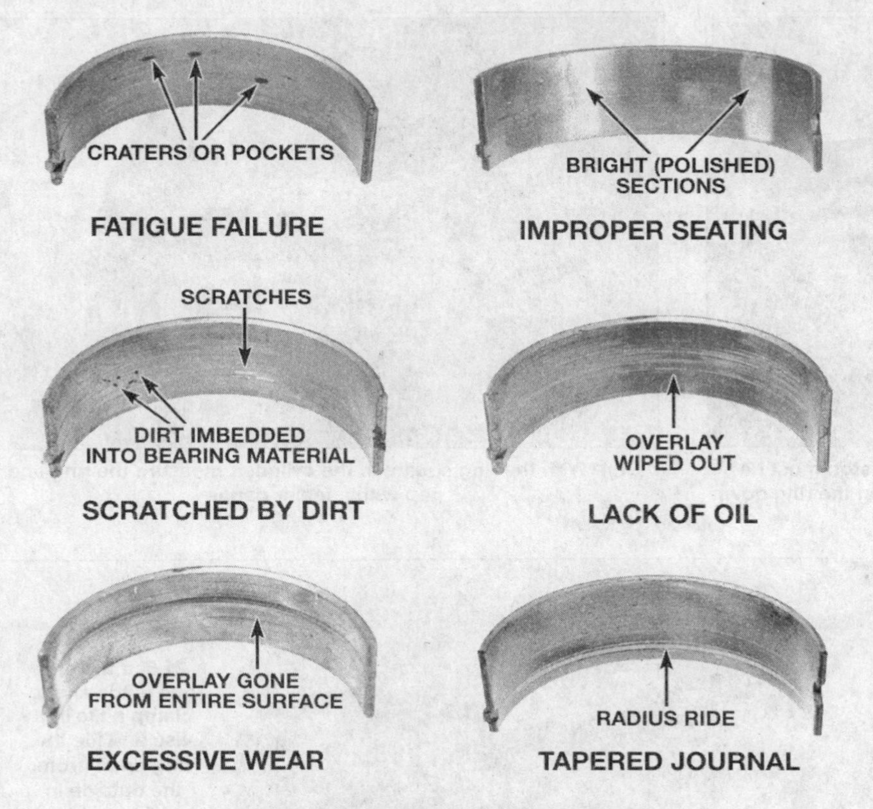

19.2 Typical indications of bearing failure

and flush them with solvent. Check the main and connecting rod bearing journals for uneven wear, scoring, pits and cracks. Check the rest of the crankshaft for cracks and other damage.

2 Using a micrometer, measure the diameter of the main and connecting rod journals and compare the results to the Specifications **(see illustration)**. By measuring the diameter at a number of points around each journal's circumference, you will be able to determine whether or not the journal is out-of-round. Take the measurement at each end of the journal, near the crank throws, to determine if the journal is tapered.

3 If the crankshaft journals are damaged, tapered, out-of-round or worn beyond the limits given in the Specifications, have the crankshaft reground by an automotive machine shop. Be sure to use the correct size bearing inserts if the crankshaft is reconditioned.

4 Refer to Section 19 and examine the main and rod bearing inserts.

19 Main and connecting rod bearings - inspection

Refer to illustration 19.2

1 Even though the main and connecting rod bearings should be replaced with new ones during the engine overhaul, the old bearings should be retained for close examination, as they may reveal valuable information about the condition of the engine.

2 Bearing failure occurs because of lack of lubrication, the presence of dirt or other foreign particles, overloading the engine and corrosion. Regardless of the cause of bearing failure, it must be corrected before the engine is reassembled to prevent it from happening again **(see illustration)**.

3 When examining the bearings, remove them from the engine block, the main bearing caps, the connecting rods and the rod caps and lay them out on a clean surface in the same general position as their location in the engine. This will enable you to match any bearing problems with the corresponding crankshaft journal.

4 Dirt and other foreign particles get into the engine in a variety of ways. If may be left in the engine during assembly, or it may pass through filters or the PCV system. It may get into the oil, and from there into the bearings. Metal chips from machining operations and normal engine wear are often present. Abrasives are sometimes left in engine components after reconditioning, especially when parts are not thoroughly cleaned using the proper cleaning methods. Whatever the source, these foreign objects often end up embedded in the soft bearing material and are easily recognized. Large particles will not embed in the bearing and will score or gouge the bearing and journal. The best prevention for this cause of bearing failure is to clean all parts thoroughly and keep everything spotlessly clean during engine assembly. Frequent and regular engine oil and filter changes are also recommended.

5 Lack of lubrication (or lubrication breakdown) has a number of interrelated causes. Excessive heat (which thins the oil), overloading (which squeezes the oil from the bearing face) and oil leakage or throw off (from excessive bearing clearances, worn oil pump or high engine speeds) all contribute to lubrication breakdown. Blocked oil passages, which usually are the result of misaligned oil holes in a bearing shell, will also oil starve a bearing and destroy it. When lack of lubrication is the cause of bearing failure, the bearing material is wiped or extruded from the steel backing of the bearing. Temperatures may increase to the point where the steel backing turns blue from overheating.

6 Driving habits can have a definite effect on bearing life. Full throttle, low speed operation in too high a gear (lugging the engine) puts very high loads on bearings, which tends to squeeze out the oil film. These loads cause the bearings to flex, which produces fine cracks in the bearing face (fatigue failure). Eventually the bearing material will loosen in pieces and tear away from the steel backing. Short trip driving leads to corrosion of bearings because insufficient engine heat is produced to drive off the condensed water and corrosive gases. These products collect in the engine oil, forming acid and sludge. As the oil is carried to the engine bearings, the acid attacks and corrodes the bearing material.

7 Incorrect bearing installation during engine assembly will lead to bearing failure as well. Tight fitting bearings leave insufficient bearing oil clearance and will result in oil starvation. Dirt or foreign particles trapped behind a bearing insert result in high spots on the bearing which lead to failure.

20 Engine overhaul - reassembly sequence

1 Before beginning engine reassembly, make sure you have all the necessary new parts, gaskets and seals as well as the following items on hand:

Common hand tools
1/2-inch drive torque wrench
Piston ring installation tool
Piston ring compressor
Short lengths of rubber or plastic hose to fit over connecting rod bolts
Plastigage
Feeler gauges
A fine-tooth file
New engine oil
Engine assembly lube or moly-base grease
RTV-type gasket sealant
Anaerobic-type gasket sealant
Thread locking compound

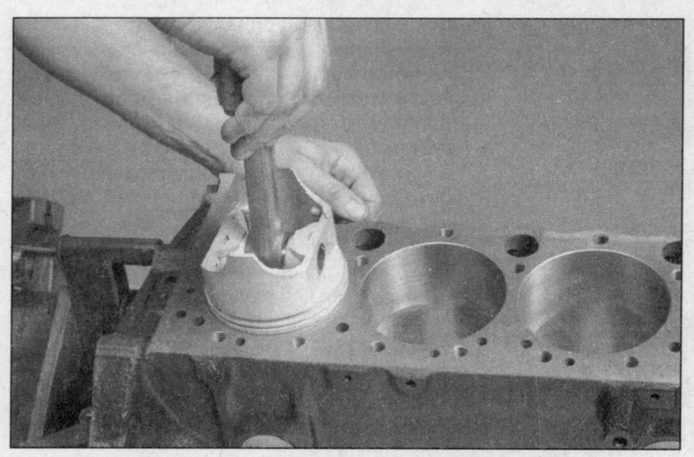

21.3 When checking piston ring end gap the piston must be square in the cylinder bore (this is done by pushing the ring down with the top of a piston as shown)

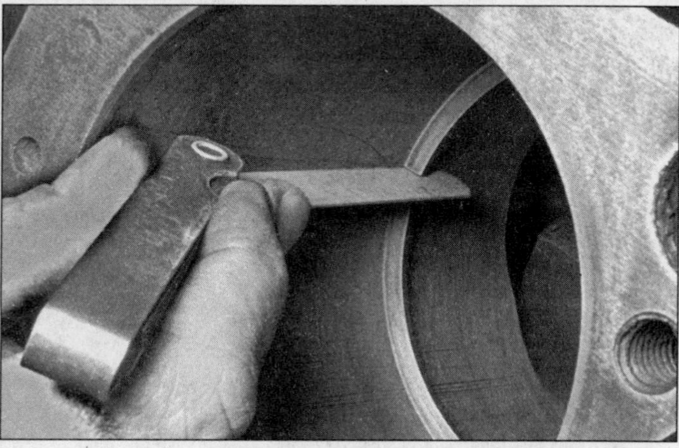

21.4 With the ring square in the cylinder, measure the ring end gap with a feeler gauge

21.5 If the end gap is too small, clamp a file in a vise and file the ring ends (from the outside in only) to enlarge the gap slightly

2 In order to save time and avoid problems, engine reassembly must be done in the following general order:

New camshaft bearings (must be done by an automotive machine shop)
Piston rings
Crankshaft and main bearings
Piston/connecting rod assemblies
Oil pump and oil strainer
Camshaft and lifters
Cylinder head(s), pushrods and rocker arms
Timing chain and sprockets (V6 and V8)
Timing gears (inline 6-cylinder)
Timing chain/gear cover
Oil pan
Intake and exhaust manifolds
Rocker arm cover(s)
Flywheel/driveplate

21 Piston rings - installation

Refer to illustrations 21.3, 21.4, 21.5, 21.9a, 21.9b and 21.12

1 Before installing the new piston rings, the ring end gaps must be checked. It is assumed that the piston ring side clearance has been checked and verified correct (Section 17).

2 Lay out the piston/connecting rod assemblies and the new ring sets so the ring sets will be matched with the same piston and cylinder during the end gap measurement and engine assembly.

3 Insert the top (number one) ring into the first cylinder and square it up with the cylinder walls by pushing it in with the top of the piston **(see illustration)**. The ring should be near the bottom of the cylinder, at the lower limit of ring travel.

4 To measure the end gap, slip feeler gauges between the ends of the ring until a gauge equal to the gap width is found **(see illustration)**. The feeler gauge should slide between the ring ends with a slight amount of drag. Compare the measurement to the Specifications. If the gap is larger or smaller than specified, double-check to make sure that you have the correct rings before proceeding.

5 If the gap is too small, it must be enlarged or the ring ends may come in contact with each other during engine operation, which can cause serious damage to the engine. The end gap can be increased by filing the ring ends very carefully with a fine file. Mount the file in a vise equipped with soft jaws, slip the ring over the file with the ends contacting the file face and slowly move the ring to remove material from the ends **(see illustration)**. When performing this operation, file only from the outside in.

6 Excess end gap is not critical unless it is greater than 0.040-inch. Again, double-check to make sure you have the correct rings for your engine.

7 Repeat the procedure for each ring that will be installed in the first cylinder and for each ring in the remaining cylinders. Remember to keep rings, pistons and cylinders matched up.

8 Once the ring end gaps have been checked/corrected, the rings can be installed on the pistons.

9 The oil control ring (lowest one on the piston) is installed first. It is composed of three separate components. Slip the spacer/expander into the groove **(see illustration)**. If an anti-rotation tang is used, make sure it is inserted into the drilled hole in the ring groove. Install the lower side rail. Do not use a piston ring installation tool on the oil ring side rails, as they may be damaged. Instead, place one end of the side rail into the groove between the spacer/expander and the ring land, hold it firmly in place and slide a finger around the piston while pushing the rail into the groove **(see illustration)**. Next, install the upper side rail in the same manner.

10 After the three oil ring components have been installed, check to make sure that both the upper and lower side rails can be turned smoothly in the ring groove.

11 The number two (middle) ring is installed next. It is stamped with a mark which must face up, toward the top of the piston. **Note:** *Always follow the instructions printed on the ring package or box - different manufacturers may require different approaches. Do not mix up the top and middle rings, as they have different cross sections.*

12 Use a piston ring installation tool and make sure that the identification mark is facing the top of the piston, then slip the ring

21.9a Installing the spacer/expander in the oil control ring groove

21.9b Install the oil ring side rails by hand - DO NOT use a ring installation tool

into the middle groove on the piston **(see illustration)**. Do not expand the ring any more than is necessary to slide it over the piston.

13 Install the number one (top) ring in the same manner. Make sure the mark is facing up. Be careful not to confuse the number one and number two rings.

14 Repeat the procedure for the remaining pistons and rings.

22 Crankshaft - installation and main bearing oil clearance check

Refer to illustrations 22.10 and 22.14

1 Crankshaft installation is the first step in engine reassembly. It is assumed at this point that the engine block and crankshaft have been cleaned, inspected and repaired or reconditioned.

2 Position the engine with the bottom facing up.

3 Remove the main bearing cap bolts and lift out the caps. Lay them out in the proper order to ensure that they are installed correctly.

4 If they are still in place, remove the old bearing inserts from the block and the main bearing caps. Wipe the main bearing surfaces of the block and caps with a clean, lint free cloth. They must be kept spotlessly clean.

5 Clean the back sides of the new main bearing inserts and lay one bearing half in each main bearing saddle in the block. Lay the other bearing half from each bearing set in the corresponding main bearing cap. Make sure the tab on the bearing insert fits into the recess in the block or cap. Also, the oil holes in the block must line up with the oil holes in the bearing insert. Do not hammer the bearing into place and do not nick or gouge the bearing faces. No lubrication should be used at this time.

6 The flanged thrust bearing must be installed in the rear cap and saddle.

7 Clean the faces of the bearings in the block and the crankshaft main bearing journals with a clean, lint free cloth. Check or clean the oil holes in the crankshaft, as any dirt here can go only one way - straight through the new bearings.

8 Once you are certain that the crankshaft is clean, carefully lay it in position (an assistant would be very helpful here) in the main bearings.

9 Before the crankshaft can be permanently installed, the main bearing oil clearance must be checked.

10 Trim several pieces of the appropriate size of Plastigage (they must be slightly shorter than the width of the main bearings) and place one piece on each crankshaft main bearing journal, parallel with the journal axis **(see illustration)**.

11 Clean the faces of the bearings in the caps and install the caps in their respective positions (do not mix them up) with the arrows pointing toward the front of the engine. Do not disturb the Plastigage.

12 Starting with the center main and working out toward the ends, tighten the main bearing cap bolts, in three steps, to the specified torque. Do not rotate the crankshaft at any time during this operation.

13 Remove the bolts and carefully lift off the main bearing caps. Keep them in order. Do not disturb the Plastigage or rotate the crankshaft. If any of the main bearing caps are difficult to remove, tap them gently from

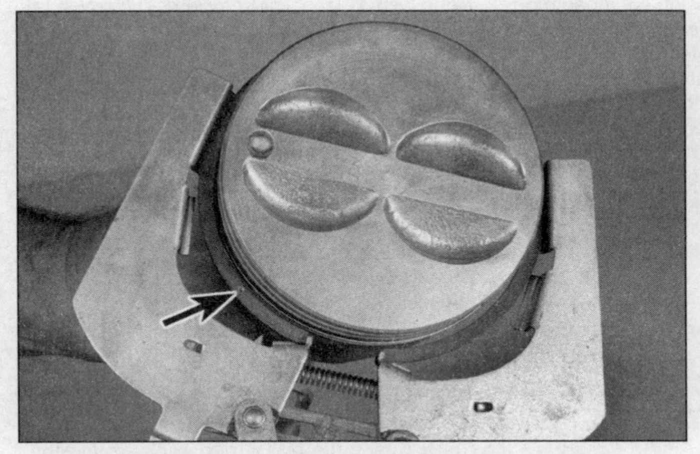

21.12 Use a ring expander to install the compression rings - make sure the mark (arrow) is facing up

22.10 Lay the Plastigage strips (arrow) on the main bearing journals, parallel to the crankshaft centerline

22.14 Compare the width of the crushed Plastigage to the scale on the container to determine the main bearing oil clearance (always take the measurement at the widest point of the Plastigage); be sure to use the correct scale - standard and metric scales are included

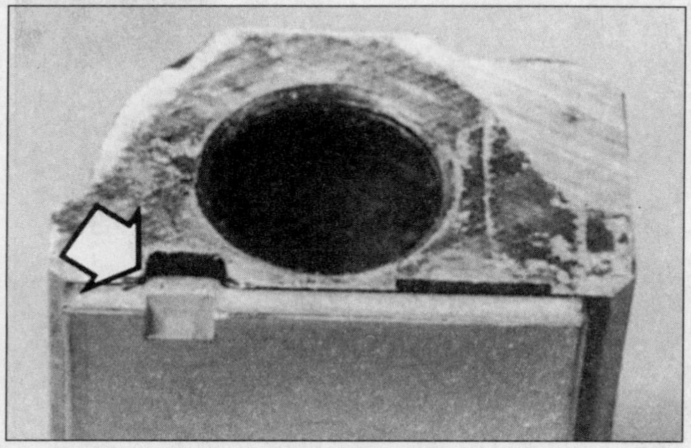

23.3 Make sure the bearing tang fits securely into the notch in the rod cap

ENGINE FRONT
A
B
A
C
D
C
PISTON TOP VIEW
24071-2C-27.5A HAYNES

23.5 Ring end gap positions

A *Oil ring rail gaps*
B *Second compression ring gap*
C *Oil ring spacer gap (position in-between marks)*
D *Top compression ring gap*

side-to-side with a soft-face hammer to loosen them.

14 Compare the width of the crushed Plastigage on each journal to the scale printed on the Plastigage container to obtain the main bearing oil clearance **(see illustration)**. Check the Specifications to make sure it is correct.

15 If the clearance is not as specified, the bearing inserts may be the wrong size (which means different ones will be required). Before deciding that different inserts are needed, make sure that no dirt or oil was between the bearing inserts and the caps or block when the clearance was measured. If the Plastigage was wider at one end than the other, the journal may be tapered (refer to Section 18).

16 Carefully scrape all traces of the Plastigage material off the main bearing journals and/or the bearing faces. Do not nick or scratch the bearing faces.

17 Carefully lift the crankshaft out of the engine. Clean the bearing faces in the block, then apply a thin, uniform layer of clean moly-base grease or engine assembly lube to each of the bearing surfaces. Be sure to coat the thrust faces as well as the journal face of the rear bearing.

18 Install the rear main oil seal (refer to Part A, Section 14, or Part B, Section 15, in this Chapter for main seal installation details).

19 Make sure the crankshaft journals are clean, then lay the crankshaft back in place in the block. Clean the faces of the bearings in the caps, then apply lubricant to them. Install the caps in their respective positions with the arrows pointing toward the front of the engine. Install the bolts.

20 Tighten all except the rear cap bolts (the one with the thrust bearing) to the specified torque. Work from the center out and approach the final torque in three steps. Tighten the rear cap bolts to 10-to-12 ft-lbs. Tap the ends of the crankshaft forward and backward with a lead or brass hammer to line up the main bearing and crankshaft thrust surfaces. Retighten all main bearing cap bolts to the specified torque, starting with the center main and working out toward the ends. On models with four bolt main bearing caps, tighten the inner bolts first, then the outer bolts, and be sure to note that different torque figures are supplied for the inner and outer bolts.

21 On manual transmission equipped models, install a new pilot bearing in the end of the crankshaft (see Chapter 8).

22 Rotate the crankshaft a number of times by hand to check for any obvious binding.

23 The final step is to check the crankshaft end play with a feeler gauge or a dial indicator as described in Section 13. The end play should be correct if the crankshaft thrust faces are not worn or damaged and new bearings have been installed.

23 Piston/connecting rod assembly - installation and rod bearing oil clearance check

Refer to illustrations 23.3, 23.5, 23.9, 23.11 and 23.13

1 Before installing the piston/connecting rod assemblies the cylinder walls must be perfectly clean, the top edge of each cylinder must be chamfered, and the crankshaft must be in place.

2 Remove the connecting rod cap from the end of the number one connecting rod. Remove the old bearing inserts and wipe the bearing surfaces of the connecting rod and cap with a clean, lint free cloth. They must be kept spotlessly clean.

3 Clean the back side of the new upper bearing half, then lay it in place in the connecting rod. Make sure the tang on the bearing fits into the recess in the rod **(see illustration)**. Do not hammer the bearing insert into place and be very careful not to nick or gouge the bearing face. Do not lubricate the bearing at this time.

4 Clean the backside of the other bearing insert and install it in the rod cap. Again, make sure the tab on the bearing fits into the recess in the cap, and do not apply any lubricant. It is critically important that the mating surfaces of the bearing and connecting rod are perfectly clean and oil free when they are assembled.

5 Space the piston ring gaps around the piston **(see illustration)**, then slip a section of plastic or rubber hose over each connecting

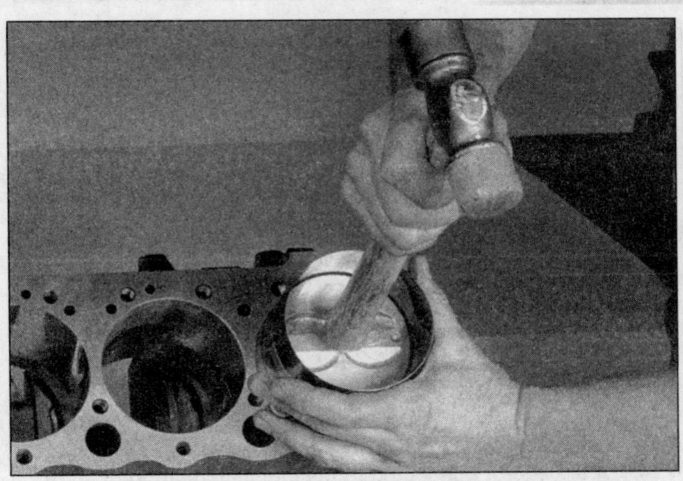

23.9 The piston can be driven (gently) into the cylinder bore with the end of a wooden hammer handle

23.11 Lay the Plastigage strips on each rod bearing journal, parallel to the crankshaft centerline

rod cap bolt.

6 Lubricate the piston and rings with clean engine oil and attach a piston ring compressor to the piston. Leave the skirt protruding about 1/4-inch to guide the piston into the cylinder. The rings must be compressed until they are flush with the piston.

7 Rotate the crankshaft until the number one connecting rod journal is at BDC (bottom dead center) and apply a coat of engine oil to the cylinder walls.

8 With the notch on top of the piston facing the front of the engine, gently insert the piston/connecting rod assembly into the number one cylinder bore and rest the bottom edge of the ring compressor on the engine block. Tap the top edge of the ring compressor to make sure it is contacting the block around its entire circumference.

9 Carefully tap on the top of the piston with the end of a wooden hammer handle while guiding the end of the connecting rod into place on the crankshaft journal **(see illustration).** The piston rings may try to pop out of the ring compressor just before entering the cylinder bore, so keep some downward pressure on the ring compressor. Work slowly, and if any resistance is felt as the piston enters the cylinder, stop immediately. Find out what is hanging up and fix it before proceeding. Do not, for any reason, force the piston into the cylinder, as you will break a ring and/or the piston.

10 Once the piston/connecting rod assembly is installed, the connecting rod bearing oil clearance must be checked before the rod cap is permanently bolted in place.

11 Cut a piece of the appropriate size Plastigage slightly shorter than the width of the connecting rod bearing and lay it in place on the number one connecting rod journal, parallel with the journal axis **(see illustration).**

12 Clean the connecting rod cap bearing face, remove the protective hoses from the connecting rod bolts and install the rod cap. Make sure the mating mark on the cap is on the same side as the mark on the connecting rod. Install the nuts and tighten them to the specified torque, working up to it in three steps. **Note:** *Use a thin-wall socket to avoid erroneous torque readings that can result if the socket becomes wedged between the rod cap and nut. Do not rotate the crankshaft at any time during this operation.*

13 Remove the rod cap, being very careful not to disturb the Plastigage. Compare the width of the crushed Plastigage to the scale printed on the Plastigage container to obtain the oil clearance **(see illustration).** Compare it to the Specifications to make sure the clearance is correct. If the clearance is not as specified, the bearing inserts may be the wrong size (which means different ones will be required). Before deciding that different inserts are needed, make sure that no dirt or oil was between the bearing inserts and the connecting rod or cap when the clearance was measured. Also, recheck the journal diameter. If the Plastigage was wider at one end than the other, the journal may be tapered (refer to Section 18).

14 Carefully scrape all traces of the Plastigage material off the rod journal and/or bearing face. Be very careful not to scratch the bearing - use your fingernail or a piece of hardwood. Make sure the bearing faces are perfectly clean, then apply a uniform layer of clean moly-base grease or engine assembly lube to both of them. You will have to push the piston into the cylinder to expose the face of the bearing insert in the connecting rod - be sure to slip the protective hoses over the rod bolts first.

15 Slide the connecting rod back into place on the journal, remove the protective hoses from the rod cap bolts, install the rod cap and tighten the nuts to the specified torque. Again, work up to the torque in three steps.

16 Repeat the entire procedure for the remaining piston/connecting rod assemblies. Keep the backsides of the bearing inserts and the inside of the connecting rod and cap perfectly clean when assembling them. Make sure you have the correct piston for the cylinder and that the notch on the piston faces to the front of the engine when the piston is installed. Remember, use plenty of oil to lubricate the piston before installing the ring compressor. Also, when installing the rod caps for the final time, be sure to lubricate the bearing faces adequately.

23.13 Measuring the width of the Plastigage to determine the rod bearing oil clearance (be sure to use the correct scale - standard and metric scales are included)

17 After all the piston/connecting rod assemblies have been properly installed, rotate the crankshaft a number of times by hand to check for any obvious binding.

18 As a final step, the connecting rod end play must be checked. Refer to Section 12 for this procedure. Compare the measured end play to the Specifications to make sure it is correct. If it was correct before disassembly and the original crankshaft and rods were reinstalled, it should still be right. If new rods or a new crankshaft were installed, the end play may be too small. If so, the rods will have to be removed and taken to an automotive machine shop for resizing.

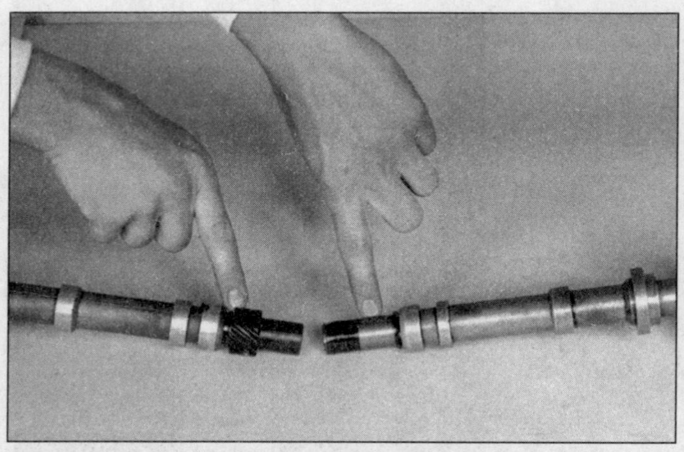

24.3 The pre-oil distributor (right) has the gear ground off and the advance weights (if equipped) removed

24.5 A drill motor connected to the modified distributor shaft drives the oil pump - make sure it turns clockwise as viewed from above

24 Pre-oiling engine after overhaul

Refer to illustrations 24.3, 24.5 and 24.6

Note: *This procedure applies only to V8 engines.*

1 After an overhaul it is a good idea to pre-oil the engine before it is installed in the vehicle and started for the first time. Pre-oiling will reveal any problems with the lubrication system at a time when corrections can be made easily and will prevent major engine damage. It will also allow the internal engine parts to be lubricated thoroughly in the normal fashion without the heavy loads associated with combustion placed on them.

2 The engine should be completely assembled with the exception of the distributor and rocker arm covers. The oil filter and oil pressure sending unit must be in place and the specified amount of oil must be in the crankcase (see Chapter 1).

3 A modified Chevrolet distributor will be needed for this procedure - a junkyard should be able to supply one for a reasonable price. In order to function as a pre-oil tool, the distributor must have the gear on the lower end of the shaft ground off **(see illustration)** and, if equipped, the advance weights on the upper end of the shaft removed.

4 Install the pre-oil distributor in place of the original distributor and make sure the lower end of the shaft mates with the upper end of the oil pump driveshaft. Turn the distributor shaft until they are aligned and the distributor body seats on the block. Install the distributor hold-down clamp and bolt.

5 Mount the upper end of the shaft in the chuck of an electric drill and use the drill to turn the pre-oil distributor shaft, which will drive the oil pump and circulate the oil throughout the engine **(see illustration)**. **Note:** *The drill must turn in a clockwise direction.*

6 It may take two or three minutes, but oil should soon start to flow out of all the rocker arm holes, indicating that the oil pump is working properly **(see illustration)**. Let the oil circulate for several seconds, then shut off the drill motor.

7 Remove the pre-oil distributor, then install the rocker arm covers. The distributor should be installed after the engine is installed in the vehicle, so plug the hole with a clean cloth.

25 Initial start-up and break-in after overhaul

1 Once the engine has been installed in the vehicle, double-check the engine oil and coolant levels.

2 With the spark plugs out of the engine and the distributor disabled by disconnecting the BAT connector (coil-in-cap models) or grounding the coil wire (separate coil models), crank the engine until oil pressure registers on the gauge.

3 Install the spark plugs, hook up the plug wires and reconnect the distributor.

4 Start the engine. It may take a few moments for the gasoline to reach the carburetor or fuel injection unit, but the engine should start without a great deal of effort.

5 After the engine starts, it should be allowed to warm up to normal operating temperature. While the engine is warming up, make a thorough check for oil and coolant leaks.

6 Shut the engine off and recheck the engine oil and coolant levels.

7 Drive the vehicle to an area with minimum traffic, accelerate at full throttle from 30 to 50 mph, then allow the vehicle to slow to 30 mph with the throttle closed. Repeat the procedure 10 or 12 times. This will load the piston rings and cause them to seat properly against the cylinder walls. Check again for oil and coolant leaks.

8 Drive the vehicle gently for the first 500 miles (no sustained high speeds) and keep a constant check on the oil level. It is not unusual for an engine to use oil during the break-in period.

9 At approximately 500 to 600 miles, change the oil and filter.

10 For the next few hundred miles, drive the vehicle normally. Do not pamper it or abuse it.

11 After 2000 miles, change the oil and filter again and consider the engine fully broken in.

24.6 Oil, assembly lube or grease will begin to flow from all of the rocker arm holes if the oil pump and lubrication system are functioning properly

Chapter 3
Cooling, heating and air conditioning systems

Contents

	Section
Air conditioning accumulator - removal and installation	14
Air conditioning compressor - removal and installation	12
Air conditioning condenser - removal and installation	13
Air conditioning system - check and maintenance	11
Antifreeze - general information	2
Coolant level check	See Chapter 1
Coolant reservoir - removal and installation	8
Cooling fan and fan clutch - removal and installation	5
Cooling system check	See Chapter 1
Cooling system servicing (draining, flushing and refilling)	See Chapter 1
Drivebelt check and adjustment	See Chapter 1
Engine oil cooler - general information	15
General information	1
Heater components (1967 through 1970 models) - removal and installation	9
Heater components (1971 through 1987 models) - removal and installation	10
Radiator - removal and installation	3
Thermostat - removal and installation	4
Underhood hose check and replacement	See Chapter 1
Water pump - check	6
Water pump - removal and installation	7

Specifications

Radiator cap rating	15 psi
Thermostat rating	195° F (91° C)
Coolant capacity*	
1967	
230 cu in	11.0 quarts
250 cu in	12.5 quarts
283 cu in, 327 cu in	17.0 quarts
1968 through 1970	
250 cu in	12.5 quarts
292 cu in	13.0 quarts
307 cu in, 327 cu in	18.0 quarts
350 cu in (C10-30)	18.5 quarts
350 cu in (K10-20)	19.0 quarts
396 cu in	24.0 quarts
1971 through 1974	
250 cu in	12.5 quarts
292 cu in	13.0 quarts
307 cu in	15.8 quarts
350 cu in	16.2 quarts
402 cu in	18.5 quarts
454 cu in	18.5 quarts
1975 through 1987	
250 cu in	15.0 quarts
292 cu in	15.0 quarts
305 cu in	18.0 quarts
350 cu in	18.0 quarts
400 cu in	20.0 quarts
454 cu in	20.0 quarts

Capacities approximate

Torque specifications	Ft-lbs
Thermostat housing bolts	20 to 30
Water pump bolts	
6-cylinder inline engine	15 to 20
V6 and V8 engines	30

Component location

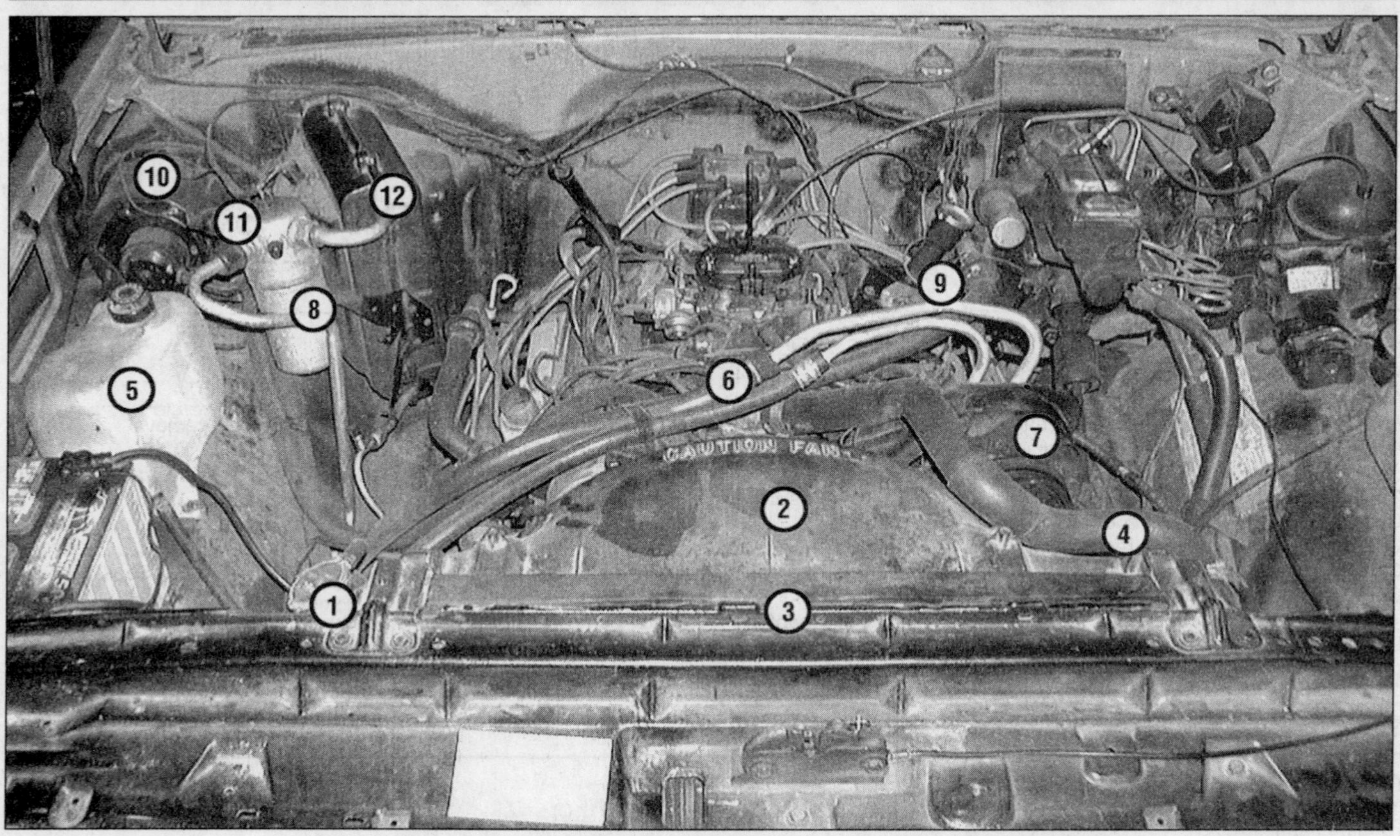

Typical V8 engine cooling, heating and air conditioning component location

1. *Radiator cap*
2. *Fan shroud*
3. *Radiator*
4. *Upper radiator hose*
5. *Coolant reservoir*
6. *Thermostat housing*
7. *Air conditioning compressor*
8. *Air conditioning accumulator*
9. *Air conditioning compressor refrigerant lines*
10. *Blower motor*
11. *Air conditioning pressure switch*
12. *Evaporator/heater core case*

1 General information

The cooling system consists of a cross-flow radiator, a thermostat and a crankshaft pulley-driven water pump. On some later models an aluminum radiator with plastic tanks is used.

The radiator cooling fan is mounted on the front of the water pump and, on later models, incorporates a fluid drive fan clutch, saving horsepower and reducing noise. On most models a fan shroud is mounted on the rear of the radiator.

The system is pressurized by a spring-loaded radiator cap, which increases the boiling point of the coolant. If the coolant temperature goes above this increased boiling point, the extra pressure in the system forces the radiator cap valve off its seat and exposes the overflow pipe. On all later models the overflow pipe leads to a coolant recovery system. This consists of a plastic reservoir into which the coolant which normally escapes down the overflow pipe is retained. When the engine cools, the excess coolant is drawn back into the radiator, maintaining the system at full capacity. This is a continuous process and provided the level in the reservoir is correctly maintained, it is not necessary to add coolant to the radiator.

Coolant in the right side of the radiator circulates up the lower radiator hose to the water pump, where it is forced through the water passages in the cylinder block. The coolant then travels up into the cylinder head, circulates around the combustion chambers and valve seats, travels out of the cylinder head past the open thermostat into the upper radiator hose on the left side of the cross-flow type radiator.

When the engine is cold the thermostat restricts the circulation of coolant to the engine. The thermostat is located in the front of the cylinder head on 6-cylinder inline engines and in the front of the intake manifold on V6 and V8 models. When the minimum operating temperature is reached, the thermostat begins to open, allowing coolant to return to the radiator.

Automatic transmission-equipped models have a cooler element incorporated into the radiator to cool the transmission fluid.

The heating system works by directing air through the heater core mounted in the dash and then to the interior of the vehicle by a system of ducts. Temperature is controlled by mixing heated air with fresh air, using a system of flapper doors in the ducts, and a heater motor. Some models are equipped with a separate auxiliary heating system for heating the rear seat area.

Air conditioning is an optional accessory, consisting of an evaporator core located under the dash, a condenser in front of the radiator, an accumulator in the engine compartment and a belt-driven compressor mounted at the front of the engine.

3.5 Typical radiator shroud mounting details

2 Antifreeze - general information

Warning: *Do not allow antifreeze to come in contact with your skin or painted surfaces of the vehicle. Flush contacted areas immediately with plenty of water. Do not store new coolant or leave old coolant lying around where it is easily accessible to children and pets - they are attracted by its sweet smell. Ingestion of even a small amount can be fatal. Wipe up garage floor and drip pan coolant spills as soon as they occur. Keep antifreeze containers covered and repair leaks in your cooling system immediately.*

The cooling system should be filled with a water/ethylene glycol based antifreeze solution which will prevent freezing down to at least -20° F (even lower in cold climates). It also provides protection against corrosion and increases the coolant boiling point.

The cooling system should be drained, flushed and refilled at least every other year (see Chapter 1). The use of antifreeze solutions for periods of longer than two years is likely to cause damage and encourage the formation of rust and scale in the system.

Before adding antifreeze to the system, check all hose connections. Antifreeze can leak through very minute openings.

The exact mixture of antifreeze to water which you should use depends on the relative weather conditions. The mixture should contain at least 50 percent antifreeze, but should never contain more than 70 percent antifreeze.

3 Radiator - removal and installation

Refer to illustrations 3.5 and 3.6

Warning: *The engine must be completely cool when this procedure is done.*

1 Disconnect the negative cable at the battery. Place the cable out of the way so it cannot accidentally come in contact with the negative terminal of the battery, as this would once again allow power into the electrical system of the vehicle.

2 Drain the cooling system as described in Chapter 1, then disconnect the upper and lower radiator hoses from the radiator.

3 If equipped with an automatic transmission, remove the transmission cooler lines from the side of the radiator - be careful not to twist the lines or damage the fittings. It is a good idea to use a flare nut wrench rather than an open end wrench. Plug the ends of the disconnected lines to prevent leakage and stop dirt from entering the system.

4 On vehicles equipped with a coolant reservoir, disconnect the overflow hose from the radiator.

5 If equipped, remove the fan shroud **(see illustration)**.

6 Remove the radiator mounting bolts or clamps and lift the radiator from the engine compartment, taking care not to contact the fan blades **(see illustration)**.

7 Prior to installation of the radiator, replace any damaged hose clamps and radiator hoses.

8 Radiator installation is the reverse of removal. When installing the radiator, make sure that the radiator seats properly in the lower saddles and that the upper clamps secure, but do not compress, the radiator core.

9 After installation, check the automatic transmission fluid level.

4 Thermostat - removal and installation

Refer to illustrations 4.4 and 4.7

Warning: *The engine must be completely cool when this procedure is done.*

1 Symptoms of a faulty thermostat are overheating, failure to warm up or slow warm-up and an inefficient heater.

2 Drain coolant (about 3 quarts) from the radiator, until the coolant level is below the thermostat housing (See Chapter 1).

3 Disconnect the upper radiator hose from the thermostat housing cover, which is located at the forward end of the cylinder head (inline engine) or intake manifold (V6 and V8 engines).

4 Remove the bolts and lift the cover off **(see illustration)**. It may be necessary to tap the cover with a soft-face hammer to break the gasket seal.

5 Note how it's installed, then remove the thermostat.

6 Use a scraper or putty knife to remove all traces of old gasket material and sealant from the mating surfaces. Make sure that no gasket material falls into the coolant passages; it is a good idea to stuff a rag in the passage. Wipe the mating surfaces with a rag saturated with lacquer thinner or acetone.

3.6 Typical radiator mounting details

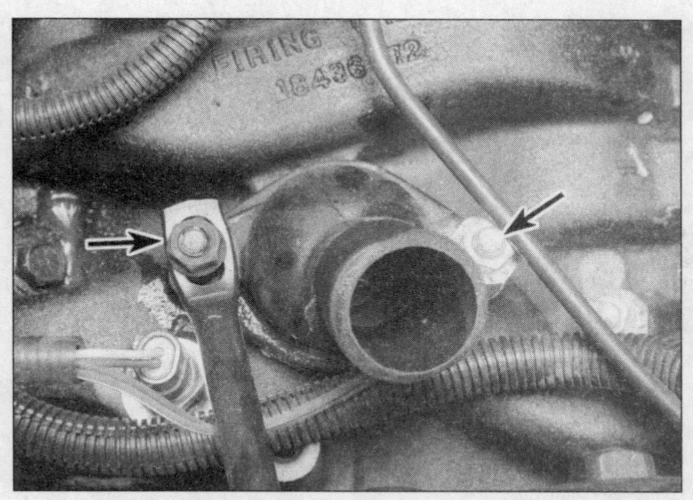

4.4 Remove the nuts/bolts (arrows) and lift off the cover (the shape of the cover varies)

7 Apply a thin layer of RTV sealant to the gasket mating surfaces of the housing and cover, then install the new thermostat in the engine. Make sure the correct end faces up - the spring is directed down into the housing manifold **(see illustration)**.

8 Position a new gasket on the housing and make sure the gasket holes line up with the bolt holes in the housing.

9 Carefully position the cover on the housing and install the bolts. Tighten them to the specified torque - do not overtighten them or the cover may be distorted.

10 Reattach the radiator hose to the cover and tighten the clamp - now may be a good time to check and replace the hoses and clamps (see Chapter 1).

11 Refer to Chapter 1 and refill the system, then run the engine and check carefully for leaks.

4.7 Lift out the thermostat - note the location of the spring

5 Cooling fan and fan clutch - removal and installation

Refer to illustration 5.4

1 The cooling fan should be replaced when the blades are damaged or bent. The fluid drive fan clutch is disengaged when the engine is cold, or at high engine speeds, when the silicone fluid inside the clutch is contained in the reservoir section by centrifugal action. Symptoms of failure of the fan clutch are continuous noisy operation, looseness leading to vibration and evidence of silicone fluid leaks.

2 Remove the fan shroud (if equipped).

3 Loosen the alternator mounting and adjustment bolts, push the alternator in towards the engine and loosen the drivebelt. Slip the drivebelt off the water pump pulley.

4 Unbolt the fan assembly and detach it from the water pump **(see illustration)**.

5 The fan clutch can be unbolted from the fan blade assembly for replacement.

6 Installation is the reverse of removal.

6 Water pump - check

Refer to illustration 6.3

1 Water pump failure can cause overheating of and serious damage to the engine. There are three ways to check the operation of the water pump while it is installed on the engine. If any one of the three following quick checks indicates water pump problems, it should be replaced immediately.

2 Start the engine and warm it up to normal operating temperature. Squeeze the upper radiator hose. If the water pump is working properly, you should feel a pressure surge as the hose is released.

3 A seal protects the water pump impeller shaft bearing from contamination by engine coolant. If this seal fails, a weep hole in the water pump snout will leak coolant **(see illustration)** (an inspection mirror can be used to look at the underside of the pump if the hole isn't on top). If the weep hole is leaking, shaft bearing failure will follow. Replace the water pump immediately.

4 Besides contamination by coolant after a seal failure, the water pump impeller shaft bearing can also be prematurely worn out by an improperly tensioned drivebelt. When the bearing wears out, it emits a high pitched squealing sound. If such a noise is coming from the water pump during engine operation, the shaft bearing has failed. Replace the water pump immediately.

5 To identify excessive bearing wear before the bearing actually fails, grasp the water pump pulley and try to force it up-and-down or from side-to-side. If the pulley can be moved either horizontally or vertically, the bearing is nearing the end of its service life. Replace the water pump.

7 Water pump - removal and installation

Refer to illustrations 7.7a, 7.7b, 7.7c, 7.8 and 7.9

Removal

1 Disconnect the negative cable at the

5.4 Cooling fan and clutch mounting details

6.3 The water pump weep hole is located on the pump snout; some pumps have two holes

7.7a 6-cylinder inline engine water pump mounting bolt locations

7.7b V6 and V8 engine water pump mounting bolt locations

battery. Place the cable out of the way so it cannot accidentally come in contact with the negative terminal of the battery, as this would once again allow power into the electrical system of the vehicle.

2 Drain the coolant (Chapter 1).

3 Remove any drivebelts associated with the water pump pulley (Chapter 1).

4 Loosen the hose clamps for the lower radiator hose and heater hose and disconnect the two hoses from the water pump.

5 Remove the fan assembly (Section 5) and water pump pulley.

6 On some models the water pump bolts may also attach the alternator or power steering brackets to the front of the engine. Where necessary, loosen the components so that the brackets can be moved aside when the water pump bolts are removed.

7 Remove the mounting bolts from the water pump and detach it from the block **(see illustrations)**. It may be necessary to grasp the pump securely and rock it back-and-forth to break the gasket seal **(see illustration)**.

Installation

8 If you are replacing the water pump, remove the heater hose extension pipe and any other components which will need to be installed on the new pump **(see illustration)**.

7.7c Grasp the water pump and rock it back-and-forth to break the gasket seal

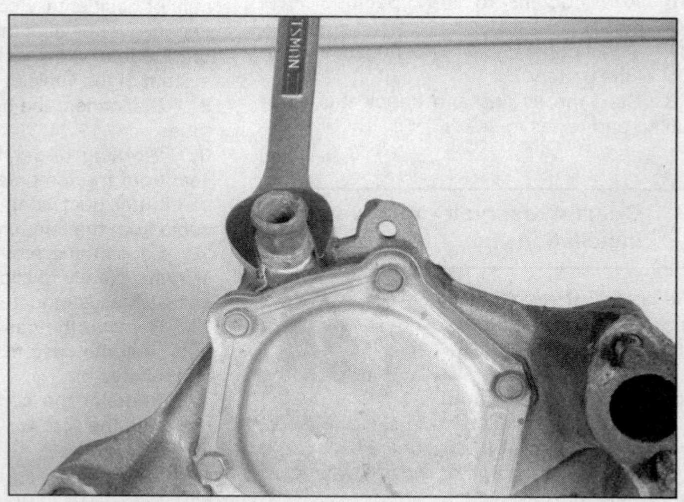

7.8 Use a wrench to carefully unscrew the heater hose pipe so it can be transferred to the new pump

7.9 Be sure to remove the old gasket(s) and clean the mating surface(s) on the block before installing the new pump

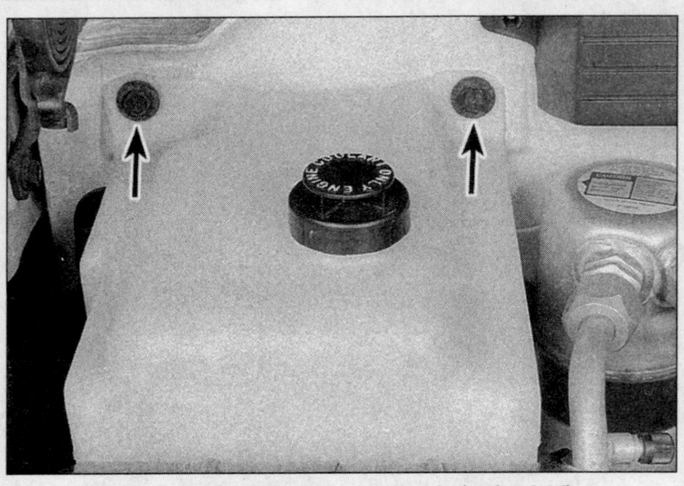

8.2 Coolant reservoir mounting bolts (typical)

Use Teflon sealing tape to prevent leakage.

9 Clean the sealing surfaces on both the block and the water pump **(see illustration)**. Wipe the mating surfaces with a rag saturated with lacquer thinner or acetone.

10 Apply a thin layer of RTV sealant to the block mounting surfaces and install a new water pump gasket(s).

11 Apply a thin layer of RTV sealant to the water pump mounting surfaces.

12 Place the water pump in position and install the bolts and studs finger tight. Use caution to ensure that the gaskets do not slip out of position. Remember to replace any mounting brackets secured by the water pump mounting bolts. Tighten the bolts to the specified torque.

13 Install the water pump pulley and fan assembly and tighten the pulley bolts securely.

14 Install the lower radiator hose, heater hose and hose clamps. Tighten the hose clamps securely.

15 Install the drivebelts (Chapter 1).

16 Add coolant to the specified level (Chapter 1).

17 Connect the cable to the negative terminal of the battery.

18 Start the engine and check the water pump and hoses for leaks.

8 Coolant reservoir - removal and installation

Refer to illustration 8.2

1 Remove the coolant overflow hose from the reservoir.

2 Remove the screws and detach the reservoir **(see illustration)**.

3 Prior to installation make sure the reservoir is clean and free of debris which could be drawn into the radiator (wash it with soap and water if necessary).

4 Installation is the reverse of removal.

9 Heater components (1967 through 1970 models) - removal and installation

Blower assembly

1 Raise the hood all the way.

2 Scribe around the right-hand hinge and then remove the hinge. Use a piece of wood or other support to hold the hood in place while the hinge is removed.

3 Disconnect the blower wire.

4 Mark the motor flange position in relation to the blower case.

5 Remove the blower mounting screws and remove the blower by carefully prying the flange from the casing.

6 Unscrew the nut which holds the fan to the blower motor shaft. Remove the fan.

Heater core

7 Disconnect the negative cable at the battery. Place the cable out of the way so it cannot accidentally come in contact with the negative terminal of the battery, as this would once again allow power into the electrical system of the vehicle.

8 Disconnect the heater hoses at the core tubes.

9 Working under the dash, remove the seal from the temperature door cable at the distributor duct adapter and disconnect the cable from the temperature door.

10 Loosen and move the right front fender skirt sufficiently to allow access to the heater case, which is mounted on the firewall.

11 Remove the case retaining screws and nuts. Pull the case from the mounting studs and remove it.

12 Remove the core from the case after removing the screws.

Controls

13 The control panel can be unbolted from the lower edge of the dash.

14 Individual control cables can be removed from the flap valves and doors.

15 Installation of heater control components is the reverse of removal.

10 Heater components (1971 through 1987 models) - removal and installation

Refer to illustrations 10.4, 10.11, 10.12, 10.15 and 10.17

1 The heater assembly is attached to the dash panel on the right side of the vehicle.

2 The blower and air inlet assembly and coolant hoses are located on the forward side of the firewall, while the heater core and air distributor duct are on the passenger side.

Blower assembly

3 Disconnect the blower lead.

4 Remove the blower mounting screws and withdraw the motor and fan **(see illustration)**. It may be necessary to cut the flange sealing strip to release the blower motor from the firewall.

5 The fan can be removed from the motor shaft after the nut has been unscrewed.

Heater core and air distributor

6 Disconnect the negative cable at the battery.

7 Place a suitable container under the heater core, disconnect the coolant hoses from the core and plug them or fasten the ends up as high as possible so the coolant does not run out. Allow the coolant in the heater core to drain into the container.

8 Remove the nuts from the distributor duct studs which project into the engine compartment.

9 Remove the glovebox and door assembly.

10 Disconnect the air, defrost and temperature door cables.

11 Remove the floor outlet and then

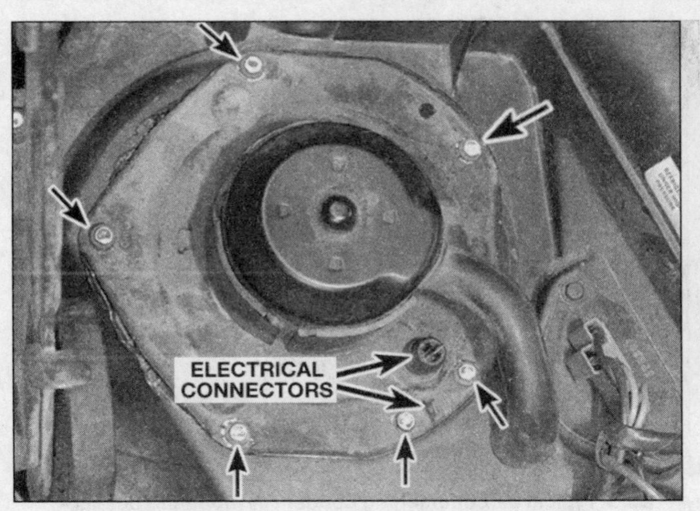

10.4 Later model blower motor installation details

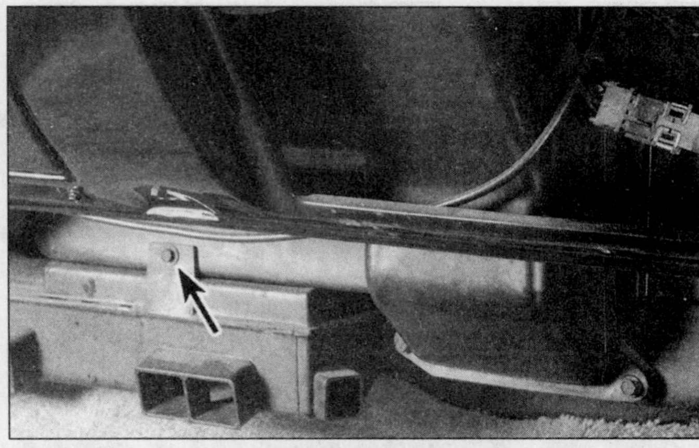
10.11 Remove the floor outlet retaining screw (arrow) and remove the outlet

10.12 Remove the heater air distributor mounting screws and lower it to the floor

10.15 Remove the heater core retaining straps (arrows)

remove the defroster duct to distributor duct screw **(see illustration)**.

12 Extract the screws retaining the heater distributor to the firewall **(see illustration)**.

13 Pull the heater assembly to the rear for access and disconnect the wiring harness.

14 Remove the heater air distributor.

15 Remove the core restraining straps and remove the core **(see illustration)**.

Controls

16 Withdraw the instrument panel bezel for access to the heater controls.

17 Disconnect the cables and unplug the blower switch connector **(see illustration)**.

18 Withdraw the control assembly through the opening above the controls.

19 Heater control component installation is the reverse of removal.

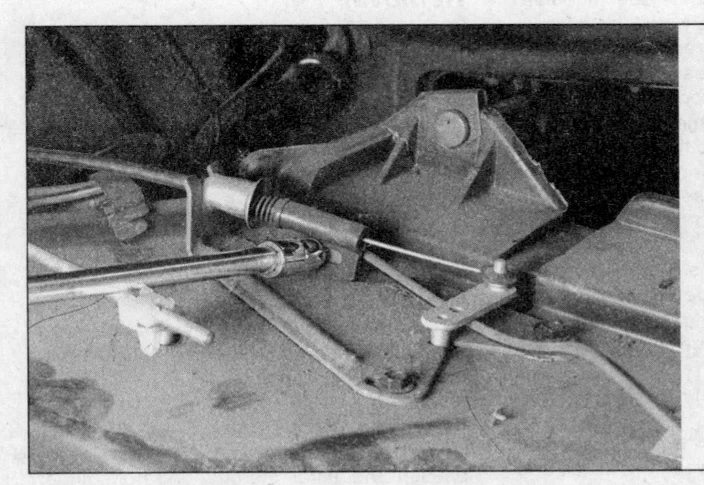
10.17 Remove the defroster and temperature door cable

11 Air conditioning system - check and maintenance

Refer to illustration 11.9

Warning: *The air conditioning system is pressurized at the factory and requires special equipment for service and repair. Any work should be left to your dealer servicing department or an automotive air conditioning shop. Do not, under any circumstances, disconnect the air conditioning hoses while the system is under pressure.*

1 The following maintenance steps should be performed on a regular basis to ensure that the air conditioner continues to operate at peak efficiency.

a) *Check the tension of the drivebelt and adjust if necessary (Chapter 1).*

b) *Check the condition of the hoses. Look for cracks, hardening and deterioration.* **Warning:** *Do not replace A/C hoses until*

11.9 **Feel the accumulator outlet and evaporator inlet with the air conditioning running to check the level of freon charge**

13.3 Typical air conditioning condenser mounting details - remove the bolts from the center support (A), then remove the condenser mounting screws (B) and carefully guide the condenser out

the system has been discharged by a dealer or air conditioning shop.

c) *Check the fins of the condenser for leaves, bugs and other foreign material. A soft brush and compressed air can be used to remove them.*

d) *Check the wire harness for correct routing, broken wires, damaged insulation, etc. Make sure the harness connections are clean and tight.*

e) *Maintain the correct refrigerant charge.*

2 The system should be run for about 10 minutes at least once a month. This is particularly important during the winter months because long-term non-use can cause hardening of the internal seals.

3 Because of the complexity of the air conditioning system and the special equipment required to effectively work on it, accurate troubleshooting and repair of the system should be left to a professional mechanic. One probable cause for poor cooling that can be determined by the home mechanic is low refrigerant charge. Should the system lose its cooling ability, the following procedure will help you pinpoint the cause.

4 Warm the engine to normal operating temperature.

5 The hood and doors should be open.

6 Set the control mode selector lever to the Norm position.

7 Set the temperature selector lever to the Cold position.

8 Set the blower control selector knob to the H position.

9 With the compressor engaged, feel the evaporator inlet pipe between the orifice and the evaporator and put your other hand on the surface of the accumulator outlet **(see illustration)**.

10 If both surfaces feel about the same temperature and if both feel a little cooler than the ambient temperature, the refrigerant level is probably okay. The problem is elsewhere.

11 If the inlet pipe has frost accumulation or feels cooler than the accumulator surface, the refrigerant charge is low.

12 If a low refrigerant charge is suspected, take your vehicle to a dealer or automotive air conditioning shop for service by a certified air conditioning technician.

12 Air conditioning compressor - removal and installation

Warning: *Have the air conditioning system discharged by a dealer service department or an air conditioning shop before beginning this procedure.*

Removal

1 Disconnect the negative cable at the battery. Place the cable out of the way so it cannot accidentally come in contact with the negative terminal of the battery, as this would once again allow power into the electrical system of the vehicle.

2 Disconnect the electrical connector from the A/C compressor and remove the line fitting bolt from the back side of the compressor.

3 Remove the drivebelt (refer to Chapter 1).

4 Remove the compressor-to-bracket bolts and nuts and lift the compressor from the engine compartment.

Installation

5 Place the compressor in position on the bracket and install the nuts and bolts finger tight. Once all the compressor mounting nuts and bolts are installed, tighten them securely.

6 Install the drivebelt (Chapter 1).

7 Connect the electrical connector to the compressor. Install the line fitting bolt to the compressor, using new O-rings lubricated with clean refrigeration oil, and tighten it securely.

8 Connect the cable to the negative terminal of the battery.

9 Take the vehicle to a dealer service department or an air conditioning shop and have the system evacuated and recharged.

13 Air conditioning condenser - removal and installation

Refer to illustration 13.3

Warning: *Have the air conditioning system discharged by a dealer service department or an air conditioning shop before beginning this procedure.*

1 Disconnect the negative cable at the battery. Place the cable out of the way so it cannot accidentally come in contact with the negative terminal of the battery, as this would once again allow power into the electrical system of the vehicle.

2 Remove the grille assembly (Chapter 11).

3 Remove the grille center and left upper fender support **(see illustration)**.

4 Disconnect the A/C line fittings. Be sure to use a back-up wrench to prevent damage to the fittings.

5 Remove the condenser-to-radiator support screws and bend the left grille support for clearance.

6 Pull the condenser forward and lower it from the vehicle.

7 Installation is the reverse of removal.

14 Air conditioning accumulator - removal and installation

Warning: *Have the air conditioning system discharged by a dealer service department or an air conditioning shop before beginning this procedure.*

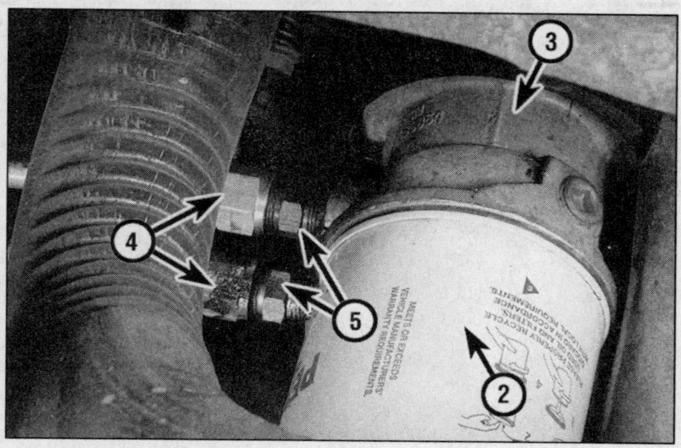

15.1 Oil cooler and related components and lines

1 Oil cooler
2 Oil filter
3 Oil filter adapter
4 Oil cooler metal lines
5 Oil cooler fittings

Removal

1 Disconnect the negative cable at the battery. Place the cable out of the way so it cannot accidentally come in contact with the negative terminal of the battery, as this would once again allow power into the electrical system of the vehicle.

2 Disconnect the accumulator inlet and outlet lines. Cap or plug the open lines immediately.

3 Remove the accumulator mounting bolts and detach the accumulator assembly.

Installation

4 Check the amount of oil in the old accumulator and add this amount plus two ounces of fresh 525 viscosity refrigerant oil to the new accumulator.

5 Place the new accumulator into position, install the mounting bolts and tighten them securely.

6 Install the inlet and outlet lines, using clean 525 viscosity refrigerant oil on the new O-rings.

7 Connect the cable to the negative terminal of the battery.

8 Have the system evacuated and recharged by a dealer service department or an air conditioning shop.

5 Engine oil cooler - general information

Refer to illustration 15.1

The engine oil cooler consists of an adapter mounted between the oil filter and engine block, the cooler (which is a radiator for engine oil) mounted in the front of the vehicle, and the connecting lines and hoses **(see illustration)**.

At regular intervals, check the hoses, lines and oil cooler for leaks and damage. Make sure everything is mounted securely and check all fittings to see if they're tight.

Notes

Chapter 4
Fuel and exhaust systems

Contents

	Section
Air filter and PCV filter replacement	See Chapter 1
Carburetor choke check	See Chapter 1
Carburetors - general information	4
Carburetor - removal and installation	5
Carburetor - servicing	6
Carburetor/throttle body mounting nut torque check	See Chapter 1
Carburetor (Rochester 1ME) - adjustments	10
Carburetor (Rochester 1MV) - adjustments	9
Carburetor (Rochester 2G series) - adjustments	11
Carburetor (Rochester 2SE/E2SE) - adjustments	15
Carburetor (Rochester 4MV) - adjustments	12
Carburetor (Rochester E4ME/E4MC) - adjustments	17
Carburetor (Rochester M and MV) - adjustments	8
Carburetor (Rochester M2M) - adjustments	16
Carburetor (Rochester M4MC/M4MCA) - adjustments	13
Carburetor (Rochester M4ME) - adjustments	14
Exhaust system check	See Chapter 1
Exhaust system - component replacement	23
Fuel filter replacement	See Chapter 1
Fuel injection system - general information	18
Fuel lines - check and replacement	7
Fuel pump - check	2
Fuel pump - removal and installation	3
Fuel system check	See Chapter 1
Fuel tank - removal and installation	21
Fuel tank - cleaning and repair	22
General information	1
Throttle Body Injection (TBI) unit - disassembly and reassembly	19
Throttle linkage - replacement	20

Specifications

Note: *The following specifications were compiled from the latest information available. If the specifications on the Vehicle Emissions Control Information label differ from those listed here, assume that the label is correct. Also, if the specifications in the rebuild kit for your particular carburetor differ from those listed here, assume that the specifications in the rebuild kit are correct.*

Fuel pump

Pressure	
6-cylinder inline engine	4.5 to 6 psi
V6 and V8 engines	5 to 9 psi
Volume (all)	1/2-pint or more in 15 seconds

Component location

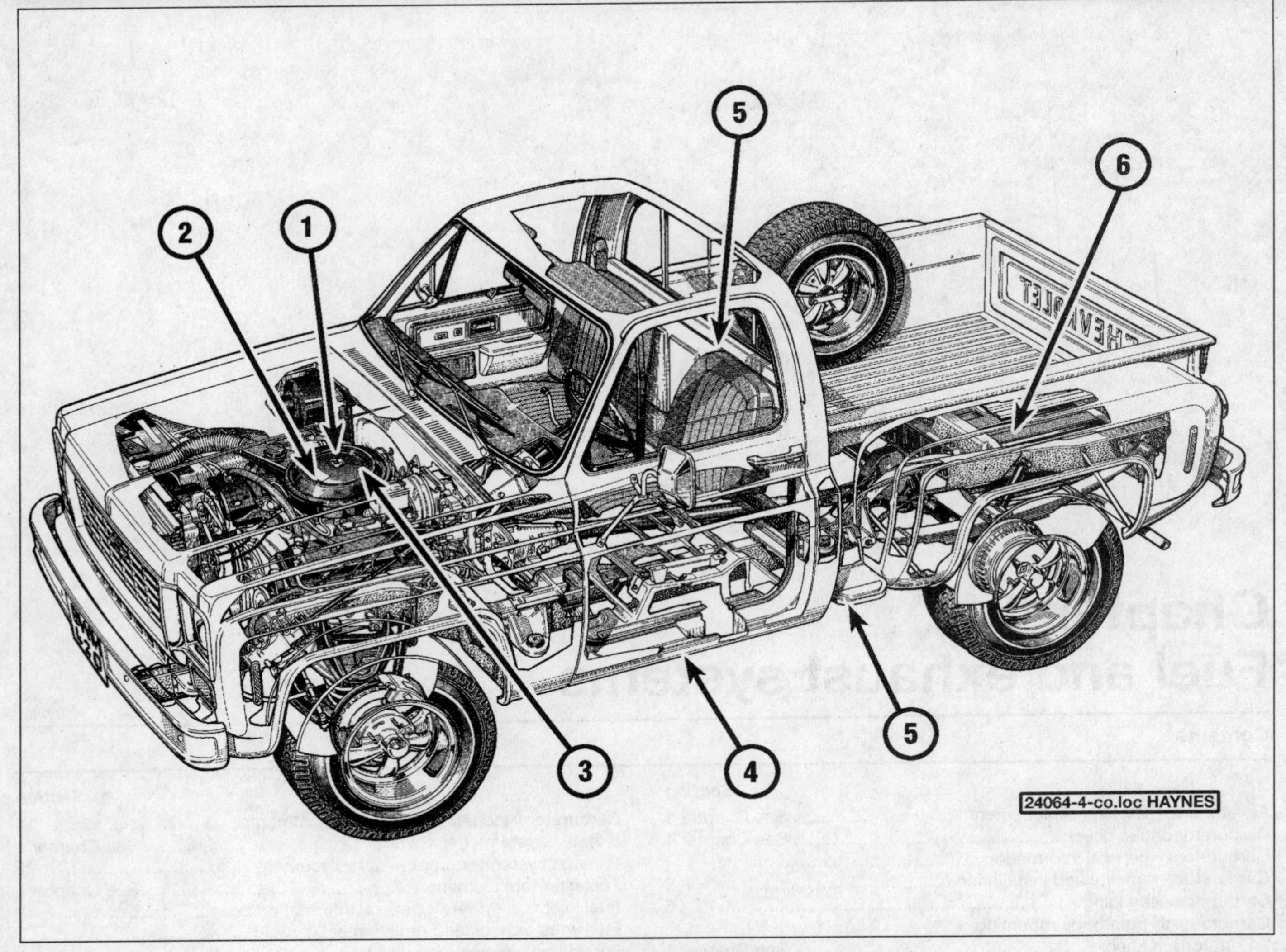

Typical fuel system component location

1 *Carburetor*
2 *Fuel filter (arrow) - carbureted models*
3 *Throttle body injection unit*
4 *Fuel filter - TBI models*
5 *Side mount fuel tanks - (A) Fuel tank switching valve located on the passenger side frame rail*
6 *Rear fuel tank*

1 General information

Fuel system

The fuel system consists of a rear-mounted gas tank, a fuel pump, a carburetor or fuel injection system and the fuel feed and return lines between the engine and the tank **(see illustrations)**. On carburetor equipped models, the fuel pump is mounted on the engine; on fuel injected models it's mounted in the gas tank.

Several types of carburetors were used over the long production life of these models. Some later models (1987 vehicles equipped with a V6 or V8 engine) were equipped with Throttle Body Injection (TBI), a type of fuel injection.

Warning: *Gasoline is extremely flammable, so extra precautions must be taken when working on any part of the fuel system. Do not smoke or allow open flames or bare light bulbs near the work area. Also, don't work in a garage if a natural gas-type appliance with a pilot light is present.*

Exhaust system

The exhaust system consists of manifolds, pipes, mufflers and, on some models, resonators which direct the exhaust gases to the rear of the vehicle. Later models are equipped with a catalytic converter which is part of the emissions system (Chapter 6).

Component location

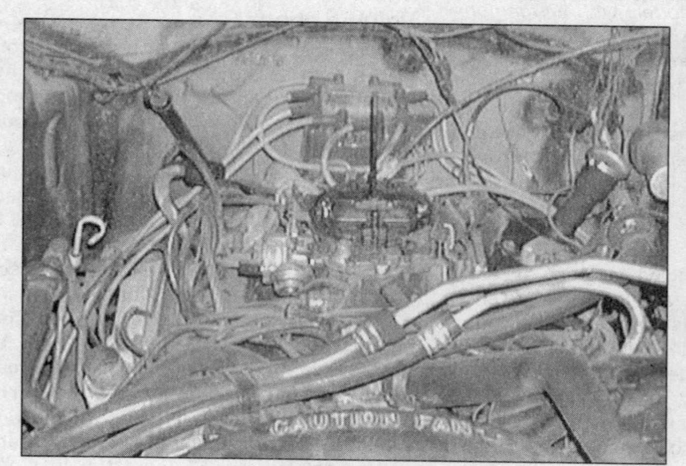

1 Carburetor

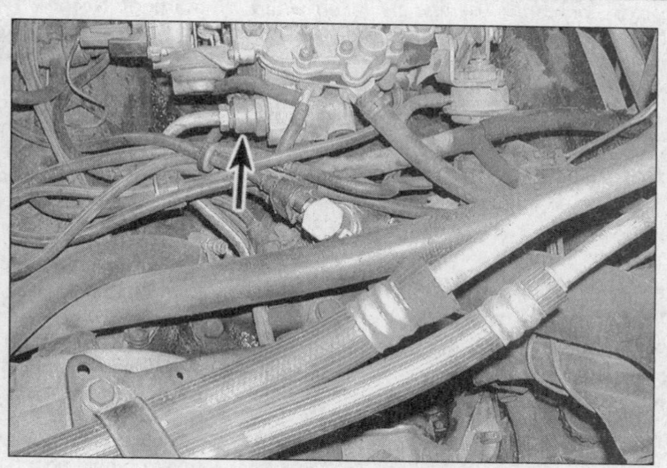

2 Fuel filter (arrow) - carbureted models

3 Throttle body injection unit

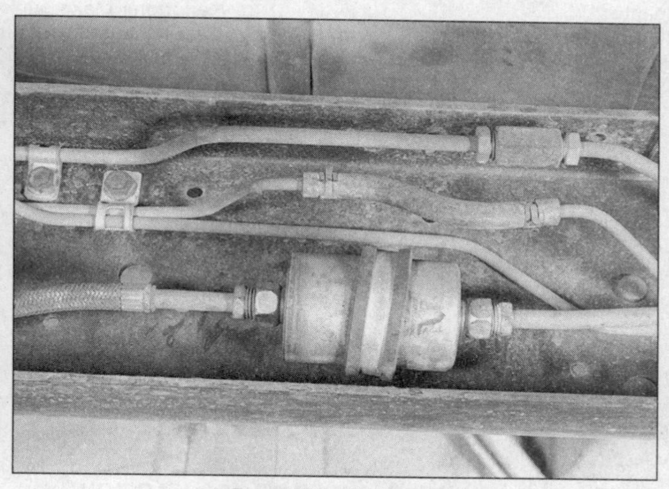

4 Fuel filter - TBI models

5 Side mount fuel tanks - (A) Fuel tank switching valve located on the passenger side frame rail

6 Rear fuel tank

2 Fuel pump - check

Warning: *Gasoline is extremely flammable, so extra precautions must be taken when working on any part of the fuel system. Do not smoke or allow open flames or bare light bulbs near the work area. Also, don't work in a garage if a natural gas-type appliance with a pilot light is present.*

Fuel injected models

1 The fuel pump on fuel injected models can be checked only with special tools and equipment - the job should be left to a dealer service department or a repair shop.

Carbureted models

2 The fuel pump is mounted on the front of the engine. It's permanently sealed and isn't serviceable or rebuildable.

Preliminary check

3 Before testing the fuel pump, check all fuel hoses and lines as well as the fuel filter (Chapter 1).

4 Remove the air cleaner assembly and disconnect the fuel line at the carburetor inlet. Detach the coil high tension wire from the distributor. Attach a jumper wire between the disconnected coil wire and a good engine ground (to prevent the engine from actually starting). **Note:** *On models with an ignition coil that's built into the distributor cap, unplug the BAT wire from the distributor cap to disable the ignition system.*

5 Hold a plastic or metal container over the end of the disconnected line and have an assistant crank the engine over. A strong spurt of gasoline should shoot from the end of the line every second revolution.

Pressure check

6 Make sure the engine has been run and brought to normal operating temperature and that the idle speed is as specified on the Vehicle Emission Control Information label.

7 Disconnect the fuel line at the carburetor (if not already done).

8 Connect a fuel pressure gauge with a flexible hose to the end of the line (the gauge must be held about 16-inches above the fuel pump, so make sure the hose is long enough). Make sure the inside diameter of the hose is no smaller than the diameter of the fuel line.

9 Start and run the engine. Note the fuel pressure on the gauge. It should be within the Specifications listed at the front of this Chapter.

Volume check

10 To check the fuel pump volume, connect a hose to the end of the fuel line and place the open end of the hose in a graduated plastic container. Volume marks should be clearly visible on the container.

11 Start the engine - it will run until the fuel in the carburetor is used up - and allow gasoline to flow into the container. At the end of 15 seconds, note the volume of fuel in the container. Compare the volume to the Specifications at the front of this Chapter.

12 If the volume is below the specified amount, attach an auxiliary fuel supply to the inlet side of the fuel pump. A small gas can with a hose forced tightly into the cap can be used as an auxiliary fuel supply. This with eliminate the possibility of a clogged tank and/or delivery line. Repeat the test and check the volume. If the volume has changed or is now normal, the fuel lines and/or tank are clogged. If the volume is still low, the fuel pump must be replaced with a new one.

Carburetor application (1967 thru 1978)

In-line engines

	230 cu in	250 cu in	292 cu in
1967 thru 1968	M	M	–
1969	–	MV	–
1970	–	M or MV	M or MV
1971	–	MV	MV
1972	–	MV	MV
1973	–	MV	MV
1974	–	MV	MV
1975	–	1 MV	1 MV
1976	–	1 MV	1 MV
1977 thru 1978	–	1 ME	1 ME

V8 engines

	283 cu in	307/327 cu in	350 cu in	396 cu in	400 cu in	454 cu in
1967 and 1968	2G	2G or 4G	–	–	–	–
1969	–	2G or 2GV	–	–	–	–
1970	–	2GV	2G,2GV or 4MV	4MV	–	–
1971	–	2GV	4MV	–	4MV	–
1972	–	2GV	4MV	–	4MV	–
1973	–	2GV	4MV	–	–	4MV
1974	–	–	2GV or 4MV	–	–	4MV
1975	–	–	2GC, M4MC or 4MV	–	4MV	4MV, M4MC or M4MCA
1976	–	–	2GC or 4MV	–	4MV	M4ME, 4MV or M4MC
		305 cu in				
1977 thru 1978	–	2GC	4MV or M4MC	–	M4MC	M4ME, 4MV or M4MC

Idle speed (1967 thru 1978)

In the following tables 'High' indicates initial setting of carburetor idle speed and 'Low' indicates final setting of carburetor idle speed

In-line engines

	230 cu in		250 cu in		292 cu in	
1967	*High*	*Low*	*High*	*Low*	*High*	*Low*
(without emission control)						
MT	550	530	550	530	–	–
AT	500	480	500	480	–	–
(with emission control)						
MT	700	680	700	680	–	–
AT	500	480	500	480	–	–
1968						
(without emission control)						
MT	550	500	500	480	–	–
AT	500	450	500	480	–	–
(with emission control)						
MT	700	680	700	680	–	–
AT	500	480	500	480	–	–
1969						
(without emission control)						
MT	–	–	500	480	–	–
AT	–	–	500	480	–	–
(with emission control)						
MT	–	–	700	680	–	–
AT	–	–	550	530	–	–
1970						
MT	–	–	750	730	700	680
AT	–	–	600	580	600	580
1971						
MT	–	–	580	550	580	550
AT	–	–	530	500	530	500
1972						
MT	–	–	800	450** 700*	775	450** 700*
AT	–	–	630	450** 600*	775	450** 700*

	230 cu in		250 cu in		292 cu in	
1973						
MT	–	–	750	450** 700*	775	450** 700*
AT	–	–	630	450** 600*	775	450** 700*
1974 (Fed)						
MT	–	–	900	850	–	–
	–	–	lean drop	950/850	–	–
AT	–	–	625	600	–	–
	–	–	lean drop	650/600	–	–
1974 (Cal)						
MT	–	–	875	850	650	600
	–	–	lean drop	950/850	lean drop	700/600
AT	–	–	–	–	–	–
	–	–	–	–	–	–
1975						
MT	–	–	850*	450**	600	450
	–	–	lean drop	950/850	lean drop	700/600
AT	–	–	600	450	–	–
	–	–	lean drop	650/600	–	–
1976 (Fed)						
MT	–	–	900	425	600	450
	–	–	lean drop	1075/900	lean drop	700/600
AT	–	–	550	425	1000	450
	–	–	lean drop	575/550	lean drop	700/600
1976 (Cal)						
MT	–	–	1000	425	600	450
	–	–	lean drop	1150/1000	lean drop	700/600
AT	–	–	1000	425	600	450
	–	–	lean drop	630/600	lean drop	700/600
1977 on						
(C10 and K10)						
MT (Fed)	–	–	750	425	–	–
MT (Cal)	–	–	850	425	–	–
(C20-30 and K20)						
MT (Fed)	–	–	600	450	600	450
MT (Cal)	–	–	–	–	–	–
(C10 and K10)						
AT (Fed)	–	–	550	425	–	–
AT (Cal)	–	–	550	425	–	–
(C20-30 and K20)						
AT (Fed)	–	–	600	450	600	450
AT (Cal)	–	–	600	450	600	450

V8 engines

	283 cu in		307/327 cu in		350 cu in		396 cu in		400 cu in		454 cu in	
1967	*High*	*Low*	*High*	*Low*	*High*	*Low*	*High*	*Low*	*High*	*Low*	*High*	*Low*
(without emission control)												
MT	550	500	550	500	–	–	–	–	–	–	–	–
AT	550	500	550	500	–	–	–	–	–	–	–	–
(with emission control)												
MT	700	680	700	680	–	–	–	–	–	–	–	–
AT	600	580	600	580	–	–	–	–	–	–	–	–

	283 cu in		307/327 cu in		350 cu in		396 cu in		400 cu in		454 cu in	
1968												
(without emission control)												
MT	–	–	550	500	–	–	–	–	–	–	–	–
AT	–	–	550	500	–	–	–	–	–	–	–	–
(with emission control)												
MT	–	–	700	680	–	–	–	–	–	–	–	–
AT	–	–	600	580	–	–	–	–	–	–	–	–
1969												
(without emission control)												
MT	–	–	550	500	–	–	–	–	–	–	–	–
AT	–	–	550	500	–	–	–	–	–	–	–	–
(with emission control)												
MT	–	–	700	680	–	–	–	–	–	–	–	–
AT	–	–	600	580	–	–	–	–	–	–	–	–
1970												
MT	–	–	700	680	600	580	600	580	–	–	–	–
AT	–	–	600	580	600	580	500	480	–	–	–	–
1971												
MT	–	–	700	600	675	600	–	–	675	600	–	–
AT	–	–	580	550	580	550	–	–	630	600	–	–
1972												
MT	–	–	1 000	450** 900*	1 000	450** 900*	–	–	675	450** 750*		
AT	–	–	650	450** 600*	630	450** 600*	–	–	675	450** 600*	–	–
1973												
MT	–	–	950	450** 900*	920	450** 900*	–	–	–	–	800	500** 700*
AT	–	–	630	450** 600*	620	450** 600*	–	–	–	–	625	500** 600*
1974 (Fed)					**Dual barrel**							
MT	–	–	–	–	950	900	–	–	–	–	825	800
	–	–	–	–	lean drop 1000/900		–	–	–	–	lean drop 850/800	
AT	–	–	–	–	625	600	–	–	–	–	615	600
	–	–	–	–	lean drop 650/600		–	–	–	–	lean drop 630/600	
					Four barrel							
MT	–	–	–	–	925	900	–	–	–	–	–	–
	–	–	–	–	lean drop 950/900		–	–	–	–	–	–
AT	–	–	–	–	615	600	–	–	–	–	–	–
	–	–	–	–	lean drop 630/600		–	–	–	–	–	–
1974 (Cal)					**Four barrel**							
MT	–	–	–	–	925	900	–	–	–	–	825	800
	–	–	–	–	lean drop 950/900		–	–	–	–	lean drop 850/800	
AT	–	–	–	–	615	600	–	–	–	–	615	600
	–	–	–	–	lean drop 630/600		–	–	–	–	lean drop 630/600	
1975					**Dual barrel**							
MT	–	–	–	–	900*	500**	–	–	740	700	–	–
	–	–	–	–	lean drop 100/900		–	–	lean drop 770/700		–	–
AT	–	–	–	–	600	500	–	–	–	–	675	650
	–	–	–	–	lean drop		–	–	–	–	lean drop 700/650	
1976 (Fed)												
MT	–	–	–	–	800	–	–	–	–	–	700	–
	–	–	–	–	lean drop 900/800		–	–	–	–	lean drop 800/700	
AT	–	–	–	–	600	–	–	–	700	–	700	–
	–	–	–	–	lean drop 650/600		–	–	lean drop 770/700		lean drop 800/700	

	283 cu in		307 cu in		350 cu in		396 cu in		400 cu in		454 cu in	
	High	*Low*	*High*	*Low*	*High*	*Low*	*High*	*Low*	*High*	*Low*	*High*	*Low*
1976 (Cal)												
MT	–	–	–	–	800	–	–	–	–	–	700	–
					lean drop						lean drop	
	–	–	–	–	900/800		–	–	–	–	800/700	
AT	–	–	–	–	600	–	–	–	700	–	700	–
					lean drop				lean drop		lean drop	
	–	–	–	–	650/600		–	–	770/700		800/700	
			305 cu in									
1977 on			*High*	*Low*								
(C10 and K10)												
MT (Fed)	–	–	700	600	800	700	–	–	–	–	–	–
MT (Cal)	–	–	–	–	–	–	–	–	–	–	–	–
(C20-30 and K20)												
MT (Fed)	–	–	–	–	875	700	–	–	770	700	–	–
MT (Cal)	–	–	–	–	800	700	–	–	–	–	–	–
(C10 and K10)												
AT (Fed)	–	–	650	500	550	500	–	–	–	–	750	700
AT (Cal)	–	–	–	–	550	500	–	–	–	–	750	700
(C20-30 and K20)												
AT (Fed)	–	–	–	–	–	–	–	–	–	–	750	700
AT (Cal)	–	–	–	–	–	–	–	–	–	–	750	700

Notes: ** Solenoid energized*
*** Solenoid disconnected*
MT – Manual transmission
AT – Automatic transmission
Fed – Federal
Cal – California
If there is a discrepancy between the idle rpm settings given here and those specified on the individual vehicle decal, the figures on the decal should be followed

Maximum CO levels for vehicles equipped with emission control carburetors

C10 and K10 models (light duty emission):

250 cu in engine	0·3%
305 cu in engine (2 barrel carburetor)	0·5%
350 cu in engine (2 barrel carburetor)	0·5%
350 cu in engine (4 barrel carburetor)	0·5%
454 cu in engine	0·5%

C20, C30, K20 and K30 models (heavy duty emission):

250 cu in engine	2·0%
292 cu in engine	0·3%
350 cu in engine (4 barrel carburetor)	0·5%
454 cu in engine	0·5%

Carburetor adjustment data (1967 thru 1978)

All dimensions are in inches

Rochester M (1967 thru 1970)

Float level	Dashpot	Metering rod	Choke rod	Fast idle rpm
$\frac{9}{32}$	$\frac{1}{16}$ to $\frac{3}{32}$ (Plunger to throttle lever)	0·140	0·150	2400

Rochester 2G/4G series

	Float level	Float drop	Accelerator pump	Choke rod	Choke vacuum break	Choke unloader	Fast idle rpm
Rochester 2G/4G 1967 thru 1970	$\frac{3}{4}$	$1\frac{3}{4}$	$1\frac{1}{8}$	N/A Manual choke	N/A Manual choke	N/A Manual choke	2100
Rochester 2GV 1969 thru 1971	$\frac{23}{32}$	$1\frac{3}{4}$	$1\frac{3}{8}$	0·60	0·140	0·215	2200 to 2400

	Float level	Float drop	Accelerator pump	Choke rod	Choke vacuum break	Choke unloader	Fast idle rpm
1972	$\frac{21}{32}$	$1\frac{9}{32}$	$1\frac{5}{16}$	0.075	0.110	0.210	2200
1973	$\frac{21}{32}$ Except No 7043108 $\frac{25}{32}$	$1\frac{9}{32}$	$1\frac{5}{16}$ Except No 7043108 $\frac{7}{16}$	0·150 Except No 7043108 0·200	0·080 Except No 7403108 0·250	0·215 Except No 7403108 0·250	2400 on high step of cam

	Float level	Float drop	Accelerator pump	Choke rod	Choke vacuum break	Choke unloader	Fast idle rpm
1974 thru 1975	$\frac{19}{32}$	$1\frac{9}{32}$	$1\frac{9}{32}$ Except Nos 7044114 7044124 $1\frac{3}{16}$	0·200 Except Nos 7044114 7044124 0·130	0·140 Except Nos 7044114 7044124 0·325	0·250 Except Nos 7044114 7044124 0·325	1600
Rochester 2GC **1975 onward**	$\frac{21}{32}$	$\frac{31}{32}$ (1975) $1\frac{9}{32}$ (1976)	$1\frac{5}{8}$ (1975) $1\frac{11}{16}$ (1976)	0·400 (1975) 0·260 (1976)	0·130	0·350 (1975) 0·325 (1976)	2000

Rochester MV series

	Float level	Choke rod	Choke vacuum break	Choke unloader	Metering rod	Fast idle rpm
1970 thru 1971	$\frac{1}{4}$	0·190 (1970) 0·180 (1971)	0·230 (1970) 0·260 (1971)	0·350	0·070	2400
1972	$\frac{1}{4}$	0·180	0·260	0·500	0·070	2400
1973	$\frac{1}{4}$	0·375	0·430	0·600	0·070	2400
1974 thru 1975 Carburetor No:						
7044021	0·295	0·275	0·350	0·500	0·080	1800 to 2400 on high step of cam
7044022	0·295	0·245	0·350	0·500	0·080	
7044321	0·295	0·300	0·375	0·500	0·080	
7044025	$\frac{1}{4}$	0·245	0·300	0·521	0·070	
7044026	$\frac{1}{4}$	0·275	0·350	0·521	0·070	

Rochester 1M series

	Float level	Choke rod	Choke vacuum break Primary	Auxiliary	Choke unloader	Metering rod	Fast idle rpm
Rochester 1MV **1975 thru 1976**							
Carburetor No: 7045002	$\frac{11}{32}$	0.260 (1975) 0.130 (1976)	0.260 0.165	0.300 (1975) 0.265 (1976)	0.325 (1975) 0.335 (1976)	0.080	2100
7045003	$\frac{11}{32}$	0.275 (1975) 0.145 (1976)	0.300	0.290	0.325 (1975) 0.335 (1976)	0.080	2100
7045004	$\frac{11}{32}$	0.245 (1975) 0.130 (1976)	0.300 0.165	0.150 (1975) 0.265 (1976)	0.325 (1975) 0.335 (1975)	0.080	2100
7045005	$\frac{11}{32}$	0.275	0.350	0.290	0.325	0.080	2100
7045302	$\frac{11}{32}$	0.245 (1975) 0.155 (1976)	0.300 0.190	0.150 (1975) – (1976)	0.275 (1975) 0.325 (1976)	0.080	2100
7045303	$\frac{11}{32}$	0.275 (1975) 0.180 (1976)	0.350 0.225	0.170 (1975) – (1976)	0.275 (1975) 0.325 (1976)	0.080	2100
7045304	$\frac{11}{32}$	0.245	0.300	0.290	0.325	0.080	2100
7045305	$\frac{11}{32}$	0.275	0.350	0.290	0.325	0.080	2100

Rochester 1ME
1977 thru 1978

Carburetor No:	Choke rod	Choke vacuum break	Choke unloader
17057001 17057303 17057005	0.125	0.150	0.325 (C10/K10) 0.275 (C20/K20)
17057002 17057004	0.110	0.135	
17057302 17057010	0.150	0.180	

	Choke rod	Choke vacuum break
17057006	0.150	0.180
17057007		
17057008		
17057009		
17057308		
17057309		
17081009	0.275	0.400
17084329		
17085009		
17085036		
17085004		

Rochester 4MV series

	Float level	Accelerator pump	Choke rod	Choke vacuum break	Choke unloader	Air valve dashpot
1970 thru 1971	$\frac{1}{4}$	$\frac{5}{16}$	0·100	0·245	0·450	0·020
1972						
Carburetor No:						
7042206	$\frac{1}{4}$	$\frac{13}{32}$	0·100	0·250	0·450	0·020
7042207	$\frac{11}{32}$	$\frac{13}{32}$	0·100	0·250	0·450	0·020
7042208	$\frac{3}{16}$	$\frac{3}{8}$	0·100	0·215	0·450	0·020
7042219	$\frac{11}{32}$	$\frac{3}{8}$	0·100	0·250	0·450	0·020
7042210	$\frac{3}{16}$	$\frac{3}{8}$	0·100	0·215	0·450	0·020
7042218	$\frac{1}{4}$	$\frac{3}{8}$	0·100	0·250	0·450	0·020
7042211	$\frac{3}{16}$	$\frac{3}{8}$	0·100	0·215	0·450	0·020
7042910	$\frac{3}{16}$	$\frac{3}{8}$	0·100	0·215	0·450	0·020
7042911	$\frac{3}{16}$	$\frac{3}{8}$	0·100	0·215	0·450	0·020
1973						
Carburetor No:						
7043202, 7043203, 7043210, 7043211	$\frac{7}{32}$	$\frac{13}{32}$	0·430	0·215	0·450	0·020
7043208, 7043215	$\frac{5}{16}$	$\frac{13}{32}$	0·430	0·215	0·450	0·020
7043200, 7043216, 7043207, 7043507	$\frac{1}{4}$	$\frac{13}{32}$	0·430	0·250 Except No 7043507 (0·275)	0·450	0·020
1974						
Carburetor No:						
7044202 and 502	$\frac{1}{4}$	$\frac{13}{32}$	0·430	0·230	0·450	0·020
7044203 and 503	$\frac{1}{4}$	$\frac{13}{32}$	0·430	0·230	0·450	0·020
7044218 and 518	$\frac{1}{4}$	$\frac{13}{32}$	0·430	0·215	0·450	0·020
7044219 and 519	$\frac{1}{4}$	$\frac{13}{32}$	0·430	0·215	0·450	0·020
7044213 and 513	$\frac{11}{32}$	$\frac{13}{32}$	0·430	0·215	0·450	0·020
7044223 and 227	0·675	$\frac{13}{32}$	0·430	0·220	0·450	0·020
7044212 and 217	0·675	$\frac{13}{32}$	0·430	0·230	0·450	0·020
7044512 and 517	0·675	$\frac{13}{32}$	0·430	0·230	0·450	0·020
7044500 and 520	0·675	$\frac{13}{32}$	0·430	0·250	0·450	0·020
7044224, 7044214 and 514, 7044215 and 515, 7044216 and 516	$\frac{11}{32}$	$\frac{13}{32}$	0·430	0·215	0·450	0·020
1975						
Carburetor No:						
7045212	$\frac{3}{8}$	0·275	0·430	0·225	0·450	0·015
7045213, 7045216	$\frac{11}{32}$	0·275	0·430	0·210	0·450	0·015
7045214, 7045215	$\frac{11}{32}$	0·275	0·430	0·215	0·450	0·015

Carburetor No	float level	Accelerator pump	Choke rod	Choke vacuum break	Choke unloader	Air valve dashpot
7045583 7045584 7045585 7045586 7045588 7045589	11/32	0·275	0·430	0·230	0·450	0·015
7045217	3/8	0·275	0·430	0·225	0·450	0·015
7045225	11/32	0·275	0·430	0·200	0·450	0·015
7045229	15/32	0·275	0·430	0·200	0·450	0·015
1976						
Carburetor No:						
7045213 7045214 7045215 7045216	11/32	9/32	0·290	0.145	0·295	0·015
7045225 7045229	11/32	9/32	0·290	0·138	0·295	0·015
7045583 7045584 7045585 7045586 7045588 7045589	11/32	9/32	0·290	0·155	0·295	0·015
17056212 17056217	3/8	9/32	0·290	0·155	0·295	0·015
1977 thru 1978						
Carburetor No:						
17056212 17056217	3/8	9/32	0·290	0·120	0·295	0·015
17057213 17057215 17057216	11/32	9/32	0·220	0·115	0·205	0·015
17057525 17057514	11/32	9/32	0·220	0·120	0·225	0·015
17057529 17057229	11/32	9/32	0·220	0·110	0·205	0·015
7045583 7045585 7045586	11/32	9/32	0·290	0·120	0·295	0·015

Rochester M4 series

Choke coil lever (all models) .... 0·120
Air valve dashpot (all models) .... 0·015
Fast idle rpm (all models except M4MC 1977 on) 1600

	Float level	Accelerator pump rod	Choke rod (fast idle cam)	Front vacuum break	Air valve wind-up	Choke unloader
Rochester M4MCA						
1975						
Carburetor No:						
7045512 7045517	17/32	0·275	0·300	0·180	9/16	0·325
Rochester M4ME						
1976	7/16	9/32	0·300	–	7/8	0·325
1977 thru 1978	3/8	9/32	Light duty emission 0·325 Heavy duty emission 0·285	–	7/8	0·325

	Float level	Accelerator pump rod	Choke rod (fast idle cam)	Front vacuum break	Air valve wind-up	Choke unloader
Rochester M4MC						
1975						
Carburetor No:						
7045202 7045203	15/32	0·275	0·300	0·180	7/8	0·325
7045218 7045219	15/32	0·275	0·325	0·180	3/4	0·325
7045220	17/32	0·275	0·300	0·200	9/16	0·325
1976						
Carburetor No:						
17056208 17056209 17056218 17056219 17056508 17056509	5/16	9/32	0·325	0·185	7/8	0·325
17056512 17056517	7/16	9/32	0·325	0·185	7/8	0·275
17057518 17056519	5/16	9/32	0·325	0·185	7/8	0·325
1977 thru 1978						
Carburetor No:						
17057202 17057204	15/32	9/32	Light duty emission 0·325 Heavy duty emission 0·285	0·160	7/8	0·280
17057218 17057222	7/16	9/32	As above	0·160	7/8	0·280
17057219	7/16	9/32	As above	0·165	7/8	0·280
17057502 17057503 17057504	15/32	9/32	As above	0·165	7/8	0·280
17057512 17057517	7/16	9/32	As above	0·165	7/8	0·240
17057518 17057519 17057522	7/16	9/32	As above	0·165	7/8	0·280
17057582 17057584	15/32	9/32	As above	0·180	7/8	0·280
17057586 17057588	7/16	9/32	As above	0·180	7/8	0·280

Carburetor application (1979 thru 1985)

In-line engines	250 cu in (4.1 liter)	292 cu in (4.8 liter)
1979	1ME or 2SE	1ME
1980 through 1982	2SE	1ME
1983	1ME,E2SE or 2SE	1ME
1984	1ME or 2SE	1ME

V8 engines	305 cu in (5.0 liter)	350 cu in (5.7 liter)	400 cu in (6.6 liter)	454 cu in (7.4 liter)
1979	M2MC	M2MC or M4M	M4M	M4M
1980	M2MC	M4M or M2MC	M4MC or M4ME	M4MC or M4ME
1981	M2ME,M4ME or M4MC	M2ME, M4ME or M4MC	----	M4M
1982	M4ME or M4MC	M4ME or M4MC	----	M4M
1983	E4ME,M4ME or M4MC	E4ME,M4ME or M4MC	----	E4M or M4M
1984	E4ME,M4ME or M4MC	E4ME,M4ME or M4MC	----	E4M or M4M
1985	M4ME,M4MC	M4ME,M4MC	----	M4ME,M4MC

Idle speed (1979 thru 1985)

In the following tables 'High' indicates the carburetor setting for fast idle in Neutral and 'Low' indicates the final carburetor idle speed. **Note:** *If there is a discrepancy between the idle rpm settings given here and those specified on the individual vehicle decal, the figures on the decal should be followed.*

In-line engines	250 cu in (4.1 liter) High	250 cu in (4.1 liter) Low	292 cu in (4.8 liter) High	292 cu in (4.8 liter) Low
1979 (Federal)				
MT	1800	750	2400	700
AT	2000	600	2400	700
1979 (California)				
MT	2100	750	2400	700
AT	2100	600	2400	700
1980 (Federal)				
MT	2000	750	2400	700
AT	2200	650	2400	700
1980 (California)				
MT	2000	750	2400	700
AT	2200	600	2400	700
1981 and 1982	Refer to emission label in the engine compartment for latest specifications			

V8 engines	305 cu in High	305 cu in Low	350 cu in High	350 cu in Low	400 cu in High	400 cu in Low	454 cu in High	454 cu in Low
1979 (Federal)								
MT	1300	600	1300	700	–	–	1600	700
AT	1600	500	1600	500	1600	500	1600	550
1979 (California)								
MT	–	–	1600	700	–	–	1600	700
AT	–	–	1600	500	1600	500	1600	550
1980 (Federal)								
MT	1300	600	1300	700	–	–	–	–
AT	1600	500	1600	500	1600	500	–	–
1980 (California)								
MT	–	–	1600	700	–	–	–	–
AT	–	–	1600	500	1600	500	–	–
1981 thru 1985	Refer to emission label in the engine compartment for latest specifications							

Carburetor adjustment data (1979 thru 1985)

Carburetor specifications – Rochester 1ME (dimensions in inches)

	1979	1980	1981	1982 thru 1985
Float level	5/16	11/32	11/32	11/32
Metering rod	0.065	0.090	0.090	0.090
Choke coil lever	0.120	0.120	–	–
Fast idle cam (choke rod)	0.275	0.275	0.275	0.275
Vacuum break	0.400	0.400	0.400	0.400
Unloader	0.521	0.520	0.520	0.520

Carburetor specifications – Rochester M2M (dimensions in inches)

	1979	1980	1981
Float level	15/32 in	7/16 in	13/32
Pump rod (inner hole)	13/32 in	9/32	5/16
Fast idle cam (choke rod)	38°	38°	38°
Front vacuum break	29°	29°	25°
Unloader	–	–	38°

Carburetor specifications – Rochester 2SE – 1979 thru 1982 (dimensions in inches)

For location of carburetor part numbers, see Carburetor Identification

	1979	1980	1981	1982
Float level	1/8	1/8	3/16	3/16
Air valve rod	0.040	2°	1°	1°
Fast idle cam (choke rod)	17°	17°	15°	15°

	1979	1980	1981	1982
Primary vacuum break	17059641 17059643 17059765 17059767 } 23.5°	17080720 17080722 } 20° 17080721 17080723 } 23.5°	17081629 – 24° 17081720 17081721 17081725 17081726 17081727 } 30°	17082482 – 23° 17082341 17082342 17082344 17082345 } 30° 17082431 17082433 } 24° 17082486 17082487 17082488 17082489 } 28°
All others	20°	22°	26°	26°
Secondary vacuum break	37°	35°	17081629 – 34° 17081720 17081721 17081725 17081726 17081727 } 37° All others 38°	17082341 17082342 17082344 17082345 } 37° 38°
Unloader	49°	41°	17081720 17081721 17081725 17081726 17081727 17081629 } 41° All others 38°	42°

Carburetor specifications — Rochester 2SE — 1983 and 1984) (dimensions in inches)

For location of carburetor part numbers, see Carburetor Identification

Float level	3/16
Air valve rod	1°
Fast idle cam (choke rod)	15°
Primary vacuum break	17083410 17083412 17083414 17083416 } 23° 17083411 17083413 17083415 17083417 17083419 17083421 17083425 17083427 } 26° 17083423 17083429 17083560 17083562 17083565 17083569 } 28°
Secondary vacuum break	38°
Unloader	42°

1984

Float level	17084348 17084349 17084350 17084351 17084352 17084353 17084354 17084355 17084410 17084412 17084425 17084427 17084560 17084562 17084569 } 11/32

Item	Carburetor number	Setting
	17084360, 17084362, 17084364, 17084366	5/32
	17084390, 17084391, 17084392, 17084393	7/16
Air valve rod		1°
Fast idle cam (choke rod)	17084348, 17084349, 17084350, 17084351, 17084352, 17084353, 17084354, 17084355, 17084360, 17084362, 17084364, 17084366	22°
	17084390, 17084391, 17084392, 17084393	28°
	17084410, 17084412, 17084425, 17084427, 17084560, 17084562, 17084569	15°
Primary vacuum break	17084348, 17084349, 17084350, 17084351, 17084352, 17084353, 17084354, 17084355, 17084360, 17084362, 17084364, 17084366, 17084390, 17084391, 17084392, 17084393	30°
	17084410, 17084412	23°
	17084425, 17084427	26°
	17084560, 17084562, 17084569	24°
Secondary vacuum break	17084348, 17084349, 17084350, 17084351, 17084360, 17084362	32°
	17084352, 17084353, 17084354, 17084355, 17084364, 17084366	35°
	17084390, 17084391, 17084392, 17084393, 17084410, 17084412	38°

	17084425 17084427	36°
	17084560 17084562 17084569	34°
Unloader	17084348 17084349 17084350 17084351 17084352 17084353 17084354 17084355 17084360 17084362 17084364 17084366	40°
	17084390 17084391 17084392 17084393 17084560 17084562 17084569	38°
	17084410 17084512	42°

Carburetor specifications — Rochester E2SE — 4.1L engine (dimensions in inches)

For location of carburetor numbers, see Carburetor Identification

1983

Float level	11/32
Air valve rod	1°
Fast idle cam (choke rod)	15°
Primary vacuum break	26°
Secondary vacuum break	38°
Unloader	42°

1984

Float level	17084356 17084357 17084358 17084359 17084632 17084633 17084635 17084636	9/32
	17084368 17084370 17084542	1/8
	17084430 17084431 17084434 17084435	11/32
	17084534 17084535 17084537 17084538 17084540	5/32
Air valve rod	1°	
Fast idle cam (choke rod)	17084356 17084357 17084358 17084359 17084368 17084370	22°
	17084430 17084431 17084434 17084435	15°
	All others — 28°	

Primary vacuum break	17084430, 17084431, 17084434, 17084435	26°
	All others	25°
Secondary vacuum break	17084356, 17084357, 17084358, 17084359, 17084368, 17084370	30°
	17084430, 17084431, 17084434, 17084435	38°
	All others	35°
Unloader	17084356, 17084357, 17084358, 17084359, 17084368, 17084370	30°
	17084430, 17084431, 17084434, 17084435	42°
	All others	45°

Carburetor specifications – Rochester M4M (1979)

For location of carburetor part numbers, see Carburetor Identification (dimensions in inches)

Float level	
17059212	7/16
17059512	13/32
17059520, 17059521	3/8
All others	15/32
Pump rod	
17059377, 17059527, 17059378, 17059528	9/32 outer hole
17059212, 17059213, 17059215, 17059229, 17059510, 17059512, 17059513, 17059515, 17059520, 17059521, 17059529	9/32 inner hole
All others	13/32 inner hole
Fast idle cam	
17059213, 17059215, 17059229, 17059513, 17059515, 17059529	37°
All others	46°
Front air valve rod	0.015
Rear air valve rod	0.015
Front vacuum break	
17059212, 17059512	24°
17059501, 17059520, 17059521	28°

17059509	30°
17059510	
17059586	
17059588	
All others	23°
Rear vacuum break	
17059363	26°
17059366	
17059368	
17059377	
17059378	
17059503	
17059506	
17059508	
17059527	
17059528	
All others	23°
Air valve spring	
17059212	3/4
17059512	
17059213	1.0
17059215	
17059229	
17059513	
17059515	
17059529	
All others	7/8
Unloader	
17059212	40°
17059213	
17059215	
17059229	
17059512	
17059513	
17059515	
17059529	
All others	42°

Carburetor specifications – Rochester M4M (1980)

For location of carburetor part numbers, see Carburetor Identification (dimensions in inches)

Float level	
17080213	3/8
17080215	
17080513	
17080515	
17080229	
17080529	
All others	15/32
Pump rod ...	9/32 inner hole
Fast idle cam	
17080213	37°
17080215	
17080513	
17080515	
17080229	
17080529	
All others	46°
Front air valve rod	0.015
Rear air valve rod	0.025
Front vacuum break	
17080212	24°
17080512	
All others	23°
Rear vacuum break	
17080290	26°
17080291	
17080292	
17080503	
17080506	
17080508	

17080212	30°
17080512	
17080213	
17080215	
17080513	
17080515	
17080229	
17080529	
All others	23°
Unloader	
17080212	40°
17080512	
17080213	
17080215	
17080513	
17080515	
17080229	
17080529	
All others	42°
Air valve spring	
17080212	3/4
17080512	
17080213	1.0
17080215	
17080513	
17080515	
17080229	
17080529	
All others	7/8

Carburetor specifications — Rochester M4M (1981)

For location of carburetor part numbers, see Carburetor Identification (dimensions in inches)

Float level	
17080212	3/8
17080213	
17080215	
17080298	
17080507	
17080512	
17080513	
17081200	15/32
17081201	
17081205	
17081206	
17081220	
17081226	
17081227	
All others	13/32
Pump rod	
17081524	5/16 outer hole
17081526	
All others	9/32 inner hole
Fast idle cam	
17080213	37°
17080215	
17080298	
17080507	
17080513	
All others	46°
Front air valve rod	0.025
Rear air valve rod	0.025
Rear vacuum break	
17081200	23°
17081201	
17081205	
17081206	
17081220	
17081226	
17081227	

17081290	24°
17081291	
17081292	
17081506	36°
17081508	
17081524	
17081526	
All others	30°
Front vacuum break	
17080212	24°
17080512	
17081200	
17081226	
17081227	
17081524	25°
17081526	
All others	23°
Air valve wind-up	
17080213	1.0
17080215	
17080298	
17080507	
17080212	3/4
17080512	
17080513	
All others	7/8
Unloader	
17080212	40°
17080213	
17080215	
17080298	
17080507	
17080512	
17070513	
17081506	36°
17081508	
17081524	38°
17081526	
All others	42°

Carburetor specifications – Rochester M4M (1982)

For location of carburetor part numbers, see Carburetor Identification (dimensions in inches)

Float level	
17080212	3/8
17080213	
17080215	
17080298	
17080507	
17080512	
17080513	
17082213	
17082513	
All others	13/32
Pump rod	
17082524	5/16 outer hole
17082526	
All others	9/32 inner hole
Fast idle cam	
17080213	37°
17080215	
17080298	
17080507	
17080513	
17082213	
17082513	
All others	46°
Air valve rod	0.025

Front vacuum break	
17082230	26°
17082231	
17082234	
17082235	
17082524	25°
17082526	
17082506	23°
17082508	
17082513	
17080513	
17082213	
17080213	
17080215	
17080298	
17080507	
All others	24°
Rear vacuum break	
17082220	34°
17082221	
17082222	
17082223	
17082224	
17082225	
17082226	
17082227	
17082290	
17082291	
17082292	
17082293	
17082230	36°
17082231	
17082234	
17082235	
17082524	
17082526	
17082293	
17082506	
All others	30°
Air valve wind-up	
17080212	3/4
17080512	
17080513	
17080213	1.0
17080215	
17080298	
17080507	
17082213	
All others	7/8
Unloader	
17080212	40°
17080213	
17080215	
17080298	
17080507	
17080512	
17080513	
17082213	
All others	39°

Carburetor specifications — Rochester M4M — 1983 (dimensions in inches)

For location of carburetor part numbers, see Carburetor Identification

5.0L engine

Float level	13/32
Pump rod	9/32 (inner)
Fast idle cam	46°
Air valve rod	0.025
Front vacuum break	—
Rear vacuum break	24°
Air valve wind-up (turns)	7/8
Unloader	39°

7.4L engine

Float level	3/8
Pump rod	9/32 (inner)
Fast idle cam	46°
Air valve rod	0.025
Front vacuum break	24°
Rear vacuum break	30°
Air valve wind-up (turns)	3/4
Unloader	40°

5.7L engine

Specification	Carburetor	Value
Float level	17080201, 17080205, 17080206, 17080290, 17080291, 17080292	15/32
	17080213, 17080298, 17080507, 17080513, 17083298, 17083507	3/8
	17082213	9/32
	All others	13/32
Pump rod		9/32 (inner)
Fast idle cam	17080213, 17080298, 17080507, 17080513, 17082213, 17083298, 17083507	37°
	All others	46°
Air valve rod		0.025 in.
Front vacuum break		23° if equipped
Rear vacuum break	17080201, 17080205, 17080206	23°
	17083290, 17083291, 17083292, 17083293	24°
	17080290, 17080291, 17080292, 17083234, 17083235	26°
	All others	30°
Air valve wind-up (turns)	17080213, 17080298, 17080507, 17080513, 17082213, 17083298, 17083507	1
	All others	7/8
Unloader	17080201, 17080205, 17080206, 17080290, 17080291, 17080292	42°
	17083234, 17083235, 17083290, 17083291, 17083292, 17083293	39°
	All others	40°

Carburetor specifications — Rochester M4M — 1984 (dimensions in inches)

For location of carburetor part numbers, see Carburetor Identification

Item	Carburetor number	Specification
Float level	17080212 17080213 17080298 17082213 17083298 17084500 17084501 17084502	3/8
	All others	13/32
Pump rod		9/32 (inner)
Fast idle cam	17080213 17080298 17080507 17080513 17082213 17083298 17083507	37°
	All others	46°
Air valve rod		0.025 in
Front vacuum break		23° if equipped
Rear vacuum break	17084226 17084227 17084290 17084292	24°
	17080212 17080213 17080298 17082213 17083298 17084500 17084501 17084502	30°
	All others	26°
Air valve wind-up (turns)	17080212 17080213 17080298 17082213 17083298 17084500 17084501	1
	All others	7/8
Unloader	17080212 17080213 17080298 17082213 17083298 17084500 17084501 17084502	40°
	All others	39°

Carburetor specifications — Rochester M4M — 1985 (dimensions in inches)

For location of carburetor part numbers, see Carburetor Identification

Item	Carburetor number	Specification
Float level	17080212 17080213 17080298 17082213 17084500 17084501 17084502 17085000 17085001	3/8
	All others	13/32
Fast idle cam	17080213 17080298 17082213 17083298 17084500 17084501	37°

Setting	Carburettor number	Value
	17080212 17084502 17085000 17085001 17085003 17085004 17085206 17085212 17085213 17085215 17085228 17085229 17085235 17085290 17085291 17085292 17085293 17085294 17085298	46°
	All others	20°
Front vacuum break (if equipped)	17080212 17084502 17085000	24°
	17080213 17080298 17082213 17083298 17084500 17084501 17085001 17085003 17085004 17085212 17085213	23°
	All others	26°
Rear vacuum break	17080212 17080213 17080298 17082213 17083298 17084500 17084501 17084502 17085000 17085001	30°
	17085205 17085208 17085210 17085216	38°
	17085209 17085211 17085217 17085219 17085222 17085223 17085224 17085225	36°
	17085226 17085227 17085228 17085229 17085290 17085292	24°
	All others	26°
Air valve wind-up (turns)	17080213 17080298 17082213 17083298 17084500 17084501	1
	17085001	1
	17080212	3/4

Air valve wind-up (continued)	17085217, 17085219, 17085222, 17085223, 17085224, 17085225 — 1/2
	All others — 7/8
Unloader	17080212, 17080213, 17080298, 17082213, 17083298, 17084500, 17084501, 17084502, 17085000, 17085001 — 40°
	17085003, 17085004, 17085213 — 35°
	17085125, 17085220, 17085221, 17085226, 17085227, 17085230, 17085231, 17085238, 17085239 — 32°
	All others — 39°

Carburetor specifications — Rochester E4M — 1983 (dimensions in inches)

For locations of carburetor part numbers, see Carburetor Identification

5.7L and 5.0L engines with a front vacuum break

Float level	7/16
Fast idle cam	20°
Front vacuum break	27°
Rear vacuum break	36°
Air valve wind-up(turns)	7/8
Unloader	36°

5.0L engine

Float level	11/28
Fast idle cam	20°
Rear vacuum break	27°
Air valve wind-up(turns)	7/8
Unloader	38°

Carburetor specifications — Rochester E4M — 1984 (dimensions in inches)

For locations of carburetor part numbers, see Carburetor Identification

Float level	17084507, 17084509, 17084525, 17084527 — 14/32
	All others — 11/32
Fast idle cam	17084205, 17084209 — 38°
	All others — 20°
Front vacuum break	17084525, 17084527 — 25°
	All others — 27°
Rear vacuum break	36° if equipped
Air valve wind-up(turns)	17084507, 17084509, 17084525, 17084527 — 1
	All others — 7/8
Unloader	17084507, 17084509, 17084525, 17084527 — 36°
	All others — 38°

Carburetor adjustment data (1986 on)

Carburetor specifications — Rochester 1ME/1MEF (dimensions in inches)

1986	Float level	Metering rod	Choke rod	Vacuum break	Unloader
Carburetor No:					
17081009	11/32	0.090	0.275	0.400	0.520
17084329					
17085009					
17085036					
17085044					
1987					
Carburetor No:					
17086101	11/32	0.090	0.120	0.275	0.520
17086102					

Carburetor specifications — Rochester M4MC/M4ME (dimensions in inches)

Float level
17085000 .... 3/8
17085001
17085003 .... 13/32
17085004
17085206
17085207
17085208
17085209
17085210
17085211
17085212
17085213
17085215
17085216
17085217
17085219
17085220
17085221
17085222
17085223
17085224
17085225
17085226
17085227
17085228
17085229
17085230
17085231
17085235
17085238
17085239
17085283
17085284
17085285
17085290
17085291
17085293
17085294
17085298

Pump rod
17085000 .... 9/32
17085001
17085003
17085004
17085206
17085208
17085209 .... 3/8
17085210 .... 9/32
17085211 .... 3/8
17085212 .... 9/32
17085213
17085215
17085216
17085217
17085219

17085220	3/8
17085221	
17085222	9/32
17085223	3/8
17085224	9/32
17085225	3/8
17085226	9/32
17085227	
17085228	
17085229	
17085230	
17085231	
17085235	
17085238	3/8
17085239	
17085283	9/32
17085284	
17085285	
17085290	
17085291	3/8
17085292	9/32
17085293	3/8
17085294	9/32
17085298	
Fast idle cam	
17085000	46°
17085001	
17085003	
17085004	
17085206	
17085208	20°
17085209	
17085210	
17085211	
17085212	46°
17085213	
17085215	
17085216	20°
17085217	
17085219	
17085220	
17085221	
17085222	
17085223	
17085224	
17085225	
17085226	
17085227	
17085228	46°
17085229	
17085230	20°
17085231	
17085235	46°
17085238	20°
17085239	
17085283	
17085284	
17085285	
17085290	46°
17085291	
17085292	
17085293	
17085294	
17085298	
Air valve rod (all)	0.025
Front vacuum break	
17085000	24°
17085001	23°
17085003	
17085004	
17085208	26°
17085209	
17085210	
17085211	

Model	Setting
17085212	23°
17085213	
17085216	26°
17085217	
17085219	
17085222	
17085223	
17085224	
17085225	

Rear vacuum break

Model	Setting
17085000	30°
17085001	
17085206	26°
17085208	38°
17085209	36°
17085210	38°
17085211	36°
17085215	26°
17085216	38°
17085217	36°
17085219	
17085220	26°
17085221	
17085222	36°
17085223	
17085224	
17085225	
17085226	24°
17085227	
17085228	
17085229	
17085230	26°
17085231	
17085235	
17085238	
17085239	
17085283	24°
17085284	26°
17085285	24°
17085290	
17085291	26°
17085292	24°
17085293	26°
17085294	
17085298	

Air valve windup (number of turns)

Model	Setting
17085000	7/8
17085001	1
17085003	7/8
17085004	
17085206	
17085208	
17085209	
17085210	
17085211	
17085212	
17085213	
17085215	
17085216	
17085217	1/2
17085219	
17085220	7/8
17085221	
17085222	
17085223	1/2
17085224	
17085225	
17085226	7/8
17085227	
17085228	
17085229	
17085230	
17085231	
17085235	

Air valve windup (number of turns) (continued)

17085238	7/8
17085239	
17085283	
17085284	
17085285	
17085290	
17085291	
17085293	
17085294	
17085298	

Unloader

17085000	40°
17085001	
17085003	35°
17085004	
17085206	39°
17085208	
17085209	
17085210	
17085211	
17085212	35°
17085213	
17085215	32°
17085216	39°
17085217	
17085219	
17085220	32°
17085221	
17085222	39°
17085223	
17085224	
17085225	
17085226	32°
17085227	
17085228	39°
17085229	
17085230	32°
17085231	
17085235	39°
17085238	32°
17085239	
17085283	
17085284	
17085285	
17085290	39°
17085291	
17085292	
17085293	
17085294	
17085298	

Carburetor specifications — Rochester E4ME (dimensions in inches)

Float level	7/16
Lean mixture screw	1.304
Idle mixture needle (number of turns)	3 (final adjustment on vehicle)
Idle air bleed valve	1.756
Choke stat lever	0.120
Choke link cam	20°
Air valve rod	0.025
Front vacuum break	
17085502	26°
17085503	
17085506	27°
17085508	
17085524	25°
17085526	
Rear vacuum break	36°
Air valve windup (number of turns)	
17085502	7/8
17085503	

17085506	1
17085508	
17085524	
17085526	
Unloader	
17085502	39°
17085503	
17085506	36°
17085508	
17085524	
17085526	

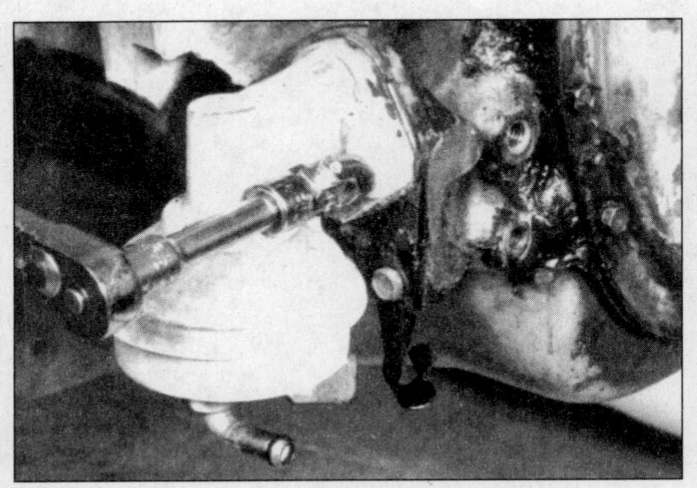

3.4a To remove the mechanical pump from the engine, remove the two mounting bolts . . .

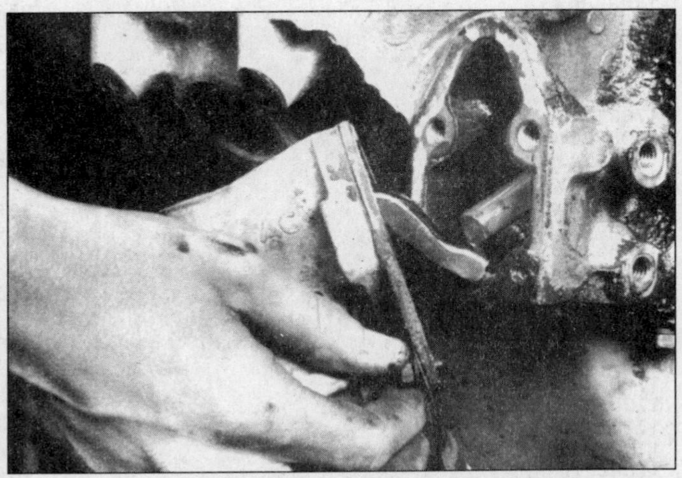

3.4b . . . then remove the pump and gasket from the block (be sure to remove all old gasket material from the pump and block mating surfaces)

3 Fuel pump - removal and installation

Warning: *Gasoline is extremely flammable, so extra precautions must be taken when working on any part of the fuel system. Do not smoke or allow open flames or bare light bulbs near the work area. Also, don't work in a garage if a natural gas-type appliance with a pilot light is present.*

Carbureted models

Refer to illustrations 3.4a, 3.4b and 3.5

Removal

1 The fuel pump is a sealed unit and cannot be serviced or rebuilt.

2 Disconnect the negative cable at the battery. **Warning:** *Place the cable out of the way so it cannot accidentally come in contact with the negative terminal of the battery - if it did, sparks could result and gasoline fumes could be ignited!*

3 Disconnect the fuel pump hoses from the fittings.

4 Remove the two bolts holding the pump to the block and detach the pump and gasket **(see illustrations)**. After noting how they are installed, the plate, gasket and rod can be removed if required.

3.5 When installing the fuel pump rod, use heavy grease to hold it in place

Installation

5 Installation is the reverse of removal. When installing the rod, make sure it's positioned with the correct end out. Heavy grease can be used to hold it in place while the fuel pump is installed **(see illustration)**.

6 Use a new gasket when installing the pump.

7 Connect the hoses (now would be a good time to install new ones).

8 Connect the battery, start the engine and check for fuel leaks.

Fuel-injected models

Removal

Refer to illustration 3.12

9 Disconnect the cable from the negative battery terminal.

10 Remove the fuel tank (Section 21).

11 The fuel pump/sending unit assembly is

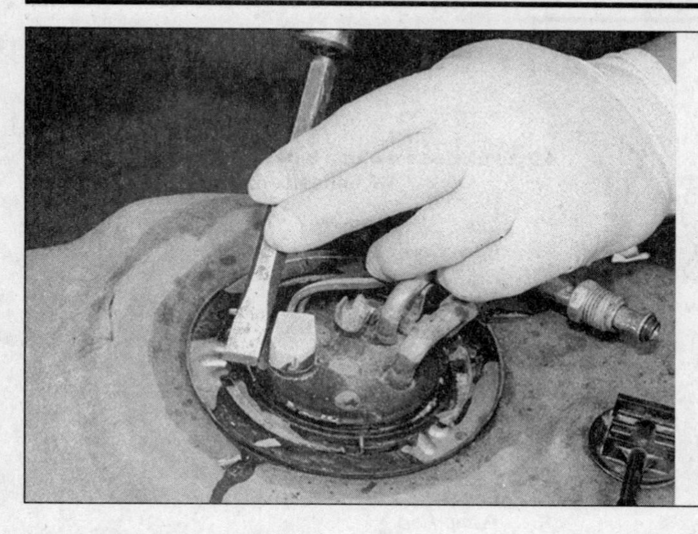

3.12 Carefully tap the lock ring counterclockwise until the locking tabs align with the slots in the fuel tank

located inside the fuel tank. It's held in place by a cam lock-ring mechanism.

12 To unlock the fuel pump/sending unit assembly, turn the lock-ring counterclockwise **(see illustration)**. If the lock-ring is too tight to release by hand, gently knock it loose with a brass or rubber hammer. **Warning:** *DO NOT use a steel hammer to knock the lock-ring loose. A spark could cause an explosion.*

13 Carefully extract the fuel pump/sending unit assembly from the tank. **Caution:** *The fuel level float and sending unit are delicate. Don't bump them into the lock-ring during removal or the accuracy of the sending unit may be affected.*

14 Check the condition of the seal around the mouth of the lock-ring mechanism. If it's dried out, cracked or deteriorated, replace it.

15 Inspect the strainer on the lower end of the fuel pump. If it's dirty, remove it, clean it with solvent and blow it out with compressed air. If it's too dirty to be cleaned, replace it.

16 If it's necessary to separate the fuel pump and sending unit, pull the fuel pump into the rubber connector and slide the pump away from the bottom support. Care should be taken to prevent damage to the rubber insulator and fuel strainer during removal. After the pump clears the bottom support, pull it out of the rubber connector for removal.

Installation

17 Insert the fuel pump/sending unit assembly into the fuel tank.

18 Turn the lock-ring clockwise until the locking cam is fully engaged by the retaining tangs.

19 Install the fuel tank (Section 21).

4 Carburetors - general information

Refer to illustrations 4.2, 4.3, 4.6, 4.7 and 4.11

Depending on model, year and engine size, carburetors of various types were used on these vehicles. On early models (1967 through 1970), some carburetors have manually-operated chokes, while later models have automatic chokes.

Rochester MV series

Carburetors in this series are a single barrel type with a triple venturi and plain tube nozzle **(see illustration)**. Fuel metering is controlled by a main well air bleed and a fixed orifice jet. During acceleration and at high engine speeds, a power enrichment system ensures top performance. On later model vehicles, an automatic choke system is incorporated. It operates via an exhaust heated coil. A throttle closing solenoid (controlled through the ignition switch) is used to ensure that the throttle valve closes completely as the ignition is turned off (to prevent dieseling).

Rochester 2G/4G series

This Rochester carburetor is a 2-barrel (2GC) or 4-barrel (4GC), side bowl design **(see illustration)**. The units installed on manual and automatic transmission equipped vehicles are similar but vary in calibration.

The main metering jets are fixed - calibration is handled by a system of air bleeds. A power enrichment valve is incorporated (power mixtures are controlled by air velocity past the boost venturi according to engine demand).

On later model vehicles, an electrically-operated throttle closing solenoid (controlled through the ignition switch) is used to ensure that the throttle valve closes completely after the ignition is turned off (to prevent dieseling). The choke is automatic and is operated by an exhaust manifold heated coil.

Rochester 4MV (Quadrajet) series

This is a 4-barrel two stage unit. The primary side uses a triple venturi system **(see illustration)**. The secondary side has two large bores and one metering system which supplements the primary main metering system and receives fuel from a common float chamber.

Rochester M4MC (Quadrajet) series

This is also a 4-barrel two stage unit and is very similar to the 4MV carburetor **(see illustration)**.

Rochester 1ME series

The Rochester 1ME carburetor is a single barrel carburetor using a triple venturi with plain tube nozzle **(see illustration 4.2)**. Fuel metering is controlled by a main well air bleed in a variable orifice jet.

During acceleration and at high engine speeds, a power enrichment system maintains top performance. An automatic choke system is incorporated, which operates via an electrically-heated choke coil. The idle system incorporates a hot idle compensator to maintain smooth engine idle during extreme hot engine operation.

A throttle closing solenoid controlled through the ignition switch is used to ensure that the throttle valve closes completely after the ignition is turned off (to prevent dieseling).

Rochester 2SE/E2SE series

This carburetor is a 2-barrel, two stage design **(see illustration)**. The primary side uses a triple venturi system. The secondary stage has a single large bore with a single tapered metering rod. An automatic choke system is incorporated, using an electrically-heated thermostatic choke coil that is mounted on the secondary side of the carburetor.

A check valve is mounted in the fuel inlet line and is used to shut off fuel flow to the carburetor and prevent fuel leaks in case of a vehicle roll-over.

Later model vehicles use the Rochester E2SE Varajet carburetor, which is very similar to the earlier 2SE model except for the changes due to increased emission restrictions. These carburetors work together with the emission systems through the use of a mixture control solenoid mounted in the float bowl.

Rochester M2M series

This is a 2-barrel unit with a triple venturi system. It is very similar to the primary side of the M4M carburetor.

The M2ME carburetors used on air conditioned vehicles have an electrically-operated air conditioner idle speed solenoid which maintains proper idle speed when the air conditioner is being used. The model M2M carburetors have an identification number stamped vertically on the left rear corner of the float bowl.

Later model vehicles use the Rochester E2M Varajet carburetor which is very similar to the earlier M2M model except for the changes due to increased emission restrictions. These carburetors work together with the emission systems through the use of a mixture control solenoid mounted in the float bowl.

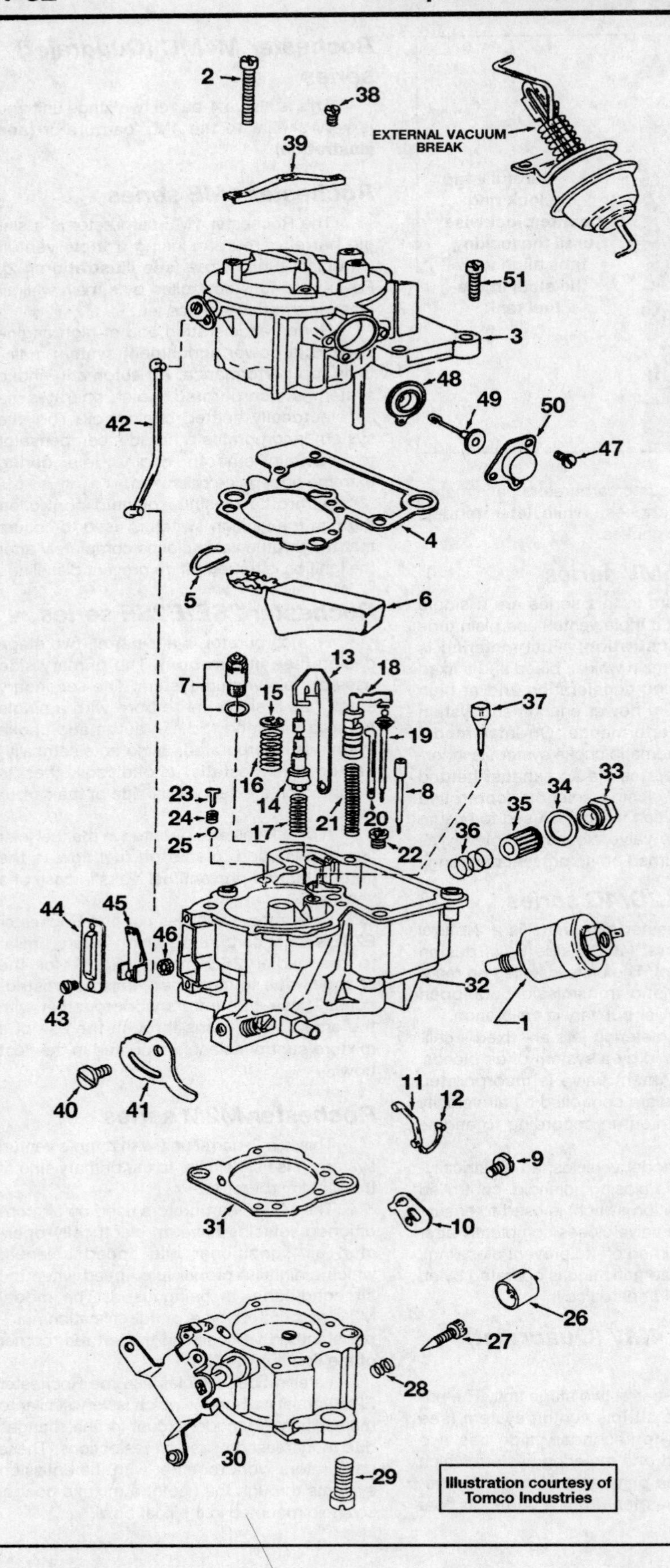

4.2 Exploded view of a typical Rochester M, MV and ME series carburetor

1 *Idle Stop Solenoid*
2 *Bowl Cover Screw & Lockwasher*
3 *Bowl Cover Assembly*
4 *Bowl Cover Gasket*
5 *Hinge Pin Float*
6 *Float Assembly*
7 *Rotary Valve & Gasket*
8 *Idle Tube Assembly*
9 *Pump & Power Link Lever Screw*
10 *Pump & Power Link Lever*
11 *Pump Link*
12 *Power Piston Rod Link*
13 *Pump Rod*
14 *Pump Cup & Stem Assembly*
15 *Pump Spring Retainer*
16 *Pump Spring*
17 *Pump Return Spring*
18 *Power Piston Assembly*
19 *Power Piston & Metering Rod Assembly*
20 *Power Piston Rod*
21 *Power Piston Spring*
22 *Main Metering Jet*
23 *Pump Discharge Spring Guide*
24 *Pump Discharge Ball Spring*
25 *Pump Discharge Check Ball*
26 *Idle Limiter Cap*
27 *Idle Adjusting Needle*
28 *Idle Adjusting Needle Spring*
29 *Throttle Body Assembly Screw & Lockwasher*
30 *Throttle Body Assembly*
31 *Throttle Body Gasket*
32 *Float Bowl Assembly*
33 *Fuel Inlet Filter Nut*
34 *Filter Nut Gasket*
35 *Fuel Filter*
36 *Fuel Filter Spring*
37 *Power valve (not used on all models)*
38 *Air Cleaner Stud Bracket Screw*
39 *Air Cleaner Stud Bracket*
40 *Fast Idle Cam Screw*
41 *Fast Idle Cam*
42 *Choke Rod*
43 *Idle Compensator Cover Screw*
44 *Idle Compensator Cover*
45 *Idle Compensator Assembly*
46 *Idle Compensator Gasket*
47 *Bowl Cover Screw*
48 *Vacuum Break Diaphragm*
49 *Vacuum Break Diaphragm Link Assembly*
50 *Diaphragm Cover*
51 *Bowl Cover Screw*

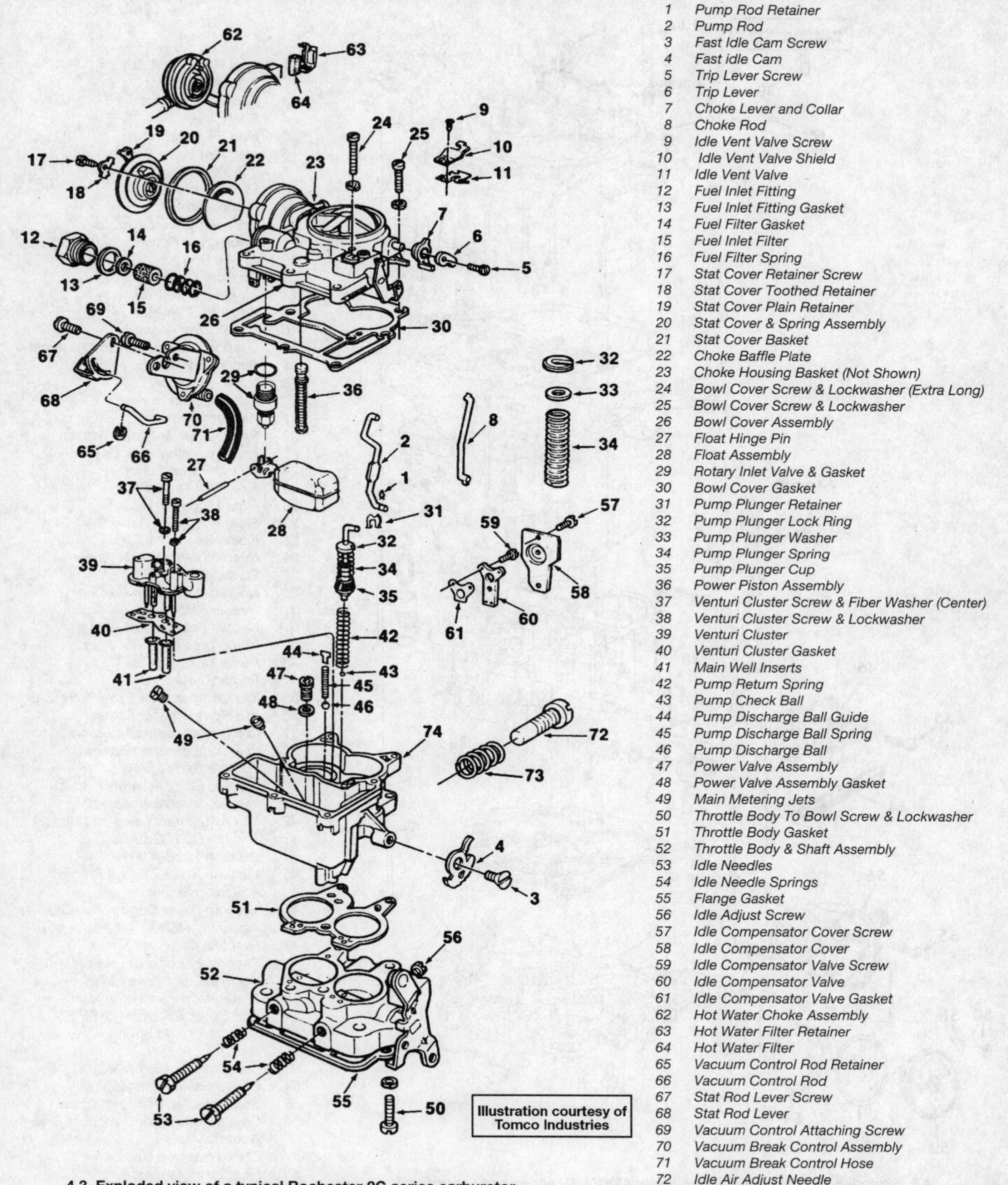

1 Pump Rod Retainer
2 Pump Rod
3 Fast Idle Cam Screw
4 Fast idle Cam
5 Trip Lever Screw
6 Trip Lever
7 Choke Lever and Collar
8 Choke Rod
9 Idle Vent Valve Screw
10 Idle Vent Valve Shield
11 Idle Vent Valve
12 Fuel Inlet Fitting
13 Fuel Inlet Fitting Gasket
14 Fuel Filter Gasket
15 Fuel Inlet Filter
16 Fuel Filter Spring
17 Stat Cover Retainer Screw
18 Stat Cover Toothed Retainer
19 Stat Cover Plain Retainer
20 Stat Cover & Spring Assembly
21 Stat Cover Basket
22 Choke Baffle Plate
23 Choke Housing Basket (Not Shown)
24 Bowl Cover Screw & Lockwasher (Extra Long)
25 Bowl Cover Screw & Lockwasher
26 Bowl Cover Assembly
27 Float Hinge Pin
28 Float Assembly
29 Rotary Inlet Valve & Gasket
30 Bowl Cover Gasket
31 Pump Plunger Retainer
32 Pump Plunger Lock Ring
33 Pump Plunger Washer
34 Pump Plunger Spring
35 Pump Plunger Cup
36 Power Piston Assembly
37 Venturi Cluster Screw & Fiber Washer (Center)
38 Venturi Cluster Screw & Lockwasher
39 Venturi Cluster
40 Venturi Cluster Gasket
41 Main Well Inserts
42 Pump Return Spring
43 Pump Check Ball
44 Pump Discharge Ball Guide
45 Pump Discharge Ball Spring
46 Pump Discharge Ball
47 Power Valve Assembly
48 Power Valve Assembly Gasket
49 Main Metering Jets
50 Throttle Body To Bowl Screw & Lockwasher
51 Throttle Body Gasket
52 Throttle Body & Shaft Assembly
53 Idle Needles
54 Idle Needle Springs
55 Flange Gasket
56 Idle Adjust Screw
57 Idle Compensator Cover Screw
58 Idle Compensator Cover
59 Idle Compensator Valve Screw
60 Idle Compensator Valve
61 Idle Compensator Valve Gasket
62 Hot Water Choke Assembly
63 Hot Water Filter Retainer
64 Hot Water Filter
65 Vacuum Control Rod Retainer
66 Vacuum Control Rod
67 Stat Rod Lever Screw
68 Stat Rod Lever
69 Vacuum Control Attaching Screw
70 Vacuum Break Control Assembly
71 Vacuum Break Control Hose
72 Idle Air Adjust Needle
73 Idle Air Adjust Needle Spring
74 Bowl Assembly

4.3 Exploded view of a typical Rochester 2G series carburetor

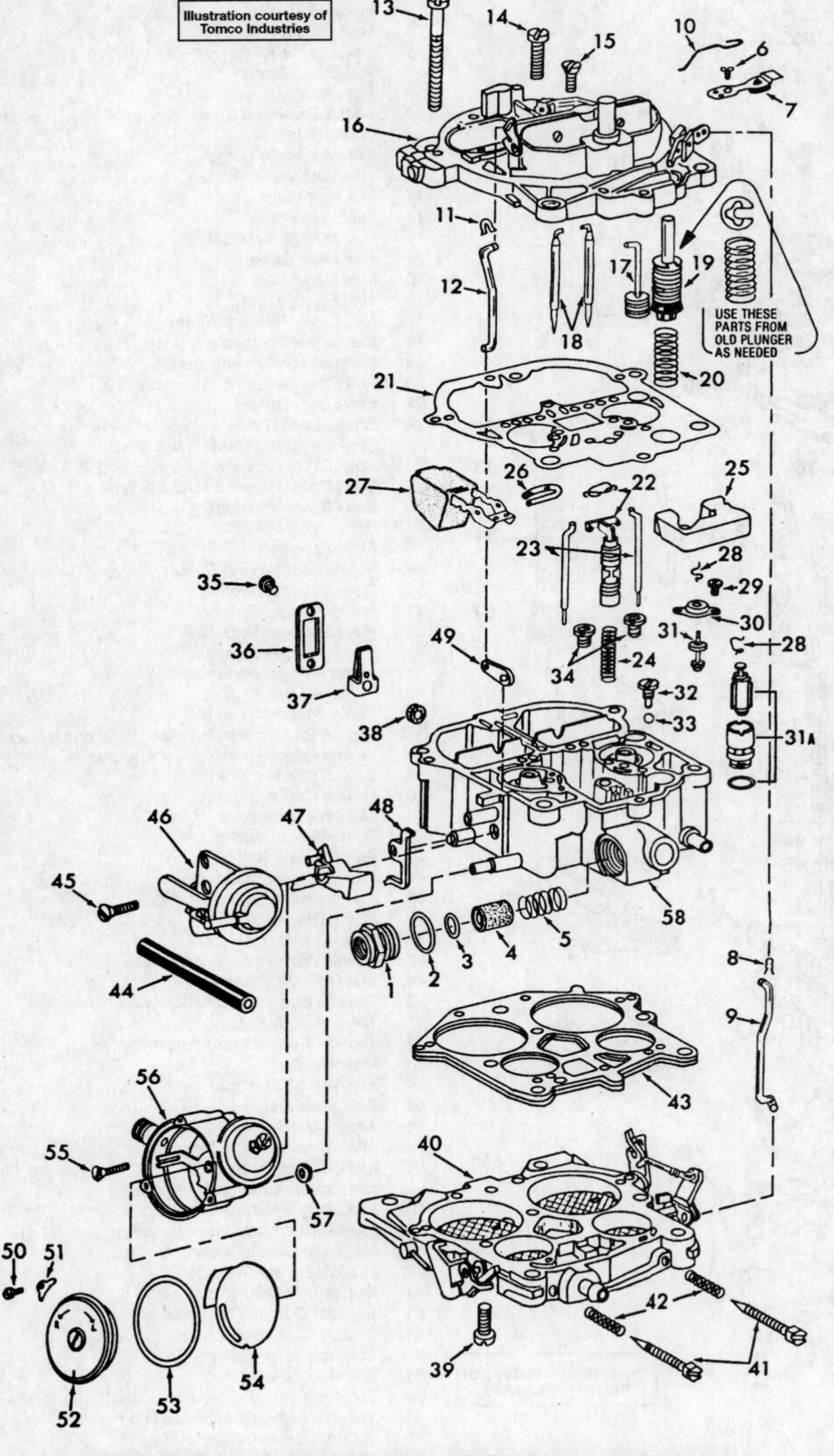

1 *Fuel Filter Nut (Inlet)*
2 *Filter Nut Gasket*
3 *Fuel Filter Gasket*
4 *Fuel Filter*
5 *Fuel Filter Spring*
6 *Idle Vent Valve Screw*
7 *Idle Vent Valve*
8 *Pump Rod Retainer*
9 *Pump Rod*
10 *Idle Vent Valve Lever*
11 *Choke Rod Retainer*
12 *Choke Rod*
13 *Air Horn Screw (4)*
14 *Air Horn Screw (3)*
15 *Air Horn Screw (2)*
16 *Bowl Cover Assembly*
17 *Dashpot Piston And Rod Assembly*
18 *Secondary Metering Rod (2)*
19 *Pump Assembly*
20 *Pump Return Spring*
21 *Air Horn Gasket*
22 *Primary Power Piston Assembly*
23 *Primary Metering Rods (2)*
24 *Power Piston Spring*
25 *Float Bowl Insert*
26 *Float Hinge Pin*
27 *Float And Lever Assembly*
28 *Float Needle Full Clip*
29 *Needle Diaphragm Retainer Screw (2)*
30 *Needle Diaphragm Retainer*
31 *Needle Diaphragm Assembly*
31A *Needle Seat Gasket Assembly*
32 *Pump Discharge Ball Plug*
33 *Pump Discharge Ball*
34 *Primary Jets (2)*
35 *Idle Compensator Cover Screw (2)*
36 *Idle Compensator Cover*
37 *Idle Compensator Assembly*
38 *Idle Compensator Gasket*
39 *Throttle Body Screw (3)*
40 *Throttle Body Assembly*
41 *Idle Adjustment Needle (2)*
42 *Idle Adjustment Needle Spring (2)*
43 *Throttle Body Gasket*
44 *Vacuum Hose (4MV)*
45 *Vacuum Break Control Bracket Attaching Screw*
46 *Vacuum Break Control And Bracket Assembly (4MV)*
47 *Fast Idle Cam*
48 *Secondary Lockout Lever (4MV)*
49 *Intermediate Choke Lever*
50 *Stat Retainer Screw (3) (4MC)*
51 *Stat Cover Retainers (3) (4MC)*
52 *Stat Cover And Spring Assembly (4MC)*
53 *Stat Cover Gasket (4MC)*
54 *Choke Baffle Plate (4MC)*
55 *Stat Housing Attaching Screw*
56 *Choke Housing And Vacuum Break Assembly (4MC)*
57 *Choke Housing Gasket (4MC)*
58 *Float Bowl Assembly*

4.6 Exploded view of a typical Rochester 4MC/4MV series carburetor

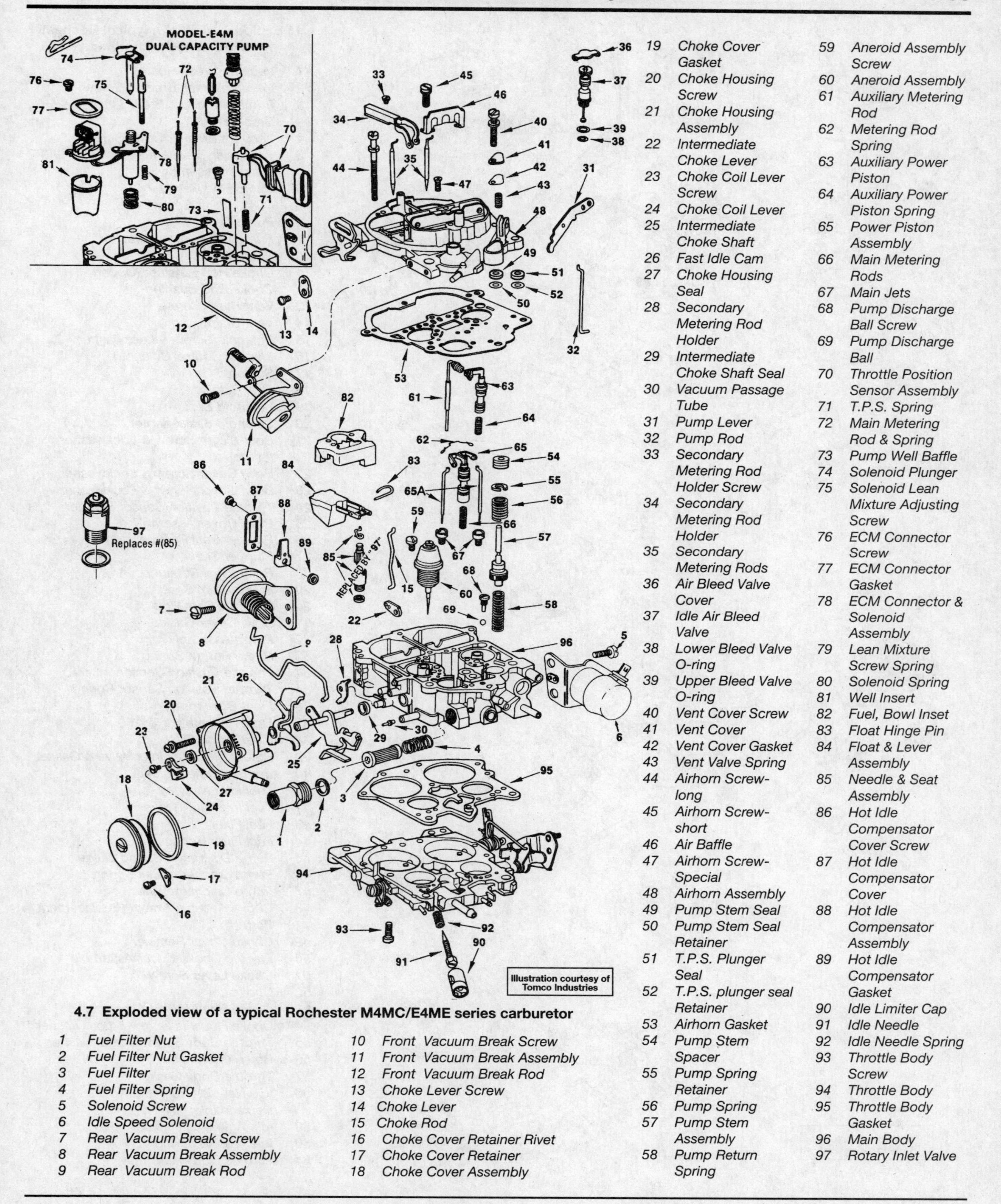

4.7 Exploded view of a typical Rochester M4MC/E4ME series carburetor

1 *Fuel Filter Nut*
2 *Fuel Filter Nut Gasket*
3 *Fuel Filter*
4 *Fuel Filter Spring*
5 *Solenoid Screw*
6 *Idle Speed Solenoid*
7 *Rear Vacuum Break Screw*
8 *Rear Vacuum Break Assembly*
9 *Rear Vacuum Break Rod*
10 *Front Vacuum Break Screw*
11 *Front Vacuum Break Assembly*
12 *Front Vacuum Break Rod*
13 *Choke Lever Screw*
14 *Choke Lever*
15 *Choke Rod*
16 *Choke Cover Retainer Rivet*
17 *Choke Cover Retainer*
18 *Choke Cover Assembly*
19 *Choke Cover Gasket*
20 *Choke Housing Screw*
21 *Choke Housing Assembly*
22 *Intermediate Choke Lever*
23 *Choke Coil Lever Screw*
24 *Choke Coil Lever*
25 *Intermediate Choke Shaft*
26 *Fast Idle Cam*
27 *Choke Housing Seal*
28 *Secondary Metering Rod Holder*
29 *Intermediate Choke Shaft Seal*
30 *Vacuum Passage Tube*
31 *Pump Lever*
32 *Pump Rod*
33 *Secondary Metering Rod Holder Screw*
34 *Secondary Metering Rod Holder*
35 *Secondary Metering Rods*
36 *Air Bleed Valve Cover*
37 *Idle Air Bleed Valve*
38 *Lower Bleed Valve O-ring*
39 *Upper Bleed Valve O-ring*
40 *Vent Cover Screw*
41 *Vent Cover*
42 *Vent Cover Gasket*
43 *Vent Valve Spring*
44 *Airhorn Screw-long*
45 *Airhorn Screw-short*
46 *Air Baffle*
47 *Airhorn Screw-Special*
48 *Airhorn Assembly*
49 *Pump Stem Seal*
50 *Pump Stem Seal Retainer*
51 *T.P.S. Plunger Seal*
52 *T.P.S. plunger seal Retainer*
53 *Airhorn Gasket*
54 *Pump Stem Spacer*
55 *Pump Spring Retainer*
56 *Pump Spring*
57 *Pump Stem Assembly*
58 *Pump Return Spring*
59 *Aneroid Assembly Screw*
60 *Aneroid Assembly*
61 *Auxiliary Metering Rod*
62 *Metering Rod Spring*
63 *Auxiliary Power Piston*
64 *Auxiliary Power Piston Spring*
65 *Power Piston Assembly*
66 *Main Metering Rods*
67 *Main Jets*
68 *Pump Discharge Ball Screw*
69 *Pump Discharge Ball*
70 *Throttle Position Sensor Assembly*
71 *T.P.S. Spring*
72 *Main Metering Rod & Spring*
73 *Pump Well Baffle*
74 *Solenoid Plunger*
75 *Solenoid Lean Mixture Adjusting Screw*
76 *ECM Connector Screw*
77 *ECM Connector Gasket*
78 *ECM Connector & Solenoid Assembly*
79 *Lean Mixture Screw Spring*
80 *Solenoid Spring*
81 *Well Insert*
82 *Fuel, Bowl Inset*
83 *Float Hinge Pin*
84 *Float & Lever Assembly*
85 *Needle & Seat Assembly*
86 *Hot Idle Compensator Cover Screw*
87 *Hot Idle Compensator Cover*
88 *Hot Idle Compensator Assembly*
89 *Hot Idle Compensator Gasket*
90 *Idle Limiter Cap*
91 *Idle Needle*
92 *Idle Needle Spring*
93 *Throttle Body Screw*
94 *Throttle Body*
95 *Throttle Body Gasket*
96 *Main Body*
97 *Rotary Inlet Valve*

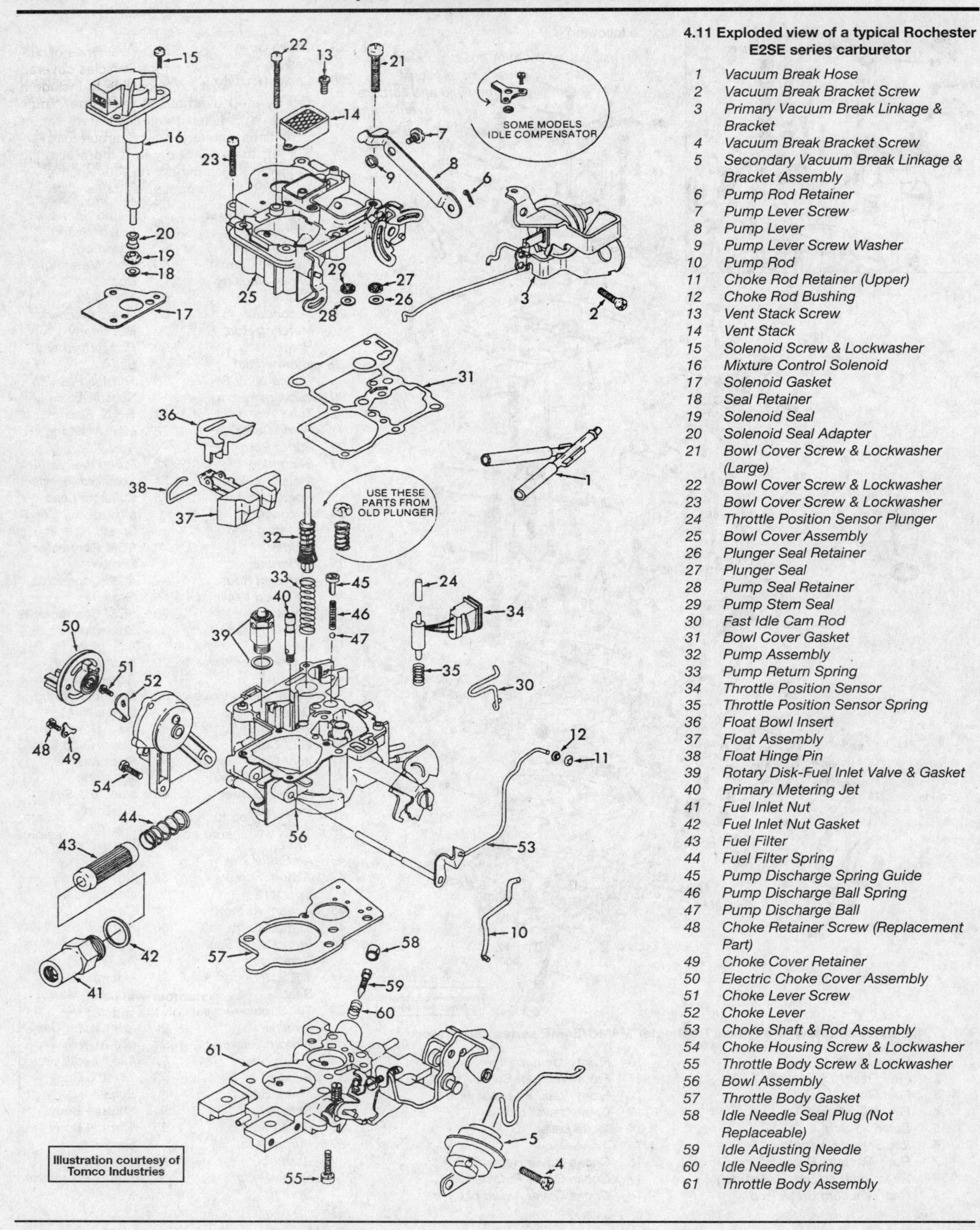

4.11 Exploded view of a typical Rochester E2SE series carburetor

1. *Vacuum Break Hose*
2. *Vacuum Break Bracket Screw*
3. *Primary Vacuum Break Linkage & Bracket*
4. *Vacuum Break Bracket Screw*
5. *Secondary Vacuum Break Linkage & Bracket Assembly*
6. *Pump Rod Retainer*
7. *Pump Lever Screw*
8. *Pump Lever*
9. *Pump Lever Screw Washer*
10. *Pump Rod*
11. *Choke Rod Retainer (Upper)*
12. *Choke Rod Bushing*
13. *Vent Stack Screw*
14. *Vent Stack*
15. *Solenoid Screw & Lockwasher*
16. *Mixture Control Solenoid*
17. *Solenoid Gasket*
18. *Seal Retainer*
19. *Solenoid Seal*
20. *Solenoid Seal Adapter*
21. *Bowl Cover Screw & Lockwasher (Large)*
22. *Bowl Cover Screw & Lockwasher*
23. *Bowl Cover Screw & Lockwasher*
24. *Throttle Position Sensor Plunger*
25. *Bowl Cover Assembly*
26. *Plunger Seal Retainer*
27. *Plunger Seal*
28. *Pump Seal Retainer*
29. *Pump Stem Seal*
30. *Fast Idle Cam Rod*
31. *Bowl Cover Gasket*
32. *Pump Assembly*
33. *Pump Return Spring*
34. *Throttle Position Sensor*
35. *Throttle Position Sensor Spring*
36. *Float Bowl Insert*
37. *Float Assembly*
38. *Float Hinge Pin*
39. *Rotary Disk-Fuel Inlet Valve & Gasket*
40. *Primary Metering Jet*
41. *Fuel Inlet Nut*
42. *Fuel Inlet Nut Gasket*
43. *Fuel Filter*
44. *Fuel Filter Spring*
45. *Pump Discharge Spring Guide*
46. *Pump Discharge Ball Spring*
47. *Pump Discharge Ball*
48. *Choke Retainer Screw (Replacement Part)*
49. *Choke Cover Retainer*
50. *Electric Choke Cover Assembly*
51. *Choke Lever Screw*
52. *Choke Lever*
53. *Choke Shaft & Rod Assembly*
54. *Choke Housing Screw & Lockwasher*
55. *Throttle Body Screw & Lockwasher*
56. *Bowl Assembly*
57. *Throttle Body Gasket*
58. *Idle Needle Seal Plug (Not Replaceable)*
59. *Idle Adjusting Needle*
60. *Idle Needle Spring*
61. *Throttle Body Assembly*

5 Carburetor - removal and installation

Warning: *Gasoline is extremely flammable, so extra precautions must be taken when working on any part of the fuel system. Do not smoke or allow open flames or bare light bulbs near the work area. Also, don't work in a garage if a natural gas-type appliance with a pilot light is present.*

1 Remove the air cleaner (see Chapter 1).

2 Disconnect the fuel and vacuum lines from the carburetor. Mark the hoses and fittings with pieces of numbered tape to ensure correct installation.

3 Mark and disconnect the wire harness connectors from the choke and carburetor.

4 Disconnect the throttle linkage and throttle return springs.

5 Disconnect the throttle valve or detent linkage (automatic transmission equipped models).

6 Remove the carburetor mounting nuts and bolts and detach the carburetor from the intake manifold. **Note:** *Quadrajet carburetors are attached to the manifold with two nuts and two long bolts. The bolts are located at the front and extend all the way through the air horn assembly, the main body and the throttle body of the carburetor.*

7 Remove the insulator and gaskets. On later models, lift off the EFE heater as well.

8 Installation is the reverse of removal, but the following points should be noted:

a) By filling the float bowl with fuel, the initial start up will be easier and there will be less strain on the battery.
b) New gaskets must be used.
c) Idle speed and mixture should be checked and adjusted if necessary.

6 Carburetor - servicing

Refer to illustration 6.7

Warning: *Gasoline is extremely flammable, so extra precautions must be taken when working on any part of the fuel system. Do not smoke or allow open flames or bare light bulbs near the work area. Also, don't work in a garage if a natural gas-type appliance with a pilot light is present.*

1 A thorough road test and check of the carburetor adjustments should be done before any carburetor service work is done. Specifications for some adjustments are listed on the Vehicle Emissions Control Information label found in the engine compartment.

2 Some performance complaints directed at the carburetor are actually a result of loose, out-of-adjustment or malfunctioning engine or electrical components. Others develop when vacuum hoses leak, are disconnected or are incorrectly routed. The proper approach to analyzing carburetor problems should include a routine check of the following:

a) Inspect all vacuum hoses and actuators for leaks and correct installation.
b) Tighten the intake manifold and carburetor mounting nuts/bolts evenly and securely (Chapters 1 and 2).
c) Perform a cylinder compression test (Chapter 2).
d) Clean or replace the spark plugs as necessary (Chapter 1).
e) Check the spark plug wires (Chapter 1).
f) Inspect the ignition primary wires (Chapter 1) and check the vacuum advance operation (Chapter 5). Replace any defective parts.
g) Check the ignition timing (Chapter 1).
h) Inspect the heat control valve in the air cleaner for proper operation (Chapter 1).
i) Service the air and fuel filters (Chapter 1).
j) Check the fuel pump (Section 2).
k) Check the PCV valve (Chapter 1).

3 Carburetor problems usually show up as flooding, hard starting, stalling, severe backfiring, poor acceleration and lack of response to idle mixture screw adjustments. A carburetor that is leaking fuel and/or covered with wet-looking deposits definitely needs attention.

4 Diagnosing carburetor problems may require that the engine be started and run with the air cleaner removed. While running the engine without the air cleaner it's possible that it could backfire. A backfiring situation is likely to occur if the carburetor is malfunctioning, but the removal of the air cleaner alone can lean the fuel/air mixture enough to produce an engine backfire. **Warning:** *DO NOT position your face directly over the carburetor throat during inspection and servicing procedures.*

5 Once it's determined that the carburetor is in need of work or an overhaul, several alternatives should be considered. If you're going to attempt to overhaul the carburetor yourself, first obtain a good quality carburetor rebuild kit. You'll also need some special solvent and a means of blowing out the internal passages of the carburetor with air.

6 Due to the fact that several types of carburetors were used on the vehicles covered by this manual, it isn't possible to include a step-by-step overhaul of each one. You'll receive a detailed, well-illustrated set of instructions with any quality carburetor overhaul kit; they will apply in a more specific manner to the carburetor on your vehicle.

7 Another alternative, which is probably the best approach, is to obtain a new or rebuilt carburetor. They are readily available from dealers and auto parts stores. The most important fact to consider when purchasing one of these units is to make sure the exchange carburetor is identical to the original. In most cases an identification number that will help you determine which type you have will be stamped on the side of the carburetor **(see illustration)**. When obtaining a rebuilt carburetor or a rebuild kit, take time to ensure that the kit or carburetor matches your application exactly. Seemingly insignificant differences can make a considerable difference in the performance of your engine.

8 If you choose to overhaul your own carburetor, allow enough time to disassemble the carburetor carefully, soak the necessary parts in the cleaning solvent (usually for at least one half day or according to the instructions listed on the carburetor cleaner) and reassemble it, which will usually take much longer than disassembly. When you're disassembling the carburetor, match each part with the illustration in the carburetor kit and lay the parts out in order on a clean work surface. Overhauls by inexperienced mechanics can result in an engine which runs poorly, or not at all. To avoid this, use care and patience when disassembling the carburetor so you can reassemble it correctly.

9 When the overhaul is complete, adjustments which may be beyond the ability of the home mechanic may be required, especially on later models. If so, take the vehicle to a dealer service department or a reputable tune-up shop for final carburetor adjustments, which will ensure compliance with emissions regulations and acceptable performance.

6.7 An identification number will be stamped on the side of the carburetor body - copy the numbers off the carburetor when purchasing a rebuild kit or a new or rebuilt carburetor

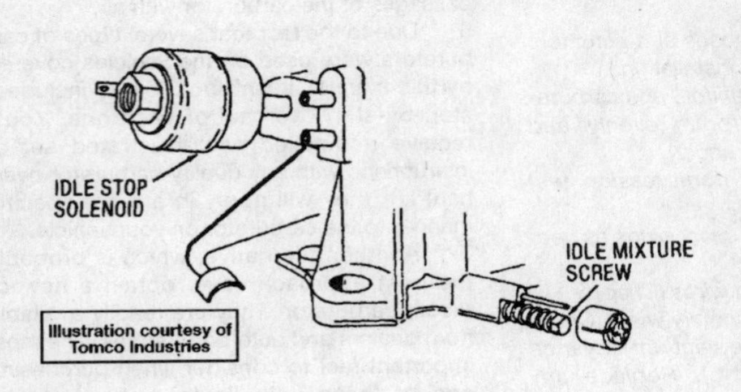

8.3 Rochester M and MV carburetor mixture screw locations

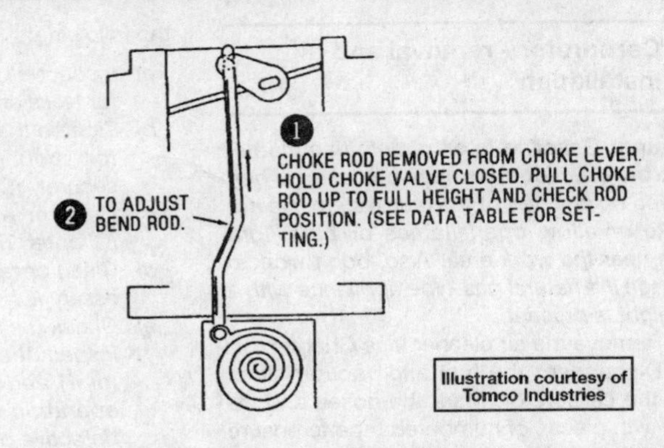

8.15 Rochester M and MV carburetor automatic choke adjustment details

7 Fuel lines - check and replacement

Warning: *Gasoline is extremely flammable, so extra precautions must be taken when working on any part of the fuel system. Do not smoke or allow open flames or bare light bulbs near the work area. Also, don't work in a garage if a natural gas-type appliance with a pilot light is present.*

1 The fuel lines on these vehicles are generally made of metal with short lengths of rubber hose connecting critical flex points such as the tank and fuel pump. The metal fuel lines are retained to the frame with various clips and brackets. They generally will require no service; however, if they are allowed to become loose from the retaining brackets, they can vibrate and eventually be worn through. If a fuel line must be replaced, leave it to a dealer or repair shop (special flaring and crimping tools are required to fabricate the lines).

2 If a short section of fuel line is damaged, rubber fuel hose can be used to replace it if it is no longer than 12-inches. Cut a length of rubber fuel hose longer than the section to be replaced and use a tubing cutter to remove the damaged portion of the metal line. Install the rubber fuel hose using two hose clamps at each end and check to make sure it isn't leaking.

3 If new fuel lines are necessary, they must be cut, formed and flared out of fuel system tubing. If you have the necessary equipment, remove the old fuel line from the vehicle and duplicate it out of new material.

4 Install the new fuel line and be careful to install new clamps and/or brackets where needed. Make sure the line is the same diameter, shape and quality as the original one. Make sure all flared ends conform to those of the original fuel line. Make sure fuel lines connected to the carburetor or other fittings are double flared. Make sure all metal particles are removed from inside the tubing before installation.

5 Always check rubber hoses for signs of leakage and deterioration.

6 If a rubber hose must be replaced, also replace the clamps.

8 Carburetor (Rochester M and MV) - adjustments

Refer to illustrations 8.3, 8.15, 8.19, 8.21, 8.27 and 8.29

Idle speed adjustment

1967 through 1972

1 Run the engine to normal operating temperature, make sure that the choke plate is completely open and have the air cleaner in position. If air conditioning is installed, switch it on during adjustment.

2 Make sure the ignition timing is correct.

3 Adjust the idle speed screw until the engine idles at the specified rpm **(see illustration)**. A tachometer should be connected to the engine to ensure accuracy.

4 Turn the mixture screw in or out to achieve the fastest, smoothest idle speed.

5 Turn the mixture screw in to obtain the specified decrease in the idle speed (lean drop).

6 Turn off the engine.

1973 and later

7 With the engine running and the solenoid lead connected, rotate the complete solenoid body to set the curb idle. This should be in accordance with the information provided on the VECI label.

8 The solenoid should be energized and the air conditioner (if installed) switched off.

9 Set the idle speed by first disconnecting the lead from the solenoid and then inserting a 1/8-inch Allen wrench into the end of the solenoid and turning it until the engine speed is as specified on the VECI label.

10 Reconnect the lead to the solenoid and shut off the engine.

Manual choke adjustment

11 Remove the air cleaner, push the choke knob all the way in and then pull it out 1/8-inch.

12 Release the choke cable clamp at the carburetor and adjust the position of the cable in the clamp until the choke plate is all the way open.

13 Tighten the cable clamp screw and then check the operation of the cable when the knob is pulled and pushed. Install the air cleaner.

Automatic choke adjustment

14 Remove the air cleaner and disconnect the choke rod from the choke lever.

15 Hold the choke plate closed with one hand and with the other, pull the choke rod down against the stop. The top of the choke rod should now be level with the bottom of the hole in the choke lever. If necessary, bend the rod until it is **(see illustration)**.

16 Reconnect the rod and install the air cleaner.

Fast idle speed adjustment

17 Make sure that the idle speed is correctly adjusted.

18 On manual choke carburetors, turn the fast idle cam to the highest position.

19 On automatic choke carburetors, set the cam follower on the highest step of the cam **(see illustration)**.

20 Bend the cam follower as necessary to achieve the specified fast idle.

Fast idle cam (choke rod) adjustment

21 After fast idle speed adjustment has been made, set the cam follower on the second step of the fast idle cam **(see illustration)**.

22 Hold the choke plate down. This should position the choke rod in the bottom end of the slot.

23 Insert a drill bit of the specified size between the lower edge of the choke plate and the air horn wall. The diameter of the drill bit should match the gap for this setting (shown in the Specifications according to the

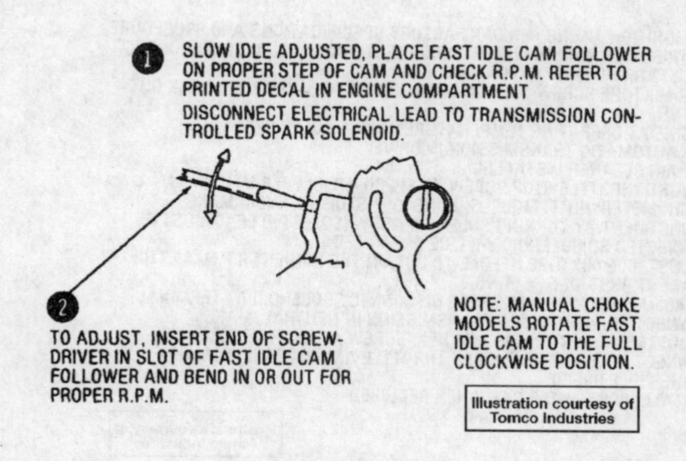

8.19 Rochester M and MV carburetor fast idle adjustment details

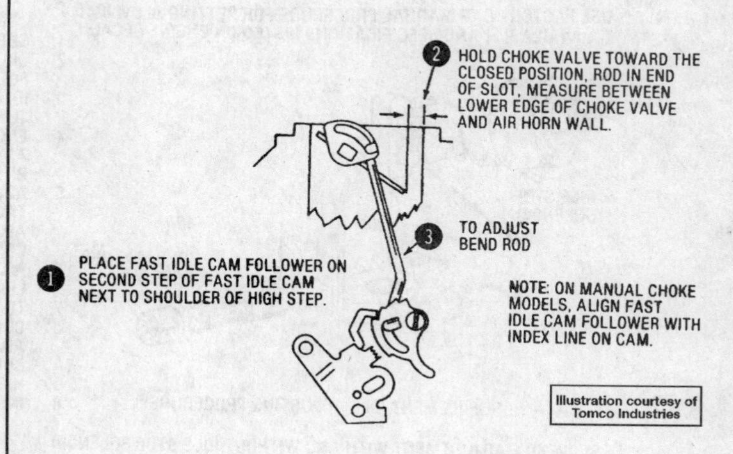

8.21 Rochester M and MV carburetor fast idle cam (choke rod) adjustment details

carburetor type).

24 Bend the choke rod as necessary to adjust the gap.

Vacuum break adjustment

25 Use a vacuum pump to seat the vacuum diaphragm.

26 If necessary, push the diaphragm plunger in until it seats.

27 Insert a drill bit of the specified size between the lower edge of the choke plate and the air horn wall **(see illustration)**.

28 Bend the connecting rod as necessary to adjust the gap.

Choke unloader adjustment

29 Press down on the choke plate. The choke rod should be in the end of the slot **(see illustration)**.

30 Move the throttle lever to hold the throttle plate wide open.

31 Insert a drill bit of the specified size between the lower edge of the choke plate and the air horn wall to measure the gap. If it differs from that specified, bend the tang on the throttle lever.

9 Carburetor (Rochester 1MV) - adjustments

Refer to illustrations 9.6, 9.24 and 9.34

Idle speed adjustment

1 Run the engine to normal operating temperature. Leave the air cleaner in position. The air conditioning must be Off on C10 and K10 models; it should be On if you have a C20, C30 or K20 model.

2 Connect a tachometer to the engine in accordance with the manufacturer's instructions.

3 On C10 and K10 models, disconnect and plug the carburetor and PCV hoses at the vapor canister.

4 On K20, C20 and larger models, disconnect the fuel tank hose from the vapor canister.

5 Place the manual transmission in Neutral. Shift the automatic transmission to Drive (C10 and K10 models) or Neutral (C20, C30 and K20 models). Make sure the parking brake is applied.

6 With the engine running, turn the idle stop solenoid (energized) in or out as neces-

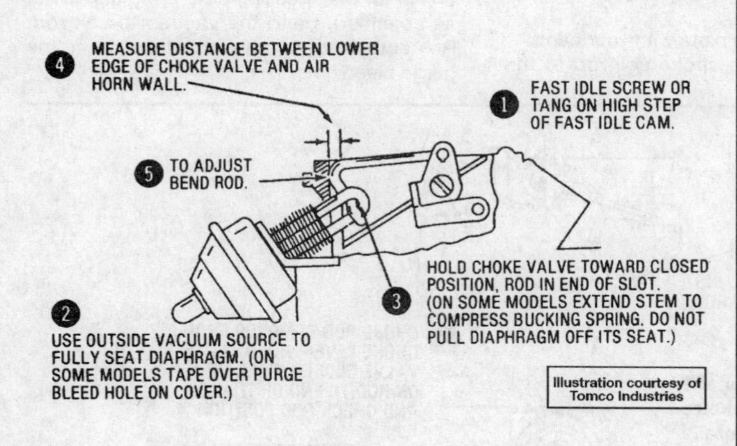

8.27 Rochester M and MV carburetor vacuum break adjustment details

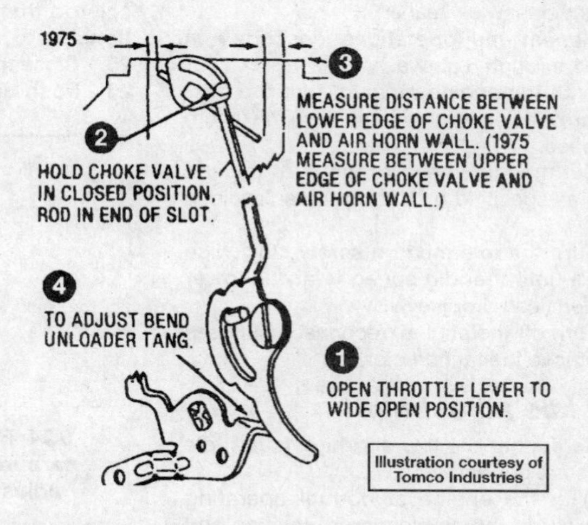

8.29 Rochester M and MV carburetor choke unloader adjustment details

USE FACTORY CAR MANUAL PROCEDURE FOR SETTING SLOW IDLE IF AVAILABLE, AND SPECIFICATIONS LISTED ON ENGINE DECAL.

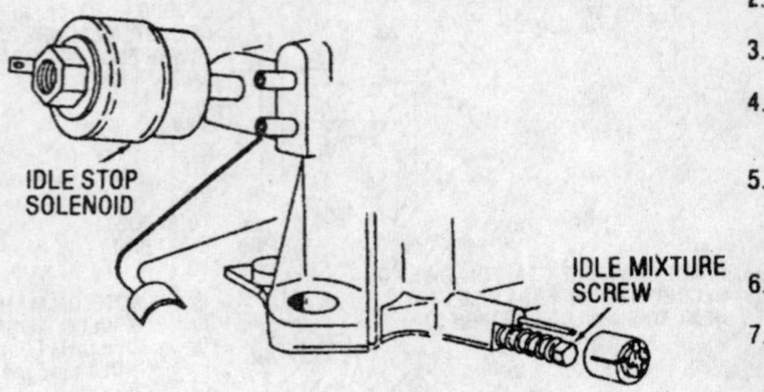

1. SET IGNITION TIMING PER CAR FACTORY SPECIFICATIONS AND PROCEDURE.
2. DISTRIBUTION VACUUM HOSE DISCONNECT AND PLUG. ALSO DISCONNECT FUEL TANK HOSE AT VAPOR CANISTER.
3. IDLE MIXTURE SCREW TURN IN UNTIL LIGHTLY SEATED THEN BACK OUT 4 TURNS.
4. ENGINE AT OPERATING TEMPERATURE, CHOKE FULLY OPEN.
   A. AUTOMATIC TRANSMISSION IN DRIVE
   B. AIR CLEANER INSTALLED
5. ADJUST THROTTLE STOP SCREW TO SPECIFIED IDLE R.P.M. USING A TACHOMETER. NOTE: MODELS USING IDLE STOP SOLENOID MAKE ADJUSTMENT BY TURNING SOLENOID ASSY. (SOLENOID LEAD MUST BE CONNECTED SO SOLENOID WILL BE ENERGIZED.)
6. ADJUST IDLE MIXTURE NEEDLE TO OBTAIN THE HIGHEST R.P.M. AT THE LEANEST BEST IDLE SETTING.
7. TO ADJUST SLOWER IDLE SPEED DISCONNECT SOLENOID AT TERMINAL CONNECTION, AUTOMATIC TRANSMISSION IN NEUTRAL, MAKE ADJUSTMENT BY TURNING 1/8" HEX. SCREW AT REAR OF SOLENOID. CONNECT SOLENOID LEAD OPEN THROTTLE AND RELEASE. RECHECK HIGHER IDLE SPEED.
8. INSTALL IDLE LIMITER CAP WHEN REQUIRED.

SUPPLEMENT IDLE ADJUSTING PROCEDURE

SLOW IDLE ADJUSTMENT WITH AND WITHOUT IDLE STOP SOLENOID

Illustration courtesy of Tomco Industries

9.6 Rochester 1MV carburetor curb idle adjustment details

sary to set the specified curb idle speed **(see illustration)**.

7 With the air conditioning Off, de-energize the idle stop solenoid by pulling off the electrical connector.

8 Turn the hex screw at the end of the solenoid body to set the low (initial) idle speed to the specified rpm.

9 Check the fast idle adjustment as described later in this Section, turn off the engine and remove the tachometer. Reconnect all the hoses.

Idle mixture adjustment

10 The idle mixture screw has a limiter cap. Normally, the screw should only be turned to the limit of the cap stop (one turn clockwise - leaner) to improve the idle quality.

11 If necessary after major overhaul, or if the cap has been broken, set the mixture screw as described below but install a new cap so future adjustment will be limited to one turn clockwise (leaner).

12 Repeat the operations described in Steps 1 through 5 above.

13 With the engine running, turn the mixture screw as necessary to achieve maximum idle speed.

14 Set the idle speed to the curb (higher) speed as specified by turning the solenoid in or out.

15 Turn the idle mixture screw clockwise (leaner) until the idle speed is at the lower specified (lean drop) level.

16 Turn off the engine, reconnect all hoses and remove the tachometer.

Fast idle adjustment

17 Verify that the low and high (curb) idle speeds are correct.

18 With the engine at normal operating temperature, the air cleaner in position and the choke fully open, disconnect the EGR signal line and plug it. If equipped, turn off the air conditioning.

19 Connect a tachometer to the engine in accordance with the manufacturer's instructions.

20 Disconnect the vacuum advance hose from the distributor. Plug the hose if the vehicle is equipped with transmission controlled spark advance.

21 Start the engine and, with the transmission in Neutral, set the fast idle cam follower tang on the high step of the cam.

22 Bend the tang as necessary to achieve the specified fast idle speed **(see illustration 8.19)**.

Fast idle cam adjustment

23 Follow the procedure in Section 8, Steps 21 through 24.

Choke coil rod adjustment

C10 and K10 models

24 Disconnect the upper end of the choke coil rod from the choke plate **(see illustration)**.

25 Close the choke plate with your hand.

26 Push up on the choke coil rod to the limit of its travel.

27 The bottom of the rod should now be level with the top of the lever. If it isn't, bend the rod.

C20-C30 and K20 models

28 Follow the procedure in Section 8, Steps 14 through 16.

Primary vacuum break adjustment

29 Set the cam follower on the highest step of the fast idle cam **(see illustration 8.27)**.

30 Seal the purge bleed hole in the vacuum break end cover with masking tape.

31 Use a vacuum pump to apply vacuum to the primary vacuum break diaphragm until the plunger is fully seated.

32 Push the choke coil rod up to the end of the slot.

33 Check the gap between the upper edge of the choke plate and the air horn wall with a drill bit of the specified size. If the gap is not as specified, bend the vacuum break rod. Remember to remove the tape from the purge bleed hole.

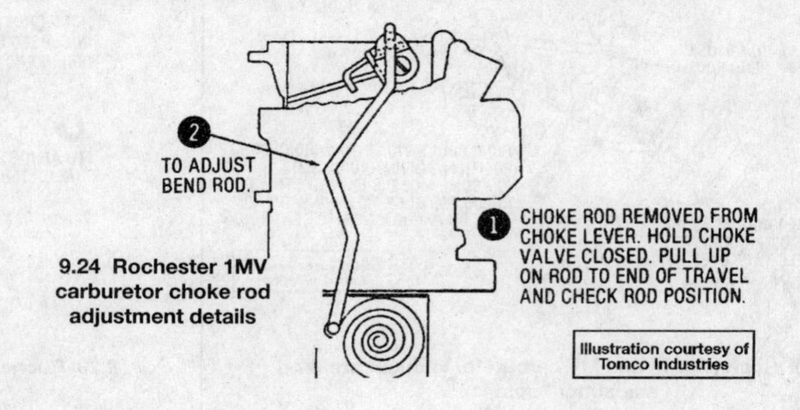

9.24 Rochester 1MV carburetor choke rod adjustment details

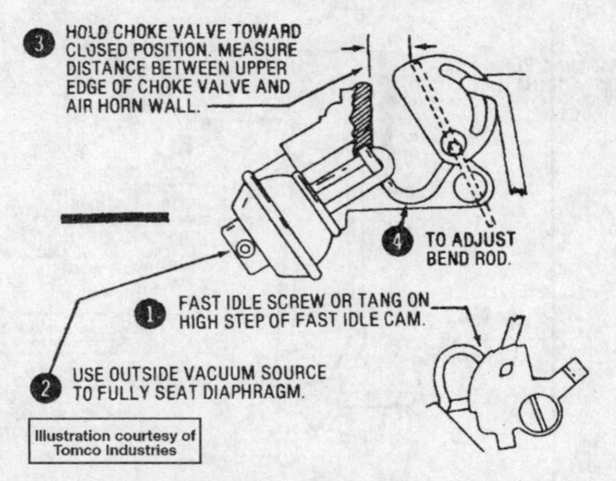

9.34 Rochester 1MV carburetor auxiliary vacuum break adjustment details

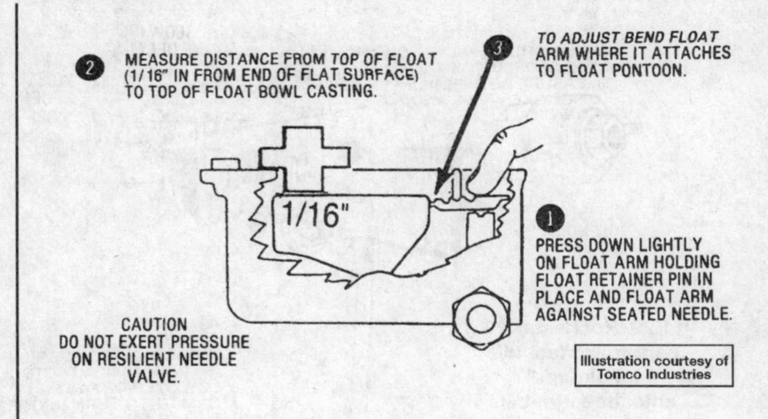

10.2 Rochester 1ME carburetor float level adjustment details

Auxiliary vacuum break adjustment

34 Set the cam follower on the highest step of the fast idle cam **(see illustration)**.

35 Apply vacuum to the auxiliary vacuum break diaphragm until the plunger is completely seated.

36 Use a drill bit to check the gap between the upper edge of the choke plate and the air horn wall.

37 If the gap is not as specified, bend the link.

Choke unloader adjustment

38 Follow the procedure in Section 8, Steps 29 through 31.

10 Carburetor (Rochester 1ME) - adjustments

Refer to illustrations 10.2 and 10.10

Idle adjustment

1 Follow the procedure described in Section 13, Steps 1 through 11, for the M4MC carburetor, but note that only one mixture control screw is used and the idle speed is set with the solenoid **(see illustration 9.6)**.

Float level adjustment

2 Push down on the end of the float arm and measure from the surface of the carburetor casting to the index point on the float **(see illustration)**.

3 If the measurement is not as specified, carefully bend the float arm to achieve the correct dimension.

Fast idle adjustment

4 Set the curb idle speed with the idle stop solenoid.

5 Set the cam follower tang on the high step of the cam **(see illustration 8.21)**.

6 Support the lever with a pair of pliers and bend the tang to obtain the specified idle speed.

Choke coil lever adjustment

7 Set the cam follower on the highest step of the fast idle cam.

8 Hold the choke plate completely closed.

9 A 0.120-inch diameter gauge must now pass through the hole in the lever and into the hole in the casting. Bend the link, if necessary, to achieve the specified clearance.

Automatic choke adjustment

10 Loosen the three screws which secure the choke housing **(see illustration)**.

11 Place the cam follower on the high step of the cam.

12 align the index marks, then retighten the screws.

Choke rod (fast idle cam) adjustment

13 Make sure the fast idle speed is as specified, then set the cam follower on the second step of the fast idle cam and against the highest step **(see illustration 8.27)**.

14 Depress the choke plate, then check the gap between the upper edge of the plate and the air horn.

Vacuum break adjustment

15 Place the cam follower on the highest step of the fast idle cam.

16 Apply vacuum to seat the diaphragm.

17 Push the choke coil lever up, then check the gap between the upper edge of the choke plate and the air horn with a drill bit of the specified size **(see illustration)**.

18 Bend the link rod to adjust the gap if necessary.

Choke unloader adjustment

19 Hold the throttle plate wide open and then depress the choke plate.

20 Measure the gap between the upper edge of the choke plate and the air horn **(see illustration 8.29)**.

21 Bend the tang to obtain the specified gap.

11 Carburetor (Rochester 2G series) - adjustments

Refer to illustrations 11.29, 11.34, 11.38, 11.42 and 11.44

Idle adjustment (1967 through 1973)

1 The procedure is similar to the one described in Section 8, Steps 1 through 6, for 1-barrel carburetors, except that the 2-barrel unit has two mixture screws. They should both be adjusted, in turn, to achieve the best idling quality.

Idle adjustment (1974 on)

2 The mixture screws on these carbure-

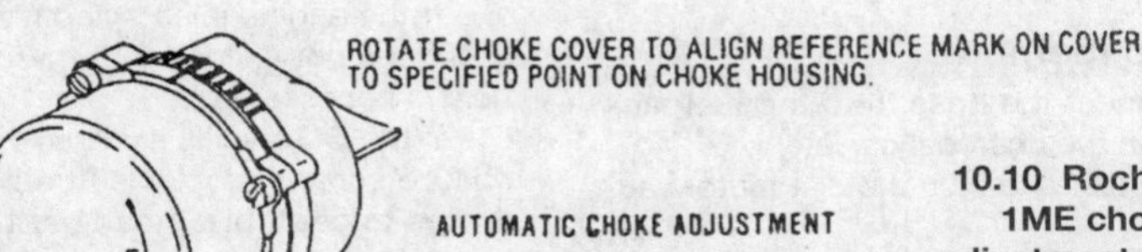

10.10 Rochester 1ME choke adjustment details

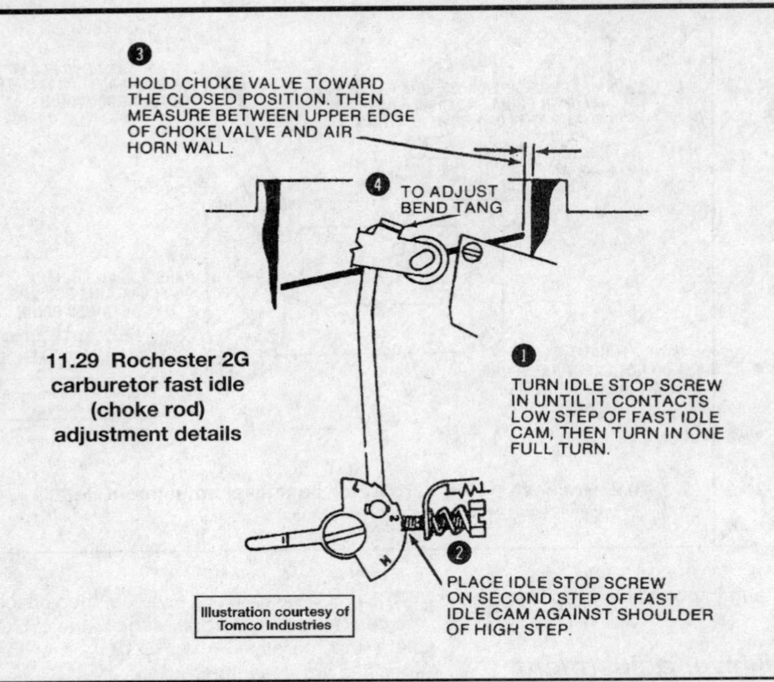

11.29 Rochester 2G carburetor fast idle (choke rod) adjustment details

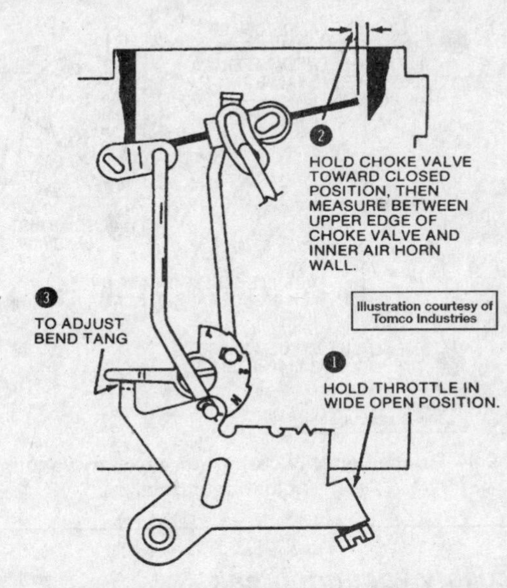

11.34 Rochester 2G carburetor choke unloader adjustment details

tors are equipped with limiter caps. The caps must remain in place and the screws adjusted (lean drop method) only to the limits of the cap stops.

3 If an idle stop solenoid is installed, first run the engine until normal operating temperature is reached. Make sure that the air cleaner is in position, the air conditioning system is Off and the parking brake is applied.

4 Disconnect the hose from the fuel tank connector on top of the charcoal canister.

5 Disconnect and plug the vacuum hose to the distributor.

6 With the engine running (manual transmission in Neutral/automatic transmission in Drive), disconnect the electrical lead from the carburetor solenoid.

7 Turn the idle speed screw until the engine idle speed is as specified.

8 Reconnect the lead to the solenoid, open the throttle momentarily, then set the specified curb idle by adjusting the screw on the solenoid.

9 Reconnect the vacuum hose and the charcoal canister hose and turn off the engine.

10 If the carburetor has been dismantled to such an extent that the mixture screw limiter caps have been removed or, if after checks of all other engine tune-up specifications, the idle mixture seems to be off, adjust it as follows in order to maintain the exhaust emission levels set by the vehicle manufacturer.

Lean drop method

11 Disconnect the hose from the fuel tank connector on the charcoal canister.

12 Disconnect and plug the distributor vacuum hose.

13 With the parking brake applied, run the engine until normal operating temperature is reached. Make sure the air cleaner is installed and the air conditioner (if applicable) is Off.

14 On vehicles with an automatic transmission, place the selector lever in Drive.

15 On vehicles with a manual transmission, disconnect the wire harness from the carburetor idle stop solenoid.

16 If not already removed, break off the tab on the idle mixture screw limiter cap.

17 Adjust the idle stop solenoid so that the engine speed is as specified (see Specifications). This is the initial idle speed.

18 Unscrew each of the idle mixture screws an equal amount until the idle speed reaches its highest point and any further movement of the screws would cause it to decrease.

19 Readjust the idle stop solenoid if necessary to bring the engine speed back to that previously set.

20 Turn the mixture screws in until the final curb idle speed is set, based on the type of transmission, as indicated in the Specifications.

21 Reconnect all disconnected components and turn off the engine.

CO meter method

22 This method can be employed if a reliable CO meter can be obtained.

23 Carry out the procedure already described in Steps 11 through 16 of this Section.

24 Install the meter probe in the tailpipe.

25 Set the initial idle speed to the Specifications by turning the idle stop solenoid.

26 If the engine idles smoothly and the CO level does not exceed 0.5, no further adjustment is necessary.

27 If the CO level is excessive, turn the idle mixture screws clockwise until the idle CO level is acceptable. Readjust the curb idle speed if necessary.

Choke rod adjustment

28 Turn the idle stop screw in until it just touches the bottom step of the fast idle cam, then screw it in exactly one full turn.

29 Position the fast idle screw on the second step of the fast idle cam against the shoulder of the high step **(see illustration)**.

30 Hold the choke plate in the closed position (using a rubber band to keep it in place) and check the gap between the upper edge of the choke plate and the inside wall of the air horn. Use a drill bit as a gauge.

31 If the gap is not as specified, bend the tang to correct it.

Choke unloader adjustment

32 Hold the throttle plates all the way open.

33 Hold the choke plate in the closed position using a rubber band to keep it in place.

34 Check the gap between the upper edge of the choke plate and the inside wall of the air horn. The gap should be as shown in the Specifications **(see illustration)**.

35 Bend the tang on the throttle lever to adjust the gap.

Vacuum break adjustment

36 Remove the air cleaner. On temperature controlled air cleaners, plug the sensor vacuum take-off port.

37 Apply vacuum to the vacuum break diaphragm until the plunger is fully seated.

38 Push the choke plate towards the closed position, then measure the gap between the upper edge of the choke plate and the air horn wall with a drill bit **(see illustration)**.

39 The gap should be as specified. If it isn't, bend the vacuum break rod.

Accelerator pump rod adjustment

40 Back out the idle speed screw.

41 Close both throttle plates completely

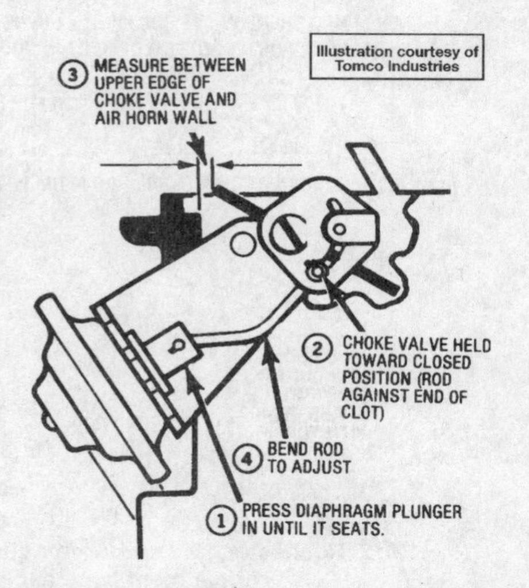

11.38 Rochester 2G carburetor vacuum break adjustment details

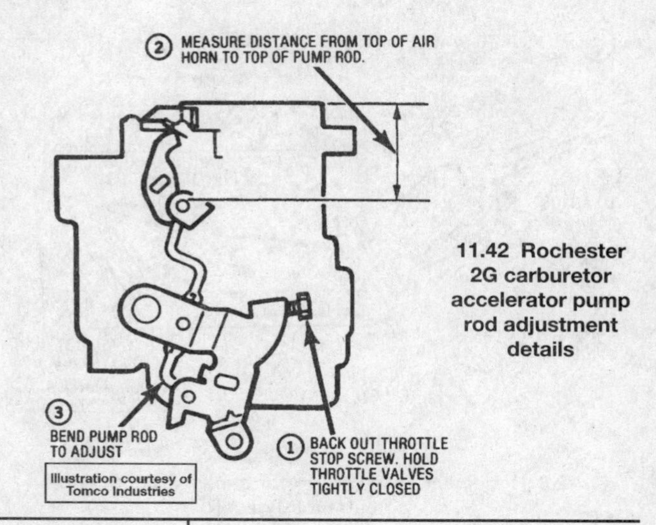

11.42 Rochester 2G carburetor accelerator pump rod adjustment details

and measure from the top surface of the air horn ring to the tip of the pump rod.

42 This measurement should be as specified. Bend the rod as required **(see illustration)**.

Choke coil rod adjustment

43 Hold the choke plate open, then, with the thermostatic coil rod disconnected from the upper lever, push down on the rod to the end of its travel.

44 The bottom of the rod should now be level with the bottom of the elongated hole in the lever. If it isn't, bend the lever by inserting a screwdriver blade into the slot **(see illustration)**.

12 Carburetor (Rochester 4MV) - adjustments

Refer to illustrations 12.2, 12.7, 12.8, 12.13 and 12.15

Idle adjustment

1 The procedure is very similar to the one for the 2G series carburetor. Refer to Section 11, Steps 2 through 27.

Fast idle adjustment

2 Place the transmission in Neutral. Position the fast idle lever on the high step of the fast idle cam **(see illustration)**.

3 Make sure that the engine is at normal operating temperature with the choke fully open.

4 On vehicles equipped with a manual transmission, disconnect the vacuum advance hose.

5 Connect a tachometer to the engine in accordance with the manufacturer's instructions.

6 Turn the fast idle screw in or out as necessary to adjust the fast idle speed.

Choke rod (fast idle cam) adjustment

7 Set the cam follower on the second step of the fast idle cam and against the high step. Rotate the choke valve towards the closed position by turning the external choke lever counterclockwise. Use a drill bit as a gauge and measure the gap between the lower edge of the choke plate (lower end) and air horn wall. Bend the choke rod if necessary to achieve the specified clearance **(see illustration)**.

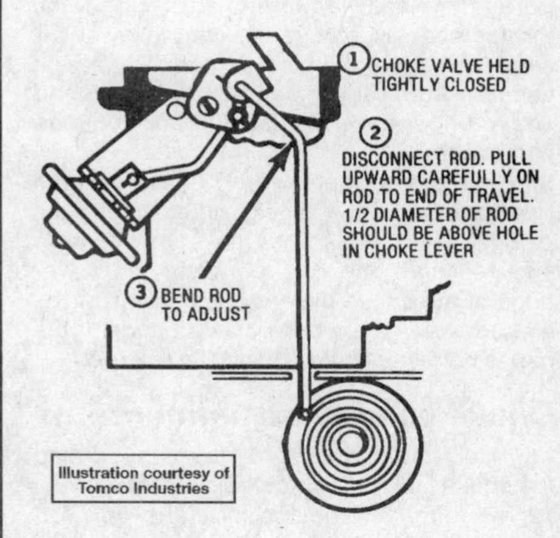

11.44 Rochester 2G carburetor choke coil rod adjustment details

ADJUST SLOW IDLE. THEN PLACE CAM FOLLOWER ON PROPER STEP OF FAST IDLE CAM AND ADJUST FAST IDLE SCREW TO PROPER R.P.M.

FAST IDLE CAM

FAST IDLE SCREW

Illustration courtesy of Tomco Industries

12.2 Rochester 4MV carburetor fast idle adjustment (typical)

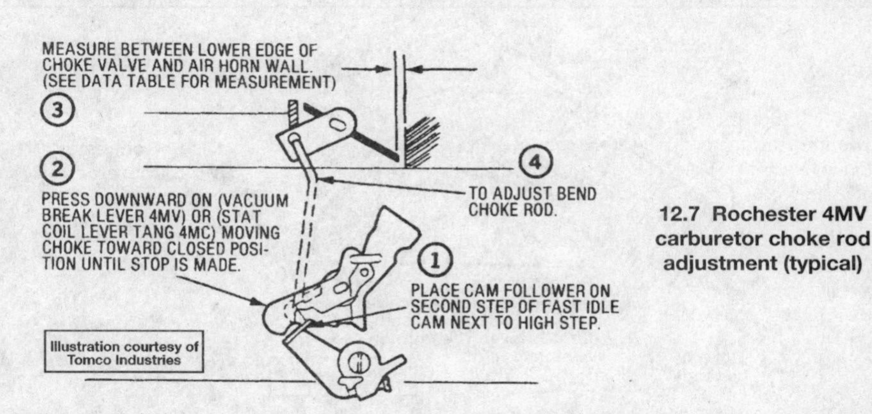

12.7 Rochester 4MV carburetor choke rod adjustment (typical)

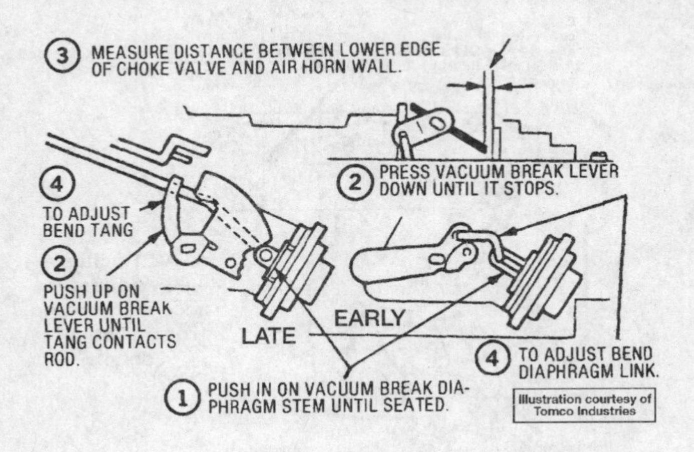

12.8 Rochester 4MV carburetor choke vacuum break adjustment (typical)

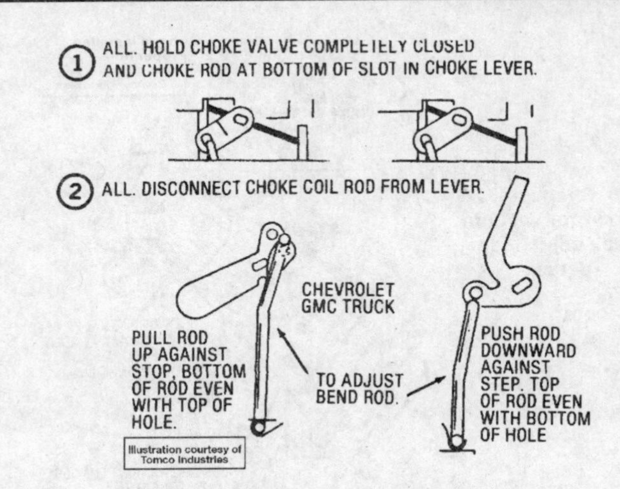

12.13 Rochester 4MV carburetor choke coil rod adjustment (typical)

Choke vacuum break adjustment

8 Seat the choke vacuum break diaphragm with a vacuum pump. Open the throttle plate slightly so the cam follower will clear the step of the fast idle cam. Rotate the vacuum break lever counterclockwise and use a rubber band to hold it in position **(see illustration)**.

9 Make sure that the end of the vacuum brake rod is in the outer slot of the diaphragm plunger.

10 Measure the gap between the lower edge of the choke plate and the inside of the air horn wall. Use a drill bit as a gauge. If the gap is not as specified, bend the link rod.

Choke coil rod adjustment

11 Hold the choke plate closed by turning the choke coil lever counterclockwise.

12 Disconnect the thermostatic coil rod and remove the coil cover, then push the coil rod down until the rod contacts the surface of the bracket.

13 Check that the coil rod fits in the notch in the choke lever. Bend the rod as necessary if it doesn't **(see illustration)**.

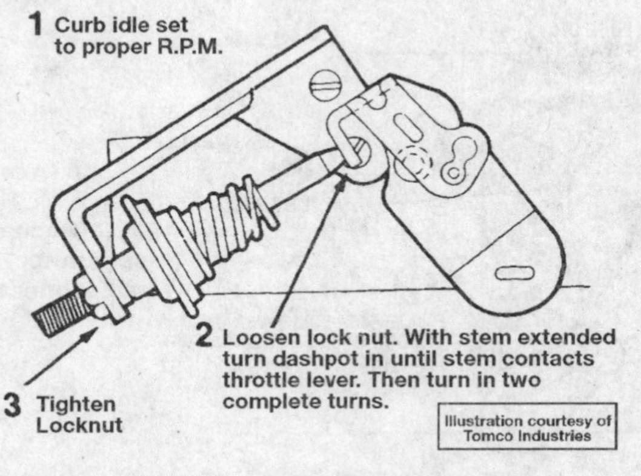

12.15 Rochester 4MV carburetor air valve dashpot adjustment (typical)

Air valve dashpot adjustment

14 Seat the choke vacuum break diaphragm, using an external vacuum source.

15 With the diaphragm seated and the air valve fully closed, measure the distance between the end of the slot in the vacuum break plunger lever and the air valve. If it's not as specified, bend the link rod as shown **(see illustration)**.

13 Carburetor (Rochester M4MC/M4MCA) - adjustments

Refer to illustrations 13.12, 13.18, 13.26, 13.30, 13.33, 13.38, 13.42, 13.48, 13.52, 13.57, 13.60 and 13.62

Mixture adjustment

1 Remove the air cleaner for access to the carburetor, keeping the vacuum hoses connected. Disconnect the remaining hoses from the air cleaner and plug them.

2 Prior to making the adjustment, the engine must be at normal operating temperature, the choke must be open and the air conditioning system (where applicable) off. The ignition timing must be correct. Make sure that the parking brake is applied.

3 Connect a tachometer to the engine in accordance with the manufacturer's instructions.

4 Remove the limiter caps from the idle mixture screws. Turn the screws in until they seat lightly, then unscrew them equally until the engine will just run.

5 Place the transmission in Neutral (manual transmission) or Drive (automatic transmission).

6 Back the mixture screws out equally 1/8-turn at a time until the maximum engine speed is obtained. Now turn the idle speed screw until the initial idle speed is as specified.

7 Repeat the adjustment to make sure that the maximum idle speed was obtained when unscrewing the mixture screws.

8 Now turn each of the idle mixture screws in 1/8-turn at a time until the idle speed is as shown for lean drop in the Specifications.

9 Check that the speed shown in the Specifications is the same as on the Vehicle Emission Control Information label. If necessary, readjust until the specified speed is achieved.

10 Reconnect the vacuum hoses, install the air cleaner and turn the engine off.

11 Install new limiter caps so future adjustment of one turn clockwise (leaner) can be done.

Pump rod adjustment

12 With the fast idle cam follower off the steps of the fast idle cam, back out the idle speed screw until the throttle valves are completely closed in the bore. Make sure that the secondary actuating rod is not restricting movement; bend the secondary closing tang if necessary, then readjust it after pump adjustment **(see illustration)**.

13 Place the pump rod in the specified hole in the lever. On all carburetors except num-

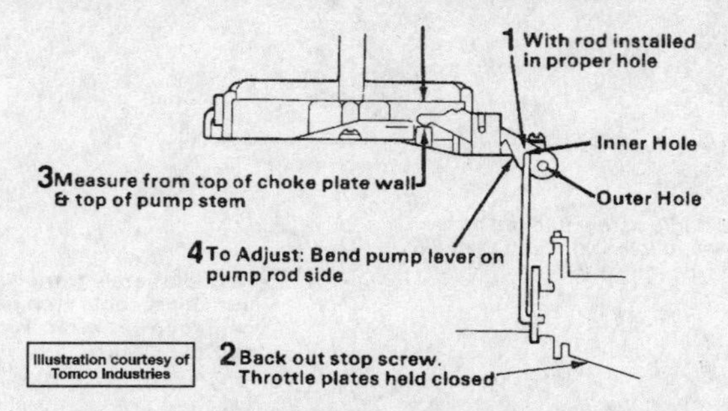

13.12 Rochester M4MC/M4MCA carburetor pump rod adjustment details

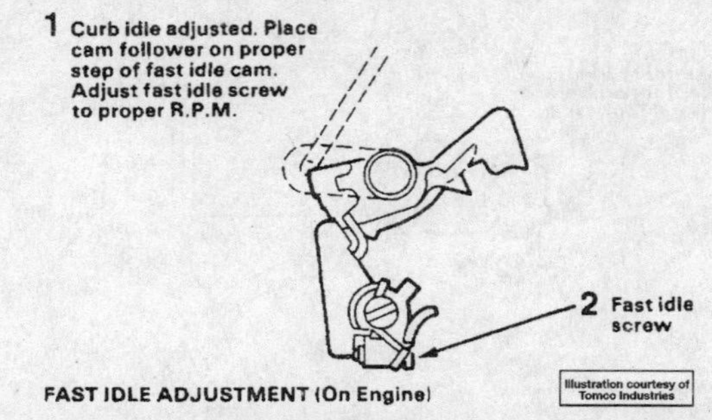

13.18 Rochester M4MC/M4MCA carburetor fast idle adjustment details

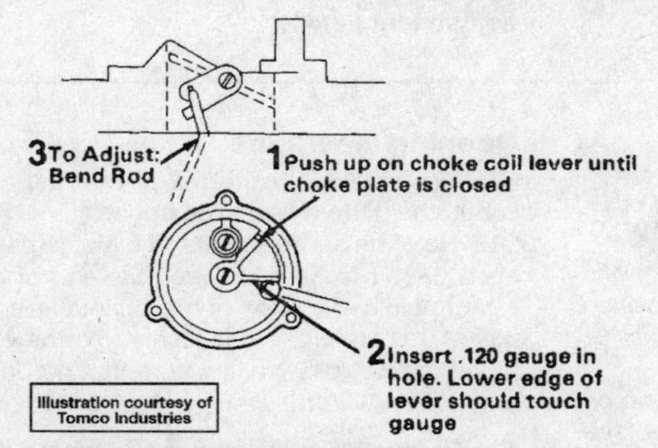

13.26 Rochester M4MC/M4MCA carburetor choke coil lever adjustment details

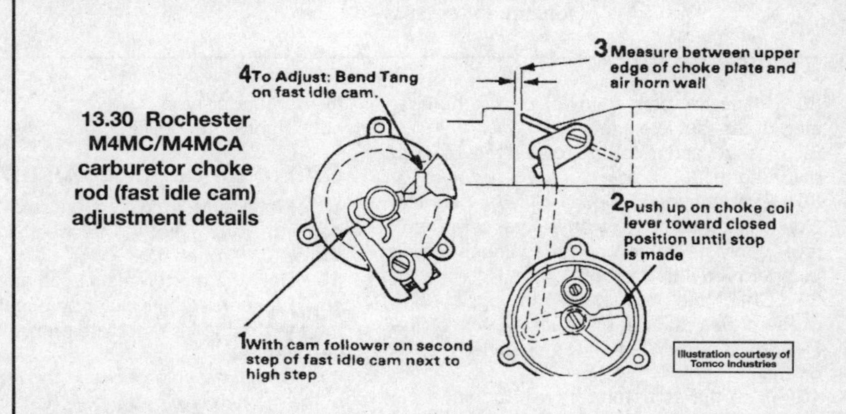

13.30 Rochester M4MC/M4MCA carburetor choke rod (fast idle cam) adjustment details

bers 17057586 and 17057588, place the rod in the inner hole. On carburetors 17057586 and 17057588, place the rod in the outer hole.

14 Measure from the top of the choke valve wall (next to the vent stack) to the top of the pump stem.

15 If necessary, adjust to obtain the specified dimension by bending the lever while supporting it with a screwdriver.

16 Adjust the idle speed.

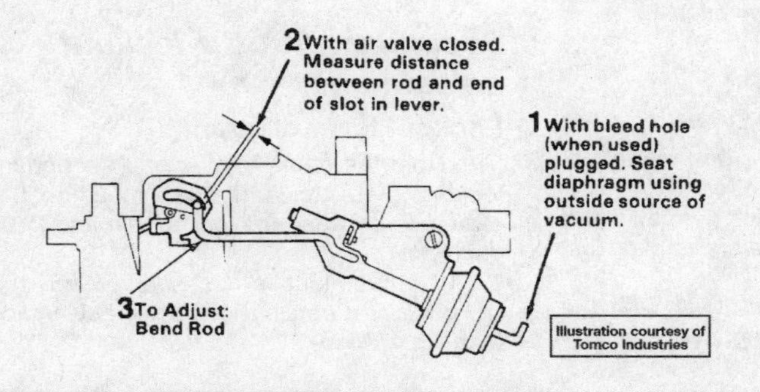

13.33 Rochester M4MC/M4MCA carburetor air valve rod adjustment details

Fast idle adjustment

17 Place the transmission in Park (automatic) or Neutral (manual).

18 Hold the cam follower on the highest step of the fast idle cam **(see illustration)**.

19 Disconnect and plug the vacuum hose at the EGR valve (if applicable).

20 Turn the fast idle screw to obtain the correct specified idle speed.

Choke coil lever adjustment

21 Remove the screws and detach the cover and coil assembly from the choke housing.

22 Push up the thermostatic coil tang (counterclockwise) until the choke plate is closed.

23 Check that the choke rod is at the bottom of the slot in the choke lever.

24 Insert a plug gauge (or the equivalent size drill bit shank) of the specified size into the hole in the choke housing.

25 The lower edge of the choke coil lever should just contact the side of the plug gauge.

26 If necessary, bend the choke rod at the point shown to adjust **(see illustration)**.

Choke rod (fast idle cam) adjustment

27 Turn the fast idle cam screw in until it contacts the fast idle cam follower, then turn it in an additional three full turns. Remove the coil cover.

28 Place the lever on the second step of the fast idle cam, against the rise of the high step.

29 Push up on the choke coil lever inside the housing to close the choke plate.

30 Measure between the upper edge of the choke plate and the air horn wall with the shank of a drill bit of the specified size **(see illustration)**.

31 If necessary, bend the tang on the fast idle cam to adjust, but make sure that the tang lies against the cam after bending.

Air valve rod adjustment

32 Using an external source of vacuum, seat the choke vacuum break diaphragm.

33 Make sure that the air valves are completely closed, then measure between the air valve dashpot and the end of the slot in the air valve lever **(see illustration)**.

34 Bend the air valve dashpot rod at the point shown, if adjustment is necessary.

Front vacuum break adjustment

35 Remove the thermostatic coil cover.

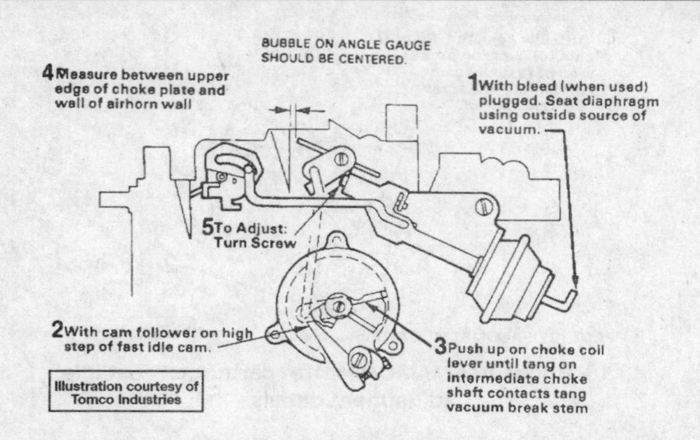

13.38 Rochester M4MC/M4MCA carburetor vacuum break adjustment details

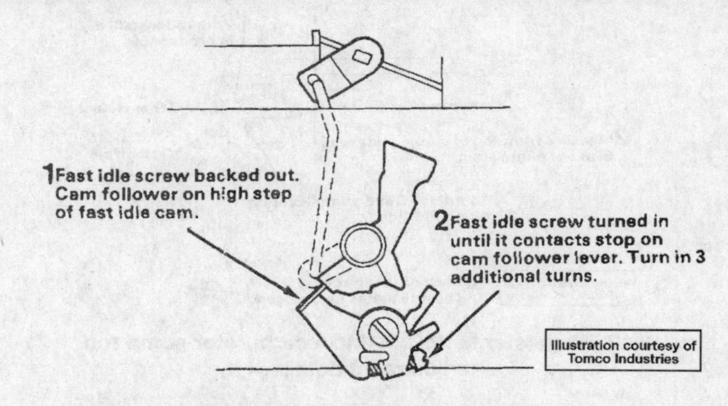

13.42 Rochester M4MC/M4MCA carburetor automatic choke coil adjustment details

36 Place the cam follower on the highest step of the fast idle cam.
37 Use a vacuum pump to seat the vacuum diaphragm.
38 Push the inner choke coil lever counterclockwise until the tang on the vacuum break lever contacts the tang on the vacuum break plunger **(see illustration)**.
39 Check the gap between the upper edge of the choke plate and the inside wall of the air horn. Use a drill bit of the specified size to do this.
40 Turn the adjuster screw to achieve the correct gap.
41 Install the cover and check the automatic choke adjustment (Steps 42 through 45).

Automatic choke coil adjustment (hot air type)

42 Place the cam follower on the highest point of the cam **(see illustration)**.
43 Loosen the three retaining screws and rotate the coil cover until the choke plate just closes.
44 Align the mark on the cover with the specified point on the housing, depending on the carburetor type.
45 Tighten the cover screws.

Choke unloader adjustment

46 Make sure that the automatic choke housing cover is set to the specified position (Steps 42 through 45).
47 Hold the throttle plate wide open.
48 Close the choke plate by pushing up on the tang of the intermediate choke lever **(see illustration)**.
49 Check the gap between the upper edge of the choke plate and the air horn inner wall. Use a suitable size drill bit as a gauge.
50 Bend the tang if necessary to adjust it.

Secondary throttle lockout adjustment

Lockout lever clearance

51 Hold the choke plates and secondary lockout plates closed, then measure the clearance between the lockout pin and lockout lever.
52 If adjustment is necessary, bend the lockout pin to obtain the specified clearance **(see illustration)**.

Opening clearance

53 Push down on the tail of the fast idle cam to hold the choke wide open.
54 Hold the secondary throttle plates partly open, then measure between the end of the lockout pin and the toe of the lockout lever.
55 If adjustment is necessary, file the end of the lockout pin, making sure that no burrs remain afterwards.

Secondary closing adjustment

56 Adjust the engine idle speed.
57 Hold the choke plate wide open, with the cam follower lever off the steps of the fast idle cam **(see illustration)**.
58 Measure the clearance between the slot in the secondary throttle plate pick-up lever and the secondary actuating rod.
59 If adjustment is necessary, bend the secondary closing tang on the primary throttle lever to obtain the specified clearance.

Secondary opening adjustment

60 Lightly open the primary throttle lever until the link just contacts the tang on the secondary lever **(see illustration)**.

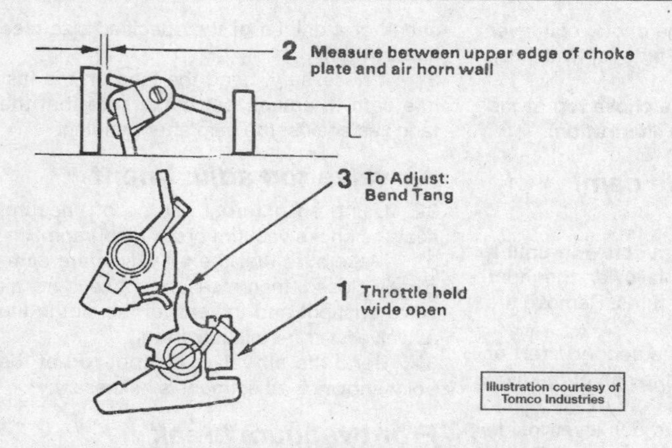

13.48 Rochester M4MC/M4MCA carburetor choke unloader adjustment details

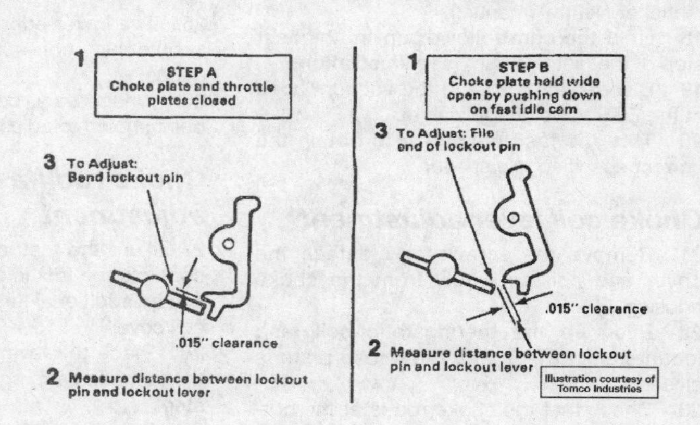

13.52 Rochester M4MC/M4MCA carburetor secondary throttle lockout adjustment details

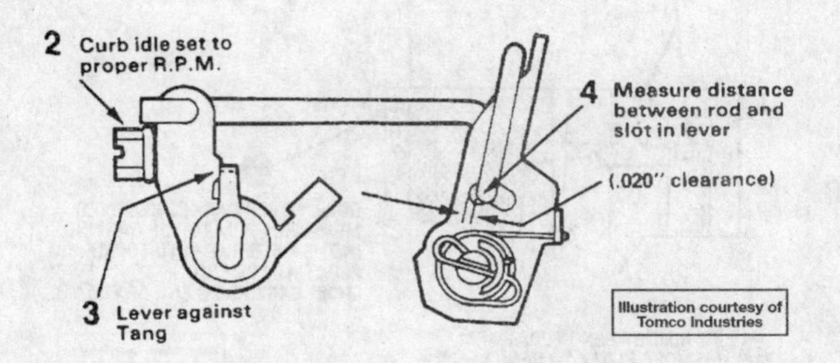

13.57 Rochester M4MC/M4MCA carburetor secondary closing adjustment details

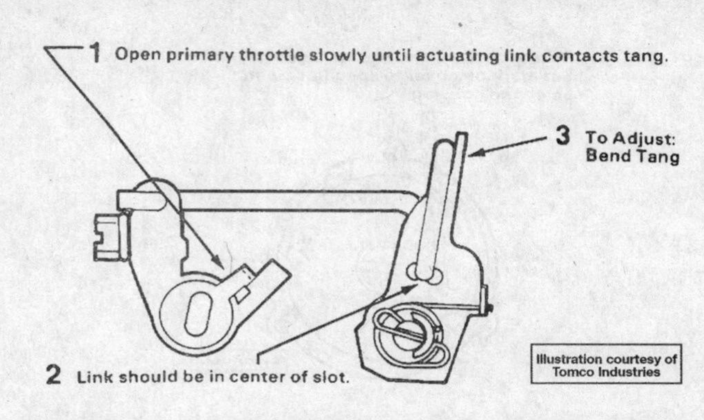

13.60 Rochester M4MC/M4MCA carburetor secondary opening adjustment details

61 Bend the tang on the secondary lever, if necessary, to position the link in the center of the secondary lever slot.

Air valve spring adjustment

62 Using an Allen wrench, loosen the lockscrew, then turn the tension adjusting screw counterclockwise until the air valve is partly open **(see illustration)**.
63 Hold the air valve closed, then turn the tension adjusting screw clockwise the specified number of turns after the spring contacts the pin.
64 Tighten the lockscrew.

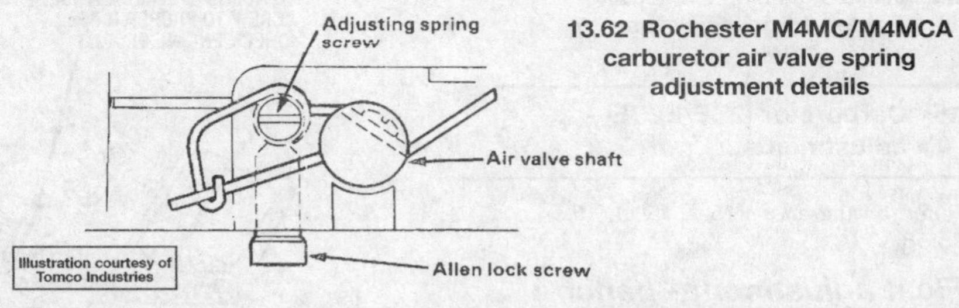

13.62 Rochester M4MC/M4MCA carburetor air valve spring adjustment details

14 Carburetor (Rochester M4ME) - adjustments

Refer to illustrations 14.7, 14.8 and 14.9

1 The Rochester M4ME carburetor is very similar to M4MC and M4MCA carburetors, with the exception of an electrically-heated (as opposed to hot air heated) choke and a rear vacuum break.
2 Except for the adjustments mentioned in the following Steps, all other on-vehicle adjustments are the same as the M4MC and M4MCA carburetors described in Section 13.

Rear vacuum break adjustment

3 Remove the thermostatic coil cover.
4 Place the cam follower on the highest step of the fast idle cam.
5 Use a vacuum pump to seat the vacuum diaphragm.
6 Push up on the choke coil lever, inside the choke housing, towards the closed choke position until the stem is pulled out and seated.
7 With the choke rod in the bottom of the slot in the choke lever, place the specified size drill bit between the upper edge of the choke plate and the air horn wall. If the dimension is not as specified, bend the vacuum break rod **(see illustration)**.

Automatic choke coil adjustment (electrically-heated type)

8 Place the cam follower on the highest step of the cam **(see illustration)**.
9 Loosen the retaining screws and rotate the coil cover until the choke plate just closes

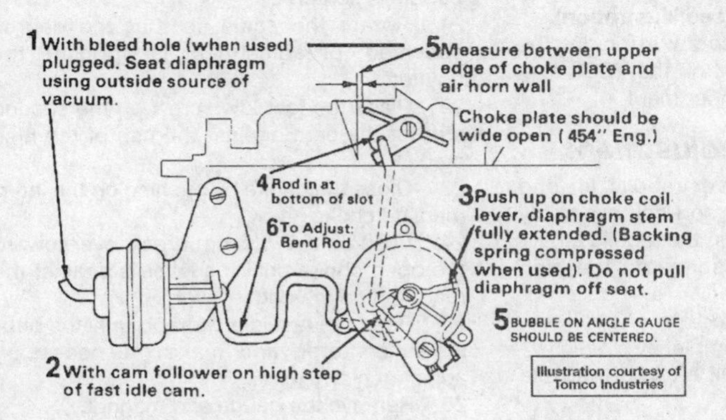

14.7 Rochester M4ME carburetor rear vacuum break adjustment details

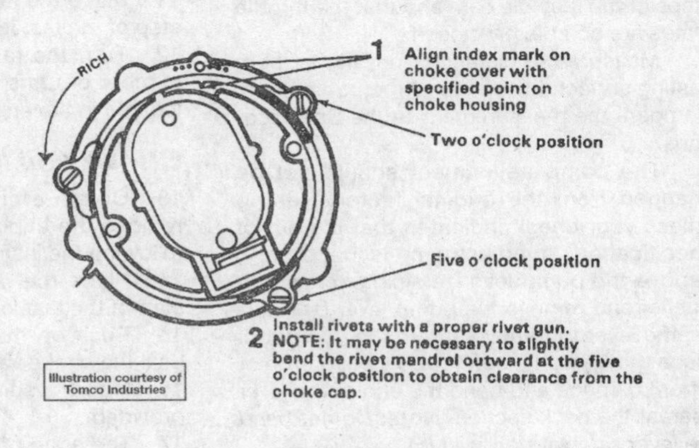

14.8 Rochester M4ME carburetor electric choke component layout

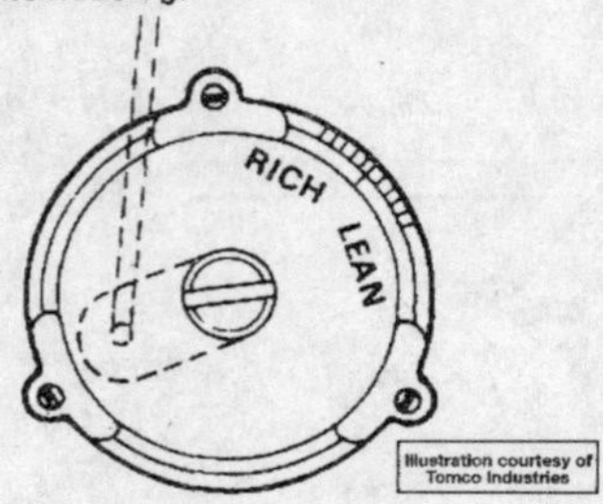

14.9 Rochester M4ME carburetor automatic choke coil adjustment details

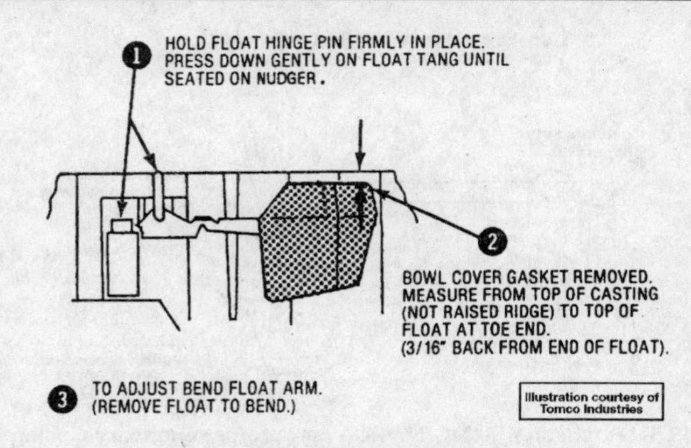

15.1 Rochester 2SE/E2SE carburetor float adjustment details

(see illustration).

10 Align the index mark on the cover with the specified mark on the housing.

11 Tighten the cover screws.

15 Carburetor (2SE/E2SE) - adjustments

Refer to illustrations 15.1, 15.11, 15.54 and 15.58

Float adjustment - bench setting

1 Hold the float retainer firmly in place while gently pushing the float against the needle **(see illustration).**

2 Measure the height of the float at the farthest point from the float hinge pin at the toe of the float.

3 If necessary, remove the float and bend the float arm up or down to adjust.

4 Visually check the float alignment after adjustment. It should not contact any other portion of the carburetor.

Pump adjustment

5 Make sure the fast idle screw is off the steps of the fast idle cam and that the throttle plates are completely closed.

6 Measure the distance from the air horn casting surface to the top of the pump stem. Compare the measurement to the Specifications.

7 The pump adjustment should not be changed from the original factory setting unless your check indicates that it's out of specification. If adjustment is necessary, remove the pump lever retaining screw and washer and remove the pump lever by rotating the lever to remove it from the pump rod. Place the lever in a soft jawed vise to protect it from damage and bend the end of the lever nearest the neck section. **Note:** *Do not bend the lever sideways or twist it.*

8 Reinstall the pump lever, washer and retaining screw and recheck the pump adjustment.

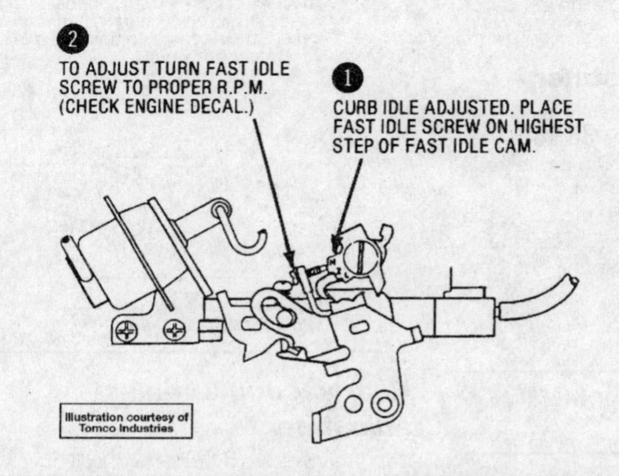

15.11 Rochester 2SE/E2SE carburetor fast idle adjustment (bench setting)

9 Tighten the retaining screw securely after the pump adjustment is correct.

10 Open and close the throttle plates, checking the linkage for freedom of movement and watching for proper pump lever alignment.

Fast idle adjustment - bench setting

11 Place the fast idle screw on the highest step of the fast idle cam **(see illustration).**

12 Turn the fast idle screw in or out the number of turns specified on the VECI label located in the engine compartment.

Choke coil lever adjustment

13 Obtain a choke cover rebuild kit and remove the choke cover and coil assembly following the instructions in the rebuild kit.

14 Place the fast idle screw on the high step of the fast idle cam.

15 Push on the intermediate choke lever until the choke valve is completely closed.

16 Insert a suitable drill bit into the hole provided.

17 The edge of the choke coil lever should just make contact with the side of the drill bit.

18 Bend the intermediate choke rod to adjust the choke coil lever as necessary.

Fast idle cam (choke rod) adjustment

19 Before proceeding, make the choke coil lever and fast idle adjustments.

20 With the choke plate completely closed, place a magnet squarely on top of it, then mount a choke valve measuring tool or an angle gauge to the magnet so that the rotating degree scale reads zero and the leveling bubble is centered.

21 Rotate the scale so that the degree specified for adjustment is opposite the pointer.

22 Place the fast idle screw on the second step of the cam against the rise of the high step.

23 Close the choke by pushing on the intermediate choke lever.

24 Push on the vacuum break lever toward the open choke until the lever is against the rear tang on the choke lever.

25 Bend the fast idle cam rod until the bubble is centered and make the necessary adjustments.

26 Remove the gauge and magnet.

Air valve rod adjustment

27 Attach a choke plate measuring gauge or angle gauge as described in Step 20.

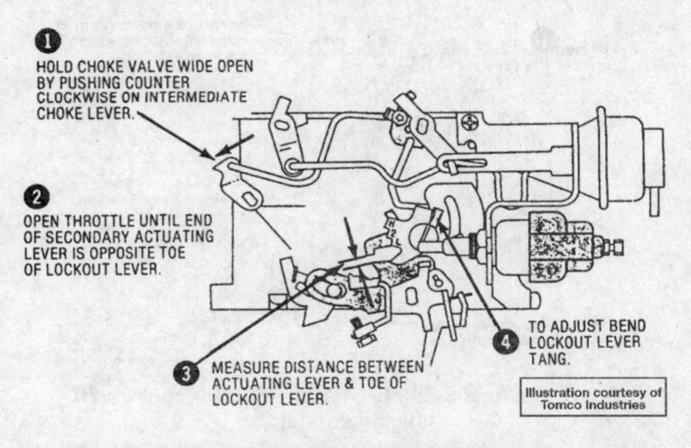

15.54 Rochester 2SE/E2SE carburetor on-vehicle fast idle adjustment details

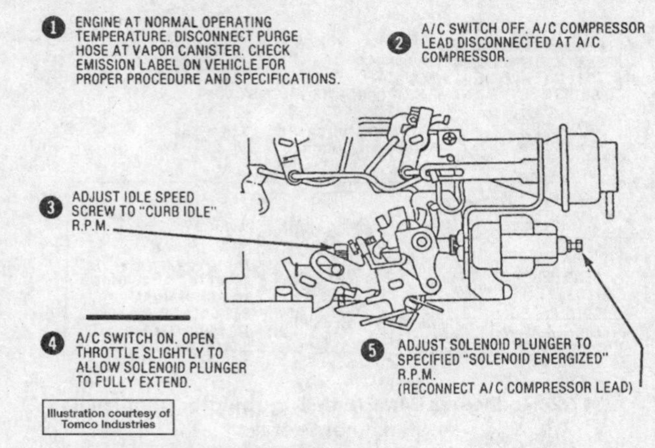

15.58 Rochester 2SE/E2SE carburetor on-vehicle idle speed adjustment details

28 Rotate the scale so that the degree specified for adjustment is opposite the pointer.
29 Using a vacuum pump, seat the vacuum diaphragm.
30 Rotate the air valve in the direction of the open air valve by applying light pressure to the air valve shaft.
31 To adjust, bend the air valve rod until the bubble on the angle gauge is centered.

Primary side vacuum break adjustment

32 Attach a choke plate measuring gauge or angle gauge as described in the fast idle cam adjustment, Step 19.
33 Rotate the scale so that the degree specified for the adjustment is opposite the pointer.
34 Seat the choke vacuum diaphragm using an outside vacuum source.
35 Hold the choke plate towards the closed position by pushing on the intermediate choke lever.
36 To adjust, bend the vacuum break rod until the bubble on the angle gauge is centered. Remove the gauge.

Secondary vacuum break adjustment

37 Mount a choke plate measuring gauge or angle gauge to the carburetor as described in the fast idle cam adjustment, Step 19.
38 Rotate the scale so that the degree specified for adjustment is opposite the pointer.
39 Using an outside vacuum source, seat the diaphragm. On models with an air bleed, plug the end cover with a piece of masking tape. Remove the tape after adjustment.
40 While reading the angle gauge, lightly push clockwise on the intermediate choke lever (in the direction of the closed choke plate) and hold in position with a rubber band.
41 To adjust, bend the vacuum break rod until the bubble in the angle gauge is centered. On later models, use a 1/8-inch Allen wrench and turn the screw in the rear cover until the bubble is centered. Remove the angle gauge and masking tape, if used.

Unloader adjustment

42 Use a choke plate measuring gauge or an angle gauge and mount it as described in the fast idle cam adjustment, Step 19.
43 Rotate the scale so that the degree specified for adjustment is opposite the pointer.
44 Install the choke thermostatic cover and coil assembly in the housing.
45 Hold the primary throttle plate wide open. Close the choke plate by pushing clockwise on the intermediate choke lever. If the engine is warm, use a rubber band to hold it in position.
46 Bend the tang to adjust the unloader until the bubble in the gauge is centered. Remove the gauge.

Secondary lockout adjustment

47 Hold the choke plate wide open by pushing counterclockwise on the intermediate choke lever.
48 Open the throttle lever until the end of the secondary actuating lever is opposite the toe of the lockout lever.
49 Insert the specified size drill bit and measure the clearance.
50 If adjustment is necessary, bend the lockout lever tang contacting the fast idle cam.

Fast idle adjustment - on vehicle

51 Set the ignition timing as per the label and prepare the vehicle for adjustments.
52 Adjust the curb idle speed if required.
53 Place the fast idle screw on the highest step of the fast idle cam.
54 Turn the fast idle screw in or out to obtain the specified fast idle rpm (see the label) **(see illustration)**.

Idle speed adjustment

Note: *On most vehicles this procedure is included on the VECI label located in the engine compartment. Consult the label before proceeding. If conflicting information exists, the label should be assumed correct.*
55 Prepare the vehicle for adjustments following the instruction on the label.
56 An assistant will be needed to keep his foot on the brake pedal while the idle is adjusted.
57 With the solenoid energized, place the selector in Drive (automatic transmission) or Neutral (manual transmission). Turn the air conditioner off, if equipped.
58 Open the throttle slightly to allow the solenoid plunger to extend fully **(see illustration)**.
59 Turn the solenoid screw to adjust the curb idle speed to the specified rpm.
60 Disconnect the solenoid wire and turn the idle speed screw to set the base idle speed to specification.
61 Reconnect the solenoid wire after the adjustment.

16 Carburetor (Rochester M2M) - adjustments

Float adjustment

1 Hold the float retainer firmly in place and gently push down on the float against the needle.
2 Measure from the top of the casting to the top of the float, at a point 3/16-inch back from the end of the float, at the toe.
3 If adjustment is necessary, remove the float and bend the float arm up or down as necessary.
4 Reinstall the float, visually check the alignment and recheck the float setting.

Pump adjustment

5 With the throttle plates completely closed, make sure the fast idle cam follower

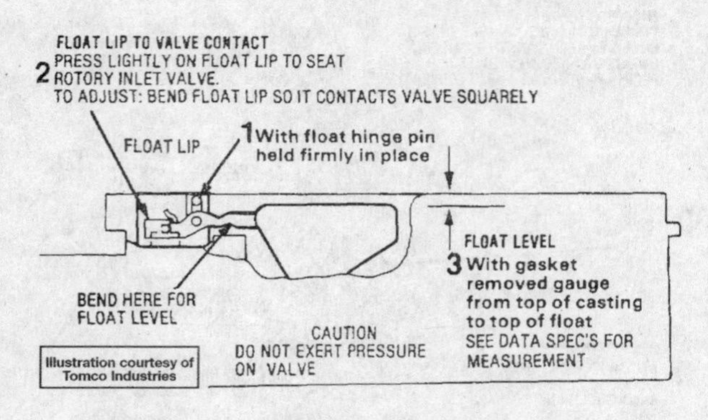

17.2 Rochester E4ME/E4MC carburetor float level adjustment details

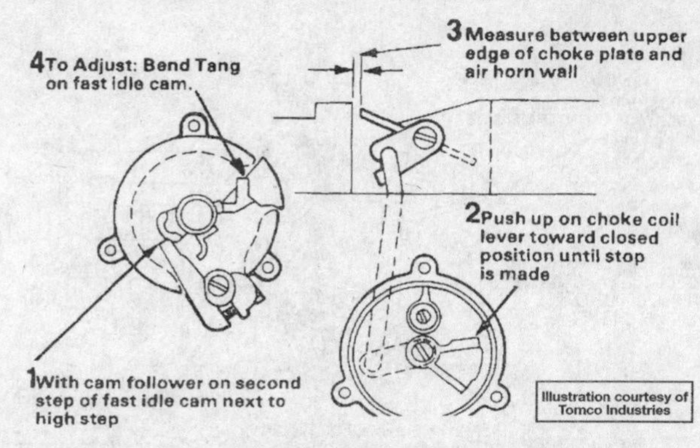

17.7 Rochester E4ME/E4MC carburetor choke rod adjustment details

lever is off the steps of the fast idle cam.

6 Insert the specified size drill bit into the specified hole in the pump lever.

7 Measure from the top of the choke plate wall, next to the vent stack, to the top of the pump stem as specified.

8 If adjustment is necessary, support the pump lever with a screwdriver while bending the lever.

Choke coil lever adjustment

9 The procedure is the same as the one for the 2SE carburetor (Section 15, Steps 13 through 18).

Fast idle adjustment - bench setting

10 Hold the cam follower on the highest step of the fast idle cam.

11 Turn the fast idle screw out until the primary throttle plates are closed. Turn the screw in to contact the lever, then turn it in an additional two complete turns.

Fast idle cam adjustment

12 This procedure is very similar to that used to adjust the 2SE carburetor (Section 15, Steps 19 through 26).

13 After Step 22, close the choke by pushing up on the choke coil lever or vacuum break lever tang. Hold in that position with a rubber band.

14 To adjust, bend the tang on the fast idle cam until the bubble on the angle gauge is centered.

15 Remove the angle gauge.

Front vacuum break adjustment

16 This procedure can be carried out in the same manner as the primary side vacuum break adjustment described in Section 15, Steps 32 through 36 for the 2SE carburetor.

Automatic choke coil adjustment (1979)

17 Place the cam follower on the highest step of the cam.

18 Loosen the three retaining screws on the choke housing.

19 Rotate the cover and coil assembly counterclockwise until the choke plate just closes.

20 Align the mark on the cover with the point on the housing that is one notch to the lean side. **Note:** *Make sure the slot in the lever engages with the coil tang.*

Unloader adjustment (1980 through 1982)

21 This procedure is very similar to the one for the 2SE carburetor (Steps 42 through 46).

22 After Step 44 hold the throttle plates wide open.

23 With the engine warm, close the choke plate by pushing up on the tang on the vacuum break lever. Hold in this position with a rubber band.

24 To adjust, bend the tang on the fast idle lever until the bubble in the angle gauge is centered. Remove the gauge.

Idle adjustment

25 The procedure is very similar to that described for the 2G carburetor. Refer to Section 11, Steps 2 through 27.

Fast idle adjustment - on vehicle

26 Place the transmission in Park or Neutral.

27 Place the cam follower on the specified step of the fast idle cam. Refer to the emission label.

28 Disconnect the vacuum hose from the EGR valve and plug the hose.

29 Start the engine and turn the fast idle adjustment screw to obtain the rpm specified on the emission label.

17 Carburetor (Rochester E4ME/E4MC) - adjustments

Refer to illustrations 17.2, 17.7, 17.12, 17.14, 17.17, 17.23, 17.28, 17.29, 17.31 and 17.37

1 Except where noted, adjustments are the same as for earlier models (Section 13). Carburetors used with the C4 system do not require adjustment of the pump rod.

Float level adjustment - bench setting

2 Hold the float retainer firmly in place, push the float down until it lightly contacts the needle and measure the float level with the gauge. The gauging point is 3/16-inch back from the toe of the float **(see illustration)**.

3 On carburetors used with the C4 system, the float should be adjusted if the height varies from that shown in the Specifications Section of this Chapter by plus-or-minus 1/16-inch.

4 If the level is too high, hold the retainer in place and push down on the center of the float until the specified setting is obtained.

5 If the level is too low on non-solenoid carburetors, remove the power piston, metering rods, plastic filler block and the float. Bend the float arm up to adjust. Reinstall the parts and visually check the alignment of the float.

6 If the level is too low on solenoid equipped carburetors, remove the metering rods and the solenoid connector screw. Count and record for use at the time of reassembly, the number of turns necessary to lightly bottom the lean mixture screw. Back the screw out and remove it, the solenoid connector and float. Bend the float arm up to adjust. Install the parts and reset the mixture screw to the recorded number of turns.

Choke rod adjustment (1983 through 1986)

7 Attach a rubber band to the green tang of the intermediate choke shaft **(see illustration)**.

8 Close the choke by opening the throttle.

9 Install a choke angle gauge and set the specified angle.

10 Place the cam follower on the second step of the cam, against the high step. If the follower does not contact the cam, turn the fast idle speed screw in until it does. The final fast idle adjustment must be made according to the information on the Emissions Control

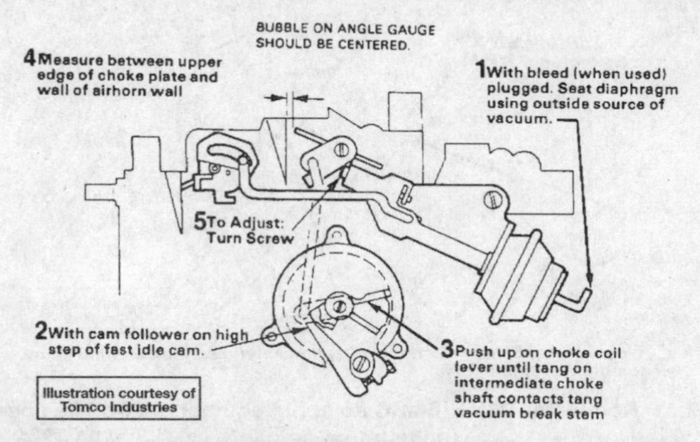

17.12 Rochester E4ME/E4MC carburetor front vacuum break adjustment details

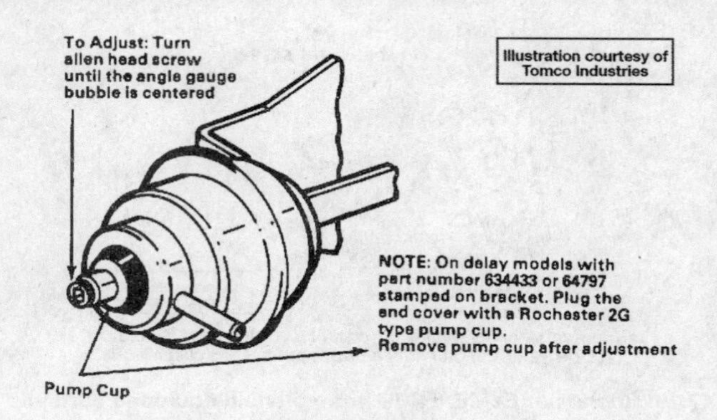

17.14 Plug the vacuum delay air bleed

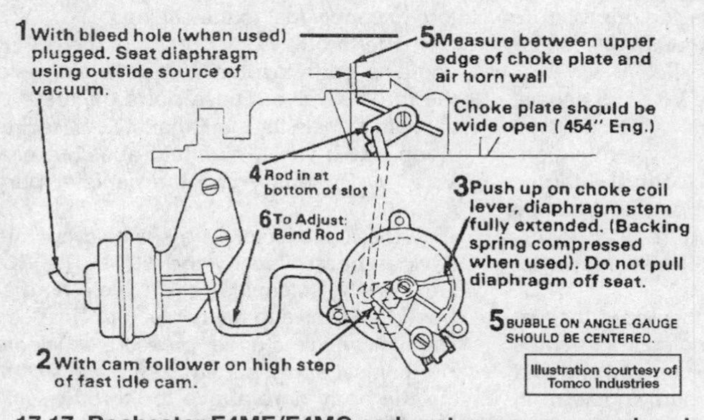

17.17 Rochester E4ME/E4MC carburetor rear vacuum break adjustment details

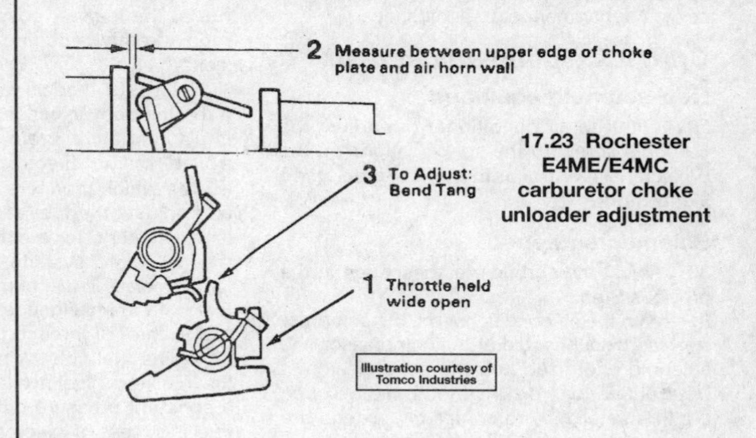

17.23 Rochester E4ME/E4MC carburetor choke unloader adjustment

Information label under the hood.
11 Bend the fast idle cam tang until the bubble is centered.

Front vacuum break adjustment (1983 through 1986)

12 Attach a rubber band to the green tang of the intermediate choke shaft and open the throttle to allow the choke to close **(see illustration)**.
13 Install a gauge tool and set the vacuum break to the specified angle.
14 Apply at least 18-inches Hg of vacuum to retract the vacuum break plunger. Plug any air bleed holes **(see illustration)**.
15 The air valve rod can sometimes restrict the plunger from retracting completely and it might be necessary to bend the rod slightly. The final rod clearance must be set after the vacuum break adjustment has been made.
16 With the vacuum applied, adjust the screw until the bubble is centered.

Rear vacuum break adjustment (1983 through 1986)

17 Attach a rubber band to the green tang of the intermediate choke shaft **(see illustration)**.
18 Open the throttle until the choke closes.
19 Set up an angle gauge tool on the carburetor and set the angle to specification.
20 Apply vacuum to retract the vacuum break plunger, making sure to plug any bleed holes.
21 The air valve rod can sometimes restrict the plunger from retracting completely. If necessary, bend the rod to allow full travel of the plunge.
22 To center the bubble, either of two methods can be used. With the vacuum applied, use a 1/8-inch Allen wrench to turn the adjustment screw. Alternatively, support the vacuum break rod and bend the rod with the vacuum applied.

Choke unloader adjustment

23 Attach a rubber band to the green tang of the intermediate choke shaft **(see illustration)** and open the throttle to allow the choke to close.
24 Install the angle gauge tool and set the angle to specification.
25 Hold the secondary lockout lever away from the pin as shown in illustration 17.23.
26 Hold the throttle lever in the wide open position and bend the fast idle lever tang until the bubble is centered.

Idle speed adjustment - preparation

27 Prior to idle speed adjustment, the engine must be at normal operating temperature and the ignition timing set to the specification on the Emissions Control Information label.
28 Some models equipped with the C4 system use an idle speed control (ISC) assembly mounted on the carburetor, which is controlled by the ECM. Do not attempt to adjust the idle on the idle speed control **(see illustration)**. Adjustment of the idle speed

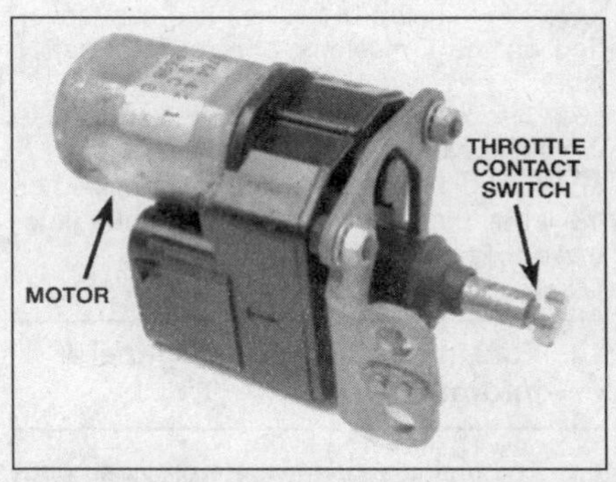

17.28 Rochester E4ME/E4MC idle speed control assembly used with the C4 system

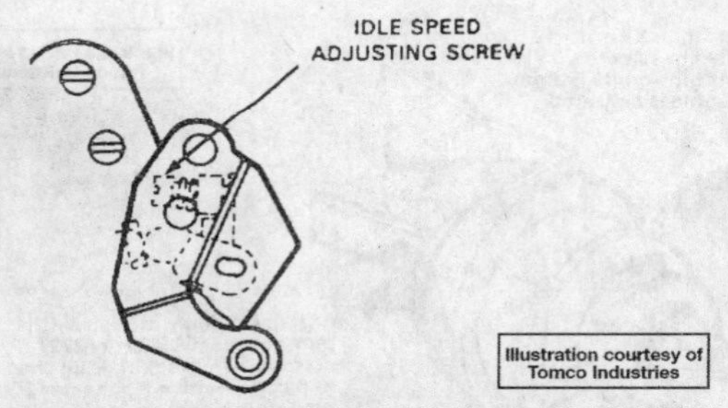

17.29 Rochester E4ME/E4MC non-solenoid equipped curb idle speed adjustment details

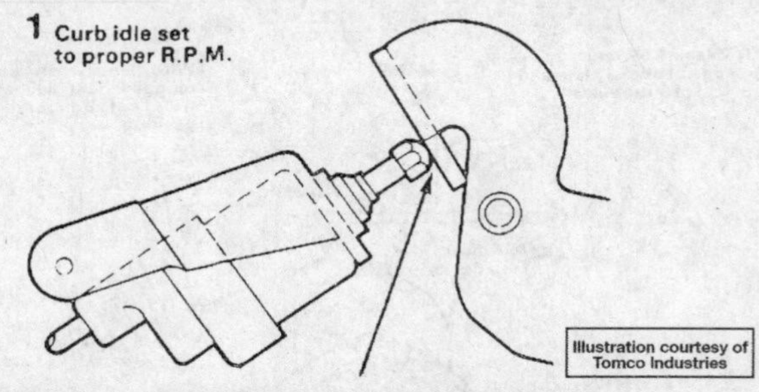

17.31 Rochester E4ME/E4MC solenoid equipped curb idle speed adjustment details

control assembly should be left to a dealer service department or a repair shop.

Curb idle speed adjustment

Non-solenoid equipped

29 With the air conditioner off, adjust the curb idle screw to the specifications on the Emissions Control Information label **(see illustration)**.

Solenoid equipped

30 Adjust the curb idle as described in the previous Step.

31 With the air conditioner on, the compressor lead disconnected at the compressor, the solenoid energized and the transmission in Neutral (manual) or Drive (automatic), open the throttle slightly to completely extend the solenoid plunger **(see illustration)**.

32 Adjust the curb idle to the specified rpm by turning the solenoid screw.

33 Reconnect the air conditioner compressor wire after adjustment.

Fast idle speed adjustment

Non-solenoid equipped

34 With the transmission in Park (automatic) or Neutral (manual), hold the cam follower on the step specified on the Emissions Control Information label. Turn the screw to obtain the correct fast idle speed.

Solenoid equipped

35 With the transmission in Park (automatic) or Neutral (manual), hold the cam follower on the step of the fast idle cam specified on the Emissions Control Information label.

36 Disconnect the vacuum hose at the EGR valve and plug it.

37 To obtain the fast idle rpm specified on the label, turn the fast idle screw **(see illustration)**.

18 Fuel injection system - general information

The fuel injection system provides optimum mixture ratios at all stages of combustion and offers immediate throttle response characteristics. It also enables the engine to run at the leanest possible fuel/air mixture ratio, greatly reducing exhaust gas emissions.

On 1987 models with a V6 or V8 engine, a throttle body injection (TBI) system is used in place of a carburetor. The system is controlled by an Electronic Control Module (ECM), which monitors engine performance and adjusts the fuel/air mixture accordingly (see Chapter 6 for a complete description of the fuel control system).

An electric fuel pump, located in the fuel tank with the fuel gauge sending unit, pumps fuel to the TBI through the fuel feed line and an in-line fuel filter. A pressure regulator in the TBI keeps fuel available to the injector at a constant pressure between 9 and 13 psi. Fuel in excess of injector needs is returned to the fuel tank by a separate line. The injector, located in the TBI, is controlled by the ECM. It delivers fuel in one of several modes (see Chapter 6 for a complete description of ECM modes of operation).

The basic TBI unit is made up of three major casting assemblies: a throttle body, a fuel meter cover and a fuel meter body. The throttle body contains the idle air control (IAC) valve (and controls air flow), a throttle position sensor (which monitors throttle angle) and two throttle valves. The fuel meter cover incorporates a built-in pressure regulator. The fuel meter body has two fuel injectors to supply fuel to the engine.

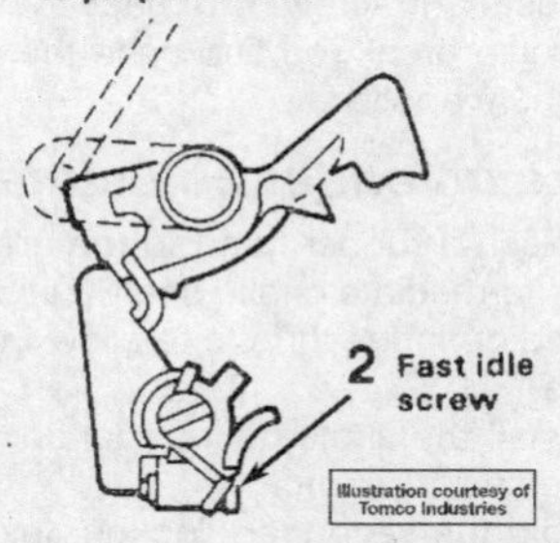

17.37 Rochester E4ME/E4MC fast idle speed adjustment details

The throttle body portion of the TBI unit contains ports located at, above and below the throttle valve. These ports generate the vacuum signals for the Exhaust Gas Recirculation (EGR) valve, manifold absolute pressure (MAP) sensor and the canister purge system.

The fuel injectors are solenoid operated devices controlled by the ECM. The ECM turns on the solenoids, which lifts a normally closed ball valve in each injector off its seat. The fuel, which is under pressure, is injected in a conical spray pattern at the walls of the throttle body bore above the throttle valve. The fuel which is not used by the injector passes through the pressure regulator before being returned to the fuel tank.

The pressure regulator is a diaphragm-operated relief valve with injector pressure on one side and air cleaner pressure on the other. The function of the regulator is to maintain a constant pressure at the injector at all times by controlling the flow in the return line. On these models the regulator has a constant bleed feature so that pressure is relieved when the engine is off. Therefore, it is not necessary to relieve the system pressure before working on the system or disconnecting fuel lines.

The purpose of the idle air control valve is to control engine idle speed while preventing stalls due to changes in engine load. The IAC valve, mounted on the throttle body, controls bypass air around the throttle valve. By moving a conical valve in, to decrease air flow, or out, to increase air flow, a controlled amount of air can move around the throttle valve. If rpm is too low, more air is bypassed around the throttle valve to increase rpm. If rpm is too high, less air is bypassed around the throttle valve to decrease rpm.

During idle, the proper position of the IAC valve is calculated by the ECM based on battery voltage, coolant temperature, engine load and engine rpm. If the rpm drops below a specified rpm, and the throttle valve is closed, the ECM senses a near stall condition. The ECM will then calculate a new valve position to prevent stalls based on barometric pressure.

19.7 The best way to remove the fuel injectors is to pry on them with a screwdriver, using a second screwdriver as a fulcrum

19.8 Remove the fuel inlet and outlet nuts from the fuel meter body

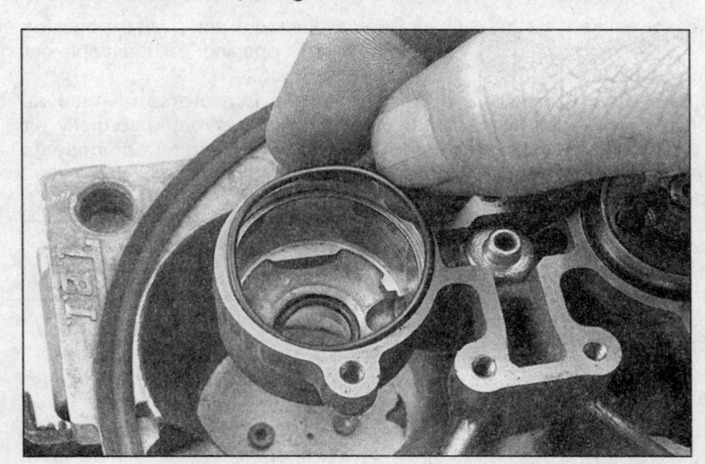

19.9 Remove the large O-ring and steel back-up washer from each injector cavity in the fuel meter body

19 Throttle Body Injection (TBI) unit - disassembly and reassembly

Refer to illustrations 19.7, 19.8, 19.9, 19.10 and 19.16

1 Disconnect the negative cable at the battery. Place the cable out of the way so it cannot accidentally come in contact with the negative terminal of the battery, as this would once again allow power into the electrical system of the vehicle.

2 Remove the air cleaner.

Disassembly

3 Remove each injector electrical connector by squeezing the two tabs together and pulling straight up.

4 Remove the fuel meter cover retaining screws and lock washers securing the fuel meter cover to the fuel meter body. Note the location of the short screws.

5 Remove the fuel meter cover. **Caution:** *Do not immerse the fuel meter cover in solvent. It might damage the pressure regulator diaphragm and gasket.*

6 The fuel meter cover contains the fuel pressure regulator. The regulator is pre-set and plugged at the factory. If a malfunction occurs, it cannot be serviced. It must be replaced as a complete assembly. **Warning:** *Do not remove the screws securing the pressure regulator to the fuel meter cover. It has a large spring inside which is highly compressed. If accidentally released, it could cause injury. Disassembly might also cause a fuel leak between the diaphragm and the regulator container.*

7 With the old fuel meter cover gasket in place to prevent damage to the casting, carefully pry each injector from the fuel meter body with a screwdriver until it can be lifted free **(see illustration)**. **Caution:** *Be careful when removing the injector to prevent damage to the electrical connector terminals, the injector fuel filter, the O-ring and the nozzle.*

8 The fuel meter body should be detached from the throttle body if it needs to be cleaned. To detach it, remove the fuel feed and return line fittings **(see illustration)** and the Torx screws that attach the fuel meter body to the throttle body.

9 Remove the old gasket from the fuel meter cover and discard it. Remove the large O-ring and steel back-up washer from the upper counterbore of each fuel meter body injector cavity **(see illustration)**. Clean the fuel meter body thoroughly with solvent and blow it dry.

10 Remove the small O-ring from the nozzle end of each injector. Carefully rotate the injector fuel filter back-and-forth and remove the filter from the base of each injector **(see illustration)**. Gently clean the filter with sol-

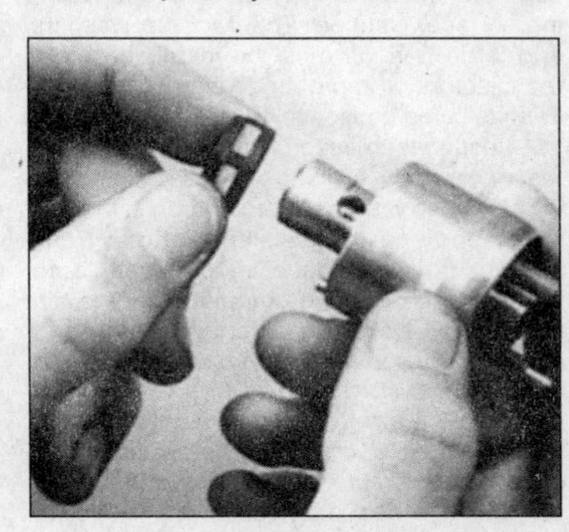

19.10 Gently rotate the fuel injector filter back-and-forth and carefully pull it off the injector nozzle

19.16 Make sure that the lug is aligned with the groove in the bottom of the fuel injector cavity

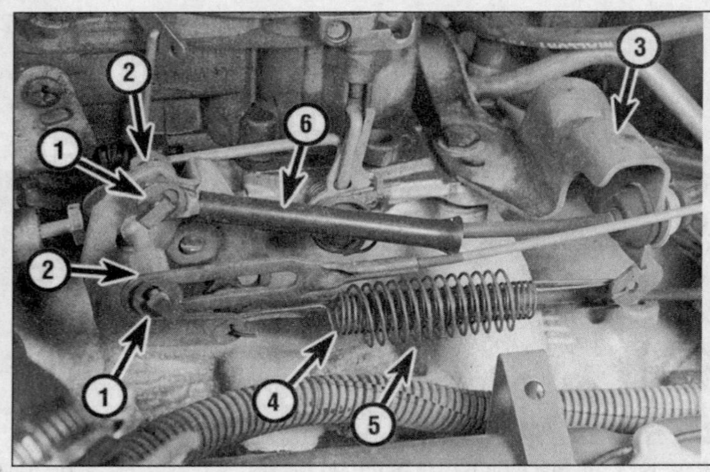

20.1 Typical early throttle cable details

1. *Fasteners*
2. *Carburetor stud*
3. *Support bracket*
4. *Inner spring*
5. *Outer spring*
6. *Cable*

vent and allow it to drip dry. It's too small and delicate to dry with compressed air. **Caution:** *The fuel injector itself is an electrical component. Do not immerse it in any type of cleaning solvent.*

11 The fuel injectors are not serviceable. If they are malfunctioning, replace them.

Reassembly

12 Install the clean fuel injector nozzle filter on the end of each fuel injector with the larger end of the filter facing the injector (the filter must cover the raised rib at the base of the injector). Use a twisting motion to position the filter against the base of each injector.

13 Lubricate new small O-rings with automatic transmission fluid. Push an O-ring onto the nozzle end of each injector until it presses against the injector fuel filter.

14 Insert the steel back-up washers into the counterbores of the fuel meter body injector cavity.

15 Lubricate new large O-rings with automatic transmission fluid and install them directly over the back-up washers. Be sure each O-ring is seated properly in the cavity, flush with the top of the fuel meter body casting surface. Caution: The back-up washers and large O-rings must be installed before the injectors or improper seating of the large O-rings could cause fuel to leak.

16 Install each injector in the cavity in the fuel meter body, aligning the raised lug on the injector base with the cast-in notch in the fuel meter body cavity **(see illustration)**. Push straight down on each injector with both thumbs until they are completely seated in each cavity. **Note:** *The electrical terminals of each injector should be approximately parallel to the throttle shaft.*

17 Install a new fuel outlet passage gasket on the fuel meter cover and a new fuel meter cover gasket on the fuel meter body.

18 Install a new dust seal in the recess in the fuel meter body.

19 Install the fuel meter cover on the fuel meter body, making sure that the pressure regulator dust seal and cover gaskets are in place.

20 Apply thread locking compound to the threads of the fuel meter cover screws. Install the screws (the two short screws go next to the injector) and tighten them to the specified torque. **Note:** *Service repair kits include a small vial of thread compound with directions for use. If not available, use Loctite 262 or equivalent. Do not use a higher strength locking compound than recommended, as it may prevent subsequent removal of the mounting screws or cause breakage of the screwhead if removal becomes necessary.*

21 Plug in the injector wire harnesses.

22 Install the air cleaner.

20 Throttle linkage - replacement

Refer to illustrations 20.1 and 20.2

1 Working under the hood, detach the linkage from the carburetor or throttle body and brackets **(see illustration)**.

2 Working inside the vehicle, disconnect the linkage or cable from the throttle pedal **(see illustration)**.

20.2 Pull the accelerator cable end from the hole in the pedal arm, then pass the cable through the slot

3 Depress the tangs and push the housing out of the firewall opening. Pull the cable out from under the hood.

4 Installation is the reverse of removal. Make sure the linkage is routed correctly - it must not contact any engine components that move or get hot.

21 Fuel tank - removal and installation

Refer to illustrations 21.15, 21.19, 21.21 and 21.29

Warning: *Gasoline is extremely flammable, so extra precautions must be taken when working on any part of the fuel system. Do not smoke or allow open flames or bare light bulbs near the work area. Also, don't work in a garage if a natural gas-type appliance with a pilot light is present.*

1 For the sake of safety and convenience, the fuel tank should be empty before removal if at all possible. If circumstances permit, run the tank dry, or almost dry, by driving the vehicle. Due to the tank shape on some models it may not be possible to siphon from the filler hole - the only alternative is to pump or siphon fuel from the fuel feed line.

2 Disconnect the negative cable at the battery. Place the cable out of the way so it cannot accidentally come in contact with the negative terminal of the battery, as this would once again allow power into the electrical system of the vehicle.

3 It may be necessary to raise the vehicle and support it securely on jackstands for clearance. **Warning:** *DO NOT work under a vehicle which is supported only by a jack.*

Early models

4 The fuel tank is either mounted behind the seat within the cab (conventional drive vehicles) or to the rear of the axle and between the side rails (four-wheel drive vehicles).

Cab mounted tank

5 Remove the seat back hinge hold-down

21.15 Remove the tank and support strap

21.19 Disconnect the tank fuel filler neck and vent hose

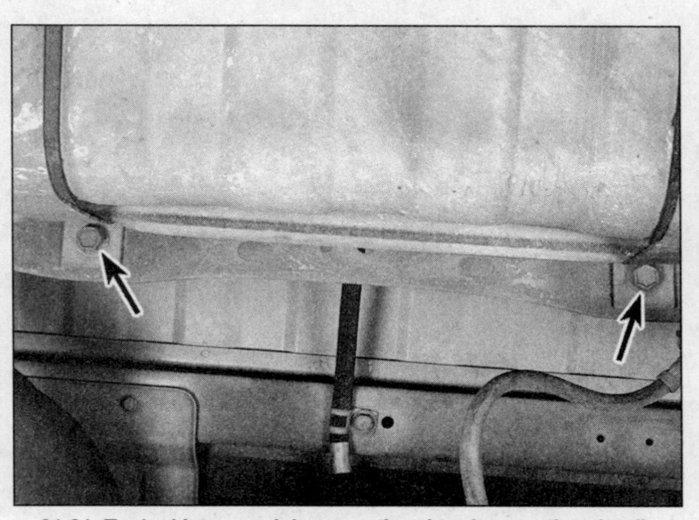

21.21 Typical later model conventional tank mounting details

21.29 Remove the strap bolts (arrows)

wing nut and tilt the seat back forward.
6 Where applicable, remove the tank cover.
7 Disconnect the fuel line, the sending unit wire and the ground cable.
8 Disconnect the filler pipe between the tank and the filler opening.
9 Remove the bolts/nuts securing the tank in position.
10 Lift the tank from the cab, disengaging the filler neck from the rubber grommet in the cab opening.
11 The sending unit can be removed by using two screwdrivers as levers to rotate it and unscrew it from the tank.
12 Installation is the reversal of removal.

Rear mounted tank

13 Disconnect the filler neck and vent tube hose.
14 Disconnect the sending unit wire at the frame.
15 Support the tank and remove the strap **(see illustration)**.
16 Lower the tank sufficiently to allow clearance to disconnect the sending unit wire, then remove the tank.
17 Remove the sending unit from the tank as described in Step 11.
18 Installation is the reverse of removal.

Later models

Sport-utility models

19 Disconnect the filler neck and vent tube hose **(see illustration)**.
20 Disconnect the sending unit wire at the frame.
21 Support the tank and disconnect the straps **(see illustration)**.
22 Lower the tank and disconnect the sending unit wire connector.
23 Remove the sending unit from the tank as described in Step 11.
24 Installation is the reverse of removal.

Pick-up models

25 The fuel tank is mounted outside the cab, outboard of the frame rail.
26 Disconnect the fuel lines, the sending unit wire and the ground strap.
27 Disconnect and remove the strap supports and clips from the fuel and vent lines.
28 Loosen the clamps at the filler neck and vent line.
29 Remove the strap bolts/nuts **(see illustration)**.
30 Remove the tank from the side frame and disengage the hose from the filler neck.
31 Remove the sending unit from the tank as described in Step 11.
32 Installation is the reverse of removal.

22 Fuel tank - cleaning and repair

1 If the fuel tank is contaminated with rust or sediment build-up, it must be removed and cleaned.
2 After the tank is removed, it should be flushed out with hot water and detergent, or, preferably, sent to a radiator shop for chemical flushing. **Warning:** *Never attempt to weld,*

solder or make any type of repairs on an empty fuel tank! Leave this work to a repair shop.

3 The use of a chemical-type sealer for on-vehicle repairs is advised only in case of an emergency; the tank should be removed and sent to a shop for more permanent repairs as soon as possible.

4 Never store a gas tank in an enclosed area where gas fumes could build up and result in an explosion or fire. Be especially careful inside garages where a natural gas appliance is located - the pilot light could cause an explosion!

23 Exhaust system - component replacement

1 The exhaust system should be inspected periodically for leaks, cracks and damaged or worn components (Chapter 1).

2 Allow the exhaust system to cool for at least three hours prior to inspecting or beginning work on it.

3 Raise the vehicle and support it securely on jackstands if necessary for clearance. **Warning:** *DO NOT work under a vehicle which is supported only by a jack!*

4 Exhaust system components can be detached by removing the heat shields (if equipped), unbolting and/or disengaging them from the hangers and removing them from the vehicle. Pipes on either side of the muffler must be removed by cutting with a hacksaw. Install the new muffler using U-bolts. If parts are rusted together, apply penetrating oil and allow it to work before attempting removal.

5 After replacing any part of the exhaust system, check carefully for leaks before driving the vehicle.

Chapter 5
Engine electrical systems

Contents

	Section
Alternator - removal and installation	13
Alternator brushes - replacement	14
Battery cables - check and replacement	4
Battery check and maintenance	See Chapter 1
Battery - emergency jump starting	3
Battery - removal and installation	2
Charging system - check	11
Charging system - general information and precautions	10
Distributor cap and rotor check and replacement	See Chapter 1
Distributor - removal and installation	6
Drivebelt check and adjustment	See Chapter 1
External voltage regulator - check and replacement	12
Ignition coil - check and replacement	9
Ignition key lock cylinder - removal and installation	See Chapter 10
Ignition module (HEI ignition) - replacement	7
Ignition pick-up coil (HEI ignition) - check band replacement	8
Ignition point replacement	See Chapter 1
Ignition switch replacement	See Chapter 10
Ignition system - check	5
Ignition system - general information and precautions	1
Ignition timing check and adjustment	See Chapter 1
Spark plug replacement	See Chapter 1
Spark plug wire check and replacement	See Chapter 1
Starter motor - in-vehicle check	16
Starter motor - removal and installation	17
Starter solenoid - removal and installation	18
Starting system - general information	15
Top Dead Center (TDC) for number one piston - locating	See Chapter 2

Specifications

Note: *The following specifications were compiled from the latest information available. If the specifications listed on the Emissions Control Information label (located in the engine compartment of your vehicle) are different, assume that the label is correct.*

Charging system

Drivebelt deflection	See Chapter 1
External voltage regulator (1967 through 1972 models)	
Charge relay	
Air gap	0.150 in
Point gap	0.030 in
Closing voltage	2.3 to 3.7 volts
Voltage regulator	
Air gap	0.067 in
Point gap	0.014 in
Voltage setting	13.8 to 14.8 volts (at 85° F)

Ignition system

Cylinder numbers	See Chapter 2
Firing order	
Inline 6-cylinder engine	1-5-3-6-2-4
V6 engine	1-6-5-4-3-2
V8 engine	1-8-4-3-6-5-7-2
Distributor rotation (all)	Clockwise
Ignition point gap	See Chapter 1
Dwell angle	See Chapter 1
Condenser capacity	0.18 to 0.23 mfd

Year and engine size (cu in)	Spark plug type	Gap	Ignition timing at idle (degrees BTDC)
1967			
230 cu in	AC-R44XLS	0.035 inch	4
250 cu in	AC-R44XLS	0.035 inch	4
283 cu in	AC-44	0.035 inch	4
327 cu in	R-43	0.035 inch	2 (220 H.P.) or 8 (185 H.P.)
1968			
250 cu in	AC-R44XLS	0.035 inch	4
307 cu in	AC-44S	0.035 inch	2
327 cu in	R-43	0.035 inch	2 (220 H.P.) or 8 (185 H.P.)
1969			
250 cu in	R44XLS	0.035 inch	4
307 cu in	R-44	0.035 inch	2
1970			
250 cu in	R-46T	0.035 inch	4
292 cu in	R-46T	0.035 inch	4
307 cu in	R-45	0.035 inch	2
350 cu in	R-44	0.035 inch	4
396 cu in	R-44T	0.035 inch	4
1971			
250 cu in	AC-R46TS	0.035 inch	4
292 cu in	AC-R44T	0.035 inch	4
307cu in	AC-R45TS	0.035 inch	2
350 cu in	AC-R45TS	0.035 inch	4
402 cu in	AC-R44TS	0.035 inch	2
1972			
250 cu in	AC-R46T	0.035 inch	4
292 cu in	AC-R44T	0.035 inch	4
307 cu in	AC-R44T	0.035 inch	4 (Manual), 8 (auto)
350 cu in	AC-R44T	0.035 inch	4 (Manual), 8 (auto)
402 cu in	AC-R44T	0.035 inch	8
1973			
250 cu in	AC-R46T	0.035 inch	6
292 cu in	AC-R44T	0.035 inch	4 (Fed), 8 (Calif)
307 cu in	AC-R44T	0.035 inch	4 (Manual), 8 (auto)
350 cu in	AC-R44T	0.035 inch	8 (Manual), 12 (auto)
454 cu in	AC-R44T	0.035 inch	10 (Federal)
1974			
250 cu in	AC-R46T	0.035 inch	8
292 cu in	AC-R44T	0.035 inch	8
350 cu in	AC-R44T	0.035 inch	8 (Fed), 12 (Calif)
454 cu in	AC-R44T	0.035 inch	10
1975			
250 cu in	AC-R46TX	0.060 inch	6 (Manual), 10 (auto)
292 cu in	AC-R44TX	0.060 inch	8
350 cu in	AC-R44TX	0.060 inch	6 (light duty), 8 (Heavy duty), 2 (Calif)
400 cu in	AC-R44TX	0.060 inch	4, 2 (Calif)
454 cu in	AC-R44TX	0.060 inch	12 (With catalytic converter), 8 (Without)
1976			
250 cu in	AC-R46TS	0.035 inch	6 (Manual), 10 (auto, Calif)
292 cu in	AC-R44T	0.035 inch	8
350 cu in	AC-R45TS	0.045 inch	6 (light duty), 8 (Heavy duty), 2 (Calif)
400 cu in	AC-R45TS	0.045 inch	4, 2 (Calif)
454 cu in	AC-R45TS	0.045 inch	12 (With converter), 8 (Without)
1977 and 1978			
250 cu in	AC-R46TS	0.035 inch	8 (Manual), 10 (Auto), 12 (Auto, Calif)
292 cu in	AC-R44T	0.035 inch	8
305 cu in	AC-R45TS	0.045 inch	8
350 cu in	AC-R45TS	0.045 inch	8 (Fed)
400 cu in	AC-R44T	0.045 inch	4, 2 (Calif)
454 cu in	AC-R44T	0.045 inch	4 (Light duty), 8 (Heavy duty)
1979			
250 cu in	R46TS	0.035 inch	10
292 cu in	R44T	0.035 inch	8
305 cu in	R45TS	0.045 inch	6
350 cu in	R44T	0.045 inch	4
350 cu in	R45TS	0.045 inch	8
400 cu in	R44T	0.045 inch	4
400 cu in	R45TS	0.045 inch	4
454 cu in	R44T	0.045 inch	4
454 cu in	R45TS	0.045 inch	8

Year and engine size (cu in)	Spark plug type	Gap	Ignition timing at idle (degrees BTDC)
1980			
250 cu in	R46TS	0.035 inch	10
292 cu in	R44T	0.035 inch	8
305 cu in	R45TS	0.045 inch	8
350 & 400 cu in (Federal)	R44T	0.045 inch	4
350 & 400 cu in (California)	R44T	0.045 inch	6
350 cu in (Federal)	R44T	0.045 inch	4
454 cu in	R44T	0.045 inch	4
1981			
250 cu in	R45TS	0.035 inch	10
292 cu in	R44T	0.035 inch	8
305 cu in	R45TS	0.045 inch	4
350 cu in (light duty)	R45TS	0.045 inch	8
350 cu in (heavy duty)	R44T	0.045 inch	4 (Fed), 6 (Calif)
454 cu in	R44T	0.045 inch	4
1982 and 1983			
250 cu in	R45TS	0.035 inch	10 (Fed), 6 (Calif)
292 cu in	R44T	0.035 inch	8
305 cu in	R44TS	0.045 inch	4 (Fed), 8 (Calif)
350 cu in (light duty)	R45TS	0.045 inch	8
350 cu in (heavy duty)	R44T	0.045 inch	6 (Fed), 4 (Calif)
454 cu in	R44T	0.045 inch	4
1984			
250 cu in	R45TS	0.035 inch	10 (Fed), 6 (Calif)
292 cu in	R44T	0.035 inch	8
305 cu in	R44TS	0.045 inch	4 (Fed), 6 (Calif)
350 cu in (light duty)	R45TS	0.045 inch	8 (Fed), 6 (Calif)
350 cu in (heavy duty)	R44T	0.045 inch	4 (Fed), 6 (Calif)
454 cu in	R44T	0.045 inch	4
1985 and 1986			
262 cu in (4.3L V6)	R43TS	0.035 inch	0 (Fed), 4 (Calif)
292 cu in	R44T	0.035 inch	8
305 cu in	R44TS	0.045 inch	4 (Fed), 6 (Calif)
350 cu in (light duty)	R45TS	0.045 inch	8 (Fed), 6 (Calif)
350 cu in (heavy duty)	R44T	0.045 inch	4 (Fed), 6 (Calif)
454 cu in	R44T	0.045 inch	4
1987			
262 cu in (4.3L V6)	R43TS	0.035 inch	0
305 cu in	CR43TS	0.035 inch	0
350 cu in (EFI)	CR43TS	0.035 inch	0
350 cu in (Carb)	R44T	0.045 inch	4 (Fed), 6 (Calif)
454 cu in (EFI)	CR43TS	0.035 inch	4
1988			
305 cu in	CR43TS	0.035 inch	0
350 cu in (EFI)	CR43TS	0.035 inch	0
350 cu in (Carb)	R44T	0.045 inch	4 (Fed), 6 (Calif)
454 cu in (EFI)	CR43TS	0.035 inch	4
454 cu in (Carb)	R44T	0.045 inch	4
1989			
350 cu in	CR43TS	0.035 inch	0
454 cu in (EFI)	CR43TS	0.035 inch	4
1990 and 1991			
350 cu in	CR43TS	0.035 inch	0
454 cu in	CR43TS	0.035 inch	4

**Refer to the Vehicle Emission Control Label (VECI) in the engine compartment; use the information there if it differs from that listed here*

1 Ignition system - general information and precautions

General information

The ignition system consists of the ignition switch, the battery, the coil, the primary (low tension) and secondary (high tension) wiring circuits, the distributor and the spark plugs.

Mechanical breaker point ignition system

On models built between 1967 and 1974, a mechanical contact breaker point ignition system was standard equipment. In this system, low voltage in the primary circuit is transformed by the coil into high voltage in the secondary circuit by the opening and closing of the contact points. The voltage is directed to the appropriate spark plug by the distributor cap and rotor. Each time the rotor aligns with one of the terminals in the cap (which are connected to the spark plug wires) the opening and closing of the contact points causes the voltage to build up and jump the gap at the spark plug. The spark at the plug ignites the fuel/air mixture in the combustion chamber. Ignition advance is handled by a combination of vacuum and mechanical advance mechanisms built into the distributor.

Component location

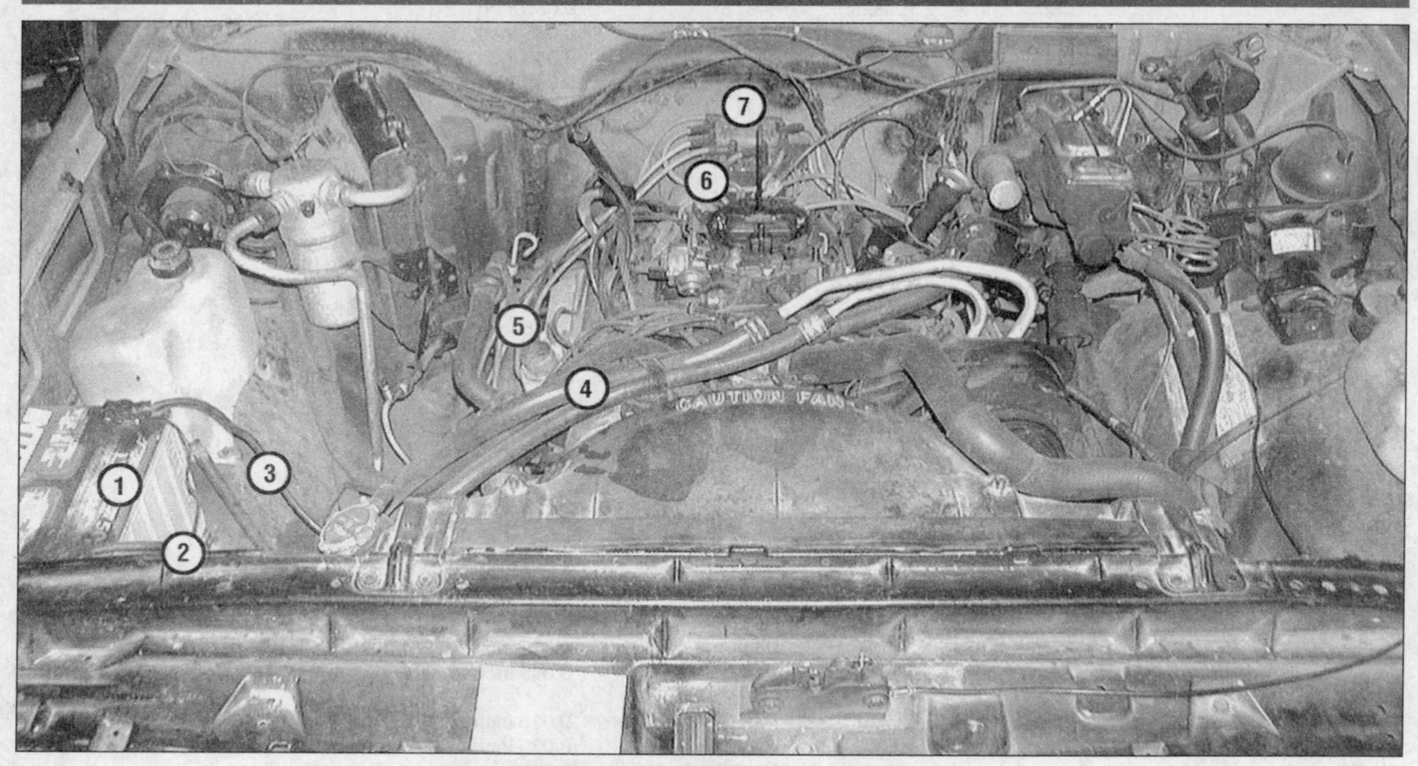

Typical V8 engine electrical component location

1. *Battery*
2. *Battery cable*
3. *Battery cable*
4. *Alternator*
5. *Starter motor (below exhaust manifold)*
6. *Distributor*
7. *High energy ignition coil (HEI)*

High Energy Ignition (HEI) system

Later models are equipped with GM's version of electronic ignition known as HEI. Some HEI distributors combine all the ignition components into one unit. The ignition coil is in the distributor cap and connects through a resistance brush to the rotor. On other HEI distributors, the coil is mounted separately. Some HEI systems (early models) utilize vacuum and mechanical advance mechanisms built into the distributor to handle ignition timing advance, while others (later models) are equipped with electronic components to handle spark advance.

Precautions

Warning: *Because of the very high voltage generated by the HEI system, extreme care should be taken whenever an operation involving ignition components is performed. This not only includes the distributor, coil, ignition module and spark plug wires, but related items such as the plug connections and test equipment.*

The secondary (spark plug) wire used with the HEI system is a carbon impregnated cord conductor encased in an 8 mm (5/16-inch) diameter rubber core with an outer silicone rubber jacket. This type of wire will withstand very high temperatures and provides an excellent insulator for the HEI's high voltage. When replacement is required, be sure to use the same type wires as the original equipment (see Chapter 1).

A silicone rubber spark plug boot forms a tight seal on each plug. The boots should be twisted 1/2-turn when removing them (for more information on spark plug wires refer to Chapter 1).

2 Battery - removal and installation

Refer to illustration 2.3

Warning: *Hydrogen gas is produced by the battery, so keep open flames and lighted cigarettes away from it at all times. Always wear eye protection when working around a battery. Rinse off spilled electrolyte immediately with large amounts of water.*

1 The battery is located in the engine compartment, near the front of the vehicle.

Removal

2 Disconnect both battery cables from the battery terminals. **Caution:** *Always disconnect the negative (-) cable first, then remove the positive (+) cable.*

3 Remove the bolt and detach the retainer **(see illustration)**.

4 Carefully lift the battery out of the carrier. **Warning:** *Always keep the battery in an upright position to prevent electrolyte spills. If you spill any on your skin, rinse it off immediately with large amounts of water.*

2.3 Battery bolt and retainer

Installation

Note: *The battery carrier and retainer should be clean and corrosion free before installing the battery. The carrier should be in good condition, to hold the battery securely and keep it level. Make certain that there are no parts in the carrier before installing the battery.*

5 Gently set the battery in position in the carrier. Don't tilt it.

6 Install the hold-down retainer and bolt. The bolt should be snug, but overtightening it may damage the battery case.

7 Install both battery cables - positive first, then negative. **Note:** *The battery posts and cable ends should be cleaned prior to connection* (see Chapter 1).

3 Battery - emergency jump starting

Refer to the *Booster battery (jump) starting* procedure at the front of this manual.

4 Battery cables - check and replacement

Refer to illustration 4.2

1 Periodically inspect the entire length of each battery cable for damage, cracked or burned insulation and corrosion. Poor battery cable connections can cause starting problems and decreased engine performance.

2 Check the cable-to-terminal connections at the ends of the cables for cracks, loose wire strands and corrosion. The presence of white, fluffy deposits under the insulation at the cable terminal connection is a sign the cable is corroded and should be replaced. Check the terminals for distortion, missing mounting bolts or nuts and corrosion **(see illustration)**.

3 If only the positive cable is to be replaced, be sure to disconnect the negative cable from the battery first. Always disconnect the negative cable first and hook it up last.

4 Disconnect and remove the cable. Make sure the replacement cable is the same length and diameter.

5 Clean the threads of the starter or ground connection with a wire brush to remove rust and corrosion. Apply a light coat of petroleum jelly to the threads to ease installation and prevent future corrosion.

6 Attach the cable to the starter or ground connection and tighten the mounting nut securely.

7 Before connecting the new cable to the battery, make sure it reaches the terminal without having to be stretched.

8 Connect the positive cable first, followed by the negative cable. Tighten the nuts and apply a thin coat of petroleum jelly to the terminal and cable connection.

Terminal end corrosion or damage.

Insulation cracks.

Chafed insulation or exposed wires.

Burned or melted insulation.

4.2 Typical battery cable problems

5 Ignition system - check

Refer to illustration 5.2

Warning: *Because of the very high secondary voltage generated by the ignition system - particularly the High Energy Ignition (HEI) system - extreme care should be taken whenever this check is performed.*

HEI system

Calibrated ignition tester method

1 If the engine turns over but won't start, disconnect the spark plug lead from any spark plug and attach it to a calibrated ignition tester for the HEI system (available at most auto parts stores).

2 Connect the clip on the tester to a ground such as a metal bracket **(see illustration)**, crank the engine and watch the end of the tester to see if bright blue, well-defined sparks occur.

3 If sparks occur, sufficient voltage is reaching the plugs to fire the engine. However, the plugs themselves may be fouled, so remove and check them as described in Chapter 1 or install new ones.

4 If no spark occurs, remove the distributor cap and check the cap and rotor as described in Chapter 1. If moisture is present, use WD-40 (or something similar) to dry out the cap and rotor, then reinstall the cap and repeat the spark test.

5 If there's still no spark, the tester should be attached to the wire from the coil (this cannot be done on vehicles with a coil-in-cap distributor) and the check should be repeated again.

6 If no spark occurs, check the primary wire connections at the coil to make sure they're clean and tight. Make any necessary repairs, then repeat the check again.

7 If sparks now occur, the distributor cap, rotor, plug wire(s) or spark plug(s) may be defective. If there's still no spark, the coil-to-cap wire may be bad. If a substitute wire doesn't make any difference, have the system checked by a dealer service department or a repair shop.

Alternative method

Note: *If you're unable to obtain an HEI tester, the following method will enable you to determine whether the ignition system has spark, but it will not tell you if there is enough voltage present to actually initiate combustion.*

8 Remove the spark plug wire from one of the spark plugs. Using an insulated tool, hold the wire about 1/4-inch from a good ground

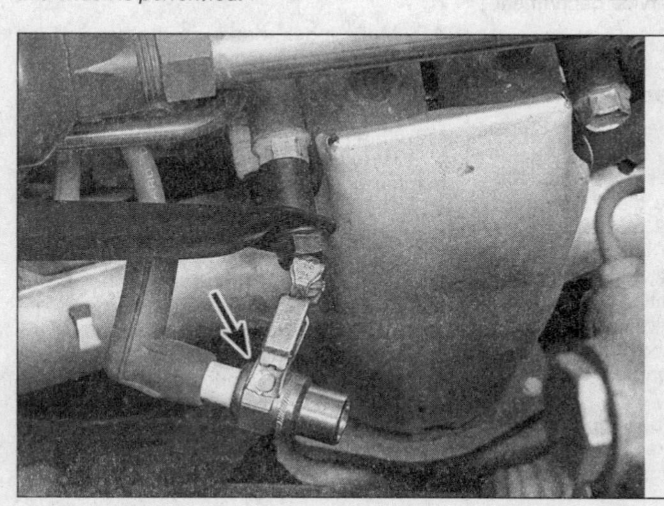

5.2 The ignition system should be checked with a spark tester that is calibrated for the HEI voltage output - if the ignition system produces a spark that will jump the tester gap, it's functioning normally and the engine should start if the plugs aren't fouled

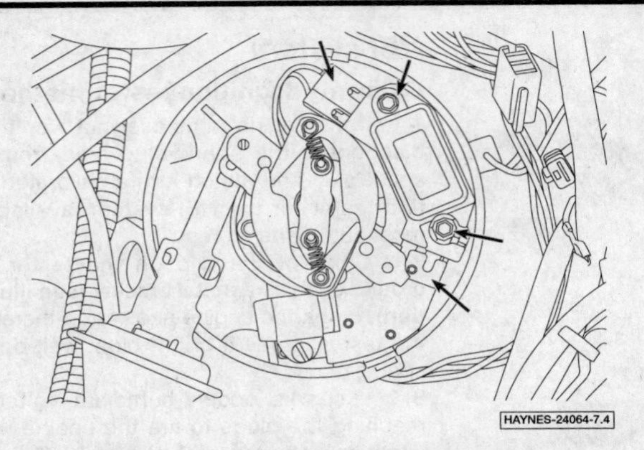

7.4 The wire connectors are attached to both ends of the module (some models) and the screws thread into the distributor body (arrows)

7.6 If you're installing a new module, apply silicone grease (included with the module) to the area where the module mounts

and have an assistant crank the engine.

9 If bright blue, well-defined sparks occur, sufficient voltage is reaching the plugs to fire the engine. However, the plugs themselves may be fouled, so remove and check them as described in Chapter 1 or install new ones.

10 If there's no spark, check another wire in the same manner. A few sparks followed by no spark is the same condition as no spark at all.

11 If no spark occurs, remove the distributor cap and check the cap and rotor as described in Chapter 1. If moisture is present, use WD-40 (or something similar) to dry out the cap and rotor, then reinstall the cap and repeat the spark test.

12 If there's still no spark, disconnect the coil wire from the distributor, hold it about 1/4-inch from a good ground and crank the engine again (this cannot be done on vehicles with a coil-in-cap distributor).

13 If no spark occurs, check the primary wire connections at the coil to make sure they're clean and tight. Make any necessary repairs, then repeat the check again.

14 If sparks now occur, the distributor cap, rotor, plug wire(s) or spark plug(s) may be defective. If there's still no spark, the coil-to-cap wire may be bad. If a substitute wire doesn't make any difference, have the system checked by a dealer service department or a repair shop.

Breaker point system

15 Follow the procedure described in Steps 8 through 13 above.

16 If sparks occur after following the instructions in Step 13, the distributor cap, rotor, plug wire(s) or spark plug(s) may be defective.

17 If there's still no spark, the coil-to-cap wire may be bad. If a substitute wire doesn't make any difference, proceed as follows.

18 Refer to Chapter 1, remove the distributor cap and check the ignition points as described.

19 If the points appear to be in good condition and the primary wires are hooked up correctly and undamaged, adjust the points as described in Chapter 1, then repeat the spark check.

20 If the primary circuit is complete (battery voltage is available at the points), the points are clean and properly adjusted and the secondary ignition circuit is functioning correctly, sparks should occur at the plug wires. If there is still no spark, have the system checked by a dealer service department or a repair shop.

6 Distributor - removal and installation

Removal

1 Disconnect the negative cable at the battery. Place the cable out of the way so it cannot accidentally come in contact with the negative terminal of the battery, as this would once again allow power into the electrical system of the vehicle.

2 Unplug the wiring harness connectors attached to the wires leading to the distributor, then refer to Chapter 1 and detach the cap and spark plug wires. Lay them aside, out of the way.

3 On models with vacuum advance, disconnect the vacuum hose from the distributor.

4 Make index marks on the base of the distributor and the engine block or manifold to ensure that you'll be able to put the distributor back in the same relative position (this will prevent drastic changes in ignition timing).

5 Refer to Chapter 2 and position the number one piston at TDC on the compression stroke. Make a mark on the edge of the distributor base directly below the rotor tip and in line with it.

6 Remove the distributor hold-down bolt and clamp, then pull straight up on the distributor to separate it from the engine. If it's stuck, squirt penetrating oil around the base and wait for it to work before attempting removal. If penetrating oil doesn't help, break the distributor free by turning it with an oil filter wrench. **Caution:** *Don't turn the crankshaft while the distributor is out of the engine.*

Installation

7 Check the condition of the O-ring or gasket at the base of the distributor. If it's damaged or deformed, oil leaks may develop.

8 To install the distributor (assuming that the crankshaft hasn't been turned), reposition the rotor about 1/8-turn clockwise, past the mark made when locating the number one piston at TDC.

9 Push the distributor into place until it's seated. The gears have to mesh properly and the oil pump driveshaft must engage in the end of the distributor shaft. The rotor should turn slightly as the gears mesh and come to rest pointing at the mark made when locating TDC for the number one piston.

10 The remaining steps are the reverse of removal. Be sure to align the marks on the distributor and block or manifold and check the ignition timing (Chapter 1).

7 Ignition module (HEI ignition) - replacement

Refer to illustrations 7.4 and 7.6

Note: *It's not necessary to remove the distributor from the engine to replace the ignition module.*

1 Disconnect the cable from the negative terminal of the battery. Remove the air cleaner to provide room to work around the distributor.

2 Remove the distributor cap and wires as an assembly and position them out of the way.

3 Remove the screws and detach the rotor.

4 Carefully detach the wires from the module terminals **(see illustration)**. If the wires are attached with a plastic connector, release the locking tang before pulling the connector off the terminals. If the wires are difficult to remove, you may find it easier to remove the module mounting screws first, detach the module and then pull off the wires. **Caution:** *Do not pull on the wires or*

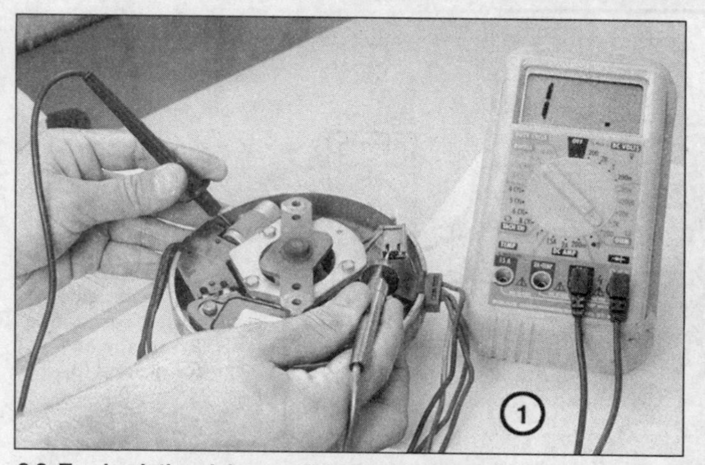

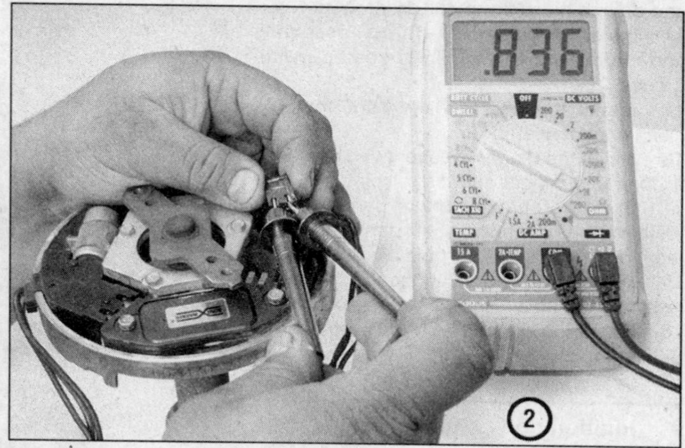

8.3 To check the pick-up coil, hook the ohmmeter leads to each wire and ground, one at a time, and note the ohmmeter reading (1) - the second check (2) requires one ohmmeter lead to be attached to each pick-up coil wire

the connectors may be damaged.

5 Remove the screws and lift out the module. **Note:** *The module can only be tested with special equipment. If you suspect that the module is malfunctioning, have it checked by an auto parts store or dealer service department.*

6 Installation is the reverse of removal. Be sure to apply the silicone dielectric grease supplied with the new module to the distributor pad **(see illustration)** - DO NOT use any other type of grease! If the grease is not used, the module will overheat and destroy itself.

8 Ignition pick-up coil (HEI ignition) - check and replacement

Refer to illustrations 8.3, 8.7, 8.9 and 8.10

1 Remove the distributor from the engine (Section 6).

2 Remove the rotor and disconnect the pick-up coil leads from the module as described in Section 7.

Check

3 Connect an ohmmeter to each terminal of the pick-up coil connector or wire and ground (one terminal at a time) **(see illustration).** The ohmmeter should indicate infinite resistance. If it doesn't, the pickup coil is defective.

4 Connect the ohmmeter between both terminals or wires of the pick-up coil connector. If a vacuum advance unit is attached to the distributor, apply vacuum from an external source and watch the ohmmeter for indications of intermittent opens (if no vacuum unit is used, flex the wires by hand). The ohmmeter should indicate one steady value within the 500 to 1500 ohm range as the wires are flexed. If it doesn't, the pick-up coil is defective.

5 If the pick-up coil fails either test, replace it.

Replacement

6 Mark the distributor shaft and gear so they can be reassembled in the same relationship.

7 Secure the distributor shaft housing in a bench vise and drive out the roll pin with a hammer and punch **(see illustration).**

8 Remove the gear and tanged washer,

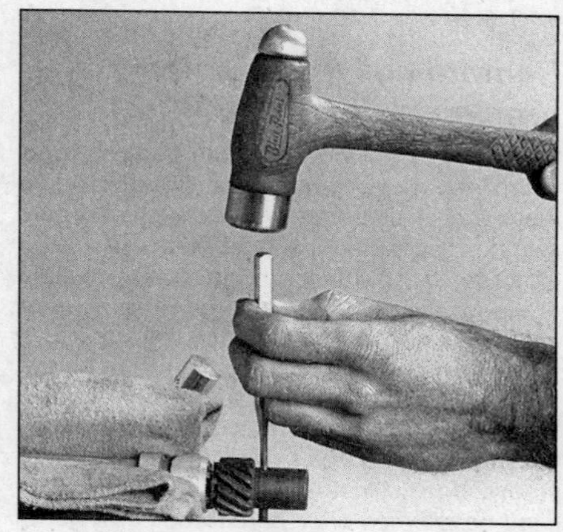

8.7 Place the distributor in a bench vise and drive out the roll pin which locks the drive gear to the bottom of the shaft

then check the shaft for burrs. If none are found, pull the shaft from the distributor.

9 Remove the three attaching screws and separate the magnetic shield **(see illustration).**

10 Remove the C-clip **(see illustration)**

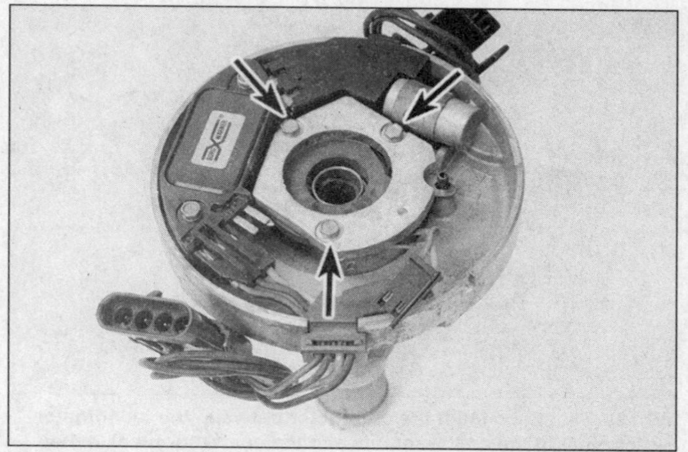

8.9 Remove the three screws (arrows) attaching the magnetic shield to the distributor base and detach the shield

8.10 The pick-up coil and pole piece assembly can be removed after carefully prying out the C-clip

and detach the pick-up coil assembly.

11 Install the new pick-up coil assembly and make sure the C-clip is seated in the groove.

12 Install the shaft. Make sure that it's clean and lubricated.

13 Install the tanged washer (with the tangs facing up), the gear and the roll pin.

14 Spin the shaft to make sure that the teeth on the distributor shaft don't touch the teeth on the pick-up coil pole piece.

15 If the teeth touch, loosen, adjust and retighten the pole piece to eliminate contact.

16 Install the distributor in the engine (see Section 6).

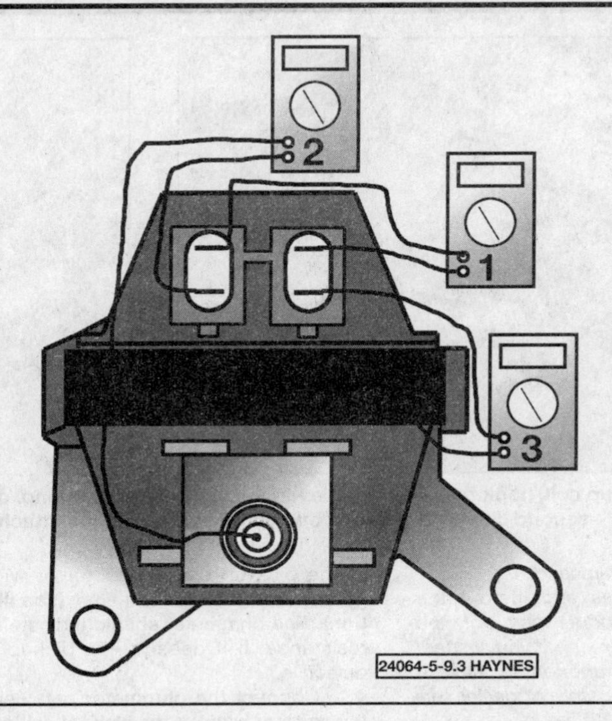

9.3 Check the coil primary resistance (1), secondary resistance (2) and the coil for ground (3)

9 Ignition coil - check and replacement

Refer to illustrations 9.3, 9.11, 9.12a, 9.12b, 9.13, 9.14 and 9.16

Remote coil (mechanical breaker point system)

1 Label the wires and terminals, then detach the wires from the coil. Using an ohmmeter, check the resistance of the primary and secondary circuits. It should be about 1 or 2 ohms for the primary circuit and 3K to 20K ohms for the secondary circuit. If the check indicates an open or short in the coil, replace it with a new one by loosening the clamp screw and sliding the coil out of the mounting bracket. Clean the bracket before installing the new coil and don't overtighten the clamp screw. **Note:** *Since electrical parts normally can't be returned once purchased, it's a good idea to have your test results confirmed by a repair shop before buying a new coil.*

Remote coil (HEI ignition system)

Check

2 Check for voltage at the coil battery terminal with a test light or voltmeter. The ignition switch must be turned to the On position. If no voltage is indicated, check for loose connections at the coil and ignition switch and make any necessary repairs.

3 Detach the coil wire harness connector and hook up an ohmmeter as shown in Test 1 **(see illustration)**. If the reading is above 1 ohm, replace the coil with a new one.

4 If the reading is 0 to 1 ohm, hook up the ohmmeter leads as shown in Test 2 **(see illustration 9.3)**. If the reading is less than 6000 ohms or more than 30,000 ohms, replace the coil with a new one.

5 If it's between 6000 and 30,000 ohms, connect the ohmmeter leads as shown in Test 3 **(see illustration 9.3)**. If the ohmmeter reading is less than infinite resistance, replace the coil. If the reading is infinity, the coil is functional.

Replacement

6 If not already done, detach the coil high tension lead and unplug the primary wire harness connector.

7 Remove the mounting nuts and separate the coil from the engine.

8 Installation is the reverse of removal.

Coil-in-cap models (HEI ignition system)

Check

9 Disconnect the cable from the negative terminal of the battery.

10 Remove the distributor cap.

11 Position the cap upside down so you can see the male spade-type terminals inside the hood protruding from the coil cap. Connect an ohmmeter to the outer two terminals

9.11 Test 1: to check the ignition coil-in-cap, first connect an ohmmeter to the tachometer and battery terminals - the reading should be zero or nearly zero

9.12a Test 2: connect the ohmmeter between the tachometer terminal of the coil assembly and the center terminal in the distributor cap . . .

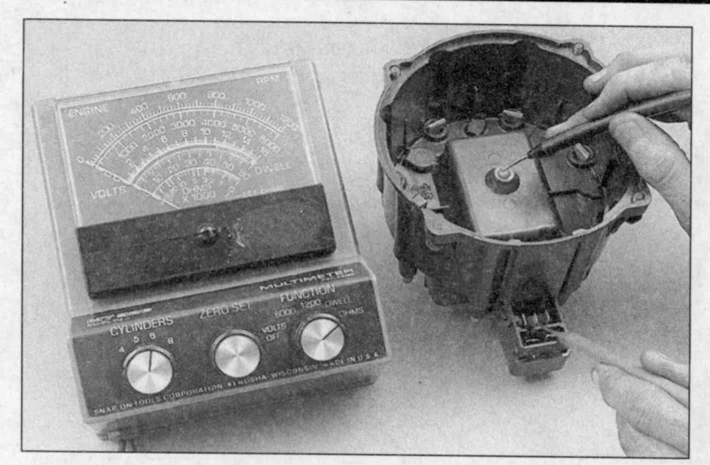

9.12b . . . then connect the ohmmeter between the ground terminal of the coil assembly and the center terminal in the distributor cap - if both readings indicate infinite resistance, replace the coil

9.13 Remove these two screws (arrows) to detach the coil cover

(see illustration). The ohmmeter should indicate very near zero resistance. If it doesn't, replace the coil.

12 Connect the ohmmeter between the TACH terminal and the center terminal of the distributor cap and note the reading on the ohmmeter **(see illustration)**. Move the ohmmeter lead to the ground terminal (center terminal on the inside row) **(see illustration)** and leave the other lead on the center terminal in the distributor cap. Be sure to use the High scale. If both ohmmeter readings indicate infinite resistance, the coil is faulty.

Replacement

13 Remove the coil cover screws **(see illustration)** and lift off the cover.

14 Carefully push the coil spade terminals through the top of the hood with a small screwdriver **(see illustration)**.

15 Remove the ignition coil mounting screws and detach the coil, with the wires, from the cap.

16 Remove the arc seal **(see illustration)**.

17 Clean the coil housing and the rest of the cap with a soft cloth and check the cap for defects. Replace it if necessary.

18 Install the new arc seal.

19 Install the new coil and tighten the mounting screws securely.

20 Position the terminals in the correct slots, then install the coil cover on the cap and tighten the screws.

21 Install the distributor cap.

22 Reconnect the cable to the negative terminal of the battery.

10 Charging system - general information and precautions

The charging system includes the alternator, voltage regulator and battery. These components work together to supply electrical power for the ignition system, lights, radio, etc. The alternator is driven by a drivebelt at the front of the engine.

The purpose of the voltage regulator is to limit the alternator's voltage to a preset value. This prevents power surges, circuit overloads, etc., during peak voltage output. On models built through 1972, the voltage regulator is mounted separately, while on later models it's contained within the alternator housing.

The charging system normally doesn't require periodic maintenance. However, the drivebelt, wires and connections should be inspected at the intervals suggested in Chapter 1.

Be very careful when working on the electrical system and be sure to note the following precautions.

a) *When making connections to the alternator from a battery, always match correct polarity.*
b) *Before using arc welding equipment to repair any part of the vehicle, disconnect the wires from the alternator and the battery terminal.*

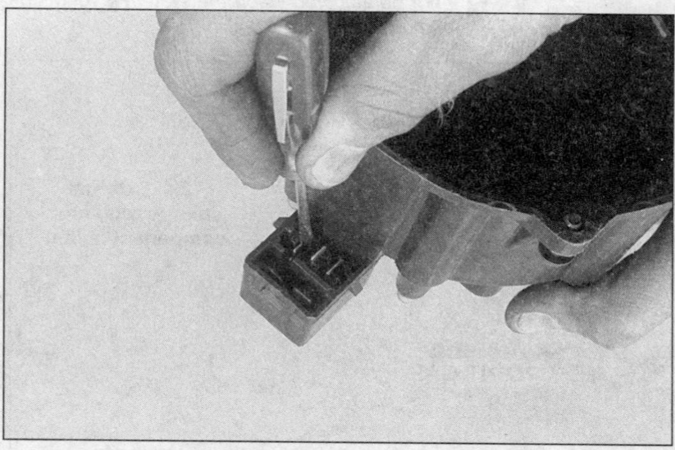

9.14 Before detaching the coil assembly from the cap, push the male spade-type connectors out of the hood with a small screwdriver

9.16 Remove the ignition coil arc seal from the bottom of the cavity in the distributor cap

11.2 A voltmeter or multimeter will be needed for charging system tests

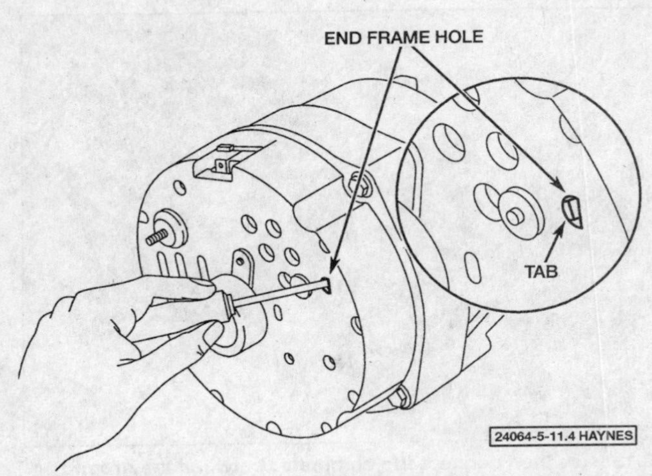

11.4 Be sure to locate the correct hole when attempting to ground the tab inside the alternator and DO NOT prolong the test or alternator damage could occur!

c) *Never start the engine with a battery charger connected.*
d) *Always disconnect both battery leads before using a battery charger.*

The charge indicator lights when the ignition switch is on, and goes out when the engine is running. If the charge indicator is on with the engine running, a charging system defect is indicated.

11 Charging system - check

Refer to illustrations 11.2 and 11.4

1 If a charging system malfunction occurs, don't immediately assume that the alternator is causing the problem. First check the following items:

a) *The battery cables where they connect to the battery. Make sure the connections are clean and tight.*
b) *The battery electrolyte specific gravity. If it's low, charge the battery.*
c) *Check the external alternator wire harness and connections. They must be in good condition.*
d) *Check the drivebelt condition and tension (Chapter 1).*
e) *Make sure the alternator mounting bolts are tight.*
f) *Run the engine and check the alternator for abnormal noise.*

2 Using a voltmeter, check the battery voltage with the engine off. It should be approximately 12-volts **(see illustration)**.

3 Start the engine and check the battery voltage again. It should now be approximately 14 to 15-volts.

4 Locate the test hole in the back of the alternator and ground the tab inside the hole by inserting a screwdriver blade and touching the tab and the case at the same time **(see illustration)**. **Caution:** *Don't run the engine with the tab grounded any longer than necessary to obtain a voltmeter reading. If the alternator is charging, it's running unregulated during this test, which may overload the electrical system and cause damage to the components.*

5 The reading on the voltmeter should be 15-volts or higher with the tab grounded.

6 If the voltmeter indicates low battery voltage, the alternator is faulty and should be replaced with a new one (Section 13).

7 If the voltage reading is 15-volts or higher and a "no charge" condition exists, the regulator or field circuit is the problem. Remove the alternator (Section 13) and have it checked further by an auto electric shop.

12 External voltage regulator - check and replacement

Refer to illustration 12.4

1 A discharged battery may be due to a malfunction in the voltage regulator. Before checking the regulator, check the following items:

a) *Drivebelt tension (Chapter 1).*
b) *Battery and cables (Chapter 1 and Section 4).*
c) *Charging circuit and connections (Section 10).*
d) *Make sure that lights and other electrical accessories have not been left on inadvertently.*

2 Detach the negative cable from the battery before removing the cover from the regulator. The regulator cover is held in place with two screws.

3 Under no circumstances should the voltage regulator contacts be cleaned - abrasive materials will destroy the contact material. Relay point and air gap adjustments can be checked with a feeler gauge to obtain approximate settings.

4 The field relay point opening may be adjusted by bending the stop. The air gap is checked with the points just touching and is adjusted by bending the flat contact spring **(see illustration)**. **Note:** *The field relay will normally operate satisfactorily even if the air gap is outside the specified limits and should not be adjusted when the system is functioning normally.*

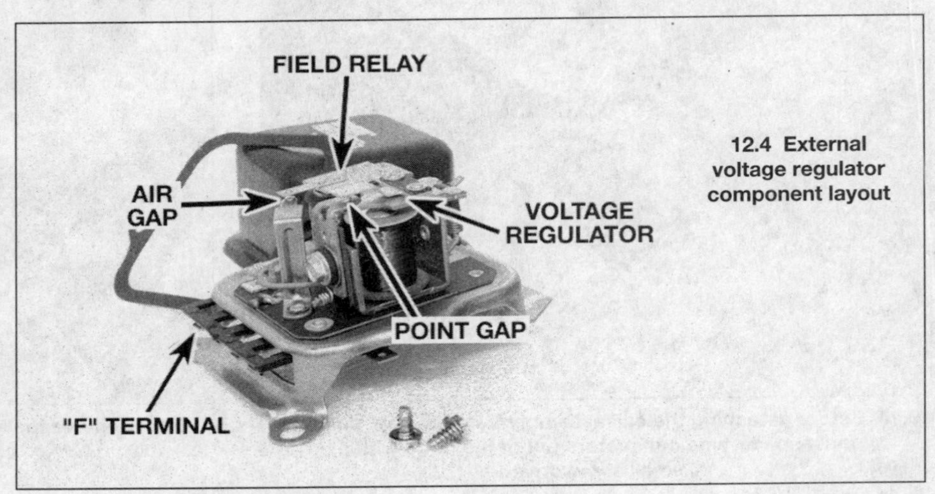

12.4 External voltage regulator component layout

13.4a To remove the drivebelt, loosen the alternator mounting bolt . . .

13.4b . . . then loosen the adjustment bolt and slip off the belt - finally, remove the mounting and adjusting bolts and remove the alternator

5 To remove the regulator, first make sure the negative battery cable has been disconnected, then detach the wires from the regulator.

6 Remove the mounting screws and detach the regulator from the vehicle.

7 Installation is the reverse of removal. Make sure that the rubber gasket is securely in place in the regulator base.

13 Alternator - removal and installation

Refer to illustrations 13.4a and 13.4b

1 Disconnect the cable from the negative terminal of the battery.

2 Loosen the bolts and remove the drivebelt (Chapter 1).

14.3a With the through bolts removed, carefully separate the drive end frame and the rectifier end frame

3 Detach the wires from the rear of the alternator. If there's any possibility of confusion later, label the wires and terminals.

4 Remove the pivot and adjustment bolts **(see illustrations)**.

5 Detach the alternator.

6 Installation is the reverse of removal. Refer to Chapter 1 for the drivebelt adjustment procedure. DO NOT hook up the battery ground cable until after the wires are reconnected to the alternator terminals.

14 Alternator brushes - replacement

Refer to illustrations 14.2, 14.3a, 14.3b, 14.4, 14.5a, 14.5b, 14.6 and 14.9

Note: *This procedure applies only to the 10-S1 series alternator used from 1973 on.*

1 Remove the alternator from the vehicle (Section 13).

2 Scribe, punch or paint marks on the front and rear end frame housings of the alternator to facilitate reassembly **(see illustration)**.

14.2 Mark the alternator end frame housings to ensure correct alignment during reassembly

3 Remove the four through-bolts holding the front and rear end frames together, then separate the drive end frame from the rectifier end frame **(see illustrations)**.

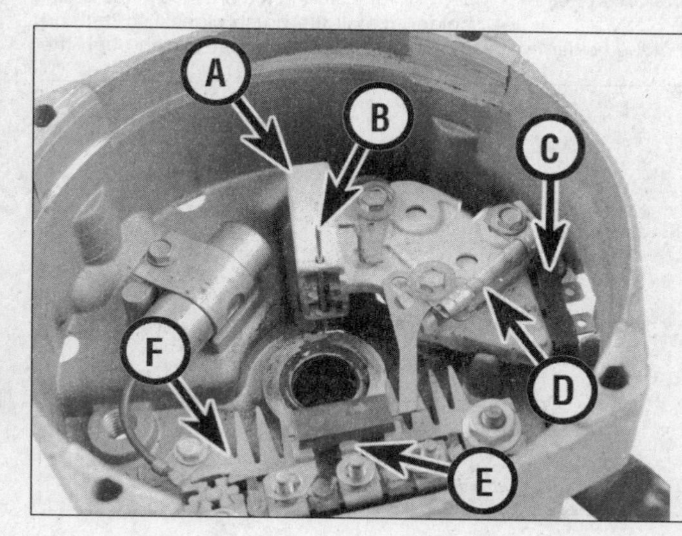

14.3b Inside a typical SI alternator

A Brush holder
B Paper clip retaining brushes (for reassembly)
C Regulator
D Resistor (not all models)
E Diode trio
F Rectifier bridge

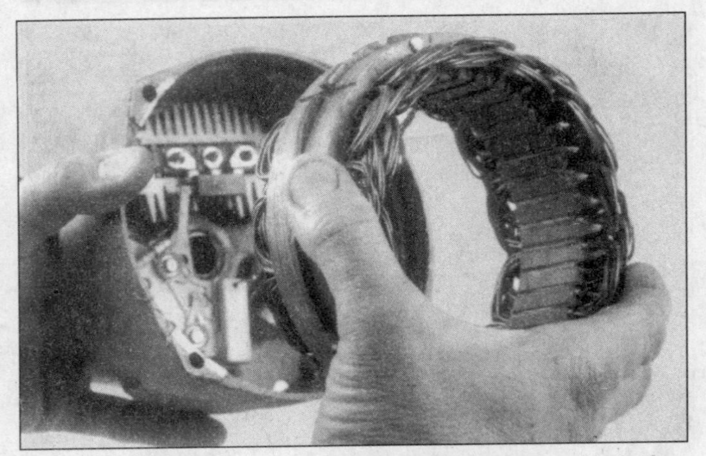

14.4 After removing the nuts holding the stator assembly to the rectifier bridge, remove the stator

14.5a Remove the screw attaching the diode trio to the end frame and remove the trio

14.5b After removing the screws that attach the brush holder and resistor (if equipped) to the end frame, remove the brush holder

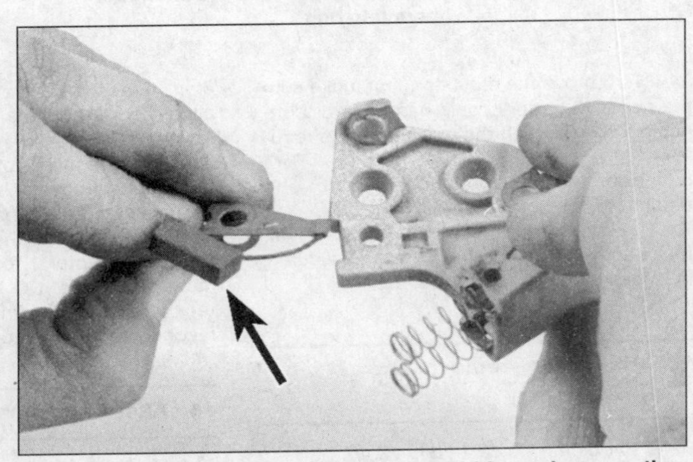

14.6 Slip the brush retainer off the brush holder and remove the brushes (arrow)

4 Remove the nuts holding the stator leads to the rear end frame (rectifier bridge) and separate the stator from the end frame **(see illustration)**.

5 Remove the screws holding the regulator/brush assembly to the end frame and detach the diode trio and the brush holder **(see illustrations)**.

6 Remove the brushes from the holder by slipping the brush retainer off **(see illustration)**.

7 Remove the springs from the brush holder.

8 Installation is the reverse of the removal procedure.

9 When installing the brushes in the brush holder, install the brush closest to the end frame first. Slip the paper clip through the rear of the end frame to hold the brush, then insert the second brush and push the paper clip in to hold both brushes while reassembly is completed **(see illustration)**. The paper clip should not be removed until the front and rear end frames have been bolted together.

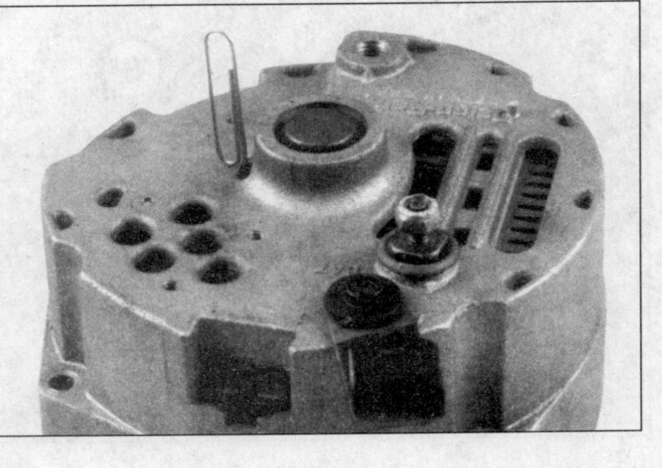

14.9 To hold the brushes in place during reassembly, insert a paper clip through the hole in the end frame nearest to the rotor shaft

15 Starting system - general information

The function of the starting system is to crank the engine. The starting system is composed of a starting motor, solenoid, battery, switch and connecting wires. The battery supplies the electrical energy to the solenoid, which then completes the circuit to the starter motor, which does the actual work of cranking the engine.

The electrical circuits are designed so the starter motor can only be operated when the clutch pedal is depressed (later models with a manual transmission) or the transmission selector lever is in Park or Neutral (automatic transmission).

Caution: *Never operate the starter motor for more than 30 seconds at a time*

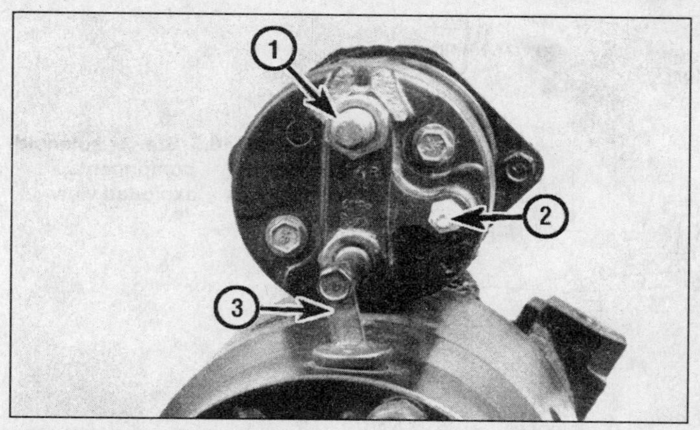

17.3 The solenoid terminals are usually different sizes to help you remember which wire goes to which terminal

1. *Battery cable terminal (+)*
2. *Ignition switch wire terminal*
3. *Starter motor strap*

17.4 Typical starter motor mounting bolts (arrows) (some starters have shims installed between the motor and the mounting pad - be sure to note how they're installed and return them to their original locations)

without pausing to allow it to cool for at least two minutes. Excessive cranking time can cause overheating and seriously damage the starter.

16 Starter motor - in-vehicle check

Note: *Before diagnosing starter problems, make sure the battery is fully charged.*

1 If the starter motor doesn't turn at all when the switch is operated, make sure the shift lever is in Neutral or Park (automatic transmission) or that the clutch pedal is depressed (manual transmission).

2 Make sure the battery is charged and that all cables, both at the battery and starter solenoid terminals, are secure.

3 If the starter motor spins but the engine isn't cranking, then the overrunning clutch in the starter motor is slipping and the motor must be removed from the engine and disassembled.

4 If, when the switch is actuated, the starter motor doesn't operate at all but the solenoid clicks, then the problem is in the battery, the main solenoid contacts or the starter motor itself.

5 If the solenoid plunger can't be heard when the switch is actuated, the solenoid itself is defective or the solenoid circuit is open.

6 To check the solenoid, connect a jumper wire between the battery positive terminal (+) and the S terminal on the solenoid. If the starter motor now operates, the solenoid is OK and the problem is in the ignition switch, Neutral start switch or wiring.

7 If the starter motor still doesn't operate, remove the starter/solenoid assembly for further testing and repair.

8 If the starter motor cranks the engine at an abnormally slow speed, first make sure the battery is charged and that all terminal connections are clean and tight. If the engine has a mechanical problem or has the wrong viscosity oil in it, it'll crank slowly.

9 Run the engine until normal operating temperature is reached, then disconnect the coil wire from the distributor cap and ground it on the engine. On vehicles with a coil-in-cap distributor, disconnect the BAT wire from the distributor cap.

10 Connect a voltmeter positive lead to the starter motor terminal of the solenoid and connect the negative lead to a good ground.

11 Operate the ignition switch and take the voltmeter readings as soon as a steady figure is indicated. Do not allow the starter motor to turn for more than 30 seconds at a time. A reading of 9-volts or more, with the starter motor turning at normal cranking speed, is normal. If the reading is 9-volts or more but the cranking speed is slow, the motor is faulty. If the reading is less than 9-volts and the cranking speed is slow, the solenoid contacts are probably burned.

17 Starter motor - removal and installation

Refer to illustrations 17.3 and 17.4

Removal

1 Disconnect the negative cable from the battery.

2 Raise the front of the vehicle and support it securely on jackstands. Apply the parking brake or block the rear tires to keep the vehicle from rolling.

3 From under the vehicle, disconnect the switch wire and battery cable from the terminals on the rear of the solenoid **(see illustration)**.

4 Remove the support bracket, if equipped, and mounting bolts and detach the starter motor **(see illustration)**. If shims are used, be sure to note how they're installed and return them to their original locations.

Installation

5 Position the starter motor and install the mounting bolts finger tight.

6 Install the support bracket, if equipped, and tighten the starter motor mounting bolts securely.

7 Attach the switch wire and battery cable to the terminals on the rear of the solenoid.

8 Lower the vehicle.

9 Connect the negative battery cable and check the starter operation.

18 Starter solenoid - removal and installation

Refer to illustrations 18.4 and 18.5

1 Remove the starter motor (Section 17).

2 Disconnect the strap from the solenoid terminal to the starter motor **(see illustration 17.3)**.

3 Remove the two screws which secure the solenoid to the starter motor.

4 Twist the solenoid 90° in a clockwise direction to disengage the flange from the starter body **(see illustration)**.

5 The solenoid is normally replaced as an

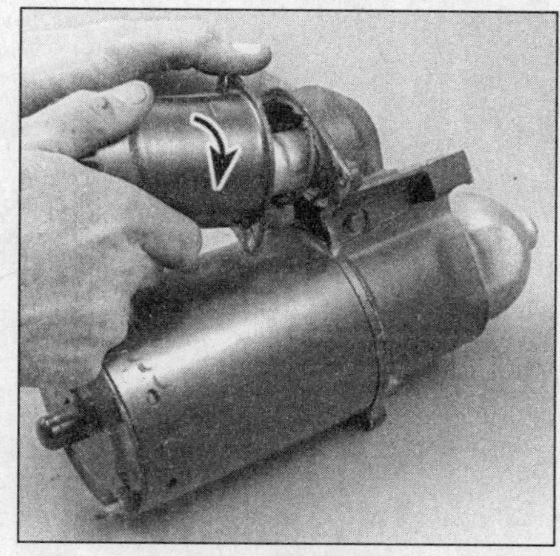

18.4 To disengage the solenoid from the starter, turn it in a clockwise direction

assembly if it's faulty. However, the end cover can be removed and the contact ring position changed to move any burned areas out of the way of the terminals and restore its function **(see illustration)**.

6 Make sure the return spring is in position on the plunger, then insert the solenoid body into the starter housing and turn the solenoid counterclockwise to engage the flange.

7 Install the two solenoid screws and connect the motor strap.

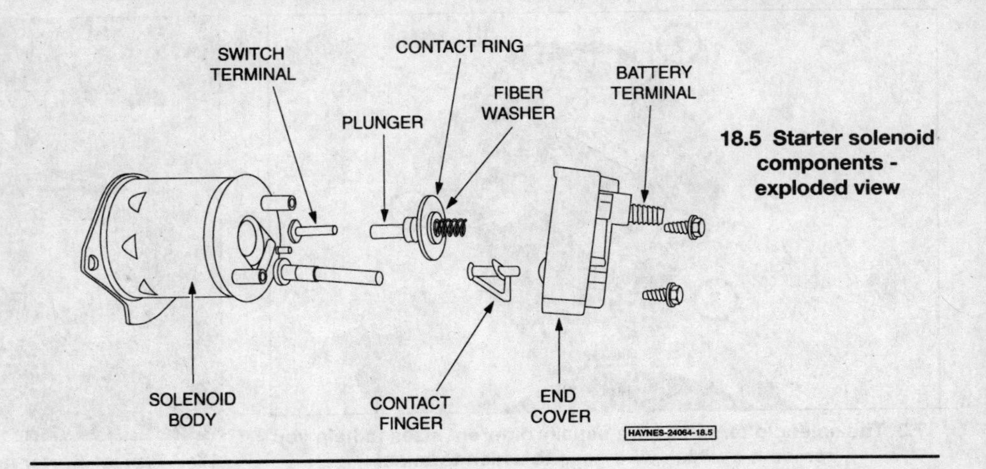

18.5 Starter solenoid components - exploded view

Chapter 6
Emissions control systems

Contents

	Section
Air Injection Reaction (AIR) system	9
Catalytic converter	17
Computer Command Control (CCC) system and trouble codes	2
Early Fuel Evaporation (EFE) system	15
Electronic Control Module/PROM removal and installation	3
Electronic Spark Control (ESC) system	7
Electronic Spark Timing (EST) system	6
Evaporative Emission Control System (EECS)	10
Exhaust Gas Recirculation (EGR) system	8
General information	1
Information sensors	4
Oxygen sensor	5
Positive Crankcase Ventilation (PCV) system	11
Pulse Air Injection Reaction (PAIR) system	13
Thermostatic Air Cleaner (THERMAC)	12
Throttle Return Control (TRC) system	14
Transmission Controlled Spark (TCS) system	16

Specifications

Torque specifications

	Ft-lbs
Catalytic converter U-bolt nuts	35
Oxygen sensor	30
Coolant temperature switch (EFE system)	8

1 General information

Refer to illustration 1.8

To prevent pollution of the atmosphere from incompletely burned and evaporating gases and to maintain good driveability and fuel economy, a number of emission control devices are incorporated on later models. They include the:

Fuel control system
Electronic Spark Timing (EST)
Electronic Spark Control (ESC) system
Exhaust Gas Recirculation (EGR) system
Evaporative Emission Control System (EECS)
Positive Crankcase Ventilation (PCV) system
Thermostatic air cleaner (THERMAC)
Air Injection Reaction (AIR) system
Early Fuel Evaporation (EFE) system
Pulse Air Injection Reaction (PAIR) system
Transmission Controlled Spark (TCS) system
Throttle Return Control (TRC) system
Catalytic converter

These systems were installed during different years for different models and engine options. Which systems were installed on which engines was also dependent upon the

Component location

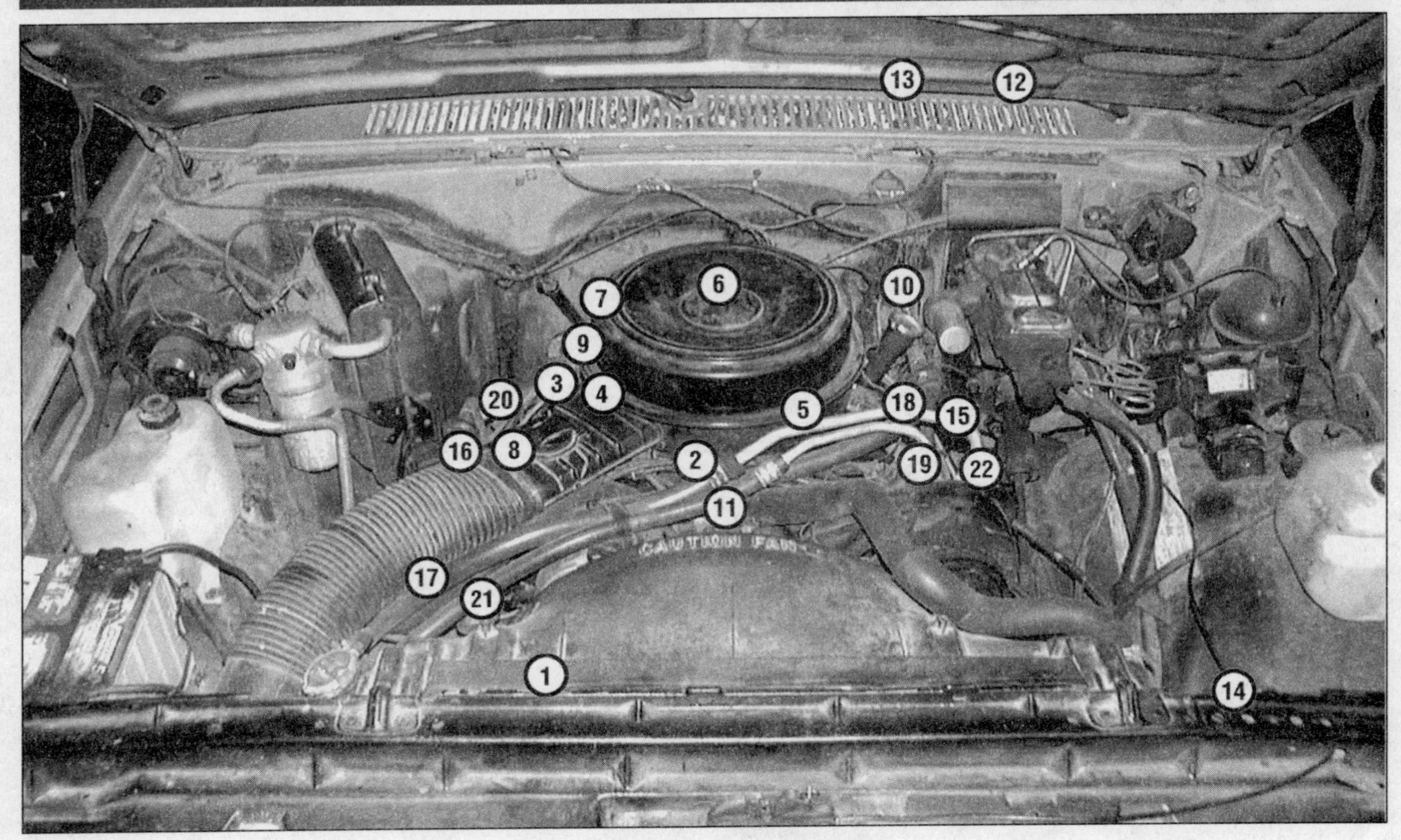

Typical V8 engine component location

1 *Typical Vehicle Emissions Control Information (VECI) label*
2 *EGR thermal vacuum switch (TVS)*
3 *Exhaust gas recirculation valve (EGR)*
4 *Electric choke*
5 *Throttle lever actuator*
6 *Thermatic air cleaner temperature sensor*
7 *Crankcase vent filter*
8 *Thermatic air cleaner vacuum diaphragm motor*
9 *Thermatic air cleaner TVS*
10 *Late model EST (timing) connector*
11 *ECM ground*
12 *Assembly Line Data Link (ALDL) connector - under dash*
13 *Service engine soon light - in instrument panel*
14 *Fuel vapor canister*
15 *Air injection pipe - left side*
16 *AIR check valve - right side*
17 *Diverter valve*
18 *AIR check valve - left side*
19 *Positive crankcase ventilation valve (PCV)*
20 *Early fuel evaporation valve*
21 *Air injection pump*
22 *Oxygen sensor - mounted in the exhaust pipe*

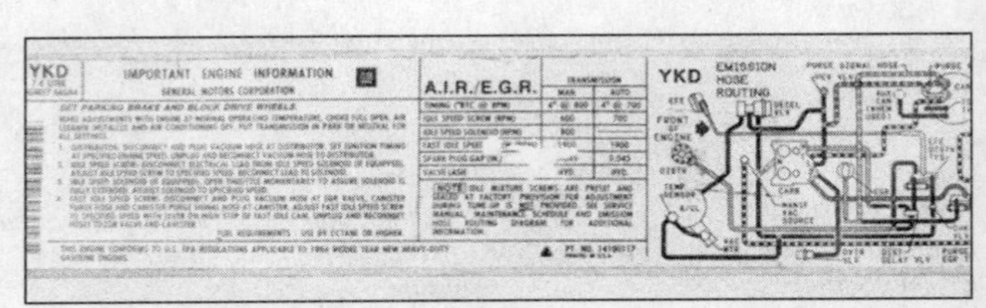

1 Typical Vehicle Emissions Control Information (VECI) label

2 EGR thermal vacuum switch (TVS) (arrow)

Component location

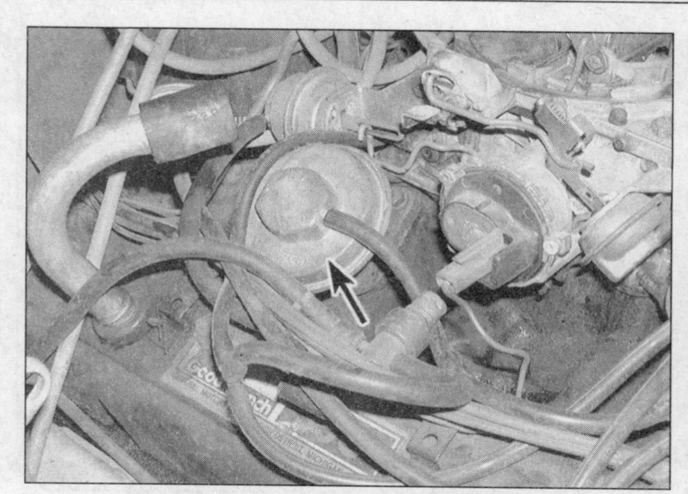

3 Exhaust gas recirculation valve (EGR) (arrow)

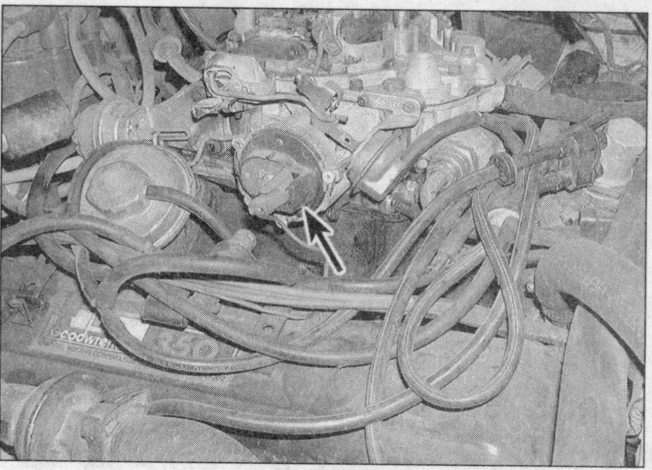

4 Electric choke (arrow)

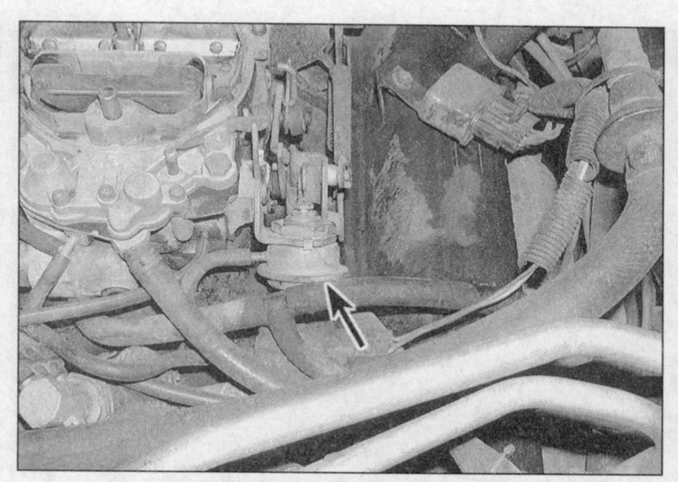

5 Throttle lever actuator (arrow)

6 Thermatic air cleaner temperature sensor (arrow)

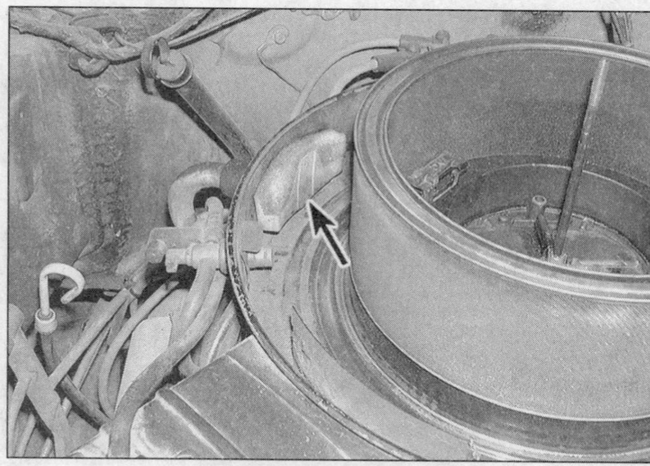

7 Crankcase vent filter (arrow)

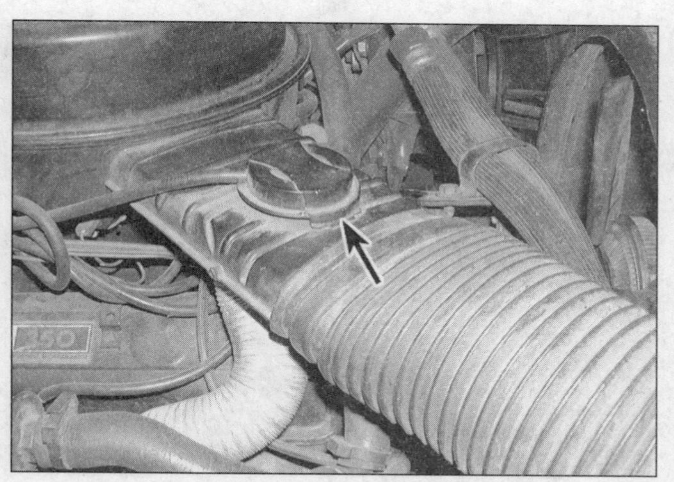

8 Thermatic air cleaner vacuum diaphragm motor (arrow)

Component location

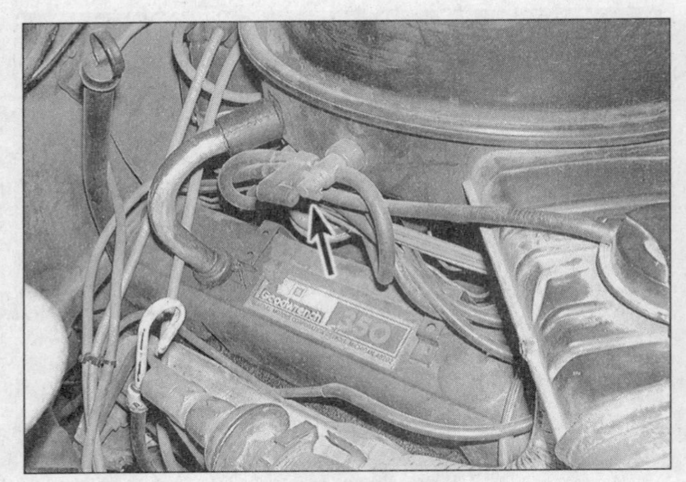

9 Thermatic air cleaner TVS (arrow)

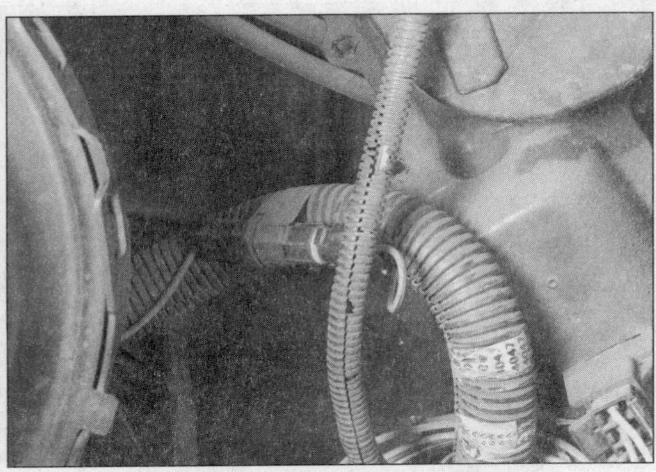
10 Late model EST (timing) connector

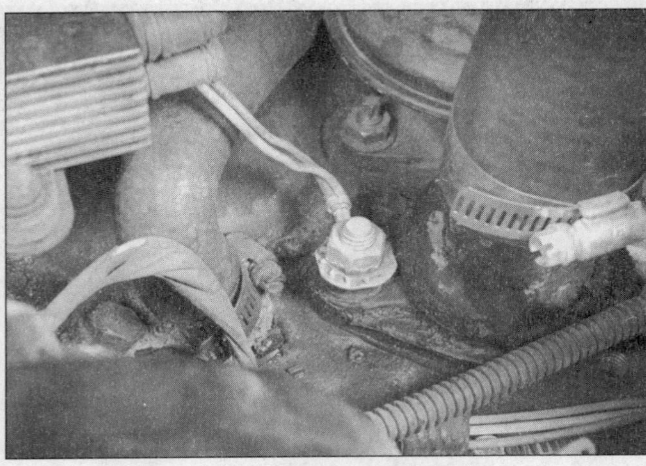
11 ECM ground

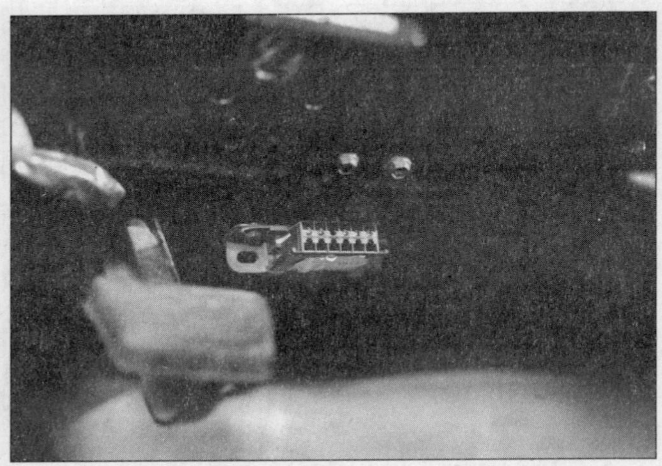
12 Assembly Line Data Link (ALDL) connector - under dash

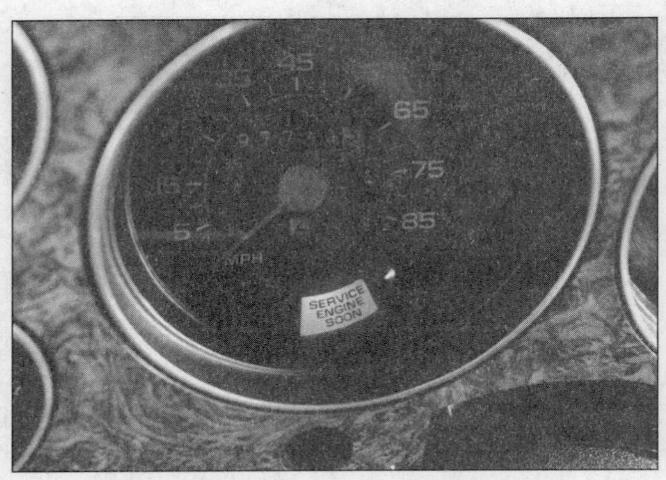

13 Service engine soon light

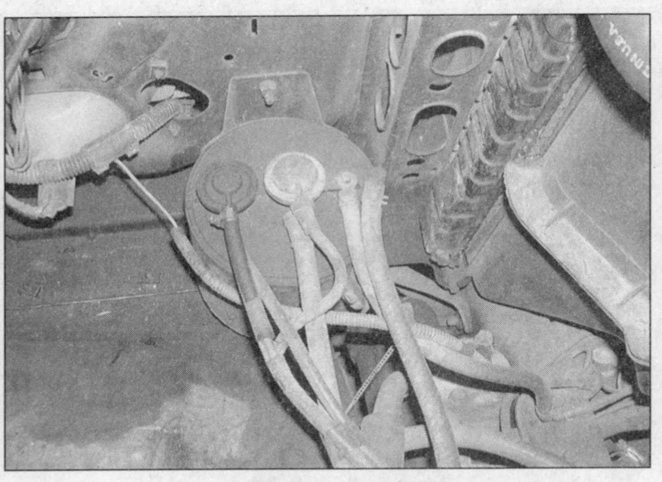
14 Fuel vapor canister

Component location

15 Air injection pipe - left side shown (arrow)

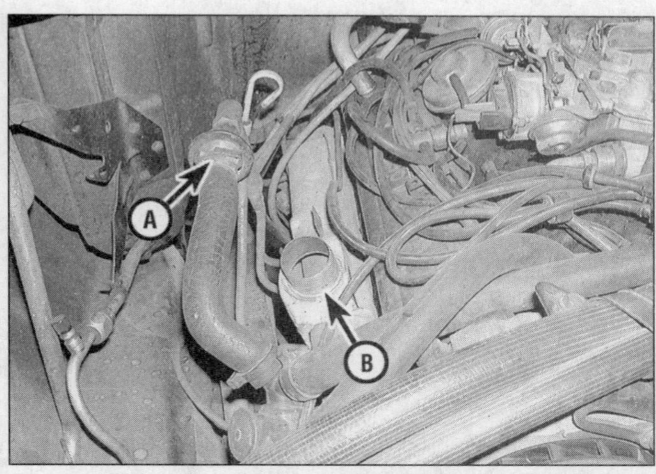

16 AIR check valve (A); Heat shield (B) - right side shown

17 Diverter valve (arrow)

18 AIR check valve (arrow) - left side shown

19 Positive crankcase ventilation valve (PCV) (arrow)

20 Early fuel evaporation valve (EFE)

Component location

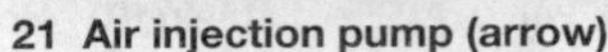

21 Air injection pump (arrow)

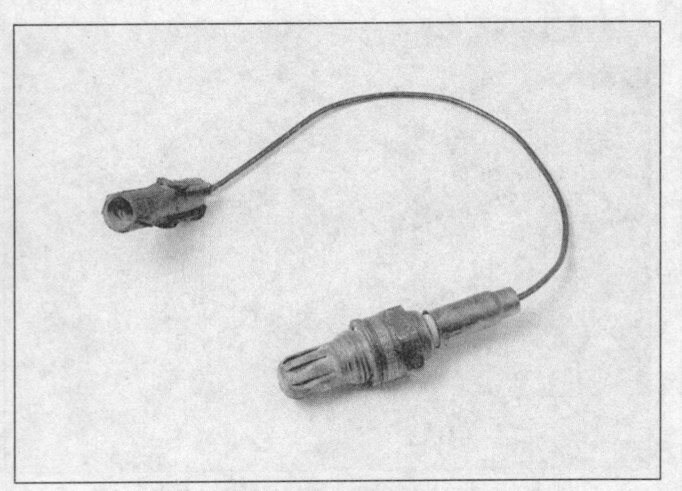

22 Oxygen sensor

area where the vehicle was first sold, with 49-state, California and Canadian models having different combinations of emission control devices.

The fuel system installed (two-barrel carburetor, four-barrel carburetor, throttle body fuel injection) also caused variances in the systems used. Refer to the systems described and illustrated in this Chapter and compare them to the equipment on your engine to determine just what systems have been installed on your vehicle.

On later models some of these systems are linked, directly or indirectly, to the Computer Command Control (CCC or C3) system.

The Sections in this Chapter include general descriptions, checking procedures within the scope of the home mechanic and component replacement procedures (when possible) for each of the systems listed above.

Before assuming that an emissions control system is malfunctioning, check the fuel and ignition systems carefully. The diagnosis of some emission control devices requires specialized tools, equipment and training. If checking and servicing become too difficult or if a procedure is beyond the scope of your skills, consult your dealer service department.

This doesn't mean, however, that emission control systems are particularly difficult to maintain and repair. You can quickly and easily perform many checks and do most (if not all) of the regular maintenance at home with common tune-up and hand tools. **Note:** *The most frequent cause of emissions problems is simply a loose or broken vacuum hose or wiring connection, so always check the hose and wiring connections first.*

Pay close attention to any special precautions outlined in this Chapter. It should be noted that the illustrations of the various systems may not exactly match the system installed on your vehicle because of changes made by the manufacturer during production or from year-to-year.

On most later model vehicles a Vehicle Emissions Control Information label is located in the engine compartment **(see illustration)**. This label contains important emissions specifications and setting procedures, as well as a vacuum hose schematic with emissions components identified. When servicing the engine or emissions systems, the VECI label in your particular vehicle should always be checked for up-to-date information.

2 Computer Command Control (CCC) system and trouble codes

Refer to illustrations 2.1a, 2.1b, 2.1c, 2.5, 2.6, 2.8 and 2.9

The Computer Command Control (CCC) system is used on later models and consists of an Electronic Control Module (ECM) and information sensors which monitor various functions of the engine and send data back to the ECM **(see illustrations)**.

The CCC system is analogous to the

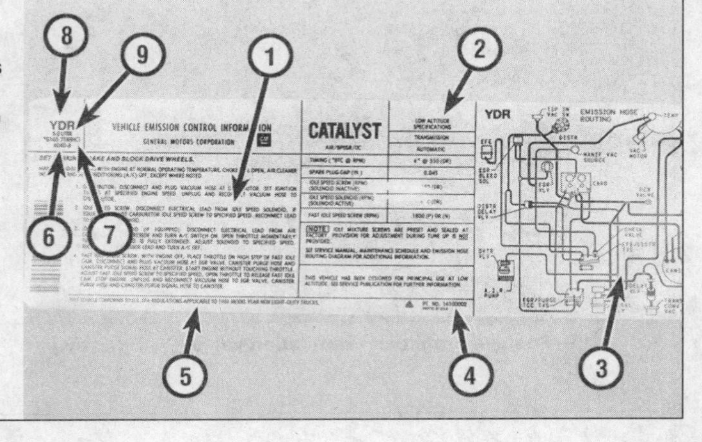

1.8 The vehicle Emission Control Information (VECI) label provides information regarding engine size, exhaust emission system used, engine adjustment procedures and specifications and an emission component and vacuum hose schematic diagram

1 *Adjustment procedures*
2 *Engine adjustment specifications*
3 *Emission component and vacuum hose routing*
4 *Label part number*
5 *Area of certification (California, Federal or Canada)*
6 *Evaporative emission system*
7 *Exhaust emission system*
8 *Label code*
9 *Engine size*

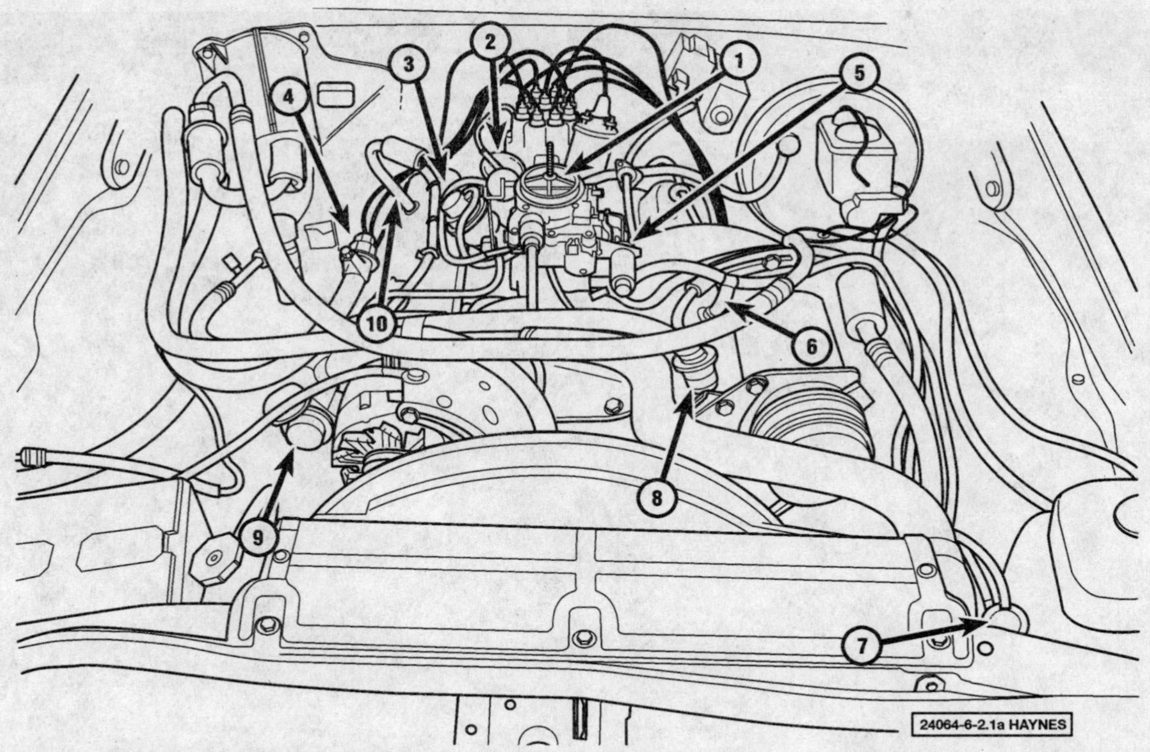

2.1a Component locations on a typical early 2bbl carburetor equipped V8 engine with California emissions package

1 *Carburetor*
2 *Vacuum advance unit*
3 *EGR valve*
4 *AIR check valve*
5 *Idle stop solenoid*
6 *PCV valve*
7 *EVAP canister*
8 *AIR check valve*
9 *Diverter valve*
10 *Thermal vacuum switches*

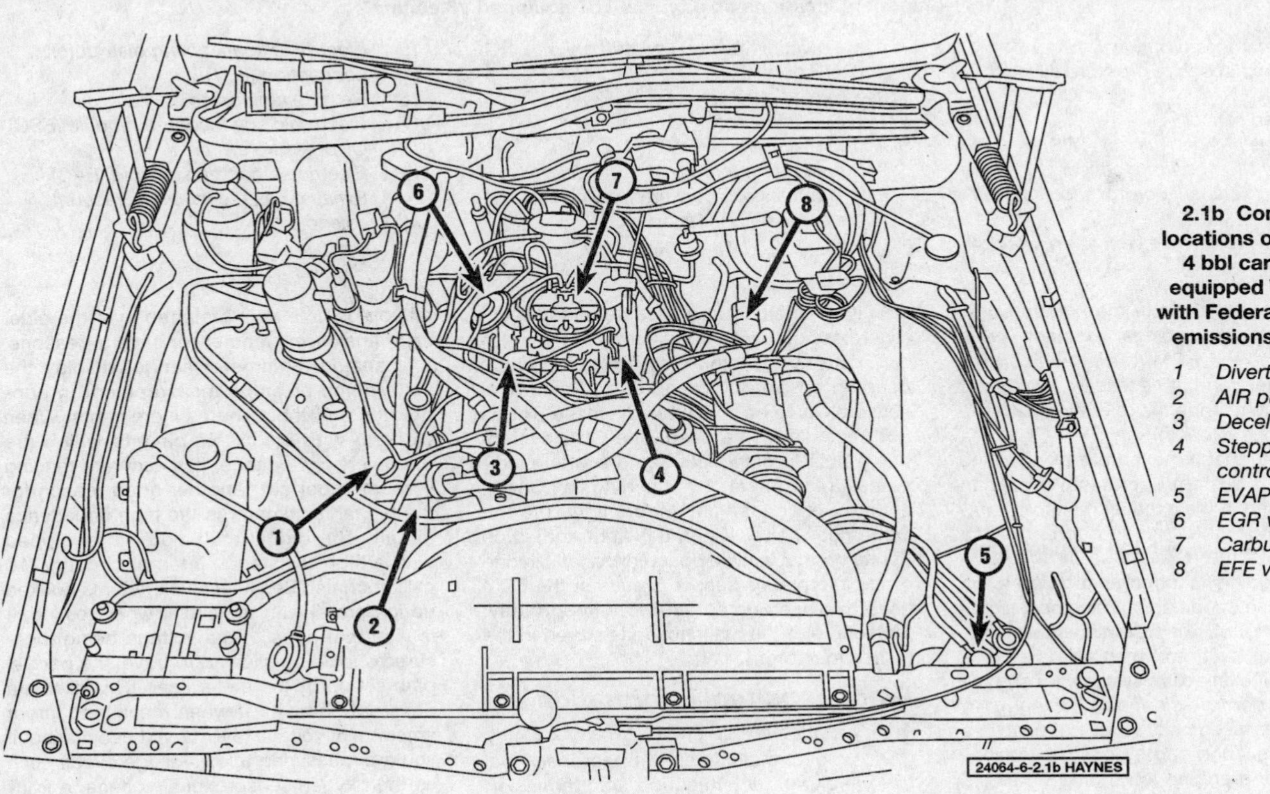

2.1b Component locations on a typical 4 bbl carburetor-equipped V8 engine with Federal (49-state) emissions package

1 *Diverter valve*
2 *AIR pump*
3 *Decel valve*
4 *Stepped speed control actuator*
5 *EVAP canister*
6 *EGR valve*
7 *Carburetor*
8 *EFE valve*

2.1c Component locations on a typical TBI-equipped V8 engine

1 *ECM harness grounds*
2 *Manifold absolute pressure sensor (MAP)*
3 *Oxygen sensor*
4 *Throttle position sensor (TPS)*
5 *Coolant temperature sensor*
6 *Electronic spark control knock sensor (ESC)*
7 *Positive crankcase vent valve (PCV)*
8 *Downshift relay (THM 400 only)*
9 *AIR pump*
10 *Fuel vapor canister (EVAP)*
11 *Fuel injector*
12 *Idle air control valve*
13 *Fuel pump relay*
14 *Transmission converter clutch electrical connection*
15 *Electronic spark timing distributor connection (EST)*
16 *Remote ignition coil*
17 *Electronic spark control module (ESC)*
18 *Oil pressure switch*
19 *Electric air control solenoid (EAC)*
20 *Exhaust gas recirculation vacuum solenoid*

central nervous system in the human body: The sensors (nerve endings) constantly relay information to the ECM (brain), which processes the data and, if necessary, sends out a command to change the operating parameters of the engine (body).

Here's a specific example of how one portion of this system operates: An oxygen sensor, located in the exhaust manifold, constantly monitors the oxygen content of the exhaust gas. If the percentage of oxygen in the exhaust gas is incorrect, indicating a too rich or too lean condition, an electrical signal is sent to the ECM. The ECM takes this information, processes it and then sends a command to the fuel injection system or carburetor mixture control solenoid, telling it to change the air/fuel mixture. This happens in a fraction of a second and it goes on continuously when the engine is running. The end result is an air/fuel mixture ratio which is constantly maintained at a predetermined ratio, regardless of driving conditions.

One might think that a system which uses an on-board computer and electrical sensors would be difficult to diagnose. This is not necessarily the case. The CCC system has a built-in diagnostic feature which indicates a problem by flashing a Check engine/Service engine soon light on the instrument panel. When this light comes on during normal vehicle operation, a fault in one of the information sensor circuits or the ECM itself has been detected. More importantly, the source of the malfunction is stored in the ECM's memory.

Diagnostic tool information

Refer to illustrations 2.5, 2.6, 2.8 and 2.9

A digital multimeter is a necessary tool for checking fuel injection and emission related components **(see illustration)**. A digital volt-ohmmeter is preferred over the older style analog multimeter for several reasons. The analog multimeter cannot display the volts-ohms or amps measurement in hundredths and thousandths increments. When working with electronic circuits which are often very low voltage, this accurate reading is most important. Another good reason for the digital multimeter is the high impedance circuit. The digital multimeter is equipped with a high resistance internal circuitry (10 million ohms). Because a voltmeter is hooked up in parallel with the circuit when testing, it is vital that none of the voltage being measured should be allowed to travel the parallel path set up by the meter itself. This dilemma does not show itself when measuring larger amounts of voltage (9 to 12 volt circuits) but if you are measuring a low voltage circuit such as the oxygen sensor signal voltage, a fraction of a volt may be a significant amount

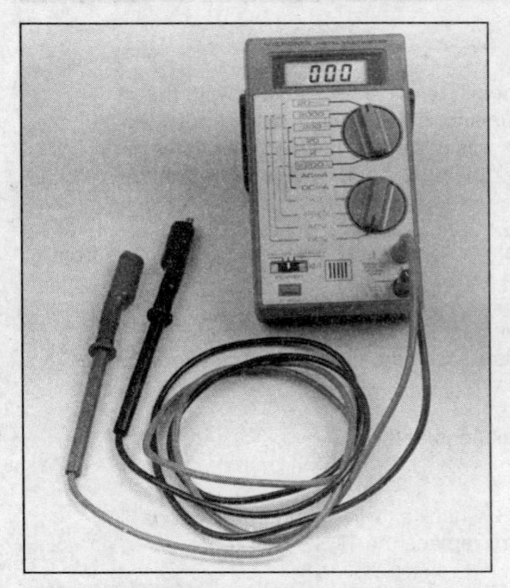

2.5 Digital multimeters can be used for testing all types of circuits; because of their high impedance, they are much more accurate than analog meters for measuring millivolts in low-voltage computer circuits

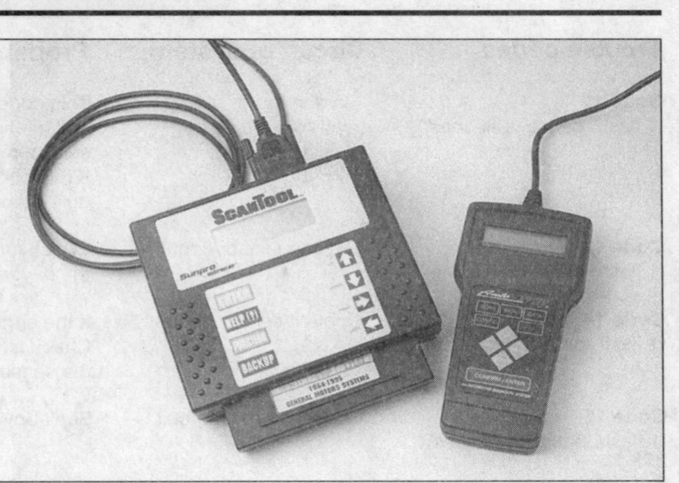

2.6 Scanners like the Actron Scantool and the AutoXray XP240 are powerful diagnostic aids - programmed with comprehensive diagnostic information, they can tell you just about anything you want to know about your engine management system, but they are expensive

when diagnosing a problem.

Hand-held scanners are the most powerful and versatile tools for analyzing engine management systems used on later model vehicles **(see illustration)**. Unfortunately, they are the most expensive. Each brand scan tool must be examined carefully to match the year, make and model of the vehicle you are working on. Often interchangeable cartridges are available to access the particular manufacturer; Ford, GM, Chrysler, etc.). Some manufacturers will specify by continent; Asia, Europe, USA, etc.

Another type of code reader and less expensive is available at parts stores **(see illustration)**. These tools simplify the procedure for extracting codes from the engine management computer by simply "plugging in" to the diagnostic connector on the vehicle wiring harness.

To retrieve this information from the ECM memory, you must use a short jumper wire to ground a diagnostic terminal. This terminal is part of a wiring connector known as the Assembly Line Data Link (ALDL) **(see illustration)**. The ALDL is located underneath the dashboard, on the left kick panel (1983 models) or just below the instrument panel to the left of the center tunnel (1984 and later models).

To use the ALDL, simply remove the plastic cover by sliding it toward you. With the connector exposed to view, push one end of the jumper wire into the diagnostic terminal (terminal B) and the other end into the terminal A (ground).

When the diagnostic terminal is grounded with the ignition on and the engine stopped, the system will enter the Diagnostic Mode. In this mode the ECM will display a "Code 12" by flashing the Check engine/Service engine soon light, indicating that the system is operating. A Code 12 is simply one flash, followed by a brief pause, then two flashes in quick succession. This code will be flashed three times. If no other codes are stored, Code 12 will continue to flash until the diagnostic terminal ground is removed.

After flashing Code 12 three times, the ECM will display any stored trouble codes. Each code will be flashed three times, then Code 12 will be flashed again, indicating that the display of any stored trouble codes has been completed.

When the ECM sets a trouble code, the Check engine/Service engine soon light will come on and a trouble code will be stored in memory. If the problem is intermittent, the light will go out after 10 seconds or when the fault goes away. However, the trouble code will stay in the ECM memory until the battery voltage to the ECM is interrupted. Removing battery voltage for 10 seconds will clear all stored trouble codes. Trouble codes should always be cleared after repairs have been completed. **Caution:** *To prevent damage to the ECM, the ignition switch must be off when disconnecting power to the ECM.*

Following is a list of the typical trouble codes which may be encountered while diagnosing the Computer Command Control System.

Also included are simplified troubleshooting procedures. If the problem persists after these checks have been made, more detailed service procedures will have to be done by a dealer service department.

2.8 Trouble code tools simplify the task of extracting the trouble codes

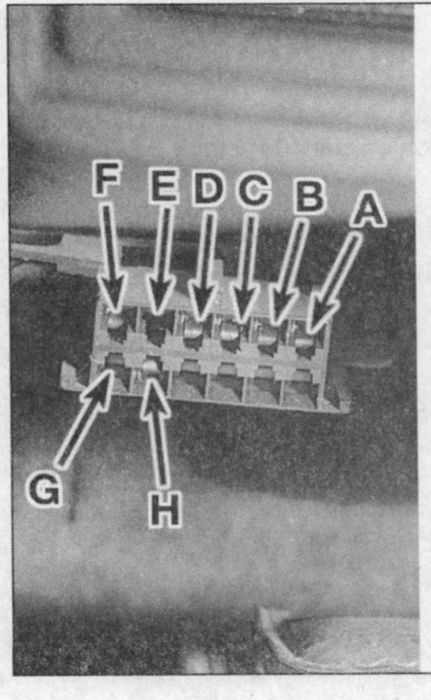

2.9 Typical late model Assembly Line Communications or Data Link (ALDL) connectors - the two terminals you will be concerned with are the A (ground) and B (diagnostic) terminals

a Ground
b Diagnostic terminal
c AIR (if used)
d Check Engine (or Service Engine Soon) light
e Serial data (special tool required - do not use)
f TCC (if used)
g Fuel pump (if used)
h Brake sensor speed input

Trouble codes	Circuit or system	Probable cause
Code 12 (1 flash, pause, 2 flashes)	No distributor reference pulse	This code will flash whenever the diagnostic terminal is grounded with the ignition turned on and the engine not running. If additional trouble codes are stored in the ECM they will appear after this code has flashed three times. If this code appears while the engine is running, no reference pulses from the distributor are reaching the ECM.
Code 13 (1 flash, pause, 3 flashes)	Oxygen sensor circuit	Check for a sticking throttle position sensor. Check the wiring and connectors from the oxygen sensor. Replace the oxygen sensor*.
Code 14 (1 flash, pause, 4 flashes)	Coolant sensor circuit	If the engine is overheating the problem must be rectified before continuing. Check all wiring and connectors associated with the coolant sensor. Replace the sensor*.
Code 15 (1 flash, pause, 5 flashes)	Coolant sensor circuit	See above, then check the wiring connections at the ECM.
Code 21 (2 flashes, pause, 1 flash)	Throttle position sensor	Check for a sticking or misadjusted TPS. Check all wiring and connections between the TPS and the ECM. Adjust or replace the TPS*.
Code 22 (2 flashes, pause, 2 flashes)	Throttle position sensor	Check the TPS adjustment. Check the ECM connector. Replace the TPS*.
Code 23 (2 flashes, pause, 3 flashes)	Mixture control solenoid	Caused by an open circuit or short to ground in the M/C solenoid circuit. Check all wires and connections.
Code 24 (2 flashes, pause, 4 flashes)	Vehicle Speed Sensor	A fault in this circuit should be indicated only when the vehicle is in motion. Disregard Code 24 if it is set when the drive wheels are not turning. Check the connections at the ECM. Check the TPS setting.
Code 32 (3 flashes, pause, 2 flashes)	EGR system	The EGR solenoid should not be energized and vacuum should not pass to the EGR valve. Diagnostic switch should close at about 2-inches of vacuum. With vacuum applied, the switch should close. Replace the EGR valve*.
Code 33 (1987 only) (3 flashes, pause, 3 flashes)	MAP sensor	Check the vacuum hoses from the MAP sensor. Check the electrical connections at the ECM. Replace the Map sensor*.
Code 34 (1984 through 1986) (3 flashes, pause, 4 flashes)	Differential pressure sensor	Check the sensor wires to make sure they are connected to the (vacuum) sensor terminals.
Code 34 (1987 only) (3 flashes, pause, 4 flashes)	MAP sensor	Code 34 will set when the signal voltage from the MAP sensor is too low. Instead the ECM will substitute a fixed MAP value and use the TPS to control fuel delivery. Replace the MAP sensor*.
Code 41 (4 flashes, pause, 1 flash)	No reference pulses to ECM	This code indicates no distributor reference pulses at certain vacuum conditions. With the engine idling, apply vacuum to the VAC or MAP sensor and pull off the vacuum hose. If there is less than a 1-volt change, there is a fault in the MAP or VAC sensor circuit. If the voltage changes more than 1-volt, check the wiring and distributor for loose or grounded connections.
Code 42 (4 flashes, pause, 2 flashes)	Electronic Spark Timing	If the vehicle will not start and run, check the wires between the HEI module and the EST and bypass terminal. Check for a bad connection at the ignition module. Check, and if necessary, replace the HEI module*.
Code 43 (4 flashes, pause, 3 flashes)	Electronic Spark Control	Check the wires from the ESC module to the ECM. Check the wires from the knock sensor to the ESC module. Check for a loose connection. Replace the knock sensor and/or ESC module*.
Code 44 (4 flashes, pause, 4 flashes)	Lean exhaust	Check the ECM wiring connection terminals. Have the fuel pressure checked. Check the oxygen sensor wire. Check the carburetor or throttle body gasket, vacuum hoses and intake manifold gaskets for leaks. Replace the oxygen sensor*.
Code 45 (4 flashes, pause, 5 flashes)	Rich exhaust	Check the evaporative charcoal canister and its components for the presence of fuel. Check the fuel pressure regulator vacuum control hose for the presence of fuel. Replace the oxygen sensor*.

Trouble codes	Circuit or system	Probable cause
Code 51 (5 flashes, pause, 1 flash)	PROM	Make sure that the PROM is properly installed in the ECM. If all pins are fully inserted in the socket, replace the PROM*. Clear the memory and recheck. If Code 51 reappears, replace the ECM*.
Code 52 (5 flashes, pause, 2 flashes)	Fuel CALPAK	Check the CALPAK PROM to insure proper installation. Replace the PROM*.
Code 53 (5 flashes, pause, 3 flashes)	System over-voltage	Code 53 will set if the voltage at ECM is greater than 17.1-volts for 2 seconds. Check the charging system.
Code 54 (1984 through 1986) (5 flashes, pause, 4 flashes)	Mixture control solenoid circuit	Code 54 sets when there is consistently high voltage at the ECM . Unplug the mixture control solenoid connector and check to make sure there is more than 10 ohms at the solenoid. If there is less than 10 ohms, the solenoid and ECM are faulty.
Code 54 (1987) (5 flashes, pause, 4 flashes)	Fuel pump circuit	Code 54 will set if the voltage at the ECM is less than 2-volts for 1.5 seconds since the last reference pulse was received. Check the fuel pump relay, circuit and connections. Check the oil pressure switch. Repair or replace faulty components*.
Code 55 (1984 through 1986) (5 flashes, pause, 5 flashes)	ECM or oxygen sensor circuits	Code 55 means high voltage at the oxygen sensor or an incorrect voltage at the ECM. Disconnect the test plug and the oxygen sensor and run the engine at idle for one minute. If the Check Engine light comes on, there is a fault in the oxygen sensor circuit. If it does not, the fault is in the ECM circuit.
Code 55 (1987) (5 flashes, pause, 5 flashes)	ECM	Be sure that the ECM ground connections are tight. If they are, replace the ECM*.

**Component replacement may not cure the problem in all cases. For this reason, you may want to seek professional advice before purchasing replacement parts.*

3 Electronic Control Module/PROM removal and installation

Refer to illustrations 3.4, 3.6a, 3.6b, 3.8, 3.10 and 3.11

Caution: *To avoid electrostatic discharge damage to the ECM, handle the ECM only by its case. Do not touch the electrical terminals during removal and installation. If available, ground yourself to the vehicle with an anti-static ground strap, available at computer supply stores.*

1 The Electronic Control Module (ECM) is located under the instrument panel.

2 Disconnect the negative cable at the battery. Place the cable out of the way so it cannot accidentally come in contact with the negative terminal of the battery, as this would once again allow power into the electrical system of the vehicle.

3 Disconnect the wire harnesses from the ECM. **Caution:** *The ignition switch must be turned to Off when removing or installing the ECM connectors.*

4 Remove the bolts and retainer and carefully detach the ECM **(see illustration)**.

5 Turn the ECM so that the bottom cover is facing up and carefully place it on a clean work surface.

6 Remove the screw(s) and lift off the PROM access cover **(see illustrations)**.

7 If you are replacing the ECM itself, the

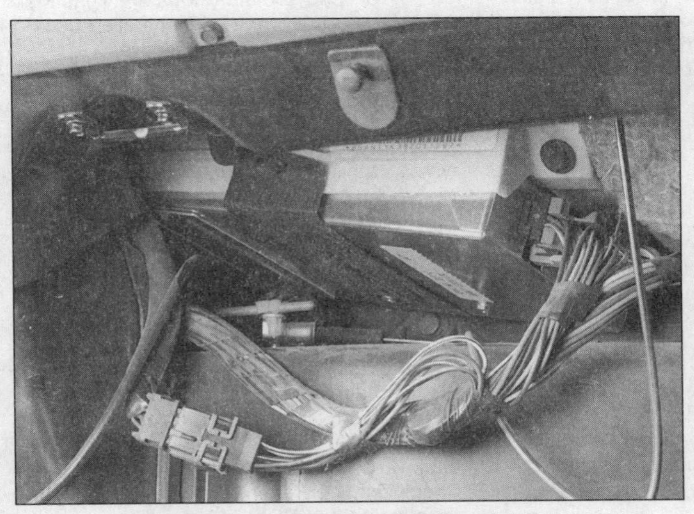

3.4 Late model ECM installation details

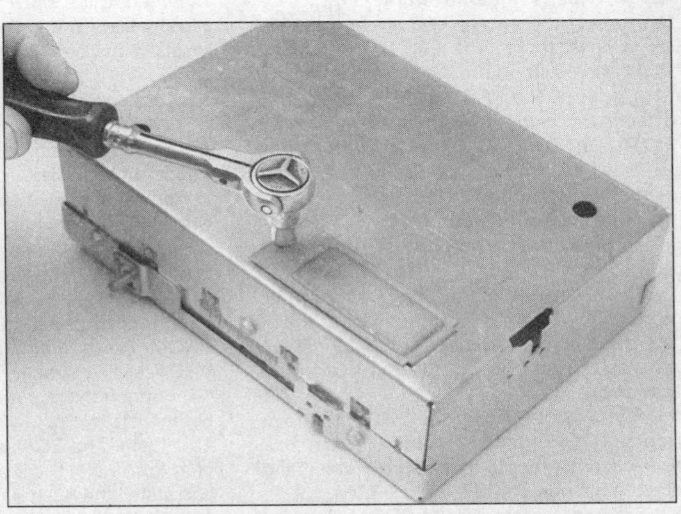

3.6a On early models the PROM cover is retained by one screw . . .

3.6b . . . while later models have two screws

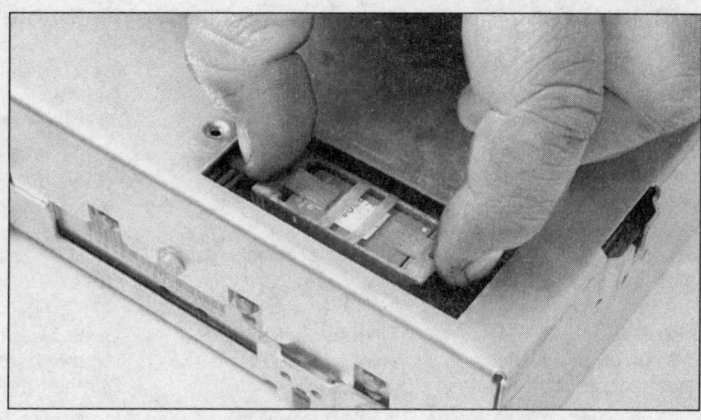

3.8 Grasp the PROM carrier at its narrow ends with the PROM removal tool (available at your dealer) and gently rock the removal tool until the PROM is disconnected

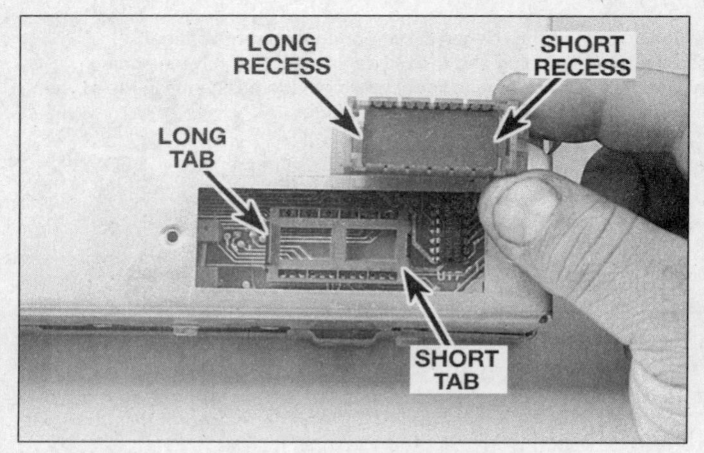

3.10 Note the reference end of the PROM carrier before setting it aside

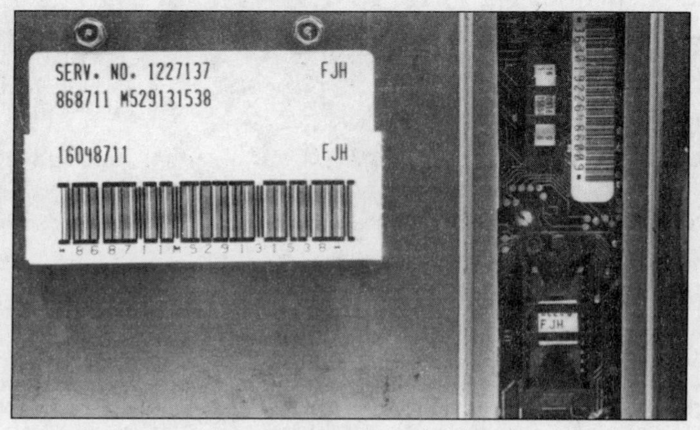

3.11 Make sure the service numbers on the ECM and the PROM are the same - otherwise, depending on what you are replacing, you either have the wrong ECM or the wrong PROM

new ECM will not contain a PROM. It will be necessary to remove the old PROM from the old ECM and install it in the new one.

8 Using a PROM removal tool (available at your dealer), grasp the PROM carrier at the narrow ends **(see illustration)**. Gently rock the carrier from end to end while applying firm upward force. The PROM carrier and PROM should lift off the PROM socket easily.

9 **Caution:** *The PROM carrier should only be removed with the special rocker-type PROM removal tool. Removal without this tool or with any other type of tool may damage the PROM or the PROM socket.*

10 Note the reference end of the PROM carrier **(see illustration)** before setting it aside.

11 If you are replacing the ECM, remove the new ECM from its container and check the service number to make sure that it is the same as the number on the old ECM **(see illustration)**.

12 If you are replacing the PROM, remove the new PROM from its container and check the service number to make sure that it is the same as the number of the old PROM.

13 Position the PROM/PROM carrier assembly squarely over the PROM socket with the small notched end of the carrier aligned with the small notch in the socket at the pin 1 end. Press on the PROM carrier until it seats firmly in the socket. If the PROM is new, make sure that the notch in the PROM is matched to the small notch in the carrier **(see illustration 3.10)**. **Caution:** *If the PROM is installed backwards and the ignition switch is turned on, the PROM will be destroyed.* Using the tool, install the new PROM carrier in the PROM socket of the ECM. The small notch of the carrier should be aligned with the small notch in the socket. Press on the PROM carrier until it is firmly seated in the socket. **Caution:** *Do not press on the PROM - press only on the carrier.* On early models, further seat the PROM by pressing down on the carrier with your fingers while using a suitable narrow blunt tool to press down on the body of the PROM. Press alternately at either end of the PROM to fully seat it.

14 Attach the access cover to the ECM and tighten the two screws.

15 Install the ECM in the support bracket, plug in the electrical connectors to the ECM and install the hush panel.

16 Start the engine.

17 Enter the diagnostic mode by grounding the diagnostic lead of the ALDL (see Section 2). If no trouble codes appear, the PROM is correctly installed.

18 If Trouble Code 51 appears, or if the Check engine/Service engine soon light comes on and remains constantly lit, the PROM is not fully seated, is installed backwards, has bent pins or is defective.

19 If the PROM is not fully seated, pressing firmly on both ends of the carrier should correct the problem.

20 It is possible to install the PROM backwards. If this occurs, and the ignition key is turned to ON, the PROM circuitry will be destroyed and the PROM will have to be replaced.

21 If the pins have been bent, remove the PROM in accordance with the above procedure, straighten the pins and reinstall the PROM. If the bent pins break or crack when you attempt to straighten them, discard the PROM and replace it with a new one.

22 If careful inspection indicates that the PROM is fully seated, has not been installed backwards and has no bent pins, but the Check engine/Service engine soon light remains lit, the PROM is probably faulty and must be replaced.

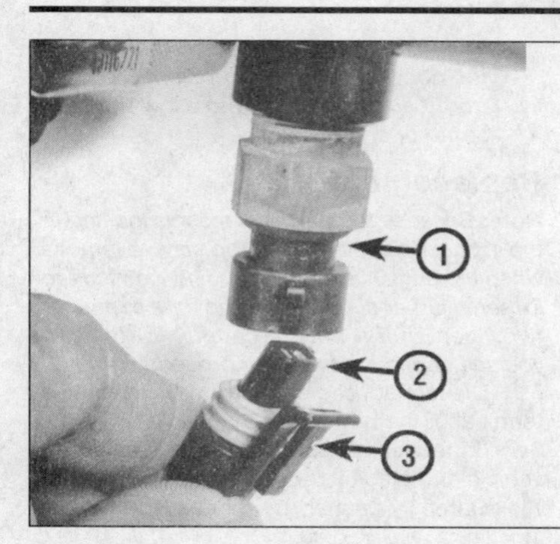

4.3 Unplug the coolant sensor plug after lifting the tab

1 *Coolant sensor*
2 *Harness connector to the ECM*
3 *Locking tab*

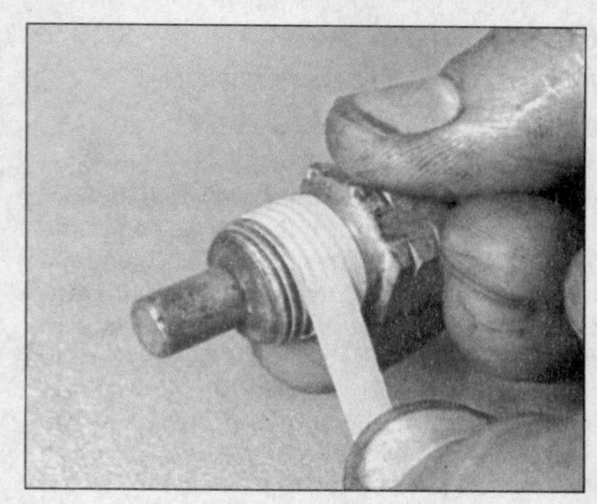

4.4 Wrap the threads of the coolant temperature sensor with Teflon tape to prevent leaks and thread corrosion

4 Information sensors

Refer to illustrations 4.3, 4.4, 4.5 and 4.7

1 ECM-equipped models use a variety of sensors to provide the ECM with information which it uses to control the engine. Depending on model, these may include:

Engine coolant temperature sensor

2 The coolant sensor is mounted in a coolant passage, usually in the front of the intake manifold. A failure in the coolant sensor circuit should set either a Code 14 or a Code 15. These codes indicate a failure in the coolant temperature circuit, so the appropriate solution to the problem will be either repair of a wire or replacement of the sensor.

3 To remove the sensor, disengage the locking tab on the connector with a small screwdriver and unplug it from the sensor **(see illustration)**. Carefully unscrew the sensor itself. **Caution:** *Handle the coolant sensor with care. Damage to this sensor will affect the operation of the entire fuel system.*

4 Before installing the new sensor, wrap the threads with Teflon sealing tape to prevent leakage and thread corrosion **(see illustration)**. Installation is the reverse of removal.

Manifold Absolute Pressure (MAP) sensor

5 The Manifold Absolute Pressure (MAP) sensor **(see illustration)** is mounted on a bracket on the intake manifold. The MAP sensor monitors the intake manifold pressure changes resulting from changes in engine load and speed and converts the information into a voltage output. The ECM uses the MAP sensor to control fuel delivery and ignition timing.

6 A failure in the MAP sensor circuit should set a Code 13 or Code 34.

Differential pressure sensor

7 The differential pressure sensor is located in the engine compartment on the firewall adjacent to the brake master cylinder **(see illustration)** and measures engine vacuum. At high vacuum the sensor sends a high voltage (almost 5-volts) signal to the ECM and at low vacuum a low voltage signal.

8 A failure in the differential pressure sensor should set a Code 34. The engine must have been run at idle for approximately two minutes before the code will be set.

Oxygen sensor

9 The oxygen sensor is mounted in the exhaust system where it can monitor the oxygen content of the exhaust gas stream.

10 By monitoring the voltage output of the oxygen sensor, the ECM will know what fuel mixture command to send to the fuel system.

11 An open in the oxygen sensor circuit should set a Code 13. A low voltage in the circuit should set a Code 44. A high voltage in the circuit should set a Code 45. Codes 44 and 45 may also be set as a result of fuel system problems.

12 Refer to Section 5 for the oxygen sensor replacement procedure.

Throttle position sensor (TPS)

13 The throttle position sensor (TPS) is located in the float bowl on carbureted models and on the end of the throttle shaft on TBI-equipped engines.

14 By monitoring the output voltage from the TPS, the ECM can determine fuel delivery based on throttle valve angle (driver demand). A broken or loose TPS can cause intermittent bursts of fuel from the injector and an unstable idle because the ECM thinks the throttle is moving.

15 A problem in any of the TPS circuits will set either a Code 21 or 22. Once a trouble code is set, the ECM will use an artificial default value for TPS and some vehicle performance will return.

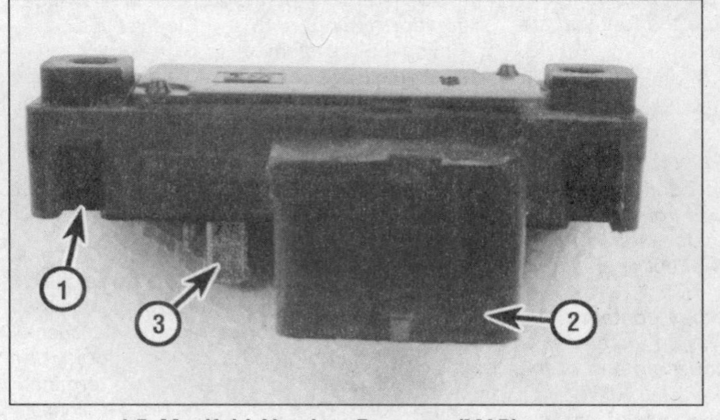

4.5 Manifold Absolute Pressure (MAP) sensor

1 *MAP sensor*
2 *ECM harness connection*
3 *Manifold vacuum connection*

4.7 The differential pressure sensor (1) is located on the firewall in the engine compartment

Park/Neutral switch (automatic transmission-equipped vehicles only)

16 The Park/Neutral (P/N) switch indicates to the ECM when the transmission is in Park or Neutral. This information is used for the Transmission Converter Clutch (TCC) and the Idle Air Control (IAC) valve operation. **Caution:** *The vehicle should not be driven with the Park/Neutral switch disconnected because idle quality will be adversely affected and a false Code 24 (failure in the Vehicle Speed Sensor circuit) may be set. For more information regarding the P/N switch, which is part of the Neutral/start and back-up light switch assembly, see Chapters 7 and 12.*

Vehicle speed sensor

17 The vehicle speed sensor (VSS) sends a pulsing voltage signal to the ECM, which the ECM converts to miles per hour. This sensor controls the operation of the Transmission Converter Clutch.

Distributor reference signal

18 The distributor sends a signal to the ECM to tell it both engine rpm and crankshaft position. See Electronic Spark Timing (EST), Section 7, for further information.

5 Oxygen sensor

Refer to illustration 5.1

General description

1 The oxygen sensor, which is located in the exhaust pipe **(see illustration)**, monitors the oxygen content of the exhaust gas stream. The oxygen content in the exhaust reacts with the oxygen sensor to produce a voltage output which varies. The ECM monitors this voltage output to determine the ratio of oxygen to fuel in the mixture. The ECM alters the air/fuel mixture ratio by controlling the pulse width (open time) of the fuel injectors, which run at a constant pressure differential between the fuel pressure and manifold pressure. Because this pressure differential is held constant, the time the ECM holds the fuel injector open directly controls the amount of fuel delivered. On carburetor equipped models a mixture control solenoid changes the fuel/air mixture by allowing more or less fuel to flow through the carburetor circuits. The MC solenoid, located in the carburetor air horn, is controlled by the ECM, which provides a ground for the solenoid. When the solenoid is energized the fuel flow through the carburetor is reduced, providing a leaner mixture. When the ECM removes the ground the solenoid de-energizes and allows more fuel to flow. A mixture of 14.7 parts air to 1 part fuel is the ideal ratio for minimizing exhaust emissions, thus allowing the catalytic converter to operate at maximum efficiency. It is this ratio of 14.7 to 1 which the ECM and the oxygen sensor attempt to maintain at all times.

5.1 Typical oxygen sensor installation

2 The oxygen sensor is like an open circuit and produces no voltage when it is below its normal operating temperature of about 600° F (315° C). During this initial period during warm-up, the ECM operates in open loop mode.

3 The proper operation of the oxygen sensor depends on four conditions:

a) ***Electrical*** *- The low voltages and low currents generated by the sensor depend upon good, clean connections which should be checked whenever a malfunction of the sensor is suspected or indicated.*
b) ***Outside air supply*** *- The sensor is designed to allow air circulation to the internal portion of the sensor. Whenever the sensor is removed and installed or replaced, make sure the air passages are not restricted.*
c) ***Proper operating temperature*** *- The ECM will not react to the sensor signal until the sensor reaches approximately 600° F (315° C). This factor must be taken into consideration when evaluating the performance of the sensor.*
d) ***Unleaded fuel*** *- The use of unleaded fuel is essential for proper operation of the sensor. Make sure the fuel you are using is of this type.*

4 In addition to observing the above conditions, special care must be taken whenever the sensor is serviced.

a) *The oxygen sensor has a permanently attached pigtail and connector which should not be removed from the sensor. Damage or removal of the pigtail or connector can adversely affect operation of the sensor.*
b) *Grease, dirt and other contaminants should be kept away from the electrical connector and the louvered end of the sensor.*
c) *Do not use cleaning solvents of any kind on the oxygen sensor.*
d) *Do not drop or roughly handle the sensor.*
e) *The silicone boot must be installed in the correct position to prevent the boot from being melted and to allow the sensor to operate properly.*

Replacement

Note: *Because the oxygen sensor is located in the exhaust pipe, it may be too tight to remove when the engine is cold. If you find it difficult to loosen, start and run the engine for a minute or two, then shut it off. Be careful not to burn yourself during the following procedure.*

5 Disconnect the cable from the negative terminal of the battery.

6 If necessary for clearance, raise the vehicle and place it securely on jackstands.

7 Carefully unsnap the pigtail lead retaining clip.

8 Disconnect the electrical connector from the sensor.

9 Note the position of the silicone boot, if equipped, and carefully unscrew the sensor from the exhaust manifold. **Caution:** *Excessive force may damage the threads.*

10 Anti-seize compound must be used on the threads of the sensor to facilitate future removal. The threads of new sensors will already be coated with this compound, but if an old sensor is removed and reinstalled, recoat the threads.

11 Install the sensor and tighten it to the specified torque.

12 Reconnect the electrical connector of the pigtail lead to the main engine wiring harness.

13 Snap the pigtail retaining clip closed.

14 Lower the vehicle and reconnect the cable to the negative terminal of the battery.

6 Electronic Spark Timing (EST) system

Refer to illustration 6.4

General description

1 To provide improved engine performance, fuel economy and control of exhaust emissions, the Electronic Control Module (ECM) controls distributor spark advance (ignition timing) with the Electronic Spark Timing (EST) system.

2 The ECM receives a reference pulse from the distributor, which indicates both engine rpm and crankshaft position. The ECM then determines the proper spark advance for the engine operating conditions and sends an EST pulse to the distributor.

Checking

3 The ECM will set EST at a specified value when the diagnostic test terminal in the ALCL/ALDL connector is grounded. To check for EST operation, the timing should be checked at 2000 rpm with the terminal ungrounded. Then ground the test terminal. If the timing changes at 2000 rpm, the EST is operating. A fault in the EST system will usually set Trouble Code 42.

6.4 The EST connector (arrow) is located in the electrical harness near the rear of the engine (V8 shown)

7.8 The Electronic Spark Control (ESC) sensor is located in the lower right side of the engine block (arrow)

Ignition timing adjustment

4 To set the initial base timing, disconnect the timing connector near the brake booster **(see illustration)**.

5 Set the timing as specified on the VECI label. This will cause a Code 42 to be stored in the ECM memory. Be sure to clear the memory after setting the timing (see Section 2).

6 For further information regarding the testing and component replacement procedures for the HEI/EST distributor, refer to Chapter 5.

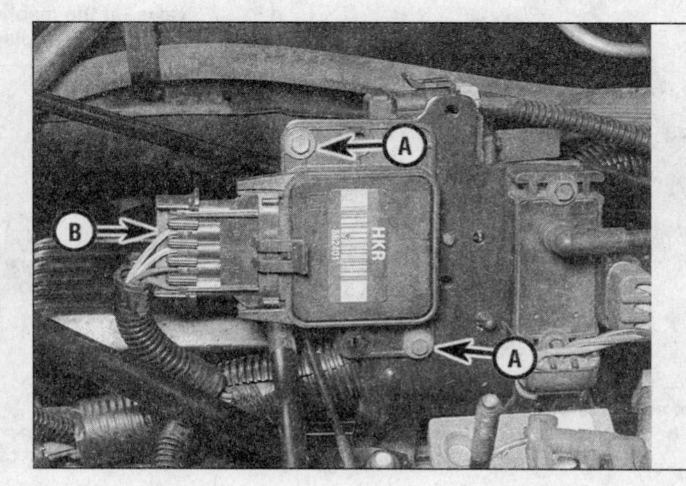

7.13 To detach the ESC module, unplug the electrical connector (B) and remove the mounting screws (A)

7 Electronic Spark Control (ESC) system

Refer to illustrations 7.8 and 7.13

General description

1 Irregular octane levels in modern gasoline can cause detonation in a high performance engine. Detonation is sometimes referred to as "spark knock," and it can severely damage the internal components of the engine.

2 The Electronic Spark Control (ESC) system is designed to retard spark timing up to 20 degrees to reduce spark knock in the engine. This allows the engine to use maximum spark advance to improve driveability and fuel economy.

3 The ESC knock sensor, which is located on the lower right side of the engine block, sends a voltage signal of 8 to 10-volts to the ECM when no spark knock is occurring and the ECM provides normal advance. When the knock sensor detects abnormal vibration (spark knock), the ESC module turns off the circuit to the ECM. The ECM then retards the EST distributor until spark knock is eliminated.

4 Failure of the ESC knock sensor signal or loss of ground at the ESC module will cause the signal to the ECM to remain high. This condition will result in the ECM controlling the EST as if no spark knock is occurring. Therefore, no retard will occur and spark knock may become severe under heavy engine load conditions. At this point, the ECM will set a Code 43.

5 Loss of the ESC signal to the ECM will cause the ECM to constantly retard the timing. This will result in sluggish performance and cause the ECM to set a Code 43.

Component replacement

ESC sensor

6 Disconnect the cable from the negative terminal of the battery.

7 If necessary for clearance, raise the vehicle and support it on jackstands. Refer to Chapter 1 and drain the cooling system.

8 Disconnect the wiring harness connector from the ESC sensor **(see illustration)**.

9 Remove the ESC sensor from the block. Coolant may flow out of the hole, so be careful not to get any in your eyes.

10 Apply thread sealant to the ESC sensor threads.

11 Installation is the reverse of the removal procedure. Be sure to refill the cooling system.

ESC module

12 Disconnect the cable from the negative terminal of the battery.

13 Disconnect the electrical connector from the module, which is located on the firewall **(see illustration)**.

14 Unscrew and remove the module.

15 Installation is the reverse of removal.

8 Exhaust Gas Recirculation (EGR) system

Refer to illustrations 8.4, 8.8, 8.17 and 8.29

General description

1 The EGR system meters exhaust gases into the intake manifold. From there the exhaust gases pass into the combustion chambers for the purpose of lowering combustion temperature, thereby reducing the amount of oxides of nitrogen (NOx) formed.

2 The main element of the system is the EGR valve, mounted on the intake manifold, which feeds small amounts of exhaust gas back into the combustion chamber.

3 The amount of exhaust gas admitted is regulated by a vacuum or backpressure controlled EGR valve in response to engine operating conditions.

4 On earlier models, a Thermal Vacuum

8.4 **The EGR valve is mounted on the intake manifold (arrows)**

8.8 **Push up on the diaphragm (arrow) when checking the EGR valve - if the diaphragm is stuck or binds in the up position, replace the valve with a new one**

8.17 **The thermal vacuum switch is threaded into the thermostat housing cover**

1 EGR *2 Vacuum supply*

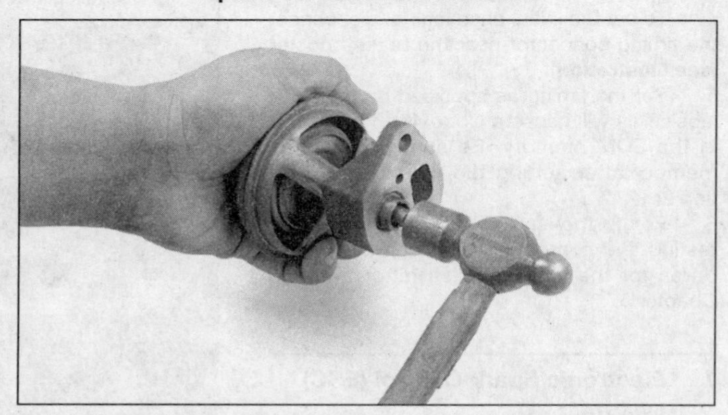

8.29 **A hammer can be used to clean the EGR valve pintle**

Switch (TVS) prevents the EGR valve from operating when the engine is cold **(see illustration)**.

5 On later models the EGR system is managed by the ECM. An ECM controlled solenoid is used in the vacuum line in order to maintain finer control of EGR flow. The ECM uses information from the coolant temperature, throttle position and manifold pressure sensors to regulate the vacuum solenoid.

6 Common engine problems associated with the EGR system are rough idle, stalling at idle after deceleration, surging at cruising speeds and stalling after cold starts. Engine overheating and detonation can also be caused by EGR system problems.

Checking

EGR valve

7 Make a physical inspection of the hoses and electrical connections to ensure that nothing is loose.

8 With the engine off, place your finger under the EGR valve and push up on the underside of the diaphragm **(see illustration)** - if the engine is hot, wear a glove to prevent burns! The diaphragm should move freely and not stick or bind. If it doesn't move or if it sticks, replace the valve with a new one.

9 With the engine running at normal operating temperature, push up on the diaphragm. The engine speed should drop.

10 If it doesn't, clean the EGR valve and passages or replace the EGR valve if necessary.

11 If the engine speed drops, increase the engine speed from idle to about 2000 rpm. The EGR valve diaphragm should move as this is done.

12 If it moves the EGR system is operating correctly.

13 If it doesn't move, attach a vacuum gauge to the vacuum hose connected to the EGR valve (use a T-fitting to connect the gauge). Start the engine and increase the speed from idle to 2000 rpm, then let it return to idle. At idle, approximately 6-inches or more of vacuum should be indicated.

14 If it is, the EGR valve is bad.

15 If it isn't check the vacuum hose to make sure it isn't leaking, restricted or disconnected.

16 Make the necessary repairs to the hose. If the hose is not a problem, check the thermal vacuum switch (non-ECM models) or EGR control solenoid (ECM models) as follows.

Thermal vacuum switch

17 Remove the carburetor-to-switch hose from the switch and check for vacuum at about 2000 rpm **(see illustration)**. The gauge should indicate approximately 10-inches of vacuum.

18 If it does, replace the switch with a new one. If it doesn't, check for a plugged or cracked hose and plugged carburetor passage.

EGR control solenoid

Note: *A Code 32 indicates a problem in the EGR system.*

19 Connect a vacuum gauge in place of the EGR valve. There should be at least ten inches of vacuum.

20 If there is not, and the vacuum hose is not damaged or leaking, the solenoid is faulty and should be replaced with a new one.

21 If there is vacuum, leave the vacuum gauge connected, turn the ignition on and, with the engine stopped, connect a test light across the solenoid harness connector terminals. The light should go on.

22 If the light does not come on, probe both harness connector terminals with a test light connected to ground. If there is no voltage at either terminal, there is an open ignition circuit

in the harness leading to the solenoid.

23 If the light does come on, ground the ALDL diagnostic and note whether the light comes on. If the light comes on, there could be a short to ground in circuit between the EGR control solenoid and the ECM (see the wiring diagrams at the end of Chapter 12). If it's okay, the ECM is faulty. **Note:** *Before replacing the ECM, use an ohmmeter to check the resistance between the solenoid terminals of the TCC and the EGR. Refer to the ECM wiring diagrams (at the end of Chapter 12) for coil terminal identification.* Replace the solenoid if the resistance measures less than 20 ohms. If the light does not come on, reconnect the solenoid, start the engine and note the vacuum gauge reading while the engine is idling. If there is vacuum replace the EGR control solenoid. If there is no vacuum the EGR control solenoid is okay.

Component replacement

EGR valve

24 Disconnect the air cleaner (Chapter 1).

25 Disconnect the EGR valve vacuum line from the valve.

26 Remove the EGR valve mounting bolts.

27 Remove the EGR valve from the manifold.

28 With the EGR valve removed, inspect the passages for excessive deposits. Scrape them clean and use a vacuum cleaner to remove the debris. **Caution:** *Do not wash the EGR valve in solvents or degreaser- permanent damage to the valve diaphragm may result. Sand blasting of the valve is also not recommended since it can affect the operation of the valve.*

29 The EGR valve can be cleaned by tapping the end of the pintle with a soft-face hammer **(see illustration)**.

30 Look for exhaust deposits in the valve outlet. Remove deposit buildup with a screwdriver.

31 Clean the mounting surfaces of the intake manifold and the valve assembly.

32 Clean the mounting surface of the valve with a wire wheel or wire brush and the pintle with a wire brush.

33 Depress the valve diaphragm and check the seat area for cleanliness and wear by looking through the valve outlet. If the pintle or the seat are not completely clean, repeat the procedure in Step 32.

34 Hold the bottom of the valve securely and try to rotate the top of the valve back-and-forth. Replace the valve if any looseness is felt.

35 Inspect the valve outlet for deposits. Remove any deposit buildup with a screwdriver or other suitable sharp tool.

36 Using a new gasket, install the old (cleaned) or new EGR valve on the intake manifold.

37 Install the EGR valve mounting bolts and tighten them securely.

38 Attach the vacuum hose to the valve.

39 Install the air cleaner (Chapter 4).

Thermal vacuum switch

40 Refer to Chapter 1 and drain about 1 quart of coolant from the system.

41 Remove the air cleaner and detach the hoses from the switch (since the switch may be used to control vacuum to other components as well, more than two hoses may be attached - label the hoses and ports to ensure correct reinstallation).

42 Unscrew the switch and remove it from the engine **(see illustration 8.17)**.

43 Apply a non-hardening sealant to the switch threads (don't get it on the end of the switch) and install the switch. Tighten it securely, then reattach the hoses and add coolant as necessary.

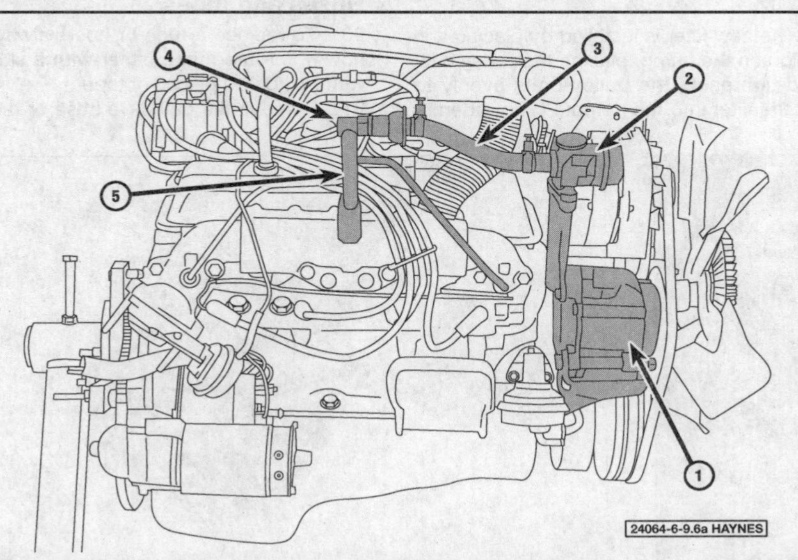

9.6a Typical Air Injection Reaction (AIR) system component layout - (Right side)

1 *Air pump*
2 *Diverter valve*
3 *High temperature hose*
4 *One way check valve*
5 *Air injection pipe*

EGR control solenoid

44 Disconnect the cable from the negative terminal of the battery.

45 Remove the air cleaner (Chapter 4).

46 Disconnect the electrical connector at the solenoid.

47 Disconnect the vacuum hoses from the solenoid.

48 Remove the mounting nut and detach the solenoid.

49 Install the new solenoid and tighten the nut securely. The remainder of the installation procedure is the reverse of removal.

9 Air Injection Reaction (AIR) system

Refer to illustrations 9.6a, 9.6b, 9.7 and 9.16

Note: *If your engine is equipped with an air pump, refer to information related to the AIR system. If no air pump is installed, refer to Section 13 for information regarding the PAIR system.*

General description

1 The AIR system helps reduce hydrocarbons and carbon monoxide levels in the exhaust by injecting air into the exhaust ports of each cylinder during cold engine operation, or directly into the catalytic converter during normal operation. It also helps the catalytic converter reach proper operating temperature quickly during warm-up. On early models, the air is injected only into the exhaust ports.

2 The AIR system uses an air pump to force the air into the exhaust stream. An air management valve, controlled by the vehicle's electronic control module (ECM) (later models) directs the air to the correct location, depending on engine temperature and driving conditions. During certain situations, such as deceleration, the air is diverted to the air cleaner to prevent backfiring from too much oxygen in the exhaust stream. One-way check valves are also used in the AIR system's air lines prevent exhaust gases from being forced back through the system.

3 The AIR system components include an engine driven air pump, air control, air switching and divert management valves, air flow and control hoses, check valves and a catalytic converter.

Checking

4 Because of the complexity of this system, it is difficult for the home mechanic to make an accurate diagnosis. If the system is malfunctioning, individual components can be checked.

5 Begin any inspection by carefully checking all hoses, vacuum lines and wires. Be sure they are in good condition and that all connections are tight and clean. Also make sure that the pump drivebelt is in good condition and properly adjusted (Chapter 1).

6 To check the pump, allow the engine to reach normal operating temperature and run

it at about 1500 rpm. Locate the hose running from the air pump **(see illustrations)** and squeeze to check for pulsations. Have an assistant increase the engine speed and check for a parallel increase in air flow. If an increase is noted, the pump is functioning properly. If not, there is a fault in the pump.

7 Each check valve can be inspected after removing it from the hose **(see illustration)**. Make sure the engine is completely cool before attempting this. Try to blow through the check valve from both ends. Air should pass through it only in the direction of normal air flow. If it is stuck either open or closed, the valve should be replaced.

8 To check the air control valve, disconnect the vacuum signal hose at the valve. With the engine running, check for vacuum at the hose. If none is felt, the line is clogged.

9 To check the deceleration valve, plug the air cleaner vacuum source, then reconnect the signal hose and listen for air flow through the ventilation pipe and into the deceleration valve. There should also be a noticeable engine speed drop when the signal hose is reconnected. If the air flow does not continue for at least one second, or the engine speed does not drop noticeably, check the deceleration valve hoses for restrictions and leaks. If no restrictions or leaks are found, replace the deceleration valve.

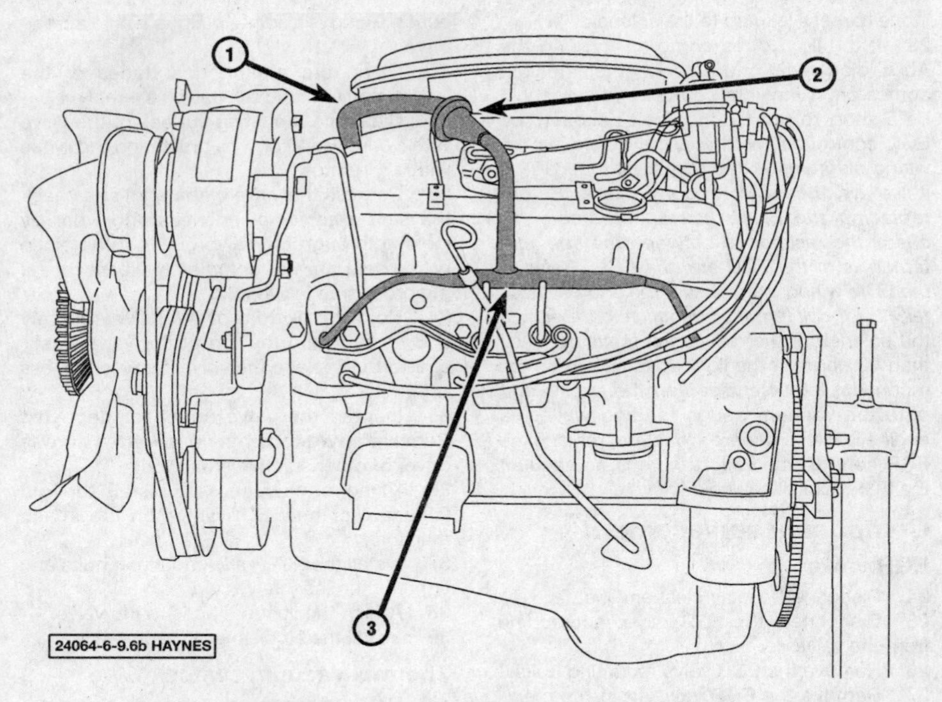

9.6b Typical Air Injection Reaction (AIR) system component layout - (Left side)

1 *High temperature hose*
2 *One way check valve*
3 *Air injection pipe*

Component replacement

Drivebelt

10 Loosen the pump mounting bolt and the pump adjustment bracket bolt.

11 Move the pump in until the belt can be removed.

12 Install the new belt and adjust it (refer to Chapter 1).

AIR pump pulley and filter

13 Compress the drivebelt to keep the pulley from turning and loosen the pulley bolts.

14 Remove the drivebelt as described above.

15 Remove the mounting bolts and lift off the pulley.

16 If the fan-like filter must be removed, grasp it firmly with a pair of needle-nose pliers **(see illustration)** and pull it from the pump. **Note:** *Do not insert a screwdriver between the filter and pump housing as the edge of the housing could be damaged. The filter will usually be distorted when pulled off. Be sure no fragments fall into the air intake hose.*

17 The new filter is installed by placing it in position on the pump, placing the pulley over it and tightening the pulley bolts evenly to draw the filter into the pump. Do not attempt to install a filter by pressing or hammering it into place. **Note:** *It is normal for the new filter to have an interference fit with the pump housing and, upon initial operation, it may squeal until worn in.*

18 Install the drivebelt and, while compressing the belt, tighten the pulley bolts securely.

19 Adjust the drivebelt tension.

Hoses and tubes

20 To replace a tube or hose, always note how it is routed first, either with a sketch or with numbered pieces of tape.

21 Remove the defective hose or tube and

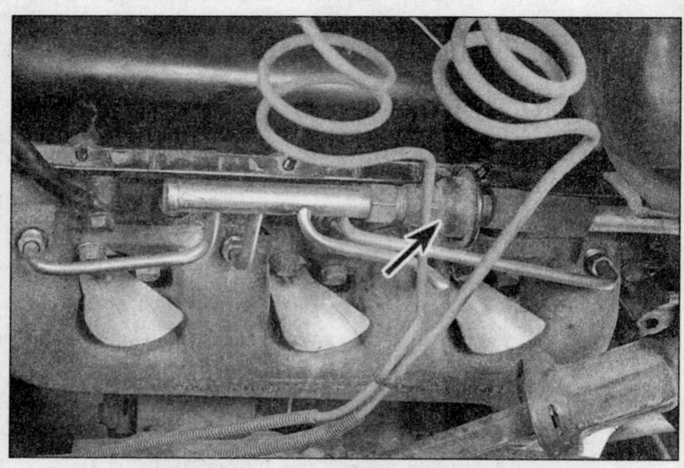

9.7 Typical one way check valve (arrow)

9.16 The filter can be pulled out of the air pump with a pair of needle-nose pliers

10.2 Typical late model charcoal - filled canister

11.1 Typical late model PCV valve (arrow)

replace it with a new one of the same material and size and tighten all connections.

Check valve

22 Disconnect the pump outlet hose at the check valve.

23 Unscrew the check valve from the pipe assembly. Be careful not to bend or twist the assembly.

24 Install a new valve after making sure that it is a duplicate of the part removed, then tighten all connections.

Air control valve

25 Disconnect the negative battery cable at the battery. Remove the air cleaner.

26 Disconnect the vacuum signal line from the valve. Disconnect the air hoses and wiring connectors.

27 If the mounting bolts are retained by tabbed lockwashers, bend the tabs back, then remove the mounting bolts and lift the valve off the adapter or bracket.

28 Installation is the reverse of the removal procedure. Be sure to use a new gasket when installing the valve.

Air pump

29 Remove the air management valve and adapter, if so equipped.

30 If the pulley must be removed from the pump it should be done prior to removing the drivebelt.

31 If the pulley is not being removed, remove the drivebelt.

32 Remove the pump mounting bolts and separate the pump from the engine.

33 Installation is the reverse of removal. **Note:** *Do not tighten the pump mounting bolts until all components are installed.*

34 Following installation, adjust the drivebelt tension (Chapter 1).

Deceleration valve

35 Disconnect the vacuum hoses from the valve.

36 Remove the screws retaining the valve to the engine bracket (if equipped) and detach the valve.

37 Install a new valve and reconnect all hoses.

10 Evaporative Emission Control System (EECS)

Refer to illustration 10.2

General description

1 This system is designed to trap and store fuel vapors that evaporate from the fuel tank, carburetor or throttle body and intake manifold.

2 The Evaporative Emission Control System (EECS) consists of a charcoal-filled canister and the lines connecting the canister to the fuel tank, ported vacuum and intake manifold vacuum **(see illustration)**.

3 Fuel vapors are transferred from the fuel tank, carburetor or throttle body and intake manifold to a canister where they are stored when the engine is not operating. When the engine is running, the fuel vapors are purged from the canister by intake air flow and consumed in the normal combustion process.

4 On later models there is a purge valve in the canister and, on some models, a tank pressure control valve. On some models the purge valve is operated by the ECM.

5 An indication that the system is not operating properly is a strong fuel odor. Poor idle, stalling and poor driveability can be caused by an inoperative purge valve, a damaged canister, split or cracked hoses or hoses connected to the wrong tubes.

Checking

6 Checking and maintenance procedures for the EECS system canister and hoses are included in Chapter 1.

7 To check the purge valve, apply a short length of hose to the lower tube of the purge valve assembly and attempt to blow through it. Little or no air should pass into the canister (a small amount of air will pass because the canister has a constant purge hole).

8 With a hand vacuum pump, apply vacuum through the control vacuum signal tube to the purge valve diaphragm. If the diaphragm does not hold vacuum for at least 20 seconds, the diaphragm is leaking and the canister must be replaced.

9 If the diaphragm holds vacuum, again try to blow through the hose while vacuum is still being applied. An increased flow of air should be noted. If it isn't, replace the canister.

Canister replacement

10 Disconnect the hoses from the canister.

11 Remove the pinch bolt and loosen the clamp.

12 Remove the canister.

13 Installation is the reverse of removal.

11 Positive Crankcase Ventilation (PCV) system

Refer to illustration 11.1

General description

1 The Positive Crankcase Ventilation (PCV) system reduces hydrocarbon emissions by circulating fresh air through the crankcase to pick up blow-by gases, which are then rerouted through a PCV valve to the carburetor or intake manifold **(see illustration)** to be burned in the engine.

2 The main components of the PCV system are vacuum hoses and the PCV valve, which regulates the flow of gases according to engine speed and manifold vacuum.

Checking

3 The PCV system can be checked quickly and easily for proper operation. It should be checked regularly because carbon and gunk deposited by blow-by gases will eventually clog the PCV valve and system hoses. The common symptoms of a plugged or clogged PCV valve include a rough idle, stalling or a slow idle speed, oil leaks, oil in the air cleaner or sludge in the engine.

4 Refer to Chapter 1 for PCV valve checking and replacement procedures.

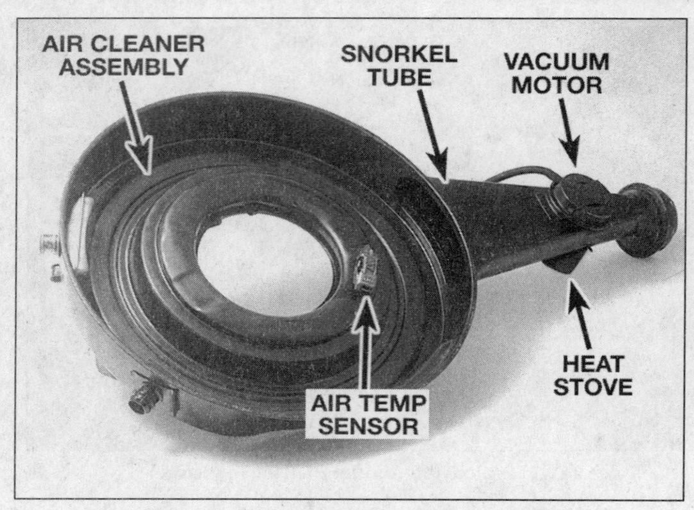

12.2 Typical THERMAC system component layout

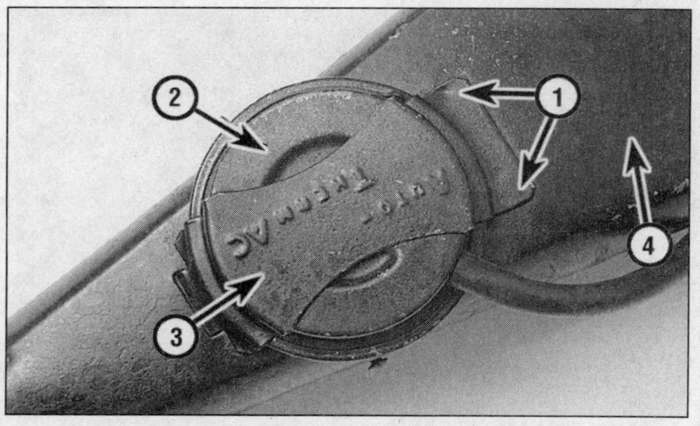

12.9 To remove the vacuum motor, drill out the spot welds and detach the retaining strap

1 Spot welds
2 Vacuum diaphragm motor
3 Retaining strap
4 Snorkel

12 Thermostatic Air Cleaner (THERMAC)

Refer to illustrations 12.2, 12.9, 12.15 and 12.20

General description

1 The thermostatic air cleaner (THERMAC) system improves engine efficiency and driveability under varying climatic conditions by controlling the temperature of the air coming into the air cleaner. A uniform incoming air temperature allows leaner air/fuel ratios during warm-up, which reduces hydrocarbon emissions.

2 The system uses a damper assembly, located in the snorkel of the air cleaner housing, to control the ratio of cold and warm air directed into the carburetor or throttle body. This damper is controlled by a vacuum motor which is, in turn, modulated by a temperature sensor in the air cleaner **(see illustration)**. On some engines a check valve is used in the sensor, which delays the opening of the damper when the engine is cold and the vacuum signal is low.

3 It is during the first few miles of driving (depending on outside temperature) that this system has its greatest effect on engine performance and emissions output. When the engine is cold, the damper blocks off the air cleaner inlet snorkel, allowing only warm air from around the exhaust manifold to enter the engine. As the engine warms up, the damper gradually opens the snorkel passage, increasing the amount of cold air allowed into the air cleaner. By the time the engine has reached its normal operating temperature, the damper opens completely, allowing only cold, fresh air to enter.

4 Because of this "cold engine only" function, it is important to periodically check this system to prevent poor engine performance when cold, or overheating of the fuel mixture once the engine has reached operating temperatures. If the air cleaner damper sticks in the "no heat" position the engine will run poorly, stall and waste gas until it has warmed up on its own. A valve sticking in the "heat" position causes the engine to run as if it is out of tune due to the constant flow of hot air to the carburetor or throttle body.

Checking

5 Refer to Chapter 1 for maintenance and checking procedures for the THERMAC system. If any of the problems mentioned above are discovered while performing routine maintenance checks, refer to the following procedures.

6 If the damper does not close off snorkel air when the cold engine is started, disconnect the vacuum hose at the snorkel vacuum motor and place your thumb over the hose end, checking for vacuum. Replace the vacuum motor if the hose routing is correct and the damper moves freely.

7 If there was no vacuum going to the motor in the above test, check the hoses for cracks, crimped areas and proper connection. If the hoses are clear and in good condition, replace the temperature sensor inside the air cleaner housing.

Component replacement

Air cleaner vacuum motor

8 Remove the air cleaner assembly from the engine and disconnect the vacuum hose from the motor.

Carbureted models

9 Drill out the two spot welds which secure the vacuum motor retaining strap to the snorkel tube **(see illustration)**.

10 Remove the motor retaining strap.

11 Lift up the motor, cocking it to one side to unhook the motor linkage at the damper assembly.

12 To install, drill a 7/64-inch hole in the snorkel tube at the center of the vacuum motor retaining strap.

13 Insert the vacuum motor linkage into the damper assembly.

12.15 Fuel injection THERMAC actuator mounting rivets (arrows)

14 Using the sheet metal screw supplied with the motor service kit, attach the motor and retaining strap to the snorkel. Make sure the sheet metal screw does not interfere with the operation of the damper door. Shorten the screw if necessary.

EFI-equipped models

15 Drill out the rivets retaining the wax pellet actuator and detach the actuator from the air cleaner housing **(see illustration)**.

16 To install, place the actuator in position in the housing, engage the spring and install new rivets.

17 Connect the vacuum hose to the motor and install the air cleaner assembly.

Air cleaner temperature sensor

18 Remove the air cleaner from the engine and disconnect the vacuum hoses at the sensor.

19 Carefully note the position of the sensor.

12.20 The sensor retainer must be pried off with a screwdriver

14.7 Throttle Return Control (TRC) plunger screw (arrow)

The new sensor must be installed in exactly the same position.

20 Pry up the tabs on the sensor retaining clip and remove the sensor and clip from the air cleaner **(see illustration)**.

21 Install the new sensor with a new gasket in the same position as the old one.

22 Press the retaining clip onto the sensor. Do not damage the control mechanism in the center of the sensor.

23 Connect the vacuum hoses and attach the air cleaner to the engine.

13 Pulse Air Injection Reaction (PAIR) system

Note: *If your engine is equipped with an air pump, refer to information related to the AIR system (Section 9). If no air pump is installed and you have a vehicle produced after the introduction of the AIR/PAIR systems, your vehicle has a PAIR system.*

General description

1 This system performs some of the functions of the AIR system, but utilizes exhaust pulses instead of an air pump to draw air into the exhaust system. Fresh air that is filtered by the air cleaner, to avoid the build-up of dirt on the check valve seat, is supplied to the system on command from the ECM (later models only). On other models, a deceleration valve is used to prevent backfiring in the exhaust system during deceleration.

2 The PAIR system consists of four pulse air valves. The combustion of the engine creates a pulsating flow of exhaust gases which are either of a positive or negative pressure. A positive pressure will force the check valve closed and no exhaust gas will flow past the valve into the fresh air supply line. A negative pressure at the pulse air valves will result in the flow of fresh air into the exhaust system. The fresh air reduces emissions by promoting a more thorough combustion.

Checking

3 Inspect the pulse air valves, pipes, grommets and hose for leaks and cracks and replace as required.

4 A simple, functional test of this system can be performed with the engine running. Disconnect the rubber hose from the air valve assembly and hold your hand over the valve's inlet hole. With the engine idling there should be constant suction at the valve. Have an assistant apply throttle. As the engine gains speed, see if the suction increases. If it doesn't, the lines are leaking or restricted or the check valve is sticking. Also make sure that air is not being blown out of the air valve - this is also an indication that the check valve is sticking open. Service or replace the components as necessary. If other PAIR problems are suspected, have a dealer or repair shop diagnose the system as they might relate to the ECM/Computer Command Control System.

Component replacement

5 To remove the pulse air valve, remove the air cleaner and disconnect the rubber hose from the plenum connecting pipe.

6 Disconnect the four pipe check valve fittings at the cylinder head and remove the check valve from the check valve pipe.

7 Assemble the check valves to the check valve pipes on a workbench before reinstalling.

8 Install the pipe check valve assemblies to the cylinder head and finger-tighten the fittings.

9 Using a large wrench or similar tool as a lever, align the check valve to the plenum grommet. Using the palm of your hand, install the check valve into the grommet. Use rubber lubricant on the grommets to ease assembly.

10 Tighten the fittings securely and reinstall the air cleaner and hose.

14 Throttle Return Control (TRC) system

Refer to illustration 14.7

General description

1 This system opens the throttle lever slightly when coasting to reduce the emission of hydrocarbons.

2 The system consists of the throttle lever actuator mounted on the carburetor, which is controlled by the solenoid control valve and the engine speed switch.

Checking

3 Inspect the hoses and electrical wires for kinks, cracks and secure connections. Check the actuator linkage for binding due to corrosion or dirt; clean as necessary.

4 To check the actuator valve, disconnect the vacuum hose and connect a vacuum pump to the valve. Apply 20 in-Hg of vacuum. If the reading on the pump gauge drops, the actuator valve is leaking and must be replaced with a new one.

5 Connect a tachometer in accordance with the manufacturer's instructions. Start the engine and have it idle at normal operating temperature, with the transmission in Neutral (manual transmission) or Park (automatic transmission).

6 Apply a vacuum of 20 in-Hg to the actuator and manually open the throttle slightly, then let it close again against the actuator plunger.

7 Record the engine rpm, which should be as specified on the VECI label in the engine compartment. If it is not, turn the idle speed adjustment screw on the plunger and repeat the check until the rpm is within specification **(see illustration)**.

8 Check for voltage between the speed switch and the control valve by attaching the voltmeter negative probe to a good engine

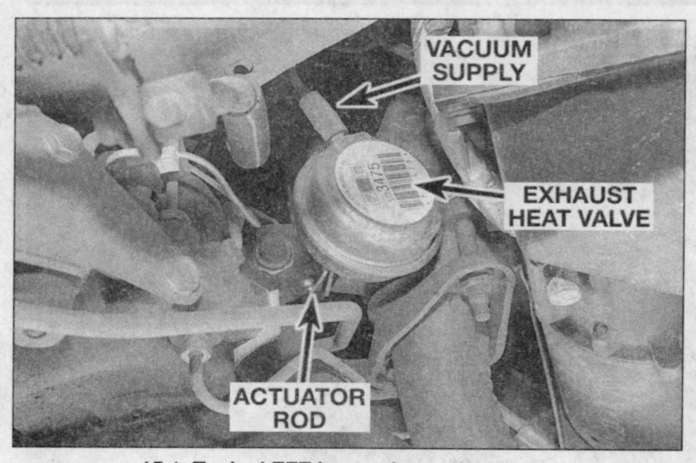

15.1 Typical EFE heat valve and actuator

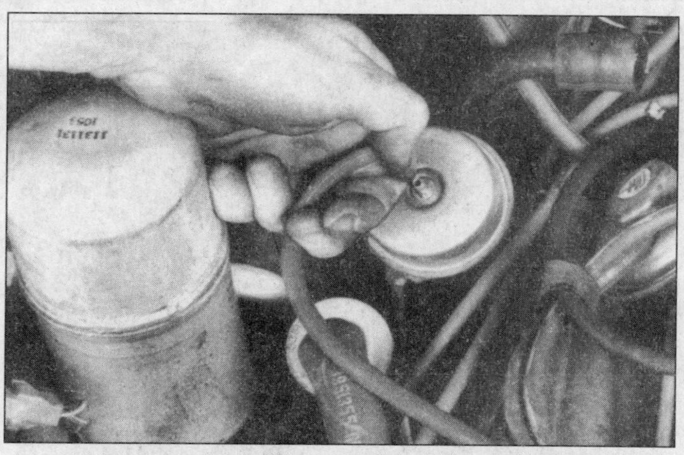

15.6 Check for a vacuum supply to the actuator with the engine idling

ground and the positive probe to the switch voltage source. The connector does not have to be unplugged for this check. Insert the probe into the connector body so it contacts the metal terminal.

9 If there is voltage at the switch and not the valve, or vice versa, there is a fault in the wiring harness. If there is no voltage at either the switch or valve, there is a fault in the wiring harness at the firewall bulkhead connector or the distributor connector.

Component replacement

Solenoid control valve

10 Disconnect the vacuum hoses and electrical connector, remove the retaining nut and detach the valve assembly from the engine.

Throttle lever actuator

11 Disconnect the vacuum hose, unscrew the large nut retaining the actuator to the bracket on the carburetor, detach the actuator from the bracket and lift it from the engine.

Speed switch

12 Unplug the electrical connector, remove the nut and washer and lift the switch off.

15 Early Fuel Evaporation (EFE) system

Refer to illustrations 15.1 and 15.6

General description

1 This system rapidly heats the intake air supply to promote evaporation of the fuel when the engine is started cold. On early models a valve in the exhaust pipe is opened and closed by a spring which is actuated by exhaust heat. Refer to Chapter 1 for maintenance operations for this type of valve. On later models a vacuum operated heat valve located in the exhaust pipe directs exhaust gas to the intake manifold to heat it when the engine is cold **(see illustration)**.

2 The components involved in the EFE's operation include the exhaust heat valve, the vacuum power actuator and rod or linkage and a coolant or oil temperature switch and vacuum hoses. **Note:** *In general, a coolant temperature switch is used on V6 and V8 engines, while the oil temperature switch is used on inline 6-cylinder engines.*

3 The coolant or oil temperature sensor is closed when the engine is cold, allowing vacuum to close the servo and direct exhaust gases to the manifold. As the engine warms up, the vacuum switch closes, cutting off vacuum to the servo so that it opens and exhaust gas is no longer directed to the manifold.

4 If the EFE system is not functioning, poor cold engine performance will result. If the system is not shutting off when the engine is warmed up, the engine will run as if it is out of tune and also may overheat.

Checking

5 With the engine cold, the transmission in Neutral (manual) or Park (automatic), start the engine. Observe the operation of the actuator rod which leads to the heat valve inside the exhaust pipe. It should immediately move the valve to the closed position. If this is the case, the system is operating correctly.

6 If the actuator rod does not move, disconnect the vacuum hose at the actuator and place your thumb over the open end **(see illustration)**. With the engine cold and at idle, you should feel a suction indicating proper vacuum. If there is vacuum at this point, replace the actuator with a new one.

7 If there is no vacuum in the line, this is an indication that either the hose is crimped or plugged or the thermal vacuum switch threaded into the water passage or the oil temperature switch threaded into the engine block is not functioning properly. Replace the hose or switch as necessary.

8 To make sure the system is disengaging once the engine has warmed up, make sure that the rod moves the exhaust heat valve to the open position as the engine reaches normal operating temperature (approximately 180 degrees).

9 If after the engine has warmed, the valve does not open, pull the vacuum hose at the actuator and check for vacuum with your thumb. If there is no vacuum, replace the actuator, If there is vacuum, replace the water or oil temperature switch.

Component replacement

Actuator and rod assembly

10 Disconnect the vacuum hose from the actuator.

11 Remove the nuts which attach the actuator to the bracket.

12 Disconnect the rod from the heat valve and remove the actuator and rod from the engine.

13 Installation is the reverse of removal.

Exhaust heat valve

14 Remove the exhaust crossover pipe, if required for clearance.

15 If the actuator is part of the heat valve, disconnect the vacuum hose. If the actuator is remotely mounted, disconnect the rod from the heat valve.

16 Remove the exhaust pipe-to-manifold nuts and tension springs.

17 Remove the heat valve from inside the exhaust pipe.

18 Installation is the reverse of removal.

Coolant temperature thermal vacuum switch

19 Refer to Chapter 1 and drain about 1 quart of coolant from the system.

20 Detach the vacuum hose from the temperature switch (located in the thermostat housing cover).

21 Unscrew the switch.

22 Apply non-hardening sealant to the threads of the new switch (don't get sealant on the end of the switch) and install it into the thermostat housing cover. Tighten it securely.

23 Reattach the vacuum hose and add coolant as required (Chapter 1).

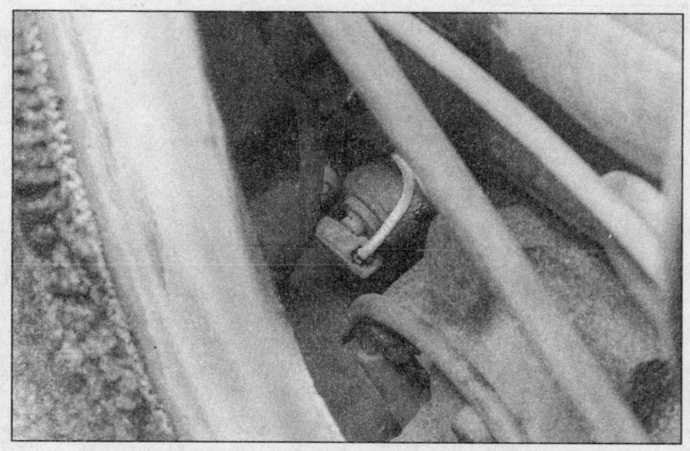
16.3 Typical TCS coolant temperature switch location in the cylinder head

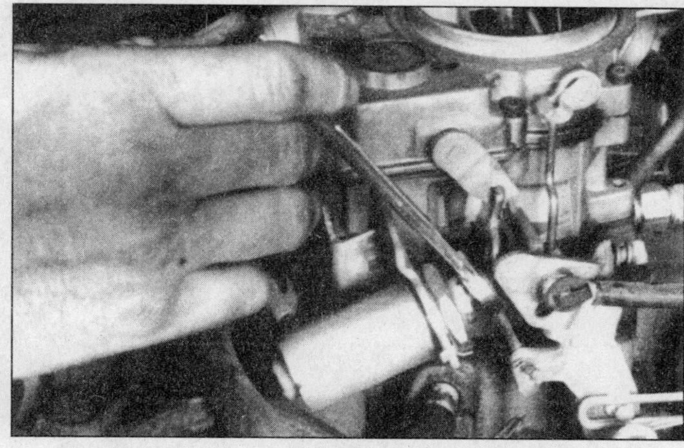
16.4 The idle stop solenoid is mounted on the carburetor

Oil temperature thermal vacuum switch

24 Detach the vacuum hoses from the temperature switch (located in the engine block).
25 Unscrew the switch.
26 Install the new switch. Tighten it securely.
27 Reattach the vacuum hoses.

16 Transmission Controlled Spark (TCS) system

Refer to illustrations 16.3 and 16.4

General description

1 The purpose of this system is to reduce the emission of certain exhaust gases by eliminating ignition vacuum advance when the vehicle is operating in low forward gears on some models.
2 The vacuum advance is controlled by a solenoid-operated switch which is actuated by a transmission switch. When the switch is in operation, the vacuum normally applied to the distributor is vented to atmosphere and ignition advance is controlled by the advance mechanism in the distributor.
3 A coolant temperature switch is wired into the solenoid circuit to prevent vacuum cut-off to the distributor at engine temperatures below 93° F **(see illustration)**.
4 In order to compensate for the retarded spark and to prevent possible engine run-on (dieseling) when the engine is turned off, an idle stop solenoid is provided on some carburetors to permit the throttle valve to close further than the normal, (slightly open) idling position of the throttle valve plate **(see illustration)**.
5 Symptoms of faults in the TCS system are:

Slow idling or dieseling when the engine is shut off (faulty idle stop solenoid)
Poor high gear performance (blown fuse)
Excessive fuel consumption (faulty temperature switch)
Backfiring during deceleration (transmission switch fault)
Hard starting when cold (vacuum advance fault)

Checking

6 Checking of this system is limited to inspecting the security of the electrical and vacuum connections.

17 Catalytic converter

Refer to illustration 17.6

General description

1 The catalytic converter is an emission control device added to the exhaust system to reduce pollutants in the exhaust gas. The converter contains platinum and rhodium, which lowers the levels of oxides of nitrogen (NOx) as well as hydrocarbons (HC) and carbon monoxide (CO).

17.6 Remove the catalytic converter U-bolts (arrows)

Checking

2 The test equipment for a catalytic converter is expensive and highly sophisticated. If you suspect that the converter on your vehicle is malfunctioning, take it to a dealer or authorized emissions inspection facility for diagnosis and repair.
3 The converter is located underneath the passenger compartment. Therefore, whenever the vehicle is raised for servicing of underbody components, check the converter for leaks, corrosion and other damage. If damage is discovered, the converter can simply be unbolted from the exhaust system and replaced.

Component replacement

4 Raise the vehicle and place it securely on jackstands.
5 Remove the U-bolt nuts from both ends of the converter.
6 Detach the converter from the exhaust pipes and lower it from the vehicle **(see illustration)**.
7 Installation is the reverse of removal.

Notes

Chapter 7 Part A
Manual transmission

Contents

	Section
Clutch pedal free play check and adjustment	See Chapter 1
Floor shift linkage - removal, installation and adjustment	4
General information	1
Manual transmission oil change	See Chapter 1
Manual transmission oil level check	See Chapter 1
Neutral start switch - removal and installation	See Chapter 8
Rear oil seal - replacement	5
Speedometer driven gear - replacement	6
Steering column shift linkage - adjustment	3
Transmission shift effort - diagnosis	2
Transmission (3-speed 76 mm) - disassembly, overhaul and reassembly	9
Transmission (3-speed 77 mm) - disassembly, overhaul and reassembly	10
Transmission (4-speed 89 mm) - disassembly, overhaul and reassembly	11
Transmission (4-speed 117 mm) - disassembly, overhaul and reassembly	12
Transmission (2-wheel drive models) - removal and installation	7
Transmission (4-wheel drive models) - removal and installation	8

Specifications

General

Transmission oil type	See Chapter 1

Torque specifications

	Ft-lbs
3-speed 76 mm transmission	
Drivegear retainer-to-case bolt	15
Side cover-to-case bolt	15
Extension-to-case bolt	45
Shift lever-to-shift shaft bolt	25
Oil filler plug	13
Transmission case-to-bellhousing bolt	75
Crossmember-to-frame bolt	55
Crossmember-to-mount bolt	40
2-3 cross-over shaft bracket retaining nut	18
1st/Reverse swivel bolt	20
Mount-to-transmission bolt	45
3-speed 77 mm transmission	
Drivegear retainer-to-case bolt	35
Cover-to-case bolt	30
Extension-to-case bolt	45
Shift lever-to-shift shaft bolt	25
Oil filler plug	15
Transmission-to-bellhousing bolt	75
Crossmember-to-frame nut	25
Crossmember-to-mount bolt	45
2-3 cross-over shaft bracket retaining nut	18
1st/Reverse swivel bolt	20
Mount-to-transmission bolt	45

Torque specifications

	Ft-lbs
4-speed 89 mm transmission	
Shift lever retaining nut	18
Extension-to-case bolt	50
Drivegear bearing retainer bolt	30
Side cover-to-case bolt	15
Back-up light switch	15
Oil filler plug	15
Transmission-to-bellhousing bolt	75
Crossmember-to-frame bolt	55
Crossmember-to-mount bolt	40
Shift lever and bracket-to-transmission	36
Control rod adjusting nuts	15
Mount-to-transmission bolts	40
4-speed 117 mm transmission	
Drivegear retainer-to-case bolt	25
Side cover-to-case bolt	20
Extension-to-case bolt	30
Lever-to-shift shaft nut	20
Oil filler plug	30
Transmission case-to-bellhousing bolt	75
Crossmember-to-frame nut	30
Crossmember-to-mount bolt	40
Transmission drain plug	20
Mount-to-transmission bolt	50

1 General information

Vehicles covered by this manual were equipped with either 3 or 4-speed manual or 3 or 4-speed automatic transmissions. All information on the manual transmission is included in this Part of Chapter 7. Information on the automatic transmission can be found in Part B of this Chapter. Information on the transfer case can be found in Part C.

Manual transmissions are identified by the number of forward gears and the distance between the mainshaft and countergear, measured at the shaft centerlines. The 3-speed 76 mm transmission has an identification number stamped on the top side of the case, below the side cover. The 3-speed 77 mm transmission has an identification number stamped on the top left side of the case. This transmission is used primarily for heavy-duty applications.

The 4-speed 89 mm transmission is fully synchronized in all forward gears and uses 4th gear mainly as an overdrive. The 4-speed 117 mm transmission is synchronized only in 2nd, 3rd and 4th gears.

Depending on the expense involved in having a transmission overhauled, it may be a better idea to consider replacing it with either a new or rebuilt one. Your local dealer or transmission shop should be able to supply information concerning cost, availability and exchange policy. Regardless of how you decide to remedy a transmission problem, you can still save a lot of money by removing and installing the unit yourself.

2 Transmission shift effort - diagnosis

Note: *Before attempting to diagnose shifting problems, make sure the clutch pedal free play is properly adjusted* (Chapter 1).

1 If excessive effort is required to shift the transmission into a particular gear, a relatively simple procedure can be used to determine if the shifter mechanism or the transmission is at fault.

2 Determine which shift rod may be binding, then disconnect it from the side of the transmission.

3 Thread two nuts onto the shift shaft extending from the transmission, lock the nuts together and use a wrench to manually shift the transmission into and out of gear.

4 Using an inch-pound torque wrench, measure the torque necessary to shift the transmission into gear. It should not exceed 72 inch-pounds.

5 Torque in excess of this amount would indicate an internal problem in the transmission. A Positraction additive may help the situation. You can obtain this special lubricant, along with additional information regarding its use, at a GM dealer.

6 If the shift effort at this location is not excessive, the problem probably lies in the shift lever or shifter. Check for binding, rust, dirt, etc.

3 Steering column shift linkage - adjustment

1 Place both the shifter levers on the transmission case in Neutral.

2 Disconnect the control rods from the levers on the steering column tube.

3 Set the levers on the steering column tube to the Neutral position and insert a gage pin (between 3/16 and 7/32-inch diameter) through the holes in the levers so the head control lever is retained in the Neutral position.

4 Connect the control rods to the column tube levers by adjusting the clamps on the shifter levers at the transmission. The levers must not move from the Neutral positions.
5 Remove the gage pin and check the gearshift operation.

4 Floor shift linkage - removal, installation and adjustment

Rod-type linkage

Removal

1 Remove the retaining screws and pull up the shift boot.
2 Remove the shift lever. On most models this can be accomplished by inserting a piece of shim stock or a small screwdriver blade into the base of the shift lever to release it.
3 Remove the retaining nuts and clips and detach the shift rods.
4 Remove the bolts and separate the shifter from the transmission.

Installation

5 Place the shifter in position and tighten the bolts securely.
6 Install the shift rods and secure them with the clips and nuts. Tighten the nuts finger tight.
7 Install the shift lever by inserting it into the shifter base until it locks.

Adjustment

8 Move the shift lever into the Neutral position.
9 With the nuts loose, move the shift rods into the Neutral position and insert a gage pin into the hole at the base of the shifter.
10 Hold the rods to keep them from moving and tighten the nuts securely.

Cover-mounted shifter

Removal

11 Remove the screws and pull the shift boot up.
12 Push down on the cap, turn the shift lever cap counterclockwise and withdraw the shift lever from the transmission.

Installation

13 Push the cap down, insert the shift lever and rotate it clockwise until it locks, then release the cap.
14 Install the shift boot.

5 Rear oil seal - replacement

Refer to illustration 5.6

1 Raise the vehicle and support it securely on jackstands.
2 Drain the transmission oil (see Chapter 1).
3 Remove the driveshaft (Chapter 8).
4 On 4-wheel drive models, remove the transfer case.

5.6 The rear oil seal can be pried out of the transmission with a long screwdriver - work around the seal, prying a little at a time, being careful not to damage the output shaft

5 On the 4-speed 117 mm transmission, remove the universal joint flange and nut for access to the oil.
6 Carefully pry the seal out of the housing with a screwdriver **(see illustration)**. Be careful not to damage the output shaft splines.
7 Clean the counterbore and check it for damage. Apply transmission oil to the lips of the new seal and coat the outer edge with a thin layer of RTV sealant.
8 Position the new seal in the bore with the open side facing IN and carefully tap it into place with a large socket and hammer (if a large socket isn't available, a piece of pipe will also work, as long as it's the correct diameter).
9 Install the driveshaft and any other components that were removed.

6 Speedometer driven gear - replacement

1 Raise the vehicle and support it securely on jackstands.
2 Disconnect the speedometer cable, remove the bolt and detach the lockplate.
3 Insert a screwdriver into the lockplate fitting and pry the gear and shaft out of the transmission.
4 Pry out the O-ring.
5 Installation is the reverse of removal. Be sure to lubricate the O-ring with transmission fluid.

7 Transmission (2-wheel drive models) - removal and installation

1 Raise the vehicle and support it securely on jackstands.
2 Drain the lubricant from the transmission (Chapter 1).
3 Disconnect the speedometer cable, back-up light switch wires and, where applicable, the throttle control switch.
4 Remove the shift controls from the transmission.
5 On vehicles equipped with a 117 mm transmission, remove the gearshift lever (Section 4). Place a clean lint-free cloth over the opening of the transmission to prevent the entry of dirt.
6 Where applicable, disconnect the parking brake lever and controls.
7 If the vehicle is equipped with a power take off, remove the unit and controls from the transmission. Place a clean lint-free cloth over the opening.
8 Remove the driveshaft (Chapter 8).
9 Support the transmission with a jack.
10 Disconnect any lines, hoses, wires or brackets which would interfere with the removal of the transmission.
11 Detach the bellhousing cover and the transmission-to-bellhousing bolts.
12 Move the transmission assembly straight back, away from the engine. Be careful to keep the transmission input shaft in alignment with the clutch plate hub. **Note:** *When removing the transmission, two or more people should take the weight of the unit. Do not allow the weight to hang on the clutch plate hub, as the disc will become distorted, seriously affecting clutch operation.*
13 When the transmission is free, lower it to the ground and remove it from under the vehicle.
14 Installation is the reverse of removal. Install and adjust the shift mechanism (Section 4). Fill the transmission with the specified lubricant (Chapter 1).

8 Transmission (4-wheel drive models) - removal and installation

Rod-type shifter transmission

1 Raise the vehicle and support it securely on jackstands.
2 Drain the transfer case and transmission lubricants (see Chapter 1).
3 Disconnect the speedometer cable and, where applicable, the throttle control switch connections.
4 Disconnect the driveshafts from the universal joint yokes at the transfer case and fasten them up out of the way.
5 Loosen and remove the bolt holding the shift lever control assembly to the adapter assembly. Push the assembly to one side and fasten it out of the way.
6 Support the transfer case with a jack and remove the bolts attaching the transfer case to the adapter.
7 Remove the bolts attaching the transfer case to the frame bracket on the right side. Separate the transfer case from the adapter.
8 Disconnect the shift control rods from the shift levers at the transmission (Section 4).

9 Support the rear of the engine with a jack or jackstand. Remove the two adapter mount bolts.

10 Remove the two top transmission-to-bellhousing bolts and install two bolts with the heads cut off in their place.

11 Remove the flywheel cover, then remove the two lower transmission-to-bellhousing bolts.

12 With the help of an assistant, slide the transmission and adapter assembly straight back on the guide pins, until the clutch gear is free of the splines in the clutch plate. **Note:** *The use of guide pins during this operation will prevent damage to the clutch plate.*

13 Carefully lower the transmission and adapter assembly to the ground and remove them from under the vehicle.

14 Installation is the reverse of the removal procedure.

Cover-mounted shifter (117 mm) transmission

15 Working in the passenger compartment, remove the screws attaching the transfer case shift lever boot retainer. Remove the retainer.

16 Remove the floor mat or carpeting.

17 Remove the screws attaching the transmission shift lever boot retainer. Slide the retainer and boot up the shift lever to remove them. Remove the gearshift lever (Section 4).

18 On GMC models, remove the center floor outlet from the heater distributor duct.

19 On vehicles equipped with a center console, the console must be removed.

20 Remove the screws attaching the transmission floor cover. To remove the cover, rotate it approximately 90-degrees to clear the transfer case shift lever.

21 Disconnect the shift lever link assembly from the transfer case shift rail connecting rod. Remove the shift lever mounting bolt and detach the shift lever control from the adapter.

22 Disconnect the back-up light wire from the switch and remove the clamps from the top cover bolt.

23 Raise the vehicle and support it securely on jackstands. Drain the transmission and transfer case lubricants (see Chapter 1).

24 Disconnect the speedometer cable from the transfer case. Also disconnect the throttle control switch.

25 Refer to Chapter 8 and disconnect the front and rear driveshafts from the transfer case. Fasten them out of the way.

26 Bend back the locktabs and remove the transmission mount-to-frame crossmember bolts. Also remove the transfer case-to-frame bracket bolts.

27 Support the transmission and transfer case assembly with a jack.

28 Remove the frame-to-crossmember bolts and detach the crossmember from the vehicle. Rotate the crossmember to clear the frame rails.

29 Remove the flywheel cover. On V8 engine equipped vehicles, remove the exhaust crossover pipe.

30 Remove the transmission-to-bellhousing bolts. **Note:** *The two upper bolts should be removed first and two bolts with the heads cut off should be installed in their places to support the transmission and prevent damage to the clutch plate hub.*

31 With the help of an assistant, slide the transmission to the rear until the main drivegear clears the clutch assembly. Lower the assembly to the ground and remove it from under the vehicle.

32 Installation of the transmission is the reverse of the removal procedure.

9 Transmission (3-speed 76 mm) - disassembly, overhaul and reassembly

Side cover overhaul

1 Shift the transmission into Neutral, raise the vehicle for access beneath and support it securely on jackstands.

2 Disconnect the control rods from the levers on the side of the transmission (Section 4).

3 Remove the cover assembly from the transmission case and allow the oil to drain.

4 Remove both shift forks from the shifter shaft assemblies and both shifter shaft assemblies from the cover.

5 Pry out the shaft O-ring seals if replacement is required.

6 Remove the detent cam spring and pivot retainer C-clip. Remove both detent cams.

7 Inspect all parts for damage and wear, and replace as necessary.

8 With the detent spring tang projecting up over the 2nd/3rd shifter shaft cover opening, install the 1st/Reverse detent cam onto the detent cam pivot pin. With the detent spring tang projecting up over the 1st/Reverse shifter shaft cover hole, install the 2nd/3rd detent cam.

9 Install the C-clip on the pivot shaft and hook the spring into the detent cam notches.

10 Install the shifter shaft assemblies carefully into the cover and the shift forks to the shifter shaft assemblies. Lift up the detent cam to allow the forks to seat properly.

11 Set the shifter levers into the Neutral detent (center position) and position the gasket on the case.

12 Carefully position the side cover, ensuring that the shift forks are aligned with their appropriate mainshaft clutch sliding sleeves.

13 Install the cover bolts. Tighten to the specified torque.

14 Add oil as necessary to the transmission.

Transmission disassembly

15 Remove the transmission as described in Section 7 or 8 and remove the side cover assembly (refer to Steps 1 through 7 if necessary).

16 Remove the drivegear bearing stem snap-ring.

17 Pull out the gear until a large screwdriver can be used to lever the drivegear bearing from its location.

18 Remove the speedometer driven gear from the rear extension, then remove the extension retaining bolts.

19 Remove the Reverse idler shaft E-ring.

20 Withdraw the drivegear, mainshaft and extension assembly together through the rear casing.

21 From the mainshaft, detach the drivegear, needle bearings and sychronizer ring.

22 Expand the snap-ring in the rear extension which retains the rear bearing and then withdraw the rear extension.

23 Using a dummy shaft, drive the countershaft (complete with Woodruff key) out of the rear of the transmission case. Carefully remove the dummy shaft and extract the countergear, bearings and thrust washers from the interior of the transmission case.

24 Drive the Reverse idler shaft out of the rear of the transmission case using a long drift.

25 The mainshaft should only be dismantled if a press or bearing puller is available, otherwise take the assembly to your Chevrolet dealer. Be careful to keep all components separated and in order for easier reassembly.

26 Remove the 2nd/3rd synchro hub snap-ring from the mainshaft. Do not mix up the synchro unit components; although identical, the components of each unit are matched in production.

27 Remove the synchro unit, 2nd speed blocker ring and 2nd speed gear from the front end of the mainshaft.

28 Depress the speedometer drivegear retaining clip and remove the gear from the mainshaft.

29 Remove the rear bearing snap-ring from its mainshaft groove.

30 Support the Reverse gear and press the mainshaft out of the rear bearing, and snap-ring from the rear end of the mainshaft.

31 Remove 1st/reverse synchro hub snap-ring from the mainshaft and remove the synchro unit.

32 Remove the 1st speed blocker ring and 1st speed gear from the rear end of the mainshaft.

Transmission overhaul

33 Clean all components in solvent and dry thoroughly. Check for wear or chipped teeth. If there has been a history of noisy gearshifts or the synchro action could easily be 'beaten' then replace the appropriate synchro unit.

34 Extract the oil seal from the rear end of the rear extension and drive in a new one with a tubular drift.

35 Clean the transmission case inside and out and check for cracks, particularly around the bolt holes.

36 Extract the drivegear bearing retainer seal and drive in a new one.

Transmission reassembly

37 Begin rebuilding the transmission by first reassembling the mainshaft. Install 2nd speed gear so that the rear face of the gear butts against the flange on the mainshaft.
38 Install the blocker ring, followed by the 2nd/3rd synchro assembly (shift fork groove nearer rear end of the mainshaft). Make sure that the notches of the blocker ring align with the keys of the synchro assembly.
39 Install the snap-ring which retains the synchro hub to the mainshaft.
40 To the rear end of the mainshaft, install the 1st speed gear, followed by the blocker ring.
41 Install the 1st/Reverse synchro unit (shift fork groove nearer the front end of the mainshaft), again making sure that the notches of the blocker ring align with the keys of the synchro unit.
42 Install the snap-ring, Reverse gear thrust washer and spring washer.
43 Install the mainshaft rear ball bearing with the outer snap-ring groove nearer the front of the shaft.
44 Install the rear bearing shaft snap-ring.
45 Install the speedometer drive gear and retaining clip.
46 Insert a dummy shaft through the countergear, and stick the roller bearings (27 at each end), needle retainer washers and the transmission case thrust washers, in position using thick grease. Note that the tangs on the thrust washers are away from the gear faces. **Note:** *If no dummy shaft is available, carefully stick the roller bearings in place, but when installing the shaft, take care that they are not dislodged.*
47 Install Reverse idler gear and shaft with Woodruff key from the rear of the transmission case. Do not install the idler shaft E-ring at this time.
48 Install the countergear assembly from the rear of the transmission case and then insert the countershaft so that it picks up the roller bearings and the thrust washers, at the same time displacing the dummy shaft or tool (if used). The countershaft should be inserted so that its slot is at its rear end when installed.
49 Expand the snap-ring in the rear extension housing and locate the housing over the rear end of the mainshaft and onto the rear bearings. Make sure that the snap-ring seats in the rear bearing groove.
50 Insert the mainshaft pilot bearings (14 of them) into the clutch gear cavity and then assemble the 3rd speed blocker ring onto the clutch drive gear.
51 Locate the clutch drive gear, pilot bearings and 3rd speed blocker ring over the front of the mainshaft. Do not install the drive gear bearing at this time. Also, make sure that the notches in the blocker ring align with the keys in the 2nd/3rd synchro unit.
52 Stick a new gasket (using grease) to the rear face of the transmission case and then, from the rear, insert the combined clutch drive gear mainshaft and rear extension. Make sure that the 2nd/3rd synchro sleeve is pushed all the way forward so that the clutch drive gear engages with the countergear anti-lash plate.
53 Install the rear extension-to-transmission case bolts. Tighten the bolts to the specified torque.
54 Install the outer snap-ring to the clutch drive gear bearing and install the bearing over the drive gear and into the front of the transmission case.
55 Install the clutch drive gear bearing shaft snap-ring.
56 Install the clutch drive gear bearing retainer and its gasket making sure that the oil return hole is at the bottom.
57 Now install the Reverse idler gear E-ring to the shaft.
58 With the synchronizer sleeves in the Neutral position, install the side cover, gasket and fork assembly (Steps 8 through 13). Tighten the bolts to the specified torque.
59 Install the speedometer driven gear in the rear extension housing.

10 Transmission (3-speed 77 mm) - disassembly, overhaul and reassembly

1 Shift the transmission into Neutral, then raise the vehicle and support it securely on jackstands.
2 Drain the oil and remove the transmission as described in Section 7 or 8.

Disassembly

3 Unbolt the access cover and remove it along with the gasket from the case.
4 Lift out the long spring that holds the detent plug in the case and, using a small magnet, remove the detent plug.
5 Unbolt the extension housing and remove it along with the gasket.
6 Apply pressure to the speedometer gear retainer and remove the speedometer drivegear retainer from the output shaft.
7 Remove the fill plug from the right side of the case. Working through the plug hole, use a 3/16-inch pin punch and drive out the countergear roll pin. Let the pin drop, it can be retrieved at a later time.
8 Insert dummy shaft tool into the hole at the front of the case. Lightly tap on the tool to push the countershaft out of the rear of the case. Allow the countergear to lie at the bottom of the case.
9 With a punch and hammer, put alignment marks in the front bearing retainer and transmission case to ensure correct reassembly. Remove the front bearing retainer and gasket.
10 Remove the snap-ring that holds the front bearing and the clutch gear shaft in place.
11 Using a puller and the necessary adapters, remove the clutch shaft front bearing.
12 Remove the snap-rings from the rear bearing and from the output shaft. It may be necessary to place a piece of barstock or a screwdriver between the 1st/Reverse sleeve and gear assembly and the transmission case. This will hold the output shaft assembly in place while removing the rear bearing.
13 Use a puller to remove the rear bearing from the output shaft.
14 Remove the set screw from the 1st/Reverse shift fork and push the shift rail out of the case.
15 Push the 1st/Reverse sleeve and gear all the way to the front and rotate the 1st/Reverse detent plug from the case.
16 Push the 2nd/3rd shift fork to the back to gain access to the set screw. Remove the set screw. Next, twist the shift rail 90-degrees with a pair of pliers to clear the bottom detent plug and, using a magnet, remove the inner lock plug.
17 Insert a 1/4-inch diameter punch through the access hole in the rear of the case to drive out the shift rail and expansion plug located in the shift rail bore at the front of the case.
18 Turn the 2nd/3rd shifter fork up and lift it out of the case.
19 Remove the bottom detent plug and its short detent spring from the case.
20 Disengage the clutch gear from the output shaft and remove the output shaft assembly. To do this, tilt the splined end of the shaft down and lift the gear end up and out of the case. At the right-rear end of the case there is a notch through which the 1st and Reverse sleeve and gear must pass.
21 Lift the clutch gear through the top of the case.
22 Remove both shifter fork shafts.
23 Remove the countergear, thrust washer and roll pin with the tool still in place.
24 Tap the end of the idler gear shaft with a hammer until the end with the roll pin clears the counter bore in the rear of the case. Remove the shaft, the Reverse idler gear and the thrust washer.
25 Remove the clutch shaft roller bearing or counter gear needle bearing or anything else that may have fallen into the case during disassembly.
26 The mainshaft should only be disassembled if a press or bearing puller is available. Be careful to keep all components separated in order to easily reassemble.
27 Remove the front output shaft snap-ring and lift out the 2nd/3rd synchronizer assembly and 2nd gear. Place alignment marks on the hub and sleeve for ease in reassembly.
28 After removing the snap-ring and tabbed thrust washer from the shaft, remove 1st gear and the blocking ring.
29 Make a note of the position of the springs and keys in the Reverse hub and place alignment marks on the hub and sleeve for correct reassembly.
30 Remove the 1st/Reverse hub retaining snap-ring then separate the sleeve gear spring and three keys from the hub.
31 Separate the hub from the output shaft using an arbor press.

Overhaul

32 Clean all components in solvent and dry thoroughly. Check the gears for wear or chipped teeth. Check the bearings for smoothness of operation inside the race. If there has been a history of noisy gear shifts or the synchro function could easily have been 'beaten' then replace the appropriate synchro unit.

33 If the bushing in the rear extension requires replacement, drive the rear seal out, then drive the bushing out using tubular drifts. Using a socket or the correct size drift, drive a new bushing in from the rear. Lubricate the inside diameter of the bushing and install the new rear seal using an appropriate tubular drift.

34 Clean the transmission case inside and out and check for cracks, particularly around the bolt holes.

35 Extract the clutch bearing retainer seal and drive in a new one.

Reassembly

36 Begin reassembling the transmission by first reassembling the mainshaft.

37 Install the 1st/Reverse synchronizer hub onto the output shaft splines. The slotted end of the hub should face the front of the shaft. Use an arbor press to complete the hub installation onto the shaft. In the rear groove, install the retaining snap-ring. **Note:** *A press must be used, do not try to drive the hub onto the shaft with a hammer.*

38 Slip the 1st/Reverse sleeve and gear half-way onto the hub with the gear end of the sleeve facing the rear of the shaft. Align the sleeve and the hub with the marks made during disassembly.

39 Place the spring in the 1st/Reverse hub making sure that the spring is bottomed in the hub and covers all three key slots. Place the three synchronizer keys in the hub with the small end of the key in the hub slot and the large end inside the hub. Push the keys all the way into the hub so that they seat on the spring, then slide the 1st/Reverse sleeve and gear over the keys until they engage in the synchronizer sleeve.

40 Install the 1st gear blocking ring to the tapered surface of the 2nd gear and install the 2nd gear onto the output shaft with the tapered surface facing the front of the shaft.

41 With the sharp edge facing out, install the tabbed thrust washer and retaining snap-ring onto the output shaft.

42 Attach the 2nd gear blocking ring to the tapered surface of the 2nd gear and install the 2nd gear onto the output shaft with the tapered surface facing the front of the shaft.

43 Install the 2nd/3rd synchronizer assembly to the output shaft. Make sure the flat portion of the synchronizer hub faces the rear. Turn the 2nd gear until the keys in the 2nd/3rd synchronizer assembly engage with the notches in the locking ring. To ease assembly, tap the synchronizer with a plastic hammer.

44 Install the retaining snap-ring on the output shaft and measure the endplay between the snap-ring and the 2nd/3rd synchronizer hub. If the endplay exceeds 0.014-inch, replace the thrust washer and all snap-rings on the output shaft assembly.

45 Smear petroleum jelly over the transmission case Reverse idler gear thrust washer and position the thrust washer in the case. Be sure to engage the thrust washer and position the thrust washer in the case. Be sure to engage the thrust washer locating tabs into the locating slots in the case.

46 With the helical cut gears toward the front of the case, install the Reverse idle gear. From the rear of the case, install the Reverse idler gear shaft, aligning the gear bore thrust washer in case bores. Align and seat the roll pin, in the shaft, into the counter bore in the rear of the case.

47 Measure the endplay between the Reverse idle gear and the thrust washer. If the play exceeds 0.018-inch, remove the idler gear and replace the thrust washer.

48 Apply a thick coat of heavy grease to the bore of the countergear and insert the dummy shaft, then load a row of 25 needle bearings into each end of the gear. Next, install one needle bearing retainer on each end of the gear to hold the needle bearings in place.

49 Cover the countergear thrust washer with petroleum jelly and position it in the case. Be sure to engage the locating tabs on the thrust washer into the locating slots in the case.

50 Slip the countershaft through the bore at the rear of the case just far enough to hold the rear thrust washer in position when the countergear is installed.

51 Insert the countershaft into the countergear, aligning the bore in the countergear with the countershaft and the front thrust washer. Before the countershaft is completely installed, make sure that the roll pin in the countershaft is aligned with the hole in the case. When the holes are aligned, remove the dummy shaft from the countergear and tap the countershaft into place.

52 Measure the countershaft endplay between the thrust washer and countergear with a feeler gauge. If the endplay exceeds 0.018-inch, remove the gear and replace the thrust washer.

53 When the correct endplay is obtained, install the roll pin in the case.

54 Place the shorter detent spring into its bore in the case. The spring should Install into place at the bottom of the 2nd/3rd shift rail bore. Next, install the lower detent plug in the detent bore, on top of the spring.

55 Install the shifter fork shafts in their case bores with the pivot lug facing up. **Note:** *Shifter fork shafts are interchangeable.*

56 Apply petroleum jelly or equivalent light grease to the clutch shaft bore and install the 15 roller bearings. Do not use heavy chassis grease as it could plug the lubricant holes.

57 Attach the blocking ring to the clutch gear and place the gear through the top of the case into position in the front case bore.

58 Place the 1st/Reverse sleeve and gear in its Neutral (centered) position on the hub. Install the output shaft assembly in the case. Be careful so the gear end of the sleeve will clear the notch in the top of the case.

59 Engage the output shaft to the clutch gear.

60 Slide the 2nd/3rd sleeve to the rear until it is in its 2nd gear position and place the 2nd/3rd shift fork in the groove of the sleeve. Be sure the shift fork set screw hole is facing up. **Note:** *The 2nd/3rd fork is the smaller of the two shift forks.*

61 Engage the 2nd/3rd shift fork in the shift fork shaft.

62 Slide the 2nd/3rd shift rail (with the tapered end facing the front of the case) through the front case bore and into the shifter fork.

63 Rotate the shift rail until the detent notches in the rail face down.

64 Depress the lower detent plug with a Phillips screwdriver and push the shift rail into the rear bore. Push the rail in until the detent plug engages the forward notch in the shift rail (the second gear position).

65 Insert the interlock plug into the detent bore. The top of the plug should be slightly below the surface of the 1st/Reverse shift rail bore.

66 Slide the 1st/Reverse synchronizer forward to the 1st gear position. Place the 1st/Reverse shifter fork (with the set screw hole facing up) in the groove of the sleeve. Engage the fork with the shifter fork shaft, then slide the 1st/Reverse shift rail through the rear case bore and shifter fork.

67 Rotate the shift rail until the detent notches in the rail face upward. Align the set screw bore in the shift rail with the set screw hole in the fork and secure with the set screw. Place the 1st/Reverse sleeve and gear into the Neutral (centered) position.

68 Install the large snap-ring on the front bearing.

69 Install the front bearing on the clutch gear shaft by hand. Drive the bearing onto the clutch shaft using an appropriate tubular drift.

70 Install the small snap-ring on the clutch gear shaft.

71 Place the bearing retainer gasket on the case, be sure the oil return hole in the case is not blocked.

72 Align the marks made during disassembly on the front bearing retainer and the transmission case and check that the oil return slot in the cap is aligned with the oil return hole in the case. Attach the bolts and tighten to the specified torque.

73 Install the large snap-ring on the rear bearing.

74 Install the rear bearing on the output shaft by hand. Drive the bearing onto the shaft and into the case with the appropriate size tubular drift. Make sure the snap-ring groove is facing the rear of the shaft.

75 Install the small snap-ring on the output shaft to hold the rear bearing in place.

76 In the hole provided in the output shaft,

engage the speedometer gear retainer. Slide the speedometer gear over the output shaft and into position with the retainer. Slide the speedometer gear over the output shaft and into position with the retainer plate facing forward.

77 Place the new extension housing gasket on the case and install the extension housing. Tighten the bolts to the specified torque.

78 Insert the expansion plug into the 2nd/3rd shift rail bore in the front of the case. When the plug is fully seated in its bore it should be approximately 1/16th-inch below the front face of the case.

79 Install the upper detent plug in the detent bore, then install the long detent spring on top of the plug.

80 Place a new gasket on the case and install the top cover.

Tighten the bolts to the specified torque.

81 Don't forget to fill the transmission with oil and tighten the fill plug.

11 Transmission (4-speed 89 mm) - disassembly, overhaul and reassembly

1 Remove the transmission from the vehicle as described in Section 7 or 8.

Disassembly

2 Disconnect the Reverse shift lever, remove the bolts and detach the side cover and shift forks. Be careful when you remove the cover from the transmission case as the Reverse detent springs and ball can fall out.

3 To gain access for the countershaft removal, remove the extension housing bolts and rotate the housing on the output shaft to expose the rear of the countertshaft. Reinstall one bolt, in the center hole on the right side, to hold the extension in an inverted position.

4 Make a hole in the countershaft expansion plug at the front of the case using a center punch or drill.

5 Using this hole, push the countershaft to the rear until the Woodruff key can be removed. After the key is removed, push the countershaft forward against the expansion plug. Using a brass drift, tap the countershaft until the plug is driven out of the case.

6 Drive the countershaft out the rear of the case using a dummy shaft. The tool will now hold the roller bearings in position within the gear bore. Lower the countershaft gear to the bottom of the case.

7 Move the extension housing back to its original position.

8 Remove the drivegear and bearing assembly through the front of the case by tapping with a brass drift.

9 Move the 3rd and Overdrive synchronizer sleeve forward, then slide the Reverse idler gear to the center of its shaft. Tap on the extension housing with a soft-faced hammer in a rearward direction. Pull the mainshaft assembly and the housing away from the case.

10 Lift the countershaft gear out of the bottom of the case.

11 To remove the Reverse idler gear shaft from the transmission, you'll need a 3/8 x 3-1/2 inch bolt with a nut and 3/8-inch drive 7/16-inch deep socket. Spin the nut on the bolt as far as possible, then insert the bolt through the rear of the deep socket. Place the bolt and socket in the case with the socket against the shaft and the head of the bolt against the case. Hold the bolt from turning and with a wrench, turn the nut against the socket. This will act as a press and remove the gear from its shaft. Remove the shaft and the Woodruff key from the transmission case.

12 Push the Reverse gearshift lever shaft in and remove it from the case. Remove the retainer and the O-ring from the case bore.

13 Remove the back-up light switch from the case.

14 To disassemble the mainshaft, a press or bearing puller will be required. If not available, your GM dealer is best equipped to carry out this operation.

15 Remove the snap-ring and slide the 3rd and Overdrive synchronizer clutch gear and sleeve assembly from the mainshaft.

16 Slide the Overdrive gear and stop ring off the mainshaft. Place alignment marks on the synchronizer parts to simplify reassembly.

17 Spread the mainshaft bearing snap-ring and remove the mainshaft from the extension housing.

18 Remove the speedometer drivegear from the mainshaft.

19 Remove the mainshaft bearing retaining snap-ring from the shaft. To remove the bearing from the mainshaft, insert a steel plate on the front side of 1st gear, then press the bearing off the mainshaft. Be careful not to damage the gear teeth.

20 Remove the bearing, bearing retainer ring, 1st gear and 1st speed stop ring from the shaft.

21 Place alignment marks on the 1st and 2nd gear synchro sleeve assembly then remove the snap-ring and slide the gear and the sleeve assembly from the mainshaft.

22 Inspect the mainshaft gear bearing surfaces for score marks and wear (or any other condition that would prevent reusing the shaft).

23 Remove the dummy shaft from the countershaft gear and carefully remove the 76 needle bearings, the thrust washers and the spacers.

24 Remove the outer snap-ring from the drivegear. If the bearing is to be replaced, use an arbor press to remove the bearings from the drivegear.

25 Remove the inner snap-ring and 16 bearing rollers from the cavity of the drivegear.

Overhaul

26 Clean all components in solvent and dry thoroughly. Check for wear or chipped teeth. If there has been a history of noisy gear shifts or the synchro function could easily be 'beaten' then replace the appropriate synchro unit.

27 Clean the transmission case inside and out and check for cracks, particularly around the bolt holes.

28 The synchronizer hubs and sliding sleeves are a matched set and should be kept together; but the keys and springs may be replaced if worn or broken. If not already marked, place alignment marks on the synchronizer hub and sleeve assembly.

29 Push the hub from the sliding sleeve and the keys will fall free. The spring can now be easily removed.

30 Place the keys in position while holding them in place, slide the sleeve onto the hub aligning the marks made before disassembly. Place the two springs in position, one on each side of the hub so all three keys are engaged by both springs.

31 Extract the oil seal from the rear of the extension housing.

32 Drive the bushing out of the housing using an appropriate size drift or socket.

33 Slide a new bushing into the rear of the extension housing and drive into place using the appropriate size drift or socket.

34 Place a new seal in the opening of the extension housing and drive it in with a tubular drift or socket.

35 Pry out the drivegear bearing retainer oil seal and replace it with a new one. drive it in with an appropriate size drift until it bottoms in the bore. **Note:** *Be sure to lubricate the inside diameter of the seal with transmission lubricant.*

36 Remove the nuts that connect the shift operating lever to the shaft. Check that the shafts have no burrs on them before removal of the shift lever. If they are free of burrs, disengage the lever from the flats on the shafts and remove them.

37 Remove the gearshift lever shafts from the side cover.

38 Pry off the E-ring from the interlock lever pivot pin and slide the interlock levers and spring from the cover.

39 If the interlock levers are cracked or worn, replace them. Slide the levers onto the pivot pin and fasten with the E-ring. Install the spring on the interlock lever hangers using pliers.

Reassembly

40 Apply grease to the bores and the housing and push each shaft into its proper bore. Next install the greased O-rings and retainers.

41 Install the operating levers (be sure 3rd/Overdrive operating lever points down) and tighten the retaining nut to the specified torque.

42 Grease the bore of the countergear and install a spacer and the dummy shaft. Center the spacer in the arbor.

43 Into each end of the gear install 19 roller bearings followed by a spacer ring and 19 more bearings and the final spacer ring.

44 Coat the countershaft thrust washers with grease and install one at the front of the countergear on the arbor with the tang side facing the case bore. Install the other washer after the countergear assembly is positioned in the bottom of the case.

45 With the outer snap-ring groove toward the front, press the drivegear bearing onto the drivegear. Be sure the bearing is seated against the shoulder of the gear.

46 Install an appropriate size snap-ring on the shaft to retain the bearing. This snap-ring is a select fit for minimum endplay.

47 Place the drivegear in a soft-jaw vise. Install the 16 bearing rollers in the cavity of the shaft. Coat the bearing rollers with grease and install the retaining snap-ring in the groove.

48 With the synchronizer cone towards the rear, slide the 2nd gear over the mainshaft and down against the shoulder on the shaft.

49 Assemble the stop-ring with the lugs aligned in the hub slots and slide the complete 1st/2nd synchronizer assembly over the mainshaft and down against the 2nd gear cone. Install a new snap-ring. Slip the next stop ring over the shaft and align the lugs into the clutch hub slots.

50 With the synchronizer cone facing the clutch sleeve gear that was just installed, place the first gear over the mainshaft and into position against the clutch sleeve gear.

51 Install the mainshaft bearing retainer ring, then slide the mainshaft bearing on. Press the bearing into position using an arbor and a suitable tool. Install an appropriate sized snap-ring on the shaft to secure the bearing. The snap-ring is a select fit for minimum endplay.

52 Slide the mainshaft into the extension housing until the bearing retaining ring engages with the slot in the extension housing. Spread the snap-ring with pliers so that the mainshaft ball bearing can move in and bottom against the thrust washer in the extension housing. Release the ring and seat it all the way around in the groove in the extension housing.

53 With the synchronizer cone facing the front, slide the Overdrive gear over the mainshaft followed by the Overdrive gear stop ring.

54 With the shift fork slot towards the rear, install the 3rd/Overdrive gear over the mainshaft followed by the Overdrive gear stop ring.

55 Apply grease to the front stop-ring and position it over the clutch gear aligning the ring lugs with the struts.

56 With the shift cover opening towards you, place the transmission case on its side.

57 Slide the countergear assembly into the case and align the tangs on the front washer with the slots in the case. Install the rear washer aligning the tangs with the slots at the rear of the case. Set the countergear in the bottom of the case, be sure the thrust washers stay in position.

58 Position a new extension gasket, coated with grease, into position on the extension.

59 Guide the mainshaft assembly into the case, being careful not to disturb the countershaft gear.

60 Rotate the extension housing until the rear of the countershaft bore is exposed. Install one bolt to hold the extension in an inverted position and prevent it from moving to the rear.

61 Insert the drivegear assembly through the front of the case and position it in the front bore. Install the outer snap-ring into the bearing groove. With everything in its proper position, tap the gear assembly lightly with a soft-faced hammer until the outer snap-ring bottoms against the case face. If it does not go into place easily, check to see if a strut, roller bearing or a stop-ring is out of position.

62 Lift the countergear assembly into position with the teeth engaged with the drivegear. Be sure the thrust washers stay in position on the ends of the arbor and that the tangs are aligned with the slots in the case.

63 Place the countershaft into the rear bore of the case, then push forward until the shaft is approximately half-way through the gear. At this time install the Woodruff key and push the shaft forward until the end is flush with the case. Remove the dummy shaft.

64 Place the Reverse shift lever shaft in the case bore, then insert the greased O-ring and retainer.

65 Remove the extension housing bolt and rotate the housing to provide clearance for installation of the Reverse idler gear in the end of the case.

66 Place the shaft in far enough to position the Reverse idler gear onto the protruding end of the shaft with the fork slot toward the rear. Simultaneously engage the slot with the Reverse shift fork.

67 Place the Woodruff key on the shaft and drive the shaft in flush with the end of the case.

68 Align the extension housing and the case and install the bolts. Tighten them to the specified torque.

69 Place the drivegear bearing retainer and gasket in position, put sealing compound on the threads of the bolts and install them. Tighten them to the specified torque.

70 Coat a new expansion plug with sealing compound and install it into the countershaft bore at the front of the case.

71 Put both synchronizer sleeves into the Neutral position and place the 1st/2nd shift fork into the groove of the 1st/2nd synchronizer sleeve; then push the Reverse idler gear to the Neutral position.

72 Turn each shift lever to the Neutral position (straight up) and install the 3rd/Overdrive shift fork into the bore under both interlock levers.

73 Apply grease to the side cover gasket and position it on the case. Next, install the Reverse detent ball, followed by the spring, into the bore in the case.

74 Begin to lower the side cover onto the case, guiding the 3rd/Overdrive shift fork into its synchronizer groove, then direct the shaft of the 1st/2nd shift fork into its bore in the side cover. Hold the Reverse interlock link against the 1st/2nd shift lever to provide clearance as the side cover is lowered into position. Use a screwdriver and raise the interlock lever, against spring tension, to allow the 1st/2nd shift fork to slip under the levers. Be sure the Reverse detent spring is positioned in the cover bore.

75 One of the eight shoulder bolts used to secure the side cover has a slightly longer shoulder which will act as a dowel to accurately locate the side cover. The remaining two bolts are standard bolts. Install the cover bolts finger tight and shift through all the gears to ensure proper operation.

76 Tighten the side cover bolts evenly to the specified torque.

77 Install the Reverse shift lever retaining nut and torque to specifications. Shift the transmission into each gear to ensure correct shift travel and smooth operation. Slight motion of the 1st/2nd shift lever toward Low gear is normal when shifting into Reverse gear.

78 The Reverse shift lever and 1st/2nd shift lever have cam surfaces which mate in Reverse position to lock the 1st/2nd lever, fork and synchronizer into the Neutral position.

79 Install the back-up light switch and tighten it to the specified torque.

12 Transmission (4-speed 117 mm) - disassembly, overhaul and reassembly

Transmission - disassembly

1 Disassemble the transmission on a clean workbench.

2 Remove the bolts attaching the transmission cover assembly to the transmission case. Remove the transmission cover assembly. **Note:** *Move the Reverse shift fork so the Reverse idler gear is partially engaged before attempting to remove the cover. The shift forks must be positioned so the rear edge of the slot in the Reverse fork is in line with the front edge of the slot in the other forks when viewed through the turret opening.*

3 Place the transmission in two gears at once in order to lock the gears. Loosen and remove the output yoke flange nut, then remove the yoke.

4 On models equipped with 4-wheel drive, use an appropriate tool to remove the mainshaft rear locknut.

5 Remove the bolts attaching the rear bearing retainer and gasket.

6 Slide the speedometer drivegear off the mainshaft.

7 At the front end of the transmission, loosen and remove the bolts retaining the drivegear bearing retainer. Remove the retainer and gasket.

8 Remove the countergear front cover screws and pry off the cover and gasket.

9 Using two large screwdrivers, carefully

pry the countergear front bearing out. Insert the screwdrivers into the groove in the cast slots in the transmission case.

10 Using snap-ring pliers, remove the countergear rear bearing snap-ring from the shaft and bearing.

11 Using a puller, remove the countergear rear bearing. Once the countergear rear bearing has been removed, the countergear assembly may rest at the bottom of the transmission case.

12 Remove the snap-ring retaining the drivegear bearing outer race to transmission case.

13 Using a brass drift, gently tap the drivegear shaft and bearing from the case. **Note:** *Align the cut-out section of the gear facing down to obtain clearance for removing the drivegear.*

14 Remove the 3rd/4th synchronizer ring.

15 Remove the rear mainshaft bearing snap-ring. Using a puller, remove the rear mainshaft bearing from the case. Slide the 1st gear thrust washer off the mainshaft.

16 Lift the rear end of the mainshaft assembly up and push it to the rear in the case bore, then swing the front end of the shaft up. Remove the mainshaft assembly from the case, then remove the 3rd/4th synchronizer cone from the shaft.

17 Slide the Reverse idler gear to the rear and move the countergear back until the front end is free of the case. In this position the countergear can be lifted from the case.

18 If necessary, the Reverse idler gear can be removed by driving the gear shaft out of the case. With the shaft removed, lift out the Reverse idler gear.

19 The transmission is now broken down into major sub-assemblies. The following Paragraphs describe how to disassemble and reassemble the various sub-assemblies.

Transmission cover

Disassembly

20 Using a small punch, drive out the pins retaining the 1st/2nd and the 3rd/4th shifter forks to the shifter shafts in the cover. Also drive out the three shifter shaft bore plugs. **Note:** *The pin retaining the 3rd/4th shifter fork to the shaft must be removed, and the shifter fork removed from the cover, before the Reverse shifter pin can be removed.*

21 Position all the shifter shafts in the Neutral position and drive the 1st/2nd and 3rd/4th shafts out of the cover and the shifter forks. **Note:** *Exercise care when removing the shafts, so that the shaft detent balls and springs and the interlock pin located in the cover are not lost as the shifter shafts are removed.*

22 Lift out the 1st/2nd and 3rd/4th shifter forks.

23 Drive out the pin retaining the reverse shifter fork to the shaft, then drive out the shifter shaft and remove the shifter fork.

Overhaul

24 Thoroughly clean all the parts with solvent and dry them with a clean lint-free cloth.

25 Inspect all parts for signs of wear or damage. Replace all damaged or worn components with new ones as required.

Reassembly

26 The shifter shafts are installed first in a prescribed sequence - Reverse first, followed by 3rd/4th, then 1st/2nd.

27 With the sequence mentioned in Paragraph 26 in mind, place the shifter fork detent springs and ball in position in their correct holes in the cover. Start the shifter shafts into the cover, depress the appropriate detent ball with a punch and push the shifter shaft into the cover. Starting with the Reverse shifter shaft, hold the fork in position and push the shaft through the bore in the fork. Align the retaining pin holes in the fork bore and the shifter shaft, then install the retaining pin. Position the fork in Neutral, then install the 3rd/4th and 1st/2nd shifter forks.

28 After the 1st/2nd shifter fork is installed, place two interlock balls between the 1st/2nd shifter shaft and the 3rd/4th shifter shaft in the crossbore of the front support boss. Install the interlock pin in the 3rd/4th (center position) shifter shaft hole and grease it to hold it in place. Continue pushing this shifter shaft through the cover bore and shifter fork until the holes in the shaft and fork align, then install a retaining pin.
Move the shaft to the Neutral position.

29 Place two interlock balls in the crossbore in the front support boss between reverse and 3rd/4th shifter shafts. Push the remaining shaft through the fork and cover bore, keeping both balls in position between the shafts, until the retaining pin holes in the fork and shaft align. Install the retaining pin.

30 Install new shifter shaft bore plugs and expand them in position.

Drivegear and shaft

Disassembly

31 If not already done, remove the mainshaft pilot bearing rollers (17) from the drivegear. Do not remove the snap-ring on the inside of the gear.

32 Remove the snap-ring securing the bearing on the stem of the gear.

33 To remove the bearing, use a special support fixture to support the bearing, then, using a press and the appropriate adapters, press the gear and shaft out of the bearing. If these tools are not available, take the assembly to a machine shop to have the bearing removed.

Overhaul

34 Wash all the parts thoroughly with solvent and dry them with a clean lint-free cloth.

35 Inspect the roller bearings for pitting and obvious wear. Inspect the bearing surface on the shaft for score marks and check the gear teeth for damage and excessive wear.

Reassembly

36 Press the bearing, with a new oil slinger, onto the shaft. **Note:** *The oil slinger should be located flush with the bearing shoulder on the gear. If the tools are not available, take the assembly to a dealer or machine shop to have the components pressed onto the shaft.*

37 Install the snap-ring to secure the bearing on the shaft.

38 Install the bearing snap-ring into the groove in the outside of the bearing.

39 Install a snap-ring in the inside of the mainshaft pilot bearing bore (if previously removed).

40 Apply a small amount of grease to the bearing surface in the shaft recess, then install the transmission mainshaft pilot roller bearings (17) and install the roller bearing retainer. **Note:** *This roller bearing retainer holds the bearing in position, and in final transmission assembly is pushed forward into the recess by the mainshaft.*

41 If not already done, remove the bolts attaching the retainer and drivegear oil seal assembly.

42 Using a screwdriver, pry out the oil seal from the retainer.

43 Using a large socket, place the new seal in position (with the lip facing up) and carefully drive it into the seal bore.

44 Install a new gasket on the retainer flange, then, when assembling the transmission, attach the assembly to the transmission case.

Mainshaft assembly

Disassembly

45 Remove 1st gear.

46 Remove the snap-ring in front of the 3rd/4th synchronizer assembly.

47 Remove the Reverse driven gear.

48 Using a press, apply force behind 2nd gear to remove the 3rd/4th synchronizer assembly, 3rd gear and 2nd gear, along with the 3rd gear bushing and thrust washer. If no press is available, take the assembly to a dealer or machine shop to have the components pressed off.

49 Remove the 2nd gear synchronizer ring and the synchronizer keys.

50 Supporting the 2nd gear synchronizer hub at its front face, press the mainshaft through, removing the 1st gear bushing and 2nd gear synchronizer hub.

51 In order to remove the 2nd gear bushing from the mainshaft, a hammer and chisel must be used to split the bushing. **Note:** *Be very careful not to nick or otherwise damage the shaft as this is done.*

Overhaul

52 Thoroughly clean all the parts with solvent and dry them with compressed air or a clean lint-free cloth.

53 The mainshaft should be inspected for evidence of scoring or excessive wear, especially at the thrust or spline surfaces. Closely examine the clutch hub and clutch sleeve for excessive wear and make sure that the sleeve slides freely on the clutch hub. Also check the fit of the clutch hub on the main-

shaft splines. **Note:** *The 3rd/4th gear clutch sleeve should slide freely on the clutch hub, but the hub should be a snug fit on the shaft splines.*
54 Inspect the 3rd gear thrust surfaces for excessive scoring and inspect the 3rd gear mainshaft bushing for excessive wear. The 3rd gear must be able to turn freely on the mainshaft bushing, but the bushing should be a press fit on the shaft.
55 Check the 2nd gear thrust washer for any signs of scoring and inspect the 2nd gear synchronizing ring for excessive wear at the thrust surface. Check the synchronizer springs for looseness or breakage.
56 Inspect the 2nd gear synchronizing ring for excessive wear and check the bronze synchronizer cone on 2nd gear for wear or damage. Closely examine the clutch gear synchronizer cone and 3rd gear synchronizer cone for wear or damage.
57 The 1st/reverse sliding gear must be a sliding fit on the synchronizer hub and should not have excessive radial or circumferential play. If the sliding gear is not free on the hub, inspect the front end of the half tooth internal splines for burrs. Burrs can be removed with an oilstone or very fine file.

Reassembly

58 Take the assembly to a dealer or machine shop and have the 2nd gear bushing, 1st/2nd gear synchronizer hub, 1st gear bushing, the synchronizer ring, 2nd gear, the 3rd gear bushing and the 3rd/4th synchronizer assembly installed on the mainshaft.
59 Install the Reverse driven gear with the fork groove facing the rear.
60 Install 1st gear on the mainshaft, against the 1st/2nd synchronizer hub. Install the 1st gear thrust washer.

Countershaft

Disassembly

61 Remove the front countergear snap-ring and lift off the thrust washer.
62 Install support blocks on the countershaft, open side to the damper. Support the assembly in a press and press the countershaft out of the clutch countergear assembly. If a press is not available, have this operation performed by a dealer or machine shop.
63 Press the shaft from the 3rd gear countergear.

Overhaul

64 Thoroughly clean all components and inspect them for any signs of damage or wear, replacing as necessary.

Reassembly

65 Take the countergear assembly to a dealer or machine shop to have the 3rd gear countergear, the clutch countergear and the clutch countergear thrust washer assemblies installed on the shaft.

Transmission - reassembly

66 Lower the countergear assembly into the transmission case and allow it to rest on the bottom of the case.
67 Place the Reverse idler gear in the transmission case with the gear teeth towards the front. Install the Reverse idler shaft (from the rear to the front), being careful to position the slot in the end of the shaft facing down. The shaft slot face must be at least flush with the transmission case.
68 Install the mainshaft assembly in the case with the rear of the shaft protruding from the rear bearing bore.
69 Position an adapter in the drivegear opening and engage the front of the mainshaft (the 1st gear thrust washer should be installed before doing this, if not already in position). Now install the snap-ring in the rear mainshaft bearing outer race and position the bearing on the shaft. Using the appropriate bearing installer or a piece of pipe, drive the bearing onto the shaft and into the transmission case bore. Align the tangs on the snap-ring with the lubrication slots in the case before driving the bearing.
70 Install the synchronizer cone on the pilot end of the mainshaft and slide it up to the clutch hub. Make sure the three cut-out sections of the 4th gear synchronizer cone align with the three clutch keys in the clutch assembly.
71 Install the snap-ring in the drivegear bearing outer race.
Index the cut-out portion of the gear teeth to obtain clearance over the countershaft gear teeth, and install the drivegear assembly in the transmission case. Raise the mainshaft to get the drivegear started and tap the bearing outer race into position.
72 Install the retainer and a new gasket. Install the attaching bolts and tighten them to the specified torque.
73 Install an adapter in the countergear front bearing bore, to support the countergear, then rotate the transmission case onto the front end.
74 Install the snap-ring in the countergear rear bearing outer race, position the bearing on the countergear and, using an appropriate driver or a piece of pipe, drive the bearing into place. Rotate the case, install the snap-ring in the countershaft rear bearing then remove the tool.
75 Tap the countergear front bearing assembly into the case.
76 Install the countergear front bearing cap (with a new gasket), install the bolts and tighten them to the specified torque.
77 Slide the speedometer drivegear over the mainshaft to the bearing.
78 Install the rear bearing retainer (with a new gasket). Be sure that the snap-ring ends are in the lubrication slot and the cut-out in the bearing retainer. Install the retainer attaching bolts and tighten them to the specified torque. On models equipped with 4-wheel drive, install the rear locknut and washer and tighten the locknut to the specified torque.
79 Install the universal joint flange.
80 Lock the transmission in two gears at once. Install the universal joint flange locknut and tighten it to the specified torque.
81 Move all the transmission gears to the Neutral position, except for the Reverse idler gear, which should be engaged approximately 3/8-inch (leading edge of the Reverse idler gear taper lines up with the front edge of 1st gear). Install the transmission cover assembly, with a new gasket. The shift forks must slide into their respective positions on the synchro clutch sleeves and the Reverse idler gear.
82 Install the transmission cover mounting bolts and tighten them to the specified torque.
83 Install the gearshift lever (Section 4) and check for satisfactory operation and selection of all the gear positions.

Chapter 7 Part B
Automatic transmission

Contents

	Section
Automatic transmission fluid and filter change	See Chapter 1
Automatic transmission fluid level check	See Chapter 1
Automatic transmission rear seal - replacement	10
Automatic transmission - removal and installation	11
Detent cable - adjustment	4
Detent switch - adjustment	5
Diagnosis - general	2
General information	1
Neutral start switch - adjustment	8
Shift linkage - check and adjustment	3
Speedometer gear - replacement	9
Throttle valve (TV) cable - check and adjustment	6
Throttle valve (TV) cable - replacement	7

Specifications

General

Transmission fluid type	See Chapter 1

Torque specifications

	Ft-lbs
Converter-to-driveplate bolts	35
Oil pan bolts	10
Transmission-to-engine bolts	35
Speedometer retainer bolt	6
Filter-to-valve body screw	10

1 General information

Since these models were introduced, fully automatic 3-speed, and on later models, 4-speed transmissions have been an option. An electronically-controlled transmission, the 4L80-E, was introduced in 1991. Diagnosis and repair of the 4L80-E must be done by a dealer service department.

Some later models are equipped with a lock-up torque converter incorporating a clutch. The clutch is designed to engage at speeds above 25 mph and provides a direct connection between the engine and the drive wheels for better efficiency and fuel economy.

Due to the complexity of the clutches and the hydraulic control system, and because of the special tools and expertise required to perform an automatic transmission overhaul, it should not be attempted by the home mechanic. Therefore, the procedures in this Chapter are limited to general diagnosis, routine maintenance, adjustment and transmission removal and installation.

If the transmission requires major repair work, it should be left to a dealer service department or an automotive or transmission repair shop. You can, however, remove and install the transmission yourself and save the expense, even if the repair work is done by a transmission specialist.

Adjustments that the home mechanic can perform include those involving the detent cable and switch, the throttle valve (TV) cable and the shift linkage.

Caution: *Never tow a disabled vehicle with an automatic transmission at speeds greater than 30 mph or distances over 50 miles.*

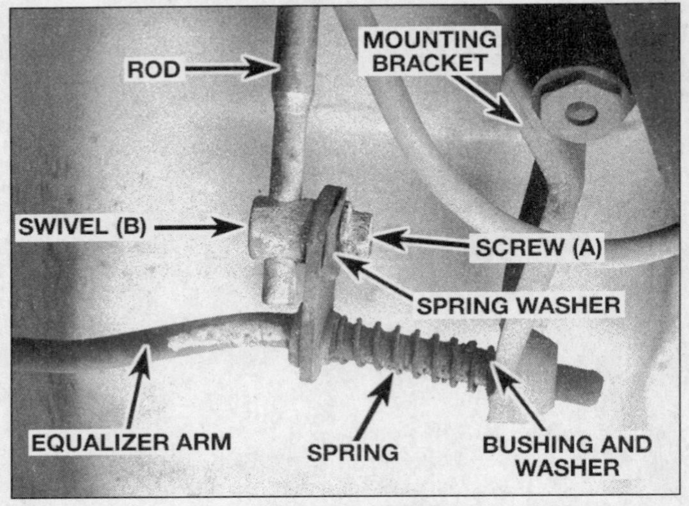

3.5 Shift linkage adjustment details (1967 through 1981 models)

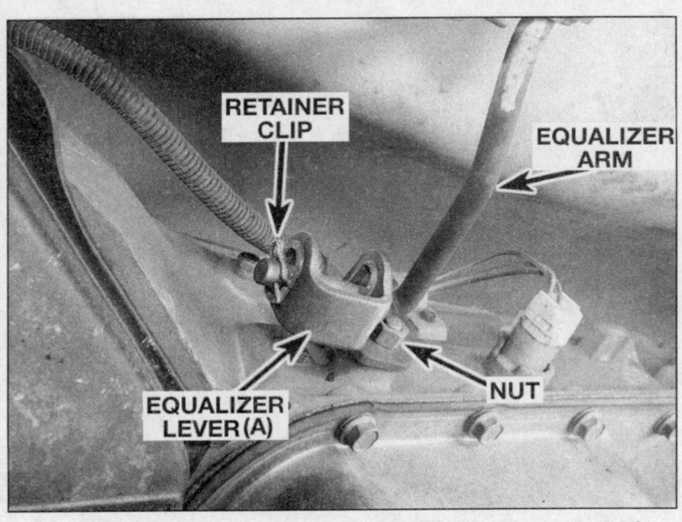

3.8 Shift linkage adjustment details (1982 and later models)

2 Diagnosis - general

1 Automatic transmission malfunctions may be caused by a number of conditions, such as poor engine performance, improper adjustments, hydraulic malfunctions and mechanical problems.

2 The first check should be of the transmission fluid level and condition. Refer to Chapter 1 for more information. Unless the fluid and filter have recently been changed, drain the fluid and replace the filter (also in Chapter 1).

3 Road test the vehicle and drive in all gears, noting discrepancies in operation.

4 Verify that the engine is not at fault. If the engine has not received a tune-up recently, refer to Chapter 1 and make sure all engine components are functioning properly.

5 Check the adjustment of the detent cable or switch and the throttle valve (TV) cable (depending on model).

6 Check the condition of all vacuum and electrical lines and fittings at the transmission, or leading to it.

7 Check for proper adjustment of the shift linkage (Section 3).

8 If at this point a problem remains, there is one final check before the transmission is removed for overhaul. The vehicle should be taken to a shop that can check the line pressure.

3 Shift linkage - check and adjustment

Refer to illustrations 3.5 and 3.8

Check

1 To check for correct linkage adjustment, pull the shift lever towards the steering wheel and select Drive by the action of the transmission detent. Do not be guided by the indicator needle, as it may be out of adjustment.

2 Release the shift lever and make sure that Low cannot be selected unless the lever is pulled.

3 Finally, pull the shift lever towards the steering wheel and let the action of the transmission detent set the lever in Neutral.

4 Release the lever and make sure that Reverse cannot be selected unless the lever is pulled. When properly adjusted, the linkage will prevent the shift lever from moving beyond the Neutral and Drive detents, unless the lever is first pulled to pass over the mechanical stop.

Adjustment

1967 through 1981 models

5 Remove the screw (A) and lock washer from the swivel (B) **(see illustration)**.

6 Set the transmission lever in the Neutral position by moving it counterclockwise to the L1 detent and then clockwise three detents (clicks) to Neutral.

7 Set the shift lever in Neutral, then attach the swivel, lock washer and screw to the lever assembly. Tighten the screw.

1982 and later models

8 Place the transmission equalizer lever (A) **(see illustration)** in the Neutral position by moving it clockwise to the Park detent and then counterclockwise two detents to the Neutral position.

9 Place the column shift lever in the Neutral gate. This is accomplished by rotating the lever until the shift lever drops into the Neutral gate notch (do not use the shift indicator pointer as a guide).

10 Connect the rod to the lever assembly.

11 Slide the swivel and clamp onto the rod and connect it.

12 Hold the column lever against the Neutral stop on the Park position side and tighten the nut securely.

All models

13 If necessary, adjust the indicator needle and the Neutral start switch.

4 Detent cable - adjustment

Refer to illustration 4.6

1 Remove the air cleaner assembly.

Early models

2 Locate and loosen the detent cable clamp screw.

3 With the choke off, position the carburetor lever in the wide open throttle position.

4 Pull the detent cable to the rear, until the wide open throttle stop in the transmission is felt. **Note:** *The cable must be pulled through the detent position to reach the wide open throttle stop.*

5 Tighten the detent cable clamp screw and check the linkage for proper operation.

Later models

6 On later models that have a snap-lock fitting rather than a cable clamp, make sure the adjuster tab is disengaged **(see illustration)** and the cable is free to slide through the fitting.

7 With the cable and fitting in the bracket and the cable end attached to the throttle lever, open the throttle valve all the way.

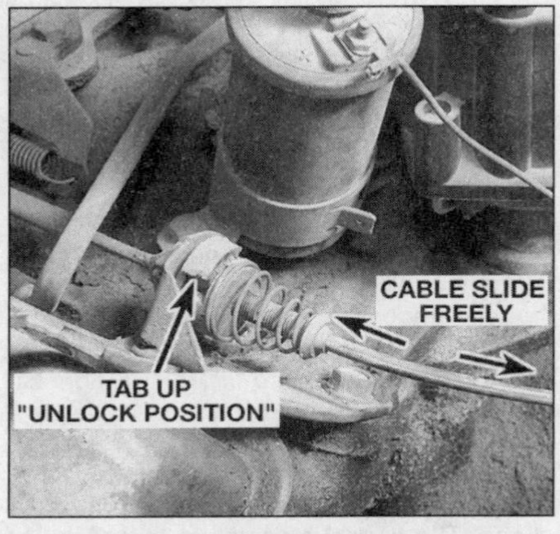

4.6 Typical snap-lock detent cable

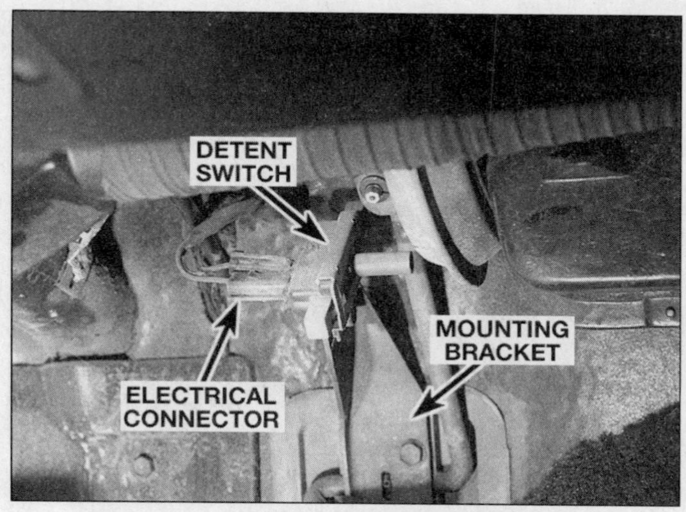

5.4 Later model detent switch adjustment details

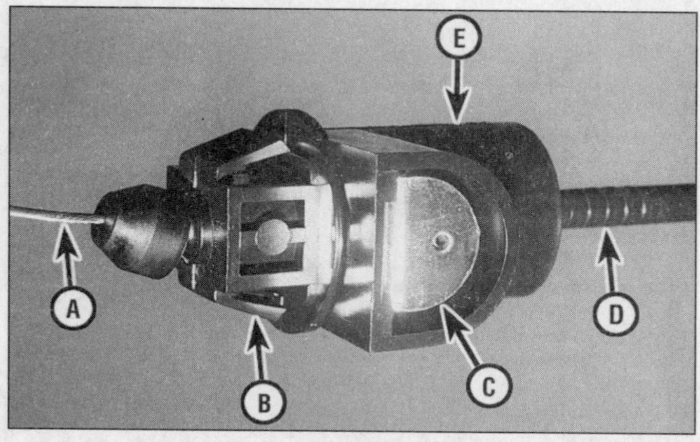

6.8 On 1981 and later models, depress the throttle valve (TV) cable release tab and pull the slider back until it rests on its stop - release the tab and open the throttle completely

A TV cable
B Locking lugs
C Release tab
D Cable casing
E Slider

8 Hold the throttle valve in the wide open position and push the snap-lock fitting tang down until it's flush with the rest of the fitting.

5 Detent switch - adjustment

Refer to illustration 5.4

1967 through 1972 models

1 Loosen the switch mounting bolts and open the throttle valve all the way (wide open throttle position).

2 Now, maintaining the wide open throttle position, move the switch until a gap of 0.200-inch (350 V8 engine) or 0.050-inch (all other engines) exists between the carburetor lever and the switch plunger.

3 When the correct gap has been obtained, tighten the switch mounting bolts.

1973 and later models

4 After installation of a new switch, press the switch plunger in all the way. The switch will adjust itself the first time the throttle is opened all the way with the accelerator pedal **(see illustration)**.

6 Throttle valve (TV) cable - check and adjustment

Refer to illustration 6.8

Check

1 The throttle valve cable used on later model transmissions should not be thought of as merely a "downshift" cable, as in earlier years. The TV cable controls line pressure, shift points, shift feel, part throttle downshifts and detent downshifts.

2 If the TV cable is broken, sticky, out of adjustment or the incorrect part for the vehicle, problems will result.

3 The check should be made with the engine running at idle speed with the selector lever in Neutral. Set the parking brake firmly and block the wheels to prevent any vehicle movement. As an added precaution, have an assistant in the driver's seat applying the brake.

4 Grab the inner cable a few inches behind where it attaches to the throttle linkage and pull the cable forward. It should easily slide through the cable terminal which connects to the throttle linkage.

5 Release the cable and it should return to its original location with the cable stop against the cable terminal.

6 If the TV cable does not operate as described, the cause is a defective or misadjusted cable or damage to the components at either end of the cable.

Adjustment

7 The engine should not be running during this adjustment.

8 Depress the readjust tab and move the slider back through the fitting away from the throttle linkage until the slider stops against the fitting **(see illustration)**.

9 Release the readjust tab.

10 Manually turn the throttle lever to the wide open throttle position, which will automatically adjust the cable. Release the throttle lever. **Note:** *Do not use excessive force at the throttle lever to adjust the TV cable. If great effort is required to adjust the cable, disconnect the cable at the transmission end and check for free operation. If it is still difficult, replace the cable. If it is now free, suspect a bent TV link in the transmission or a problem with the throttle lever.*

11 After adjustment, check for proper operation as described in Steps 3 through 6 above.

7 Throttle valve (TV) cable - replacement

Refer to illustrations 7.4 and 7.6

1 Disconnect the negative cable at the battery. Place the cable out of the way so it cannot accidentally come into contact with the terminal, which would once again allow current flow.

2 Remove the air cleaner assembly.

3 Depress the readjust tab on the TV cable and push the slider through the fitting, away from the throttle lever **(see illustration 6.8)**.

4 Separate the TV cable terminal from the throttle lever **(see illustration)**.

7.4 To detach the throttle valve cable from the carburetor linkage, pull the cable forward and pass the post through the large hole (arrow)

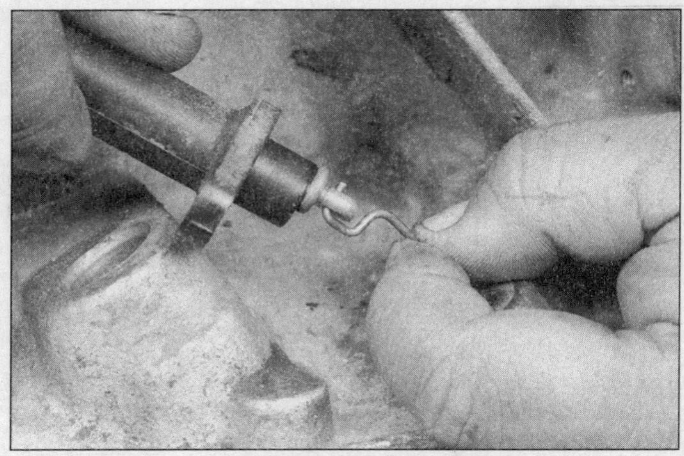

7.6 Pull up on the TV cable and disengage it from the link

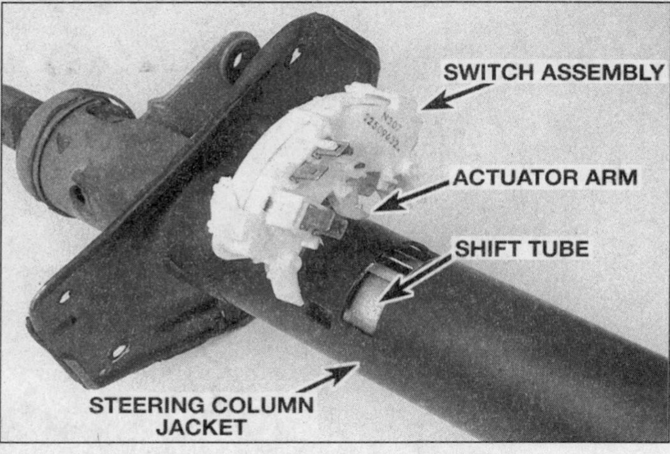

8.3 Neutral start switch details

5 Squeeze the locking tabs and separate the cable assembly from the bracket.
6 Remove the bolt that holds the TV cable to the transmission. Pull up on the cable and disconnect it from the link. Remove any routing clips or straps that may hold the TV cable in place **(see illustration)**.
7 Replace the seal in the transmission case hole. Connect the new cable to the link and bolt the cable to the transmission case.
8 Route the cable and reinstall any straps or clips used to secure it.
9 Guide the cable through the bracket until the locking tabs engage.
10 Connect the cable terminal to the throttle lever.
11 Adjust the cable by pushing in on the readjust tab. Push the slider through the fitting away from the throttle body until the slider stops against the fitting.
12 Release the readjust tab. Open the throttle lever as far as possible to automatically adjust the cable.
13 Check the cable for binding or sticking.

8 Neutral start switch - adjustment

Refer to illustration 8.3

1 Place the shift lever in the Neutral position.
2 Loosen the screws attaching the switch to the steering column.
3 Using a 0.098-inch diameter gauge pin, rotate the switch alternately clockwise and counterclockwise until the pin can be inserted into the switch and the steering column (to a depth of 3/8-inch) **(see illustration)**. A 3/32-inch drill bit should work as a gauge pin.
4 While holding the switch in this position, tighten the switch mounting screws.
5 When a satisfactory adjustment has been carried out, remove the gauge pin.

9 Speedometer gear - replacement

1 This operation is carried out without removing the transmission from the vehicle.
2 The procedure is identical for both manual and automatic transmissions - see Chapter 7, Part A.

10 Automatic transmission rear seal - replacement

1 Raise the vehicle and support it safely on jackstands.
2 Support the transmission with a jack.
3 Remove the driveshaft (see Chapter 8).
4 On 4-wheel drive models, remove the transfer case.
5 Using a long screwdriver, pry the old seal out of the end of the transmission.
6 Compare the new seal with the old one to make sure they are the same.
7 Drive the new seal into position using a large socket or a piece of pipe which is the same diameter as the seal.
8 Once in position, coat the lips of the new seal with automatic transmission fluid.
9 Reinstall the various components in the reverse order of removal, referring to the necessary Chapters as needed.

11 Automatic transmission - removal and installation

Refer to illustration 11.9

4-wheel drive models

1 Disconnect the negative battery cable from the battery.
2 Raise the vehicle and support it safely on jackstands.
3 Remove the transfer case shift lever and rod (see Chapter 7C).
4 Disconnect the vacuum modulator line and the speedometer drive cable at the transmission and fasten them out of the way.
5 Disconnect the manual control lever rod and the detent cable from the transmission.
6 Disconnect the front and rear driveshafts at the transfer case (see Chapter 8).
7 Support the transmission with a jack.
8 If equipped, remove the strut-to-transmission bolt.
9 Remove the transmission-to-adapter case bolts **(see illustration)** and position a jack or jackstand under the transfer case.
10 Remove the transfer case-to-frame bracket bolts and remove the transfer case.
11 On V8 engine equipped vehicles, remove the exhaust system crossover pipe.
12 Remove the bolts securing the rear crossmember in position and lift it away from the transmission.
13 Remove the torque converter underpan and mark the driveplate-to-torque converter relationship for reinstallation in the same position, then remove the driveplate-to-converter bolts.
14 Support the engine at the oil pan rail with a jack.
15 Lower the transmission until the upper transmission housing-to-engine bolts can be reached using a universal socket with a long extension bar. Remove the upper bolts.
16 Remove the remainder of the transmission housing-to-engine bolts.
17 To remove the transmission, move it slightly to the rear and down. Have an assistant to assist in this operation. Lower the transmission to the ground and remove it from beneath the vehicle. **Note:** *Watch the converter when moving the transmission back. If it does not move with the transmission, pry it off the driveplate before proceeding. When removing the transmission, keep the front end up to prevent the converter from falling out.*
18 Installation is the reverse of removal. Tighten all bolts to the specified torque.

2-wheel drive models

19 Disconnect the negative battery cable from the battery.
20 Raise the vehicle and support it securely on jackstands.
21 Disconnect the vacuum modulator line and the speedometer drive cable at the transmission. Tie this line and cable up out of the way.

11.9 Remove the transmission-to-transfer case adapter mounting bolts

22 Disconnect the manual control lever rod and the detent cable from the transmission.
23 Refer to Chapter 8 and disconnect the driveshaft.
24 Support the weight of the transmission with a jack.
25 Disconnect the rear engine mount at the transmission extension, then remove the transmission support crossmember.
26 Remove the torque converter cover, mark the driveplate-to-converter relationship to ensure correct realignment during installation, then remove the bolts.
27 Support the engine at the oil pan rail with a jack or jackstand.
28 Slightly lower the rear of the transmission, so the upper transmission housing-to-engine bolts can be reached with a universal socket and a long extension bar. Remove the upper bolts.
29 Remove the remainder of the transmission housing-to-engine bolts.
30 With the help of an assistant, remove the transmission by moving it to the rear and down. Lower the transmission to the ground and remove it from under the vehicle. **Note:** *Watch the converter when moving the transmission back. If it doesn't move with the transmission, pry it off the driveplate before proceeding. When removing the transmission, keep the front end up to prevent the converter from falling out.*
31 Installation is the reverse of removal. Tighten all bolts to the specified torque.

Notes

Chapter 7 Part C
Transfer case

Contents

	Section
General information	1
Transfer case (Dana 20) - disassembly, overhaul and reassembly	7
Transfer case (New Process 203) - disassembly, overhaul and reassembly	4
Transfer case (New Process 205) - disassembly, overhaul and reassembly	5
Transfer case (New Process 208) - disassembly, overhaul and reassembly	6
Transfer case oil change	See Chapter 1
Transfer case oil level check	See Chapter 1
Transfer case - removal and installation	3
Transfer case shift linkage - check and adjustment	2

Specifications

Rear output shaft endplay	
New Process 203	0.000 to 0.005 inch
New Process 205	0.002 to 0.027 inch
Dana 20	0.001 to 0.005 inch
Front output shaft endplay (Dana 20)	0.001 to 0.005 inch
Idler gear endplay (New Process 205)	0.000 to 0.002 inch

Torque specifications

	Ft-lbs
New Process 203 transfer case	
Adapter-to-transfer case attaching bolt	38
Adapter-to-transmission attaching bolt	40
Transfer case bracket-to-frame nut (upper)	50
Transfer case shift lever attaching nut (lower)	65
Transfer case shift lever attaching nut	25
Transfer case shift lever rod swivel lock nut	50
Transfer case shift lever locking arm nut	150
Skid plate attaching bolt retaining nut	45
Crossmember support attaching bolt retaining nut	45
Adapter mount bolt	25
Intermediate case-to-range box bolt	30
Front output bearing retainer bolt	30
Output shaft yoke nut	150
Front output rear bearing retainer bolt	30
Differential assembly screw	45
Rear output shaft housing bolt	30
Poppet ball retainer nut	15
Power takeoff cover bolt	15
Front input bearing retainer bolt	20
Filler plug	25

Torque specifications

	Ft-lbs
New Process 205 transfer case	
Idler shaft locknut	200
Idler shaft cover bolts	18
Front output shaft front bearing retainer bolt	30
Front output shaft yoke locknut	200
Rear output shaft bearing retainer bolt	30
Rear output shaft yoke locknut	150
Power take off cover bolt	15
Front output shaft rear bearing retainer bolt	30
Drain and filler plugs	30
Transfer case-to-frame bolt	130
Transfer case-to-adapter bolt	25
Adapter mount bolt	25
Transfer case bracket-to-frame nut (upper)	30
Transfer case bracket-to-frame nut (lower)	65
Adapter-to-transmission bolt (manual transmission)	22
Adapter-to-transmission bolt (automatic transmission)	35
Transfer case control mounting bolt	100
New Process 208 transfer case	
Shift lever-to-shifter assembly nut	10 to 20
Knob assembly-to-shifter assembly nut	19 to 30
Shifter assembly-to-transfer case bolt	89 to 103
Shift arm-to-case nut	10 to 15
Detent retainer bolt	20 to 25
Indicator switch	18
Adapter-to-transmission bolt	19 to 30
Adapter-to-transfer case bolt	19 to 30
Rear case-to-front case bolt	23
Rear retainer-to-rear case bolt	23
Operating lever-to-range sector locknut	18
Yoke nut	120
Lockplate-to-front case	30
Lubrication filler plug and drain plug	18
Skid plate-to-crossmember nut	39 to 52
Support strut rod-to-transmission end bolt	30 to 39
Support strut rod-to-transfer case end bolt	111 to 148
Dana 20 transfer case	
Shift rail set screw	15
Front output shaft rear cover bolt	30
Front output shaft front bearing retainer bolt	30
Front output shaft yoke locknut	150
Intermediate shaft lockplate bolt	15
Rear output shaft housing bolt	30
Rear output shaft yoke locknut	150
Case bottom cover bolt	15
Transfer case-to-adapter bolt	45
Transfer case-to-frame bolt	45
Adapter mount bolt	25

1 General information

Four-wheel drive models are equipped with a transfer case mounted on the rear of the transmission. Drive is transmitted from the engine, through the transmission and the transfer case to the front and rear axles by driveshafts.

2 Transfer case shift linkage - check and adjustment

Refer to illustration 2.3

Check

1 Whenever the vehicle is raised, check the transfer case shift linkage for freedom of operation, corrosion, damaged connections or loose bolts, correcting as necessary.

Adjustment

2 The New Process 203 and 208 transfer case shift linkages are adjustable.

New Process 203

3 Align the gage holes in the levers with

2.3 To adjust the New Process 203 transfer case shift linkage, align the gage holes and insert a drill bit

3.4 Skid plate and crossmember mounting details

the gage hole in the shifter mechanism and insert a gage pin or appropriate size drill bit, so the shift lever will be locked in Neutral **(see illustration)**.

4 Place the transfer case shift arms in Neutral (pointing straight down in the 6 o'clock position).

5 With both the swivel and locknuts loosely assembled to the rod, rotate the swivel until the ends of the rod will enter both levers at the same time and then secure with the retainer.

6 Tighten the locknuts against the swivel, making sure the arm doesn't move.

7 Repeat Steps 5 and 6 for the other rod, lever and arm.

8 Remove the gage pin.

New Process 208

9 Place the shift lever in the 4 HI position.

10 Push the lower shift lever forward against the 4 HI stop.

11 Place the rod swivel in the shift lever hole.

12 Hang the 0.200-inch thick gage tool behind the swivel.

13 Run the rod nut up against the gage with the shifter against the 4 HI stop.

14 Remove the gage and then push the swivel rearward against the rod nut.

15 Run the front rod nut against the swivel and tighten it securely.

3 Transfer case - removal and installation

Refer to illustrations 3.4 and 3.7

1 Raise the vehicle and support it securely on jackstands.

2 Drain the transfer case oil into a suitable container.

3 Disconnect the speedometer drive cable, back-up light switch and the throttle control system (TCS) switch.

4 Position a jack or jackstand under the transmission, then remove the crossmember supports and, where necessary, remove the skid plate(s) **(see illustration)**.

3.7 Typical transfer case mounting details

5 Disconnect the front and rear driveshafts and fasten them out of the way (refer to Chapter 8 for further information).

6 Disconnect the shift lever rod from the shift rail link. On full-time four wheel drive models, disconnect the shift levers at the transfer case.

7 Using a suitable cradle or stand, support the transfer case and remove the bolts attaching the transfer case to the transmission adapter **(see illustration)**.

8 Carefully move the transfer case to the rear until the input shaft clears the adapter, then lower the assembly from the vehicle.

9 Installation is a reversal of removal, making sure that all nuts and bolts are tightened to the specified torque. Be sure to fill the transfer case with the correct grade and quantity of oil (see Chapter 1).

4 Transfer case (New Process 203) - disassembly, overhaul and reassembly

Disassembly

1 Before attempting to disassemble the transfer case, ensure that all external surfaces are clean and free of grime. If the transfer case oil was not drained prior to removal, do it now.

2 Position the transfer case on a work bench or suitable work table.

3 Use a tool to hold the rear output shaft flange and loosen the retaining nut. **Note:** *Tap the dust shield rearward on the shaft (away from the bolts) to obtain the necessary clearance to remove the bolts from the flange and allow installation of the output shaft holding tool.*

4 Remove the bolts retaining the front output shaft front bearing retainer and remove the bearing retainer and gasket from the transfer case.

5 With the help of an assistant, position the assembly on wood blocks.

6 Remove the bolts retaining the rear output shaft assembly to the transfer case and disengage the assembly from the case.

7 Carefully slide the carrier unit from the shaft. **Note:** *A 1-1/2 to 2-inch diameter water hose clamp may be installed on the input shaft at this time to prevent the loss of bearings when removing the input shaft assembly from the range box.*

8 Raise the shift rail and drive out the roll pin retaining the shift fork to the rail.
9 Remove the shift rail poppet ball plug. Lift out the spring and poppet ball from the case. A small magnet will simplify removal of the ball.
10 Push the shift rail down, lift up on the lockout clutch and remove the shift fork from the clutch assembly.
11 Remove the bolts retaining the front output shaft rear bearing retainer to the transfer case. Tap on the front of the shaft or carefully pry the retainer away from the case. Remove the retainer from the shaft. Recover any roller bearings which may fall from the rear cover.
12 If it is necessary to replace the rear bearing, support the cover and press the bearing from the cover. When installing a new bearing, position it squarely and press it into the cover until it is flush with the opening.
13 From the lower side of the case, carefully pry out the output shaft front bearing.
14 Disengage the front output shaft from the chain and remove the shaft from the transfer case.
15 Remove the bolts attaching the intermediate chain housing to the range box and lift the intermediate housing from the range box.
16 Remove the chain from the intermediate housing.
17 Remove the lockout clutch, drive gear and input shaft assembly from the range box. Be careful not to lose the roller bearings (see Note in Step 7).
18 Pull up on the shift rail and disconnect the rail from the link.
19 Remove the input shaft assembly from the range box by lifting it up and away.
20 At this point the transfer case is disassembled into its subassemblies. Each of the subassemblies should then be disassembled for cleaning and inspection.

Cleaning and inspection

Bearings: Wash all the bearings and rollers in a suitable solvent. Inspect the bearings and rollers for evidence of chipping, cracks or worn areas. Replace any damaged or worn components.
Shafts and gears: Thoroughly clean all shafts and gears. Inspect the shaft splines and bearing surfaces for signs of chipped teeth or excessive wear.
Case, cover and housings: Clean the transfer case, cover and housings thoroughly with solvent and inspect them for cracks or damage.

Overhaul

Differential carrier assembly

21 Remove the bolts from the carrier assembly and separate the carrier sections.
22 Lift the pinion gear and spider assembly from the carrier.
23 Remove the pinion thrust washers, pinion roller washers, pinion gears and roller bearings from the spider unit.
24 Clean and inspect all components, replacing worn or broken parts as necessary.
25 Begin reassembly by loading the roller bearings to the pinion gears (132 rollers are required, 33 in each pinion). Use petroleum jelly to retain them in position.
26 Install the pinion roller washer, pinion gear, roller washer and thrust washer on each leg of the spider.
27 Place the spider assembly into the front half of the carrier, with the undercut surface of the spider thrust surface facing downward or toward the gear teeth.
28 Position the carrier halves together, install the retaining bolts and tighten them to the specified torque.

Lockout clutch assembly

29 Remove the front side gear from the input shaft assembly and remove the thrust washer, roller bearings (123) and spacers from the front side gear bore. Note the position of the spacers to aid reassembly.
30 Using suitable snap-ring pliers, remove the snap-ring retaining the drive sprocket to the clutch assembly. Slide the drive sprocket from the front side gear.
31 Remove the lower snap-ring, then remove the sliding gear, spring and spring cup washer from the front side gear.
32 Clean and inspect all components for signs of wear, replacing as necessary.
33 Install the spring cup washer, spring and sliding clutch gear on the front side gear.
34 Install the snap-ring that retains the sliding clutch to the front side gear.
35 Load the roller bearings (123) and spacers in the front side gear. Use petroleum jelly to retain the roller bearings in position.
36 Install the thrust washer in the gear end of the front side gear.
37 Slide the drive sprocket onto the clutch splines and install the retaining ring.

Input shaft assembly

38 Slide the thrust washer and spacer from the shaft.
39 Using snap-ring pliers, remove the snap-ring retaining the input bearing retainer assembly to the shaft and remove the bearing retainer assembly from the shaft.
40 Support the low speed gear (large gear) and tap the shaft from the gear and thrust washer.
41 Using a screwdriver, pry behind the open end of the large snap-ring, retaining the input bearing in the bearing retainer, and remove the snap-ring from the retainer. Tap the bearing from the retainer.
42 Remove the pilot roller bearings (15) from the end of the input shaft.
43 Remove and discard the O-ring from the end of the shaft.
44 Clean and inspect all components, replacing as necessary.
45 Begin reassembly by positioning the bearing to the retainer and tapping or pressing it into place. The ball loading slots should be toward the concave side of the retainer.
46 Install the large snap-ring to secure the bearing in the retainer. **Note:** *The snap-ring is a selective fit; to provide the tightest fit, four thicknesses (A, B, C or D) are available.*
47 Install the low speed gear onto the shaft, with the clutch end toward the gear end of the shaft.
48 Position the thrust washers on the shaft, aligning the slot in the washer with the pin in the shaft. Slide or tap the washer into position.
49 Position the input bearing retainer on the shaft and install a new snap-ring. **Note:** *The snap-ring is a selective fit. To provide the tightest fit, four thicknesses (A, B, C or D) are available.*
50 Slide the spacer and thrust washer onto the shaft. Align the spacer with the location pin.
51 Install the roller bearings (15) in the end of the shaft and use a heavy grade grease to retain them in position.
52 Install a new O-ring on the end of the shaft.

Range selector housing (range box) disassembly

53 Remove the poppet plate spring, plug and gasket.
54 Disengage the sliding clutch gear from the input gear and remove the clutch fork and sliding gear from the case.
55 Remove the shift lever assembly retaining nut and the upper shift lever from the shifter shaft.
56 Remove the shift lever snap-ring and lower lever.
57 Push the shifter shaft assembly downward and remove the lockout clutch connector link. **Note:** *The long end of the connector link engages the poppet plate.*
58 Remove the shifter shaft assembly from the case and separate the inner and outer shifter shafts. Remove and discard the O-rings.
59 Inspect the poppet plate for damage. If it is necessary to remove the poppet plate, drive the pivot shaft from the case. The poppet plate and spring should be removed from the bottom of the case.
60 Remove the input gear bearing assembly.
61 Remove the large snap-ring from the bearing outer diameter.
62 Tap the input gear and bearing from the case.
63 Remove the snap-ring retaining the input shaft bearing to the shaft and remove the bearing from the input gear.
64 From the intermediate case side, remove the countershaft from the cluster gear and the case. Use a bar of suitable diameter for this. Remove the cluster gear assembly from the case. **Note:** *Recover the roller bearings (72) from the gear case and shaft.*
65 Remove the cluster gear thrust washers from the case.
66 Clean and inspect all components, replacing as necessary.

Range selector housing (range box) reassembly

67 Using an alignment tool and a heavy grade grease, install the roller bearings (72) and spacers in the cluster gear bore.
68 Using a heavy grade grease, position the countershaft thrust washers in the case. Engage the tab on the washers with a slot in the case thrust surface.
69 Position the cluster gear assembly in the case and install the countershaft, through the front face of the range box, into the gear assembly. The countershaft face with the flat should face forward, and must be aligned with the case gasket.
70 Install the bearing (without snap-ring) on the input gear shaft, positioning the snap-ring groove outward, and install a new retaining ring on the shaft. Position the input gear and bearing in the housing. **Note:** *The snap-ring is a selective fit. To provide the tightest fit, four thicknesses (A, B, C or D) are available.*
71 Install the snap-ring in the outside diameter of the bearing.
72 Align the oil slot in the retainer with the drain hole in the case and install the input gear bearing retainer, gasket and retaining bolts. Tighten the bolts to the specified torque.
73 If removed, install the poppet plate and pivot pin assembly in the housing. Use a suitable sealant on the pin.
74 Install new O-rings on the inner and outer shifter shafts. Lubricate the O-rings and assemble the inner shaft in the outer shaft.
75 Push the shifter shafts into the housing, engaging the long end of the lockout clutch connector link to the outer shifter shaft before the shaft assembly bottoms.
76 Install the lower shift lever and retaining ring.
77 Install the upper shift lever and shifter shaft retaining nut.
78 Install the shift fork and sliding clutch gear. Push the fork up into the shifter shaft assembly to engage the poppet plate, sliding the clutch gear forward onto the input shaft gear.
79 Install the poppet plate spring, gasket and plug in the top of the housing. Check the spring engagement with the poppet plate.
80 Remove the bearing retainer attaching bolts, retainer and gasket from the housing.
81 Using suitable snap-ring pliers, remove the snap-ring retaining the bearing on the shaft.
82 Using a screwdriver or other suitable tool, pry the bearing from the case and remove it from the shaft.
83 Carefully inspect the input gear for burrs, scoring, heat discoloration, etc. Also inspect the condition of the seal in the bearing retainer. Replace components as necessary.
84 Install the bearing with the snap-ring on the input gear shaft. Position the bearing to the case and tap into place with a soft faced mallet. **Note:** *The snap-ring is a selective fit. To provide the tightest fit, four thicknesses (A, B, C or D) are available.*
85 Install the snap-ring to retain the bearing on the shaft.
86 Position a new gasket and the bearing retainer to the housing. Install the retaining bolts, tightening them to the specified torque.
87 Remove the bearing retainer attaching bolts, retainer and gasket.
88 Pry the seal out of the retainer.
89 Position the new seal to the retainer, and install it using a length of tubing of suitable diameter.
90 Position the bearing retainer and gasket on the housing. Install the attaching bolts, tightening them to the specified torque.

Rear output shaft housing assembly disassembly and reassembly

91 Remove the speedometer driven gear from the housing.
92 If not removed during transfer case disassembly, remove the flange nut and washer. Remove the flange from the shaft.
93 Tap on the flange end of the pinion, with a soft faced mallet, to remove the pinion from the carrier. If the speedometer gear is not on the pinion shaft, reach into the carrier and remove it from the housing.
94 Pry out the old seal from its bore.
95 Using a screwdriver, pry behind the open ends of the snap-ring and remove the snap-ring retaining the rear bearing in the housing.
96 Pull or tap the bearing from the housing.
97 To remove the front bearing, insert a long drift through the rear opening, and drive the bearing from the housing.
98 Begin reassembly by positioning the rubber seal in the bearing bore. Use grease to hold it in place. Position the roller bearing in the bore and press it into place until it bottoms in the housing.
99 Position the rear bearing to the case and tap it into place.
100 Install the snap-ring to retain the bearing in the case. **Note:** *The snap-ring is a selective fit. To provide the tightest fit four thicknesses (A, B, C or D) are available.*
101 Position the rear seal to the bore and drive it into place, using a length of tubing of suitable diameter, until it is approximately 1/8 to 3/16-inch below the housing face.
102 Install the speedometer drive gear and shims on the output shaft and install the shaft into the carrier through the front opening.
103 Install the flange, washer and retaining nut. Leave the nut loose until shim requirements are determined.
104 Install the speedometer driven gear in the case.

Front output shaft bearing retainer seal replacement

105 Pry or drive out the existing seal from the retainer bore.
106 Apply a suitable sealer to the outer diameter of the new seal.
107 Position the seal to the retainer bore and, using a piece of suitable diameter tubing, tap the seal into the retainer.

Front output shaft rear bearing replacement

108 Remove the rear cover from the transfer case.
109 Support the rear cover and press the bearing from the cover.
110 Position the new bearing to the outside face of the cover and, using a suitable piece of wood to cover the bearing, press it into the cover until it is flush with the opening.
111 Position the gasket and cover to the transfer case and tap into place.
112 Install the cover retaining bolts, tightening them to the specified torque.

Reassembly

113 Place the range box on blocks, with the input gear side toward the bench.
114 Position the range box to transfer case housing gasket on the input housing.
115 Install the lockout clutch and drive sprocket assembly on the input shaft assembly. **Note:** *A 2-inch hose clip may be installed on the end of the shaft to prevent loss of bearings from the clutch assembly.*
116 Install the input shaft, lockout clutch and drive sprocket assembly in the range box, aligning the tab on the bearing retainer with the notch in the gasket.
117 Connect the lockout clutch shift rail to the connector link, and position the rail in the housing. Rotate the shifter shaft while lowering the shift rail into the housing, to prevent the link and rail from being disconnected.
118 Install the drive chain the chain housing, positioning the chain around the outer wall of the housing.
119 Install the chain housing on the range box, engaging the shift rail channel of the housing to the shift. Position the chain on the input drive sprocket.
120 Install the front output sprocket in the case, engaging the drive chain to the sprocket. Rotate the clutch drive gear to assist in positioning the chain on the drive sprocket.
121 Install the shift fork to the clutch assembly and the shift rail, then push the clutch assembly fully into the drive sprocket. Install the roll pin retaining the shift fork to the shift rod.
122 Install the front output shaft bearing.
123 Install the front output shaft bearing retainer, gasket and retaining bolts.
124 Install the front output shaft flange, gasket seal, washer and retaining nut. Tap the dust shield back in place after installing the bolts in the flange.
125 Install the front output shaft rear bearing retainer, gasket and retaining bolts. **Note:** *If the rear bearing was removed, position the new bearing to the outside face of the cover and press it into the cover until it is flush with the opening.*
126 Install the differential carrier assembly on the input shaft. The carrier bolt heads

should face the rear of the shaft.

127 Install the rear output housing assembly, gasket and retaining bolts. Load the bearings in the pinion shaft.

128 Check the rear output shaft endplay. To do this, install a dial indicator on the rear housing so that it contacts the end of the output shaft. Holding the rear flange, rotate the front output shaft and determine the highest point of gear run-out on the rear shaft. Zero the dial indicator and, with the rear shaft set at this point, pull up on the end of the shaft to determine the endplay.

129 Remove the dial indicator and install a shim pack onto the shaft, in front of the rear bearing, to control the endplay to within 0 to 0.005-inch. Hold the rear flange and rotate the front output shaft to check for binding of the rear output shaft.

130 Install the lockout clutch shaft rail poppet ball and spring and screw the plug into the case.

131 Install the poppet plate spring, gasket and plug, if not installed during reassembly of the range box.

132 Install the shift levers on the range box shifter shaft (if not left on the linkage in the vehicle).

133 Tighten all bolts, locknuts and plugs to the specified torque.

134 Fill the transfer case to the proper level with the correct grade of oil. Install and tighten the filler plug to the specified torque.

5 Transfer case (New Process 205) - disassembly, overhaul and reassembly

Disassembly

Rear output shaft and yoke

1 Loosen the rear output yoke nut.

2 Loosen and remove the rear output shaft housing bolts and remove the housing and retainer assembly from the transfer case.

3 Remove the yoke retaining nut and the yoke from the shaft, then remove the shaft assembly from the housing.

4 Using a suitable pair of pliers, remove and discard the housing bearing snap-ring.

5 At the inside of the housing, remove the thrust washer and washer pin.

6 Remove the tanged bronze washer.

7 Remove the gear needle bearings (32 per row), spacer and second row of needle bearings.

8 Remove the tanged bronze washer from the shaft.

9 Remove the pilot rollers (15), retaining ring and washer.

10 Remove the oil seal retainer, ball bearings, speedometer gear and spacer. If required, the housing bearing may be pressed or drifted out.

11 Remove the oil seal from the retainer.

12 Discard all gaskets and place the parts in a safe place for later inspection.

Front output shaft

13 Remove the yoke locknut, lift off the washer and remove the yoke.

14 Remove the front bearing attaching bolts and remove the retainer.

15 Remove the front output shaft bearing retainer attaching bolts.

16 Using a soft faced mallet, tap on the end of the output shaft to remove the shaft, gear assembly and the rear bearing retainer from the case.

17 Remove the sliding clutch from the output high gear, washer and bearing which are still in the case.

18 Using suitable snap-ring pliers, remove the gear retaining ring from the shaft.

19 Remove the thrust washer and pin from the shaft.

20 Slide off the front output low gear, needle bearings (32 per row) and the spacer from the front output shaft.

21 If it is necessary to replace the front output shaft rear bearing, support the cover and press the bearing from the cover. A new bearing is simply pressed into the case, but ensure that the force is applied to the outside diameter of the bearing only.

Shift rail and fork assemblies

22 Remove the two poppet nuts on top of the transfer case. Lift out the two poppet springs and using a magnet, remove the poppet balls.

23 Using a 1/4-inch diameter punch, drive the cup plugs into the case.

24 Position both shift rails in Neutral. Using a long narrow punch, drive the shift fork pins through the shift rails into the case.

25 Remove the clevis pins and the shift rail connector link.

26 Remove the shift rails, upper (range) rail first, followed by the lower (4-wheel) rail.

27 Remove the shift forks and the sliding clutch from the case.

28 Remove the front output high gear, washer and bearing from the case.

29 Remove the shift rail cup plugs and pins that were driven into the case earlier.

Input shaft assembly

30 Using suitable snap-ring pliers, remove the snap-ring in front of the bearing. Use a soft faced hammer to tap the shaft out of the rear of the case. Tap the bearing out of the front of the case.

31 Tip the case onto the power takeoff face and remove the two interlock pins from inside the case.

Idler gear

32 Remove the idler gear shaft nut.

33 Remove the idler shaft rear cover attaching bolts and lift off the cover.

34 Using a suitable drift and a soft faced mallet, drive out the idler gear shaft.

35 Remove the idler gear by way of the output shaft opening.

36 Remove the two bearing cups from the idler gear.

Overhaul

37 Thoroughly wash all components in solvent. Carefully inspect all bearings and rollers for evidence of chipping, cracks or worn spots. Inspect the shaft splines and gears for chipped teeth or excessive wear. Replace worn or damaged components as necessary.

Reassembly

Idler gear

38 If previously removed, press the two bearing cups into the idler gear.

39 Assemble the two bearing cones, spacer, shims and idler gear to the dummy shaft J-23429 and check the endplay of the idler gear, which should be 0.000 to 0.002-inch. Adjust or replace shims as necessary to achieve this endplay.

40 Install the idler gear assembly with the dummy shaft into the transfer case. Install the large end first, through the front output bore opening.

41 Install the idler shaft from the large bore side and drive it through the idler gear with a soft faced hammer.

42 Install the washer and a new locknut on the idler shaft. Check for the correct endplay again and check that the assembly rotates freely. Tighten the idler shaft locknut to the specified torque.

43 Install the idler shaft cover (with a new gasket) in position and screw in the attaching bolts. Tighten the bolts to the specified torque. The flat on the idler shaft cover must be located adjacent to the front output shaft rear cover.

Shift rail and fork assemblies

44 Press the two rail seals, metal lip outwards, into the case.

45 Install the interlock pins through the large bore or power takeoff cover opening.

46 Start the front output drive shift rail into its bore in the case from the back, slotted end first, with the poppet notches upward.

47 Install the shift fork (long end inward) into the shift rail. Push the rail, through the shift fork, into the neutral position.

48 Install the input shaft bearing and shaft into the case.

49 Start the range rail into its bore in the case from the front, poppet notches upward.

50 Install the sliding clutch onto the shift fork. Place the assembly over the input shaft in the case. Position the fork to receive the range shift rail. Push the range shift rail through the shift fork and into the neutral position.

51 Install new lock pins through the holes at the top of the case and drive them into the forks. **Note:** *When installing the range rail lock pin, place the case on the power takeoff opening.*

Front output shaft

52 Install two rows of needle bearings (32 in each row), separated by the spacer, in the front low output gear and retain them in posi-

6.3 **Remove the nut securing the front yoke and remove the front yoke. Discard the yoke seal, the washer and the yoke nut**

6.7a **Remove the bolts that attach the rear retainer to the rear case. Tap on the retainer with a plastic mallet and remove the retainer and pump housing as an assembly**

6.7b **Tap on the retainer with a rubber mallet to loosen it**

tion with a sufficient amount of grease.

53 Place the front output shaft, spline end down, in a soft jaw vise. Install the front low gear over the shaft, with the clutch gear facing down, and install the thrust washer pin and thrust washer. Retain these components in position with a new snap-ring. **Note:** *The snap-ring should be positioned so that the opening is opposite the pin.*

54 Position the front wheel high gear and washer into the case. Install the sliding clutch in the shift fork, then put the fork and rail in the four wheel drive (4 HI) position, with the clutch teeth in mesh with the teeth of the front wheel high gear.

55 Align the washer, high gear and the sliding clutch with the bearing bore. Insert the front output shaft and low gear assembly through the high gear assembly.

56 Using a piece of tubing, install a new seal in the bearing retainer. Install the front output bearing and retainer into the case.

57 Clean and grease the roller in the front output rear bearing retainer. Install the retainer, using a new gasket. Apply thread sealant to the attaching bolts before tightening them to the specified torque.

58 Install the front output yoke, washer and locknut. Tighten the locknut to the specified torque.

Rear output shaft

59 Install two rows of needle bearings (32 in each row), separated by the spacer, into the output low gear. Use sufficient grease to retain the needle bearings in position.

60 Install the thrust washer onto the rear output shaft, tang down in the clutch gear groove. Install the output low gear onto the shaft with the clutch teeth facing down.

61 Install the thrust washer over the gear with the tab pointing up and away from the gear. Install the washer pin and the large thrust washer over the shaft and pin. Rotate the washer until the tab fits into the slot approximately 90-degrees away from the pin. Install the snap-ring to retain these components and check the endplay, which should be between 0.002 and 0.027-inch.

62 Grease the pilot bore or the rear output shaft and install the needle bearings (15). Install the thrust washer and retain in position with a new snap-ring.

63 Clean, grease and install a new bearing in the retainer housing using a suitable bearing driver.

64 Install the housing onto the output shaft assembly. Install the spacer and speedometer gear, then install the bearing.

65 Install the rear bearing retainer seal using a piece of suitable diameter tubing.

66 Install the bearing retainer assembly onto the housing with one or two gaskets, depending on the clearance. Tighten the retainer attaching bolts to the specified torque.

67 Install the yoke, washer and locknut on the output shaft. Tighten the yoke locknut to the specified torque.

68 Position the range rail in the high position and install the output shaft and retainer assembly on the transfer case.

69 Install the power takeoff cover and gasket.

70 Install the cup plugs at the rail pin holes.

71 Install the shift rail cross link, clevis pins and lock pins.

6 Transfer case (New Process 208) - disassembly, overhaul and reassembly

Refer to illustrations 6.3, 6.7a, 6.7b, 6.10, 6.11, 6.12, 613, 6.14, 6.15, 6.16. 6.17, 6.18, 6.19, 6.21, 6.22a, 6.22b, 6.23, 6.24a, 6.24b, 6.25a, 6.25b and 6.26

Disassembly)

1 Remove the transfer case (Section 3).

2 Remove both the drain plug and the fill plug.

3 Remove the nut securing the front yoke and remove the front yoke **(see illustration)**. Discard the yoke seal, the washer and the yoke.

4 Put the transfer case on its end, placing the front case on wooden blocks. It may be necessary to cut V-notches in the blocks in order to clear the mounting studs in the front of the case.

5 Unscrew the lock mode indicator switch and washer from the top of the front case.

6 From the bottom of the front case, remove the detent bolt, spring and poppet.

7 Remove the bolts that attach the rear retainer to the rear case. Tap on the retainer with a plastic mallet and remove the retainer

6.10 Note the position of the oil pump and that the side facing the case interior has a recess in it, then remove the pump from the mainshaft

1 *Oil pump gear*
3 *Pump housing*
2 *Speedometer gear*

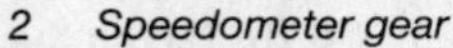

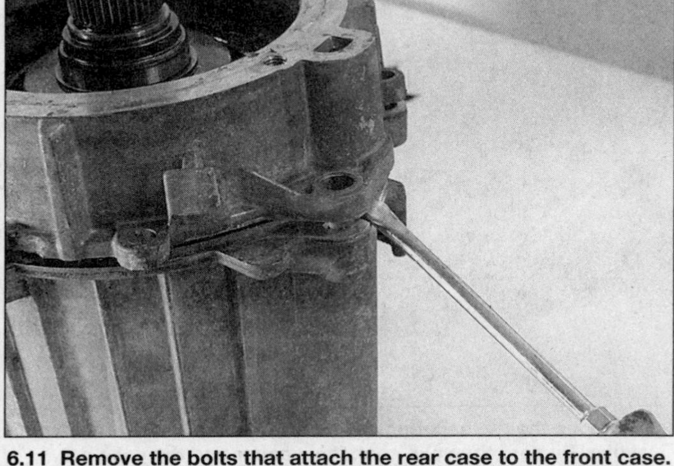

6.11 Remove the bolts that attach the rear case to the front case. Insert screwdrivers into the slots that are cast in the case ends and gently pry upward. Do not attempt to pry the cases apart at any point on the mating surfaces. Separate the rear case from the front case

6.12 Remove the bushing, retainer and spring from the shift rod

6.13 Note the position of the rear thrust bearings and races on the front output shaft and remove the complete assembly

6.14 Remove the snap-ring that retains the driven sprocket

6.15 Remove the snap-ring that retains the drive sprocket and remove the spacer washer

6.16 Lift evenly on both sprockets to remove the drive and driven sprockets and drive chain as an assembly. The mainshaft roller bearings may fall out of the driven sprocket

6.17 Lift out the front output shaft along with the front thrust bearing assembly

6.18 Remove the synchronizer blocker ring from the mainshaft

6.19 Lift the synchronizer mode fork bushing, mode fork and bracket as an assembly away from the mainshaft. Be careful, as the synchronizer keys may fall free from the hub

and pump housing as an assembly **(see illustrations)**.

8 Separate the pump housing from the retainer and pry the pump seal from the housing. Throw away the seal.

9 Remove the speedometer drive gear from the mainshaft.

10 Note the position of the oil pump and that the side facing the case interior has a recess in it, then remove the pump from the mainshaft **(see illustration)**.

11 Remove the bolts that attach the rear case to the front case. Insert screwdrivers into the slots that are cast in the case ends and gently pry upward. Do not attempt to pry the cases apart at any point on the mating surfaces. Separate the rear case from the front case **(see illustration)**.

12 Remove the bushing, retainer and spring from the shift rod **(see illustration)**.

13 Note the position of the rear thrust bearings and races on the front output shaft and remove the complete assembly **(see illustration)**.

14 Remove the snap-ring that retains the driven sprocket **(see illustration)**.

15 Remove the snap-ring that retains the drive sprocket and remove the spacer washer **(see illustration)**.

16 Lift evenly on both sprockets to remove the drive and driven sprockets and drive chain as an assembly. The mainshaft roller bearings may fall out of the driven sprocket **(see illustration)**.

17 Lift out the front output shaft along with the front thrust bearing assembly **(see illustration)**.

18 Remove the synchronizer blocker ring from the mainshaft **(see illustration)**.

19 Lift the synchronizer mode fork bushing, mode fork and bracket as an assembly away from the mainshaft **(see illustration)**. Be careful, as the synchronizer keys may fall free from the hub.

20 Remove the shift rail **(see illustration)**.

21 Pull the mainshaft out of the case with the synchronizer hub and snap-ring attached.

22 Remove the annulus gear snap-ring and

6.21 Pull the mainshaft out of the case with the synchronizer hub and snap-ring attached

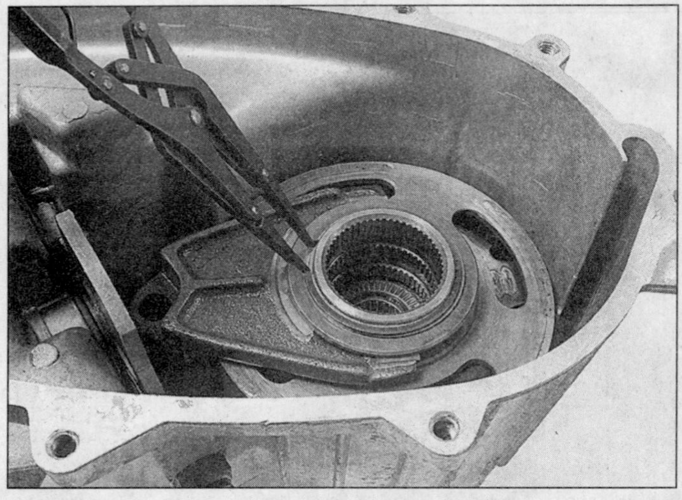
6.22a Remove the annulus gear snap-ring and thrust washer from the case

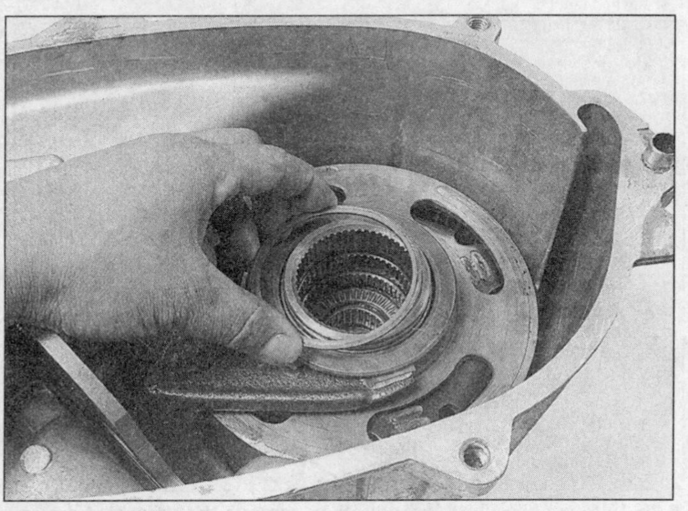
6.22b Remove the annulus gear thrust washer

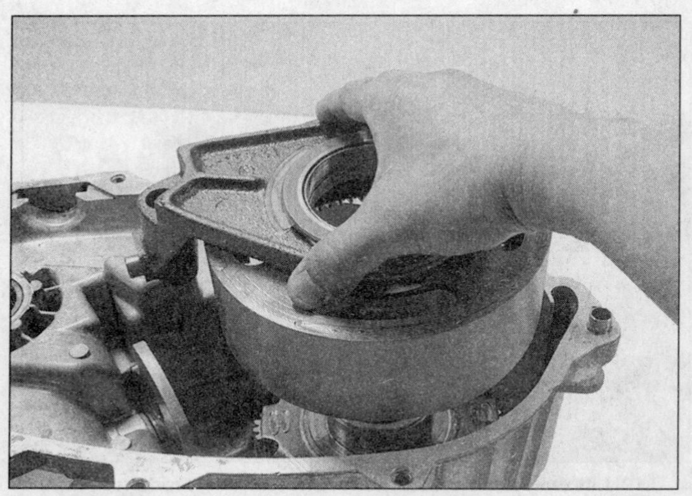
6.23 Remove the annulus gear and range fork as an assembly. To disengage the fork lug from the range sector turn the fork counterclockwise

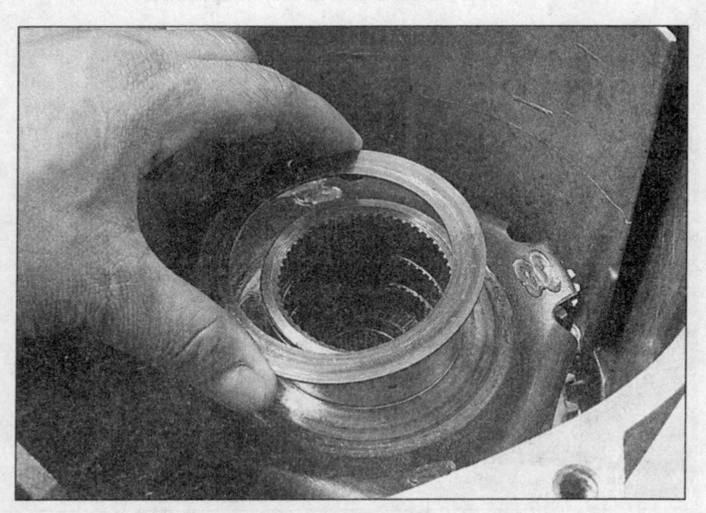
6.24a Remove the planetary thrust washer . . .

6.24b . . . and remove the planetary assembly from the case

6.25a After removing the mainshaft thrust bearing from the input gear, lift the gear straight up and out of the case

6.25b Lift the input gear strait up and out of the case

6.26 Note the position of the input gear thrust bearing and race and remove it

thrust washer from the case **(see illustrations)**.

23 Remove the annulus gear and range fork as an assembly. To disengage the fork lug from the range sector turn the fork counterclockwise **(see illustration)**.

24 Lift out the planetary thrust washer and remove the planetary assembly from the case **(see illustrations)**.

25 After removing the mainshaft thrust bearing from the input gear, lift the gear straight up and out of the case **(see illustrations)**.

26 Note the position of the input gear thrust bearing and race and remove them **(see illustration)**.

27 Remove the nut and washer attaching the range sector operating lever, then remove the lever and sector shaft seal and seal retainer.

28 Remove the range sector from the case.

29 Remove the output shaft seals from the front and rear case seal bores.

30 Check the lockplate for looseness or wear and inspect to see if it's broken or cracked. If so, remove the bolts that attach the lockplate to the front case and throw them away.

31 Lift the lockplate from the case.

32 Apply sealant around the bolt holes of both the case and the new lockplate and position the lockplate in the case, aligning the bolt holes.

33 Put thread sealant on the threads of the new lockplate attaching bolts and tighten them to the specified torque.

Overhaul

34 Thoroughly wash all components in a suitable cleaning solvent. Carefully inspect all bearings and rollers for evidence of chipping, cracks or worn spots. Inspect the shaft splines and gears for chipped teeth or excessive wear. Replace worn or damaged components as necessary.

35 The bearings used in the transfer case all have bearing oil feed holes. After replacing any bearings, check that the bearing position does not obstruct or block the oil feed hole.

36 Drive the rear output bearing out of the retainer using a hammer and a brass drift. Pry the rear seal out using a screwdriver or brass drift.

37 With the shielded side of the new bearing facing the interior of the case, drive it into position using an appropriate size tubular drift.

38 Install the snap-ring that holds the bearing in place.

39 Insert the new rear seal using an appropriate size drift.

40 Remove the front output shaft bearing using an appropriate size tubular drift.

41 Install the new bearing, positioning it so the oil feed hole is not covered, and drive it into place with an appropriate size drift.

42 Attach a C-clamp to the rear case using a block of wood beneath the case so as not to distort the case. Mount the C-clamp firmly in a vise. Using a slide hammer and the appropriate adapter, remove the front output shaft rear bearing.

43 Using an appropriate size tubular drift, install the new bearing. Be sure the oil feed hole is not covered and that the bearing is seated flush with the edge of the case bore to allow room for the thrust bearing assembly.

44 Remove the C-clamp from the case.

45 Remove both the front and rear input bearings simultaneously using an appropriate size tubular drift.

46 Install the rear bearing first, then install the front bearing using an appropriate size tubular drift. Be sure the oil feed holes to the bearing are not covered and that the bearing is mounted flush with the case.

47 If the mainshaft pilot bearing cannot be removed by hand, a slide hammer or a similar internal-type blind hole bearing puller must be used.

48 Insert the new pilot bearing by hand, if possible. Otherwise, use an appropriate size drift. Be sure the bearing is seated flush with the edge of the oil hole and if installation tools were used, check that the bearing is positioned so that it will not block the oil feed hole.

Reassembly

49 During reassembly lubricate all the components with either petroleum jelly or automatic transmission fluid.

50 Place the input gear race and thrust bearing in position in the front case and install the input gear.

51 Place the mainshaft thrust bearing in the input gear.

52 Place the range sector shaft seal and seal retainer into the end of the front case.

53 Slip the range sector into position.

54 Place the operating lever into position on the range selector shaft and install the shaft washer and lock nut. Tighten to the specified torque.

55 Slide the planetary assembly over the input gear until it is fully seated and meshed with the gear.

56 Place the planetary thrust washer on top of the planetary hub.

57 If the inserts were removed from the range fork, replace them now. Engage the range fork into the annulus gear and place the annulus gear over the planetary assembly. The range fork lug should be inserted into the range sector slot.

58 Install the snap-ring that retains the annulus gear.

59 Line up the shaft bores in the range fork and in the case and slip in the shift rail.

60 Check that the mainshaft thrust bearing is properly seated in the input gear, then install the mainshaft.

61 With the synchronizer keys in position, install the synchronizer and mode fork as an assembly.

62 Install the synchronizer blocker ring to the mainshaft.

63 Thoroughly coat the mainshaft with petroleum jelly and place the first bearing

retainer into position. Install one row of 60 needle bearings on the mainshaft, then install the second bearing retainer. Finally, put in another row of 60 needle bearings.

64 Install the front thrust bearing assembly for the front output shaft in the front case. The sequence is thick race, thrust bearing, thin race.

65 Place the front output shaft into its bore in the front case.

66 Position the drive and driven sprockets in the chain and align the sprockets with the shafts. Install the assembly with the tooth side of the drive sprocket facing the case interior.

67 Place the spacer on the drive sprocket and secure it with a snap-ring.

68 Install the snap-ring onto the driven sprocket.

69 Place the rear thrust bearing assembly (thin race, thrust bearing, thick race) into position on the front output shaft.

70 With the recessed side of the oil pump facing downward, toward the case interior, place the oil pump gear on the mainshaft.

71 Install the speedometer drive gear on the mainshaft.

72 Position the magnet in the front case if it was removed.

73 Install the bushing, spring and retainer on the shift rail.

74 Apply sealant to the mating surface of the front case and place the rear case in position on it. Be sure the rear thrust bearing assembly for the front output shaft is seated in the rear case.

75 Align the case bolt holes and alignment dowels and install the bolts. The two bolts installed at the opposite ends of the case should have flat washers. Tighten the bolts in a criss-cross pattern to the specified torque.

76 Install a new seal into the pump housing. Apply petroleum jelly to the pump housing tabs and insert the housing into the rear retainer.

77 Apply sealant to the mating surface of the rear retainer. Install the retainer to the case, aligning the index marks. Tighten the bolts to the specified torque.

78 Coat the lip of a new oil seal with petroleum jelly and install it in the retainer bore.

79 Install the washer and indicator switch in the top of the front case and tighten to the specified torque.

80 Spread a small amount of sealant on the detent retainer bolt and insert the detent poppet, spring and bolt. Tighten the bolt to the specified torque.

81 Install the drain plug with a new gasket to the specified torque.

82 Install a new output shaft seal in the front of the case.

83 Install the front yoke, yoke seal washer and yoke nut. Tighten the nut to the specified torque.

84 Pour six pints of Dexron type automatic transmission fluid into the transfer case. Install and tighten the fill plug to the specified torque.

7 Transfer case (Dana 20) - disassembly, overhaul and reassembly

Disassembly

1 Before disassembly begins, thoroughly clean all external surfaces of the transfer case with a suitable solvent.

2 Remove the cover from the bottom of the transfer case.

3 Remove the intermediate gear shaft lockplate at the rear side of the transfer case.

4 Using a dummy shaft, drive the intermediate shaft out of the rear of the case as shown.

5 Remove the intermediate gear and thrust washers from the case.

6 Remove the bolts attaching the rear output shaft sub-assembly to the transfer case and remove the complete unit from the case.

7 Remove the locknut and washer from the front output shaft yoke (if not removed during unit removal from vehicle) and tap the yoke from the output shaft.

8 Loosen the dog set screw from the rear wheel drive shift fork. Rotate the shift rail to preload the poppet ball and pull the shift rail out of the housing. Remove the shift fork and clutch gear from the case. Remove the poppet ball and spring from the retainer.

9 Remove the front output shaft rear cover and shims.

10 Using a soft faced hammer, tap on the front of the front output shaft, removing the shaft from the rear of the case. Remove the front washer, front bearing, gear washer and the output gear from the case as they clear the shaft. This procedure will also remove the rear bearing cup. **Note:** *The front bearing cup can be removed, if required, by using a suitable puller or driver.*

11 Loosen the dog set screw at the front wheel drive shift fork. Swing the fork and gear toward the cover opening and lift out the gear. Rotate the shift rail to preload the poppet ball and pull the shift rail out of the retainer. Remove the fork as it clears the shift rail and remove the poppet ball and spring from the retainer.

12 Support the inner race of the rear bearing and press the bearing from the shaft.

13 Remove the front output shaft bearing retainer and gasket.

14 Using a suitable puller or pry tool, remove the shift rail lip seals.

15 Remove the speedometer driven gear from the rear retainer.

16 Support the yoke assembly in a soft jawed vise and remove the yoke locknut and washer.

17 Remove the assembly from the vise and, using a soft faced hammer, tap the yoke from the shaft.

18 Support the rear face of the retainer and press the output shaft from the retainer.

19 Pry the seal from the housing bore.

20 Remove the tapered bearing.

21 Using a suitable drift, drive the bearing cup from the housing rear bore.

22 Using a suitable brass drift, remove the bearing cup from the housing front bore.

23 Remove the shims and the speedometer gear from the shaft.

24 Using suitable press plate, remove the front bearing from the shaft.

25 Collapse and remove the pilot bearing from the output shaft pilot bore.

Overhaul

26 Thoroughly clean all components in a suitable solvent and dry with lint-free cloth or dry compressed air.

27 Carefully inspect all bearings and rollers for evidence of chipping, cracks, or worn spots, replacing as necessary.

28 Closely examine all shafts, splines and gears for signs of chipped teeth or excessive wear, replacing as necessary.

29 Inspect the case, cover and bearing cups for any signs of cracking or damage, replacing as necessary.

Reassembly

30 Press a new pilot bearing into the bore of the rear output shaft. Using a suitable piece of tubing. **Note:** *The bearing identification should face outward.*

31 Using a suitable guide, press the front bearing onto the shaft.

32 Install the front bearing cup in the front housing bore, using a suitable guide tool, until the cup seats in the housing.

33 Install the rear bearing and cup in the rear housing bore, using a suitable guide tool, until the cup seats in the housing.

34 Install the seal in the housing rear bore. Use a suitable piece of tubing for this.

35 Insert the output shaft into the front bore of the housing and position the yoke on the rear of the shaft. Support the front of the shaft and press the assembly together, seating the bearing on the shaft.

36 Support the yoke assembly in a soft jawed vise and install the washer and locknut. **Note:** *The locknut must be tightened to the specified torque before installing the driveshaft, during unit installation.*

37 Install the speedometer driven gear and the lockplate in the housing.

38 Position the front bearing cup and tap it into the bore until it is flush with the inner face.

39 Position the retainer and gasket to the case and install the attaching bolts.

40 If the shift rail seals were removed, position new seals in the retainer rail bores and press them into place with a block of wood or other suitable tool.

41 Install the poppet ball and spring (red) for the front wheel drive shift rail in the retainer.

42 Depress the poppet ball and push the shift rail through the retainer into the case. Install the shift fork on the shift rail and tighten the dog set screw. **Note:** *The shift fork set screw boss should face the front of the case.*

43 Install the poppet ball and spring (yellow) for the rear wheel drive shift rail in the retainer. **Note:** *Position the front wheel drive shift rail in neutral, so that interlocks do not interfere with the second rail installation.*

44 Depress the poppet ball and install the shift rail through the retainer into the case. Install the shift fork on the rail and tighten the set screw.

45 Support the front output shaft and press the rear bearing onto the shaft. Use a suitable guide tool for this operation.

46 Place the front wheel drive clutch gear and drive gear in the shift fork with the collar toward the rear of the case.

47 Install the front output shaft through the rear of the case into the clutch gear, drive gear and spacer. Position the rear bearing cup to the case and tap it to within 1/8-inch of the seat.

48 Position the front bearing on the shaft. Supporting the rear of the shaft, press the bearing into position using a suitable guide tool. Tap the front of the shaft rearward to reposition the rear bearing cup.

49 Install the front output shaft rear bearing cover and shim pack.

50 Install the rear wheel drive clutch gear in the case and the shift fork. The gear shift collar should face the rear of the case.

51 Using a piece of suitable tubing, install the front output shaft retainer seal.

52 If the shift rail cups are damaged, or have been removed, position a new cup in the case bore and install by tapping on the end of the cup.

53 Install the front output shaft yoke, washer and locknut. **Note:** *The yoke locknut must be tightened to the specified torque before installing the driveshaft during unit installation.*

54 Using a dummy shaft, install the needle bearings and spacers in the intermediate gear. Position the thrust washers in the case, with the tang in the groove on the case and supporting the intermediate gear in the case. Install the intermediate shaft through the rear of the case. **Note:** *The intermediate shaft is a press fit into the case front bore. Align the lockplate slot in the shaft with the bolt hole, before installing the shaft in the front bore. Take care not to damage the shaft O-ring.*

55 Install the intermediate shaft lockplate and bolt, and tighten to the specified torque.

56 Install the rear output shaft bearing assembly. Tighten the retaining bolts to the specified torque. **Note:** *Take care when engaging the input shaft to prevent damage to the pilot bearings.*

57 Install the rear output shaft yoke, washer and locknut (if not previously done). **Note:** *The rear yoke locknut must be tightened to the specified torque before checking endplay, and before installing the rear driveshaft.*

58 Shift the transfer gears and check for satisfactory gear engagement and shift rail movement. The four wheel drive rail will have the greater poppet ball spring tension.

59 Position the bottom cover and gasket to the case. Install the cover bolts and tighten them to the specified torque.

60 After a major overhaul or rebuild, the front and rear output shafts should be checked for endplay. The endplay must be between 0.001 and 0.005-inch, and shims should be added or removed to achieve this.

Notes

Chapter 8
Clutch and driveline

Contents

	Section
Axle assembly - removal and installation	23
Axle bearing (semi-floating axle) - replacement	20
Axles - general information	17
Chassis lubrication	See Chapter 1
Clutch - adjustment	3
Clutch components - removal, inspection and installation	4
Clutch cross shaft (later mechanical linkage equipped vehicles) - removal and installation	8
Clutch operated Neutral start switch - replacement	12
Clutch - description and check	2
Clutch hydraulic system - bleeding	11
Clutch idler lever and shaft assembly (early models) - removal and installation	5
Clutch master cylinder - removal and installation	9
Clutch pedal (early models) - removal and installation	6
Clutch pedal free play check and adjustment	See Chapter 1
Clutch pedal (later models) - removal and installation	7
Clutch slave cylinder - removal and installation	10
Differential oil change	See Chapter 1
Differential oil level check	See Chapter 1
Driveshaft and universal joints - description and check	13
Driveshaft center support bearing - removal and installation	15
Driveshaft - removal and installation	14
Front drive axle (K-Series) with drum brakes - axleshaft removal and installation	26
Front drive axle (K-Series) with disc brakes - axleshaft removal and installation	27
Front wheel bearings (four-wheel drive) - check, repack and adjustment	28
Front wheel drive hub - removal and installation	25
Full-floating axleshaft - removal and installation	21
General information	1
Hub/drum assembly and wheel bearings (full-floating axle) - removal, installation and adjustment	24
Oil seal (semi-floating axle) - replacement	19
Pinion oil seal (all models) - replacement	22
Semi-floating rear axleshaft - removal and installation	18
Universal joints - removal, overhaul and installation	16

Specifications

Clutch

Pedal free play	See Chapter 1
Clutch plate lining thickness (minimum)	1/16 in

Torque specifications	Ft-lbs
Pressure plate-to-flywheel bolts	11 to 18
Fork lever pivot ball stud	25
Flywheel housing-to-engine bolts	30

Driveshaft

Torque specifications	Ft-lbs
Driveshaft-to-front axle bolt	15
Driveshaft-to-transfer case flange bolt	74
Driveshaft-to-rear axle (strap type) bolt	12 to 17
Driveshaft-to-rear axle (U-bolt type) bolt	18 to 22
Center bearing support-to-hanger bolt	20 to 30
Hanger-to-frame bolt	40 to 50

Axles

Torque specifications	Ft-lbs
Full-floating axleshaft-to-hub flange bolts	115
Differential cover bolt	10 to 20
Differential lock screw	25
Rear hub bearing outer locknut	65

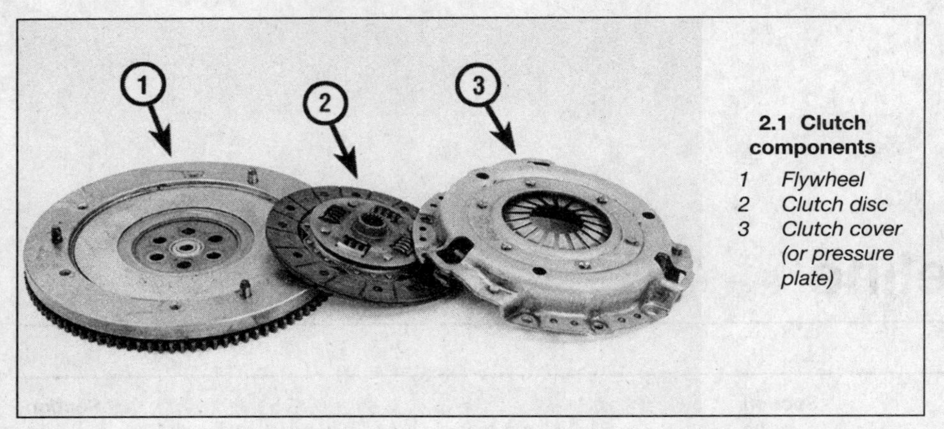

2.1 Clutch components

1 *Flywheel*
2 *Clutch disc*
3 *Clutch cover (or pressure plate)*

1 General information

The Sections in this Chapter deal with the components from the rear of the engine to the rear wheels (except for the transmission and transfer case, which is dealt with in Chapter 7) and forward to the front wheels on four-wheel drive models. In this Chapter, the components are grouped into three categories: clutch, driveshaft and axles (differentials and axleshafts). Separate Sections within this Chapter cover checks and repair procedures for components in each of these three groups.

Since nearly all these procedures involve working under the vehicle, make sure it is safely supported on sturdy jackstands or a hoist where the vehicle can be safely raised and lowered.

2 Clutch - description and check

Refer to illustration 2.1

All models equipped with a manual transmission feature a single dry plate, diaphragm spring-type clutch **(see illustration)**.

Clutch actuation is handled by a mechanical linkage on earlier models and a hydraulic system on 1984 and later models.

When the clutch pedal is depressed, the linkage acts on the clutch fork in the bellhousing. The clutch fork moves the release bearing into contact with the pressure plate fingers, disengaging the clutch driven plate from the flywheel.

Clutch component terminology can sometimes be a problem since common names often differ from the terminology used by the manufacturer. For example, the driven plate is also called the clutch disc or plate, the clutch release bearing is sometimes called a throwout bearing and the clutch fork is often called a release lever.

Some preliminary checks should be performed to diagnose a clutch problem.

a) *To check clutch "spin-down" time, run the engine at idle speed with the transmission in Neutral (clutch pedal up - engaged). Depress the clutch pedal, wait nine seconds and shift the transmission into Reverse. No grinding noise should be heard. A grinding noise would indicate component failure in the pressure plate assembly or the clutch plate.*
b) *To check for complete clutch release, run the engine (with the brakes applied to prevent vehicle movement) and hold the clutch pedal approximately 1/2-inch from the floor. Shift the transmission between 1st and Reverse gear several times. If the shifts are not smooth, component failure is indicated.*
c) *Visually check the clutch pedal bushing at the top of the clutch pedal to make sure there is no binding or excessive wear.*
d) *Working under the vehicle, check to see if the clutch fork is securely attached to the ball/stud.*

3 Clutch - adjustment

1967 through 1983 models are equipped with a clutch linkage requiring periodic checks and adjustment. Refer to Chapter 1 for the clutch pedal free play check and adjustment procedure on these models. 1984 through 1987 models have hydraulic clutch actuation, which automatically compensates for clutch wear.

4 Clutch components - removal, inspection and installation

Refer to illustrations 4.4a, 4.4b, 4.9, 4.10, 4.16 and 4.20

Warning: *Dust produced by clutch wear and deposited on clutch components contains asbestos, which is hazardous to your health. Do not blow it out with compressed air and do not inhale it. Do not use gasoline or solvents to remove the dust. Brake system solvent should be used to flush the dust into a drain pan. After the clutch components are wiped clean with a rag, dispose of the contaminated rags and solvent in a covered container.*

Removal

1 Access to the clutch components is normally accomplished by removing the transmission and bellhousing, leaving the engine in the vehicle. If the engine is being removed for major repairs, then the opportunity should be taken to check the clutch for wear and replace worn components as necessary. The following procedures are described with the engine in the vehicle.

2 Refer to Chapter 7 and remove the transmission, and, on four-wheel drive models, the transfer case. Disconnect the clutch fork rod and pull back the spring. Support the engine with a hoist from above or a jack under the oil pan while the transmission is out of the vehicle. If a jack is used under the engine, make sure a piece of wood is positioned between the jack and the oil pan to spread the load. **Caution:** *On some models the oil pump pick-up is very close to the bottom of the oil pan. If the pan is bent or distorted in any way, engine oil starvation could occur.*

3 Remove the bellhousing.

4 To support the clutch plate during removal, install a clutch alignment tool through the clutch plate hub **(see illustrations)**.

5 Carefully inspect the flywheel and pressure plate for indexing marks. The marks usually are an "X", an "O" or a white letter. If none are visible, use a center punch to make some of your own so the pressure plate and flywheel can be installed in the same relative positions.

6 Turning each one 1/2-turn at a time, slowly loosen the pressure plate-to-flywheel bolts. Work in a criss-cross pattern and loosen the bolts a little at a time until all spring pressure is relieved. Hold the pressure

4.4a A clutch alignment tool (shown here) . . .

4.4b . . . or an old transmission input shaft must be used to support the clutch plate prior to loosening, or to center the plate prior to tightening the pressure plate bolts

4.9 The clutch plate lining, springs and splines (arrows) should be checked for wear

4.10 Pry the clutch fork clip (arrow) up with a screwdriver to release the fork from the ballstud

plate to keep it from falling and completely remove the bolts, then separate the pressure plate and clutch plate from the flywheel.

Inspection

7 Ordinarily, when a problem develops in the clutch it can be attributed to wear of the clutch driven plate assembly. However, all components should be inspected at this time.

8 Inspect the flywheel carefully for score marks, cracks, evidence of overheating (which will appear as blue spots) and damaged ring gear teeth. If damage or wear are noted, remove the flywheel and take it to an automotive machine shop to see if it can be resurfaced, otherwise a new one will be required.

9 Inspect the lining on the clutch plate. There should be at least 1/16-inch of lining above the rivet heads. Check for loose rivets, distortion, cracks, broken springs and any obvious damage **(see illustration)**. As mentioned above, the clutch plate is normally replaced each time it is removed, so if there is any doubt about its condition, install a new one.

10 Remove the clutch fork from the ballstud by pressing it out of the ballstud clip with a screwdriver **(see illustration)**.

11 Check the release bearing (throwout bearing) for roughness and excessive wear by turning it with your hands. The release bearing is usually replaced each time it is removed (the cost is relatively low compared to the labor required to gain access to it).

4.16 When installing the clutch and pressure plate, make sure the friction surfaces are clean and grease-free

12 If the ballstud is worn, it can be replaced by unscrewing it from the bellhousing with a wrench and installing a new one.

13 The clutch fork retainer clip may be removed by prying it from the fork with a small screwdriver.

14 Check the machined surface of the pressure plate for score marks, cracks and evidence of overheating (blue spots on the friction surface). If the pressure plate is scored or otherwise damaged, take it to an automotive machine shop to see if they can resurface it. Also check for distorted and cracked springs and release fingers. If a new pressure plate is required, rebuilt units are available.

Installation

15 Before installation, carefully clean the flywheel and pressure plate machined surfaces. Do not get oil or grease on the clutch friction surfaces. Handle the parts only with clean hands.

16 Attach the clutch plate and pressure plate to the flywheel **(see illustration)**. Make sure the clutch plate is installed with the cor-

4.20 Lubricate the clutch release bearing recess (arrow) and the fork groove with high temperature grease

rect side against the flywheel (most replacement clutches are marked "flywheel side" or something similar to avoid confusion). Also make sure the index marks are aligned to ensure correct clutch/pressure plate balance.

17 Holding the clutch plate in position with an alignment tool **(see illustrations 4.4a and 4.4b)**, tighten the pressure plate-to-flywheel bolts finger tight only, working around the pressure plate.

18 If an alignment tool has not been used up to this point, center the clutch plate at this time. Move the plate until it is centered so the splines of the transmission input shaft will pass though the plate and into the pilot bearing.

19 Slowly, one turn at a time, tighten the pressure plate-to-flywheel bolts. Work in a criss-cross pattern to prevent distortion of the pressure plate as the bolts are tightened to the specified torque. Remove the alignment tool.

20 Using high temperature grease, lubricate the entire inner surface of the release bearing. Make sure the groove inside is completely filled. Also lubricate the ballstud and the clutch fork groove in the release bearing **(see illustration)**.

21 Install the clutch fork in the bellhousing, making sure the fork engages with the ballstud. Attach the release bearing to the clutch fork.

22 Install the bellhousing.

23 Install the transmission.

24 Line up the pushrod with the clutch fork and then attach the return spring.

25 On non-hydraulic clutch models adjust the clutch (Chapter 1).

5 Clutch idler lever and shaft assembly (early models) - removal and installation

1 Disconnect the negative cable at the battery. Place the cable out of the way so it cannot accidentally come in contact with the negative terminal of the battery, as this would once again allow power into the electrical system of the vehicle.

2 Disconnect and remove the clutch fork return spring.

3 Disconnect the pedal pushrod and the clutch fork pushrod from the idler lever and shaft assembly.

4 Loosen the shaft ballstud nut and slide the stud out of the frame bracket slot.

5 Clean and inspect all components for wear, replacing as necessary.

6 Begin reassembly by installing the idler lever and shaft assembly on the ballstud.

7 Connect the pedal pushrod and clutch fork pushrod to the idler lever and shaft assembly.

8 Adjust the clutch linkage as described in Chapter 1.

6 Clutch pedal (early models) - removal and installation

1 Disconnect the negative cable at the battery. Place the cable out of the way so it cannot accidentally come in contact with the negative terminal of the battery, as this would once again allow power into the electrical system of the vehicle.

2 Disconnect and remove the clutch pedal return spring.

3 Remove the nut, bolt and washers securing the lever in position. Remove the lever.

4 Remove the clutch pedal from the sleeve assembly and remove the brake pedal return spring.

5 Remove the bolt retaining the sleeve assembly to the brake pedal support.

6 Remove the bushings and and the sleeve assembly.

7 Check all components for wear and replace if necessary.

8 Lubricate the shaft bushings with grease.

9 Installation is the reverse of removal. When finished, the clutch pedal free play should be checked and adjusted as described in Chapter 1.

7 Clutch pedal (later models) - removal and installation

Removal

1 Disconnect the negative cable at the battery. Place the cable out of the way so it cannot accidentally come in contact with the negative terminal of the battery, as this would once again allow power into the electrical system of the vehicle.

2 Working beneath the vehicle, disconnect the clutch pushrod at the cross shaft.

3 Refer to Chapter 10 and remove the steering column covers.

4 With the steering column covers removed, remove the screws retaining the pushrod boots to the bulkhead.

5 Disconnect the parking brake release handle from the instrument cluster.

6 On vehicles equipped with air conditioning, remove the duct from the lower left side of the instrument cluster.

7 Disconnect the Neutral start switch from the clutch pedal arm.

Mechanical clutch

8 Remove the bolts attaching the lower section of the clutch pedal arm to the upper arm. **Caution:** *Maintain pressure on the clutch pedal because when the lower bolt is removed the arm will snap up.*

9 Remove the lower arm and pushrod from the vehicle.

10 Remove the clutch pedal return spring.

11 Remove the pedal pivot shaft retaining nut.

12 Using a metal rod, push out the pivot shaft, allowing the rod to support the brake pedal components in position when the pivot shaft is removed.

13 Remove the clutch pedal assembly from the support assembly.

14 Remove the pedal bushings and spacer from the pedal arm.

15 Clean and inspect all components for wear, replacing where necessary.

Hydraulic clutch

16 Disconnect the master cylinder pushrod from the clutch pedal.

17 Remove the clutch pedal pivot nut, bolt or stud and remove the clutch pedal. On models which use a stud, insert a long bolt or rod into the clutch assembly bracket while withdrawing the stud.

Installation

18 Installation is the reverse of removal. Lubricate all contact surfaces with chassis grease.

19 When finished, check, and if necessary, adjust the clutch pedal free play (Chapter 1).

8 Clutch cross shaft (later mechanical linkage equipped vehicles) - removal and installation

1 Disconnect and remove the clutch release fork return spring.

2 Disconnect the pedal-to-cross shaft operating rod from the cross shaft.

3 Disconnect the release fork-to-cross shaft operating rod from the release fork.

4 Loosen the cross shaft ballstud nut, then slide the ballstud from the slot in the bracket.

5 Pull the cross shaft from the ballstud nut at the engine.

6 Installation is the reverse of removal after applying grease to the ballstud. After installation, adjust the pedal free play (Chapter 1).

9 Clutch master cylinder - removal and installation

1 Remove the lower steering column covers and, if equipped, the lower left side air conditioning duct.
2 In the passenger compartment, remove the clip and disconnect the master cylinder pushrod from the pedal.
3 In the engine compartment, disconnect and plug the fluid lines at the master cylinder.
4 Remove the nuts and detach the master cylinder from the firewall.
5 Installation is the reverse of removal. Be sure to use a new gasket.
6 Bleed the clutch hydraulic system (see Section 11).

10 Clutch slave cylinder - removal and installation

Refer to illustration 10.3

1 Disconnect the negative cable at the battery. Place the cable out of the way so it cannot accidentally come in contact with the negative terminal of the battery, as this would once again allow power into the electrical system of the vehicle.
2 Raise the vehicle and support it securely on jackstands.
3 Disconnect and plug the master cylinder-to-slave cylinder fluid line **(see illustration)**.
4 Remove the nuts and detach the slave cylinder from the bellhousing.
5 Installation is the reverse of removal.
6 Bleed the clutch hydraulic system (Section 11).

11 Clutch hydraulic system - bleeding

1 The hydraulic system should be bled whenever any part of the system has been removed or if the fluid level has been allowed to fall so low that air is drawn into the master cylinder. The procedure is similar to bleeding a brake system.
2 Fill the master cylinder reservoir with new brake fluid of the specified type (see Chapter 1). **Caution:** *Do not reuse any of the fluid coming from the system during the bleeding operation or use fluid which has been inside an open container for an extended period of time.*
3 Raise the vehicle and support it securely on jackstands.
4 Remove the slave cylinder and tip it up at a 45° angle so the bleeder valve is up.
5 Remove the dust cap which fits over the bleeder valve and push a length of plastic hose over the fitting. Place the other end of the hose into a clear container with about two inches of brake fluid. The hose end must be in the fluid at the bottom of the container.

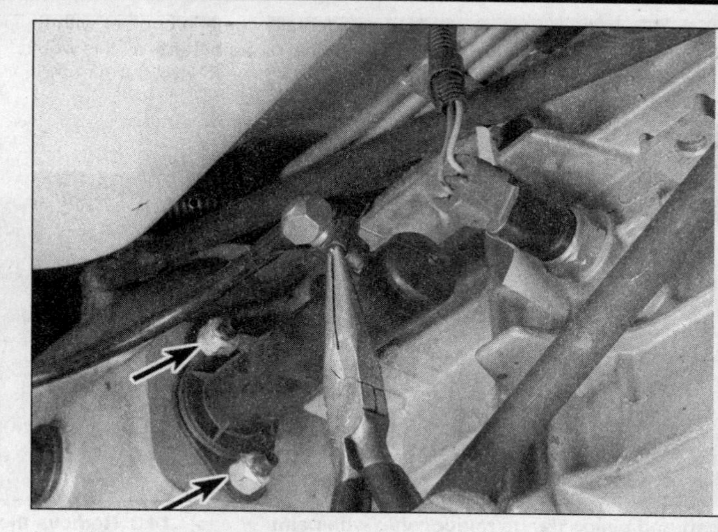

10.3 Clutch slave cylinder mounting (typical)

6 Have an assistant depress the clutch pedal and hold it. Open the bleeder valve on the slave cylinder, allowing fluid to flow through the hose. Close the bleeder valve when no more fluid flows from the hose. Once closed, have your assistant release the pedal.
7 Continue this process until all air is evacuated from the system, indicated by a solid stream of fluid being ejected from the bleeder valve. Keep a close watch on the fluid level inside the master cylinder reservoir. If the level drops too low, air will be sucked back into the system and the process will have to be started all over again.
8 Install the slave cylinder and lower the vehicle. Check carefully for proper operation before driving the vehicle.

12 Clutch operated Neutral start switch - replacement

1 Some models are equipped with a clutch operated Neutral start switch which is mounted on the clutch pedal support and allows the engine to be started only with the clutch pedal fully depressed.
2 Disconnect the negative cable at the battery. Place the cable out of the way so it cannot accidentally come in contact with the negative terminal of the battery, as this would once again allow power into the electrical system of the vehicle.
3 Remove the retaining bolt and detach the switch from the clutch pedal bracket.
4 Unplug the electrical connector from the switch.
5 Install a new switch in the reverse order. Check to be sure that the engine can be started only when the clutch pedal is fully depressed.

13 Driveshaft and universal joints - description and check

The driveshaft runs between the transmission and the rear axle (differential) and, on four-wheel drive models, between the transfer case and the front axle.

All driveshafts used to drive the rear wheels have needle bearing type universal joints. One-piece shafts have a splined sliding sleeve at the front connecting to the output shaft of the transmission, while two-piece shafts have a central slip joint.

The purpose of these devices is to accommodate, by retraction or extension, the varying shaft length caused by the movement of the rear axle as the rear suspension deflects. On four-wheel drive vehicles, due to the extent of the front driveshaft angle, a constant velocity joint is used at the transfer case end of the driveshaft.

On two-piece shafts, the shaft is supported near the center by a center support bearing mounted in a bracket attached to the frame crossmember.

Some universal joints can be lubricated, but other types are sealed for life. The constant velocity joint used on four-wheel drive vehicles must be lubricated at regular intervals (see Chapter 1).

Since the driveshaft is a balanced unit, it is important that no undercoating, mud, etc. be allowed to build up on it. When the vehicle is raised for service it is a good idea to clean the driveshaft and inspect it for obvious damage. Also make sure that the small weights used to originally balance the driveshaft are in place and securely attached.

Whenever the driveshaft is removed, it must be reinstalled in the same relative position to preserve the balance.

Problems with the driveshaft are usually indicated by a noise or vibration while driving the vehicle. A road test should verify if the problem is the driveshaft or another vehicle component.

a) *On an open road, free of traffic, drive the vehicle and note the engine speed (rpm) at which the problem is most evident.*
b) *With this in mind, drive the vehicle again, this time keeping the transmission in each gear for an extended time and running the engine up to the rpm noted.*

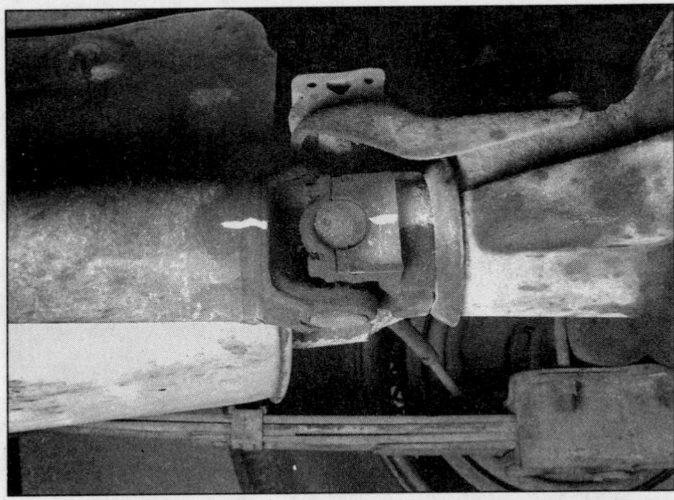

14.2 Mark the driveshaft and pinion flange relationship with paint

14.3 Remove the front driveshaft mounting nuts (arrows)

c) If the noise or vibration occurs at the same engine speed regardless of which gear the transmission is in, the driveshaft is not to blame, since the speed of the driveshaft varies in each gear.

d) If the noise or vibration decreased or was eliminated, visually inspect the driveshaft for damage, material on the shaft which would affect balance, missing weights and damaged universal joints.

To check for worn universal joints:

a) On an open road, free of traffic, drive the vehicle slowly until the transmission is in high gear. Let off on the throttle, allowing the vehicle to coast, then accelerate. A clunking or knocking noise indicates worn universal joints.

b) Drive the vehicle at about 10 to 15 mph, then place the transmission in Neutral, allowing the vehicle to coast. Listen for abnormal driveline noises.

14 Driveshaft - removal and installation

Refer to illustrations 14.2, 14.3, 14.9a and 14.9b

1 Raise the vehicle and support it securely on jackstands.

2 Using chalk or white paint, mark the relationship between the various installed components to ensure correct alignment during installation **(see illustration)**. It is very important that the front driveshaft-to-front axle and transfer case and rear driveshaft-to-rear axle companion flange be marked in this way.

Front driveshaft (four-wheel drive)

3 Disconnect the front U-joint by removing the retainers and unbolt the rear U-joint from the transfer case flange **(see illustration)**.

4 Collapse the driveshaft sufficiently to allow the front U-joint to be disengaged from the front axle.

5 Withdraw the driveshaft from the vehicle by lowering it at the rear between the transfer case and the frame. Support the driveshaft and make sure the cap assemblies do not fall from the open ends of the U-joint. Wrap tape around the caps if necessary, to hold them in place.

6 To install, extend the driveshaft to its full length and then compress it to half the full travel.

7 Slide the slip yoke end of the driveshaft toward the front axle and fit the U-joint caps into the tabs on the flange. Install the retainers and bolts. Tighten the bolts securely.

8 Extend the driveshaft, seat it against the transfer case flange and install the bolts. Tighten the bolts securely.

Rear driveshaft

9 Unbolt the U-bolts or straps, whichever

14.9a U-bolt type driveshaft-to-axle flange mount

14.9b Strap type driveshaft-to-axle flange mount

15.2 Remove the center support bearing mounting bolts (arrows)

16.3 Remove outer-type universal joint snap-rings with a small pair of pliers

type is used **(see illustrations)**. It is a good idea to tape the universal joint bearing cups to the trunnions to keep them from falling off.

10 On vehicles which have a two-piece shaft, remove the bolts from the center bearing bracket.

11 Push the shaft forward slightly to disengage it from the rear axle, then, while lowering it, withdraw the shaft assembly to the rear. The splined front sliding sleeve section will be drawn off the transmission output shaft during the removal operation so have a plastic bag or sheet of plastic film and a large rubber band handy to seal the rear of the transmission, preventing fluid or lubricant loss.

12 To install a one-piece shaft, reverse the removal operations.

13 To install a two-piece driveshaft, slide the front part of the shaft into the transmission. On two-wheel drive models the center joint spline has an alignment key, so it is not possible to install it incorrectly. On four-wheel drive vehicles equipped with two-piece driveshafts, the method of installation is as follows.

14 Side the front part of the shaft into the transmission and bolt the center bearing bracket to the crossmember. Slide the grease cap and gasket (if equipped) onto the rear splines and rotate the shaft until the front universal joint trunnion is in a vertical position.

15 Support the rear section of the shaft and align the universal joint yoke trunnions in the same vertical attitude as the front one.

16 Rotate the rear section of the shaft four splines toward the left side of the vehicle and connect the rear shaft to the front shaft. This alignment operation is known as phasing.

17 Connect the rear of the shaft to the rear axle flange, then tighten the grease cap on the slip joint.

18 Tighten the support bearing bolts and the U-joint nuts or bolts securely.

15 Driveshaft center support bearing - removal and installation

Refer to illustration 15.2

1 Remove the rear driveshaft (Section 14).

2 Remove the bolts which retain the bearing assembly to the bracket **(see illustration)**.

3 Separate the front driveshaft and lower the bearing assembly from the vehicle.

4 Installation is the reverse of removal.

16 Universal joints - removal, overhaul and installation

Refer to illustrations 16.3, 16.4, 16.5, 16.6, 16.17, 16.20, 16.21a, 16.21b, 16.23 and 16.31

Note: *Always purchase a universal joint service kit for your model vehicle before beginning this procedure. Also, read through the entire procedure before beginning work.*

1 Remove the driveshaft (Section 14).

Outer snap-ring type

2 Place the driveshaft on a bench equipped with a vise.

3 Remove the snap-rings, using a small pair of pliers **(see illustration)**.

4 Support the cross (trunnion) on a short piece of pipe or a large socket and use another socket to press out the cross by closing the vise **(see illustration)**.

5 Press the cross through as far as possible, then grip the bearing cup with locking pli-

16.4 To remove the U-joint from the driveshaft, use a vise as a press - the small socket will push the cross and bearing cup into the large socket

16.5 Grip the bearing cap with locking pliers and remove it from the yoke

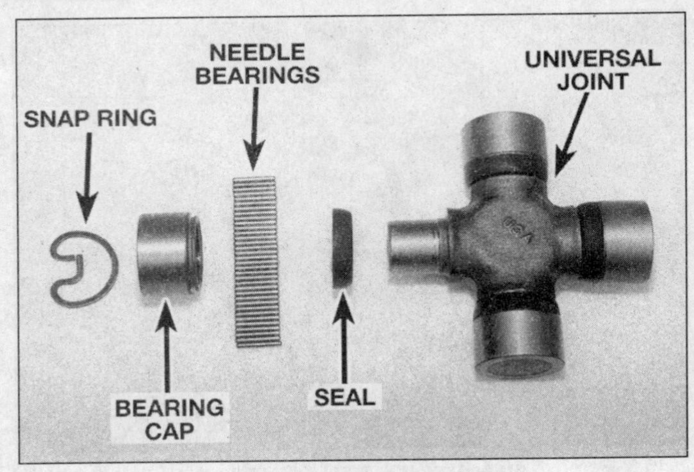

16.6 Outer snap-ring type U-joint

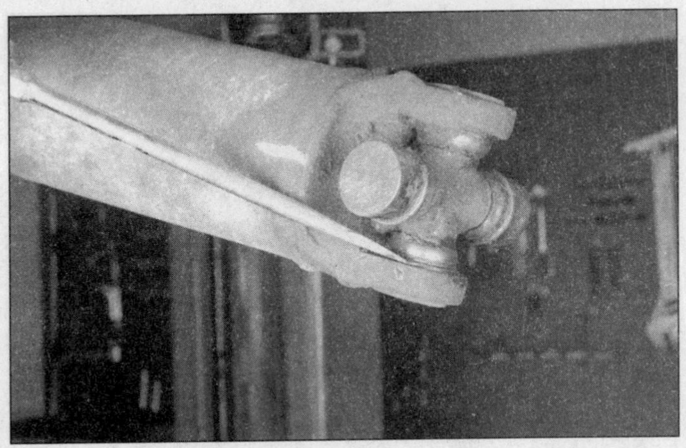
16.17 Remove inner-type snap-rings from the U-joint by tapping them off with a screwdriver and hammer

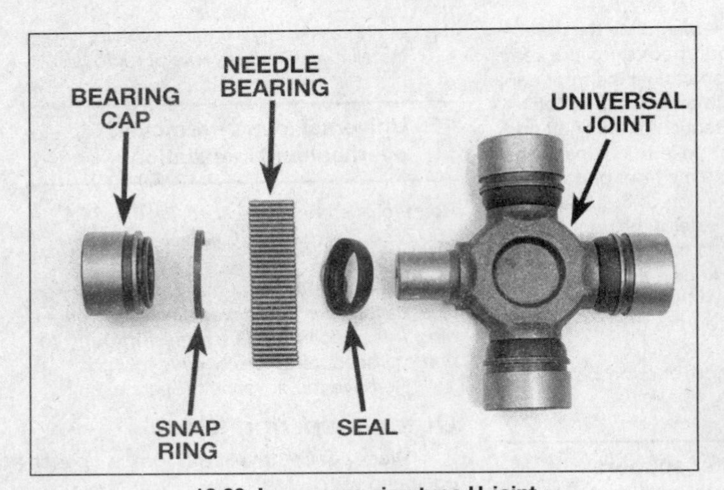

16.20 Inner snap-ring type U-joint

16.21a Press the injected plastic-type joint bearing cup into place and install the snap-ring in the groove

ers and remove it **(see illustration)**.

6 A universal joint repair kit will contain a new trunnion, seals, bearings, cups and snap-rings **(see illustration)**.

7 Inspect the bearing cup housing in the driveshaft for wear and damage.

8 If the bearing cup housings in the yoke are so worn that the cups are a loose fit in the yokes, the driveshaft itself will have to be replaced with a new one.

9 Make sure that the dust seals are properly located on the trunnion so that the cavities face the trunnion.

10 Using a vise, press one bearing cup into the yoke approximately 1/4-inch.

11 Use thick multi-purpose grease to hold the needle rollers in place in the cup.

12 Insert the trunnion into the partially installed bearing cup, taking care not to dislodge the needle rollers.

13 Stick the needle bearings into the opposite cup, hold the trunnion in correct alignment and press both cups into place by slowly and carefully closing the jaws of the vise.

14 Use a socket slightly smaller in diameter than the caps to press the caps into the yoke. Press in one side, install the snap-ring, then press the other side to shift the trunnion assembly tight against the installed snap-ring, and install the other snap-ring.

15 Repeat the operations for the remaining two bearing cups.

Injected plastic (inner snap-ring) type

16 If the joint has never been rebuilt, press out the bearing cups as described in Steps 5 and 6.

17 If the joint has been previously rebuilt, remove the snap-rings (bearing retainers) inboard of the yokes **(see illustration)**.

18 If this is the first time the joint is being rebuilt it will not be necessary to remove the snap-rings; the pressing operation will shear the molded plastic retaining material.

19 Remove the trunnion (cross) and clean all plastic material from the yoke. Use a small punch to remove the plastic from the injection holes.

20 Reassembly is the same as for the outer snap-ring joint described in Steps 9 through 14, except that the snap-rings are installed inside the yoke ears **(see illustration)**.

21 When installing the bearing cup, press it in until the snap-ring can be installed **(see illustration)**. If difficulty is encountered, strike the yoke sharply with a hammer. This will spring the yoke ears slightly and allow the snap-ring groove to move into position **(see illustration)**.

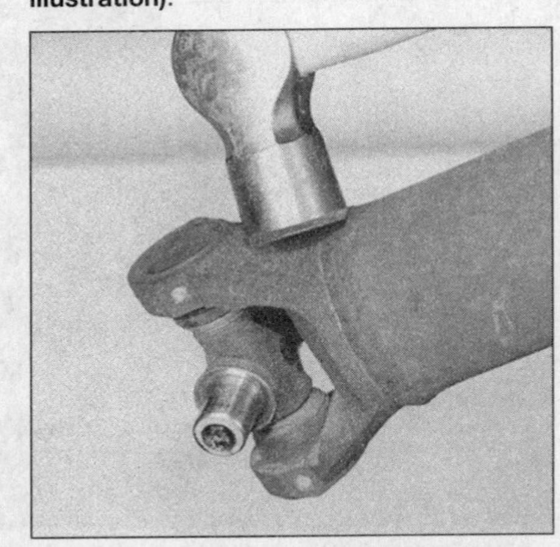
16.21b Strike the yoke with a hammer to reposition the retainer groove

16.23 Alignment marks must be made before beginning disassembly of the CV joint (arrows)

16.31 Install the centering ball by tapping it into place with a brass drift and hammer

Double cardan constant velocity (CV) joint

22 The CV joint can be disassembled for inspection after obtaining an inspection kit containing two bearing cups and retainer. An overhaul kit will be required if damage or wear is evident.

23 Use a hammer and punch to mark the relative positions of the flange yoke and coupling before beginning disassembly **(see illustration)**.

24 Disassemble the joint as described in Steps 3 through 6.

25 Disengage the flange yoke and trunnion (cross) from the centering ball. Pry the seal from the ball socket and remove the washers, springs and the 3 ball seats.

26 Clean the ball seat insert bushing and inspect it for wear. Replace the flange yoke and trunnion assembly if wear is excessive.

27 Clean the seal, ball seats, spring and washers and inspect them for wear.

28 Remove all plastic material from the groove in the coupling yoke (if this type is used).

29 Inspect the centering ball for wear and damage.

30 If the ball is damaged or worn, use a puller to draw it off the stud or take it to a dealer or repair shop for removal and installation of a new one.

31 Place a new ball onto the stud and tap it into position until it seats firmly on the stud shoulder **(see illustration)**. It is extremely important that the ball surface not be damaged during installation.

32 Using grease provided in the repair kit, lubricate all the parts and insert them in the ball seat cavity in the following order: spring, small washer, three ball seats (with the largest opening out to receive the ball), large washer and the seal.

33 Lubricate the seal lip and press it into the cavity with the lip facing in. Fill the cavity with the provided grease.

34 Attach the flange yoke to the centering ball, making sure that the alignment marks are correctly positioned.

35 Install the bearing caps as described in Steps 9 through 14.

17 Axles - general information

Four-wheel drive front axles are very similar in construction to rear axles. All axles are hypoid gear types with cast carriers and pressed in axle tubes. Some models are equipped with limited slip or locking differentials designed to give better traction when one wheel has more traction than the other.

Front axles are equipped with steering knuckles at the ends of the axle tubes and universal joints at the ends of the axleshafts, allowing the front wheels to steer. Several types of front hubs were used over the long production life of these models, including locking, automatic and full-time four-wheel drive.

Two types of axleshafts are used; semi- and full-floating. Semifloating axleshafts are supported at the outer end of the axle by bearings pressed into the outer end of the axle tube and are retained in the differential splines. Full-floating axleshafts ride in the differential splines at their inner ends and roller bearings mounted in the end of the axle tube at the outer end. They are held in place by a hub mounted flange which can be unbolted, allowing the axleshaft to be removed with the wheel in place.

18 Semi-floating rear axleshaft - removal and installation

Refer to illustrations 18.3, 18.4, 18.6 and 18.7

Removal

1 Raise the rear of the vehicle, support it securely on jackstands and remove the wheel and brake drum. On locking differential equipped models, remove both rear wheels and brake drums.

2 Remove the differential cover and allow the oil to drain into a container (Chapter 1, Section 37).

Conventional differential

3 Remove the lock screw and pull out the pinion shaft **(see illustration)**.

18.3 Remove the lock screw and pull the pinion shaft out of the differential

18.4 With an assistant pressing the axleshaft into the differential, the C-lock can be pulled out of the groove in the end of the axleshaft (semi-floating axle)

18.6 Withdraw the pinion shaft until it rests on the case (locking differential)

4 Have an assistant push in on the outer flanged end of the axleshaft while you remove the C-lock from the groove in the inner end of the shaft **(see illustration)**.

Locking differential

5 Rotate the differential for access, support the pinion shaft so it won't fall into the case, then remove the lock screw.

6 Withdraw the pinion shaft part way. Rotate the differential until the shaft touches the case, providing enough clearance for access to the C-locks **(see illustration)**.

7 Use a screwdriver to rotate the C-lock until the open end points in **(see illustration)**.

8 With the C-lock in position so it will pass through the end of the thrust block, push the axleshaft in and remove the C-lock. Repeat the operation for the opposite axleshaft.

All models

9 With the C-lock removed, withdraw the axleshaft, taking care not to damage the oil seal. Some models have a thrust washer in the differential; make sure it does not fall out when the axleshaft is removed.

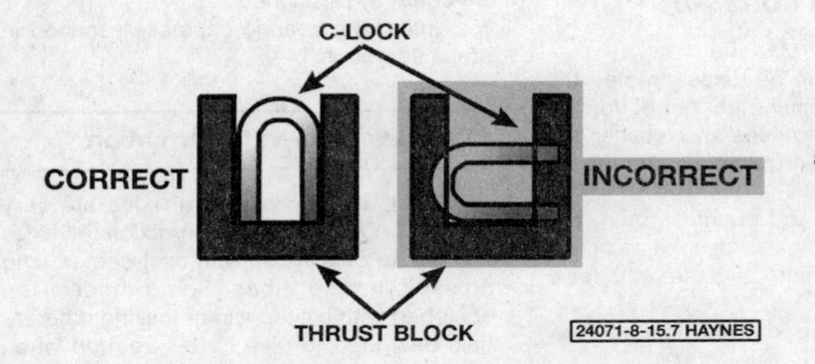

18.7 The C-lock must be positioned as shown before the axleshaft can be removed (locking differential)

Installation

All models

10 To install, carefully insert the axleshaft into the housing and seat it securely in the differential.

Conventional differential

11 Install the C-lock in the axleshaft groove and pull out on the flange to lock it.

12 Insert the pinion shaft, align the hole in the shaft with the lock screw hole and install the lock screw. Tighten the lock screw to the specified torque.

Locking differential

13 Install the C-locks with the pinion shaft still partially withdrawn, making sure the C-locks are positioned as shown in **illustration 18.7**.

14 Withdraw the axleshaft carefully until the C-lock is clear of the thrust block.

15 After the C-locks are installed, push the pinion shaft into place with the groove lined up with the lock screw hole and then install the lock screw. Tighten the lock screw to the specified torque.

16 Install the cover and fill the differential with the specified oil.

17 Install the brake drums and wheels and lower the vehicle.

19.2 The end of the axleshaft can be used to pry the old seal out of the axle housing

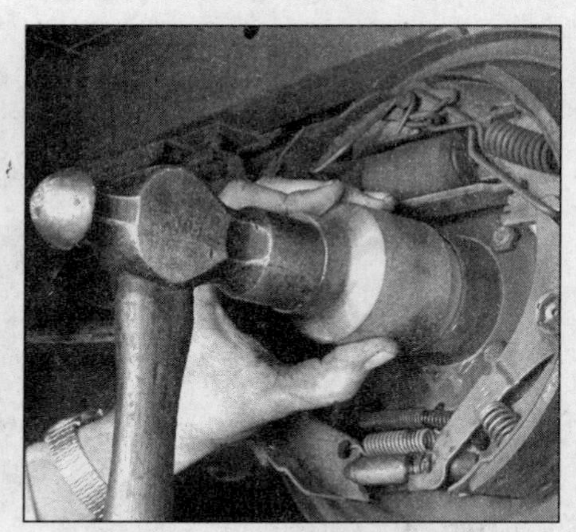

19.3 Use a seal installer tool, large socket or piece of pipe to tap the seal evenly into place

19 Oil seal (semi-floating axle) - replacement

Refer to illustrations 19.2 and 19.3

1 Remove the axleshaft (Section 18).

2 Pry out the old oil seal from the end of the axle housing using a large screwdriver or the inner end of the axleshaft as a lever **(see illustration)**.

3 Apply high melting point grease to the oil seal recess and tap the new seal evenly

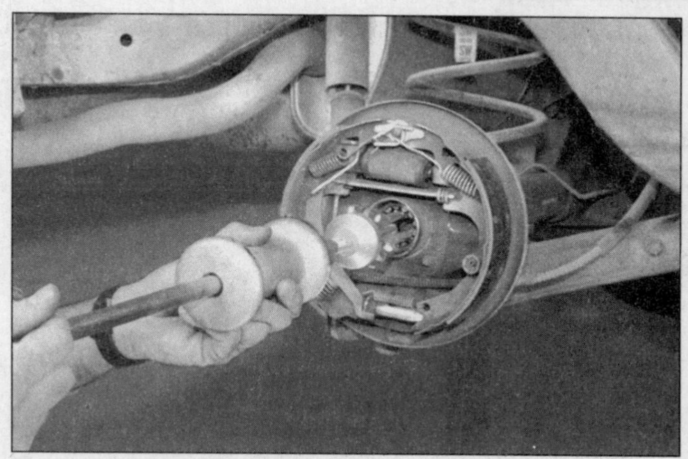

20.3 A slide hammer is required when removing the semifloating axleshaft wheel bearing

20.4 Use a special bearing installer tool, large socket or piece of pipe to tap the bearing evenly into the axle housing

21.1 Remove the bolts which attach the axleshaft flange to hub-wheel installed

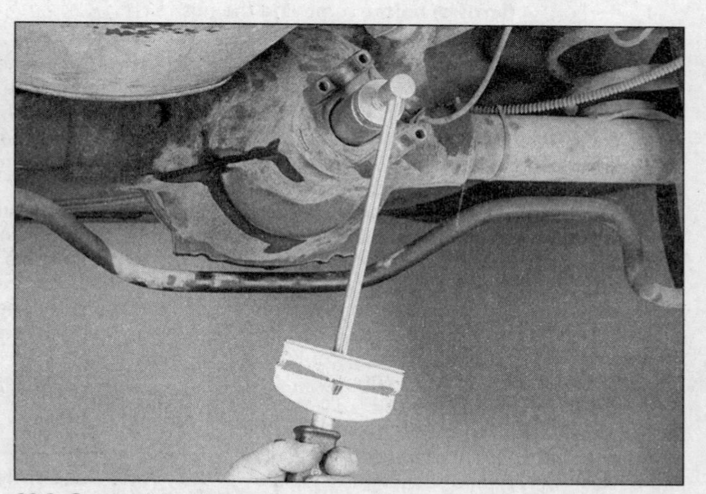

22.3 On semi-floating axles, use an in-lbs torque wrench to check the torque necessary to move the pinion

into position using a hammer and seal installation tool **(see illustration)**, large socket or piece of pipe so that the lips are facing in and the metal face is visible from the end of the axle housing. When correctly installed the face of the oil seal should be flush with the end of the axle housing.

20 Axle bearing (semi-floating axle) - replacement

Refer to illustrations 20.3 and 20.4

1 Remove the axleshaft (Section 18) and the oil seal (Section 19).

2 A bearing puller which engages behind the bearing will be required for this job.

3 Attach a slide hammer and extract the bearing from the axle housing **(see illustration)**.

4 Clean out the bearing recess and drive in the new bearing using a piece of pipe positioned against the outer bearing race **(see illustration)**. Make sure that the bearing is tapped in to the full depth of the recess and that the numbers on the bearing are visible from the outer end of the axle housing.

5 Discard the old oil seal and install a new one, then install the axleshaft.

6 On some early models, the bearings are pressed onto the axleshaft itself. These will have to be taken to a dealer or repair shop for removal as they are very tight and a retaining collar must be cut off.

21 Full-floating axleshaft - removal and installation

Refer to illustration 21.1

1 Remove the bolts which attach the axleshaft flange to the hub. There is no need to remove the wheel or jack up the vehicle **(see illustration)**.

2 Tap the flange with a soft faced hammer to loosen the shaft and then grip the rib in the face of the flange with a pair of self-locking pliers. Twist the shaft slightly in both directions and then withdraw it from the housing. Place a drip pan under the outer end of the axle to catch any lubricant which might leak out while the axle is removed.

3 Installation is the reverse of removal, taking care to hold the axleshaft level in order to engage the splines at the inner end with those in the differential side gear. Always use a new gasket on the flange and keep both the flange and hub mating surface free from grease or oil.

22 Pinion oil seal (all models) - replacement

Refer to illustrations 22.3, 22.4, 22.6 and 22.10

1 Raise the rear of the vehicle and support it securely on jackstands.

2 Disconnect the driveshaft and fasten it out of the way.

3 On semi-floating axles, use an inch-pound torque wrench to check the torque required to rotate the pinion and record it for use later **(see illustration)**.

4 Scribe or punch alignment marks on the

22.4 Mark the relative positions of the pinion, nut and flange (arrows) before removing the nut

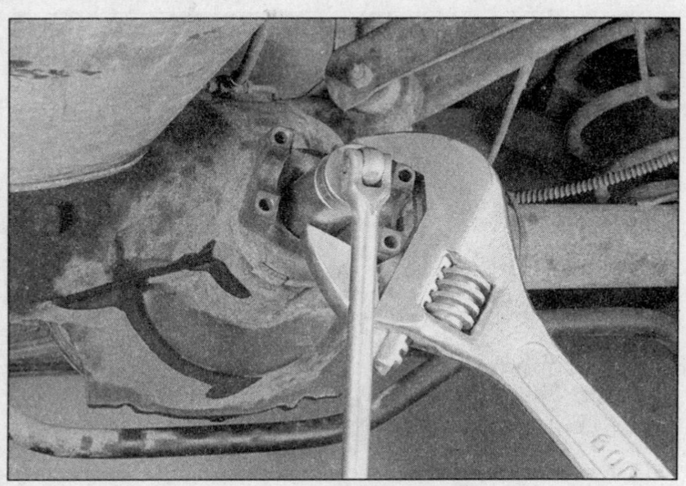

22.6 A large adjustable wrench can be used to immobilize the pinion flange while you loosen the locknut

pinion stem, nut and flange **(see illustration)**.

5 Count the number of threads visible between the end of the nut and the end of the pinion and record it for use later.

6 A special flange holding tool or a large wrench must now be used to keep the companion flange from moving while the self-locking pinion nut is removed **(see illustration)**.

7 Remove the pinion nut.

8 Withdraw the companion flange. It may be necessary to use a two or three jaw puller engaged behind the flange to draw it out. Do not attempt to pry behind the flange or hammer on the end of the pinion shaft.

9 Pry out the old seal and discard it.

10 Lubricate the lips of the new seal with high-temperature grease and tap it evenly into position with a large socket, making sure it enters the housing squarely and to its full depth **(see illustration)**.

11 Align the mating marks made before disassembly and install the companion flange. If necessary, install and tighten the pinion nut to draw the flange into place. Do not try to hammer the flange into position.

12 Apply non-hardening sealant to the ends of the splines which are visible in the center of the flange so oil will be sealed in.

13 Install the washer (if equipped) and pinion nut. Tighten the nut carefully until the original number of threads are exposed and the marks are lined up.

14 On semi-floating axles, measure the torque required to rotate the pinion and tighten the nut in small increments until it matches the figure recorded in Step 5. In order to compensate for the drag of the new oil seal, the nut should be further tightened so that the rotational torque of the pinion exceeds that recorded by no more than 1 to 5 in-lbs.

15 Connect the driveshaft and lower the vehicle.

23 Axle assembly - removal and installation

Front axle

1 Disconnect the driveshaft from the front axle and fasten it out of the way.

2 Raise the front of the vehicle enough to remove the weight from the front springs.

3 Disconnect the lower shock absorber mounts.

4 Disconnect the steering link rod from the steering arm.

5 Disconnect and plug the brake hoses.

6 Disconnect the axle vent hose.

7 Remove the U-bolt nuts and detach the axle from the springs.

8 Raise the vehicle to allow the axle to be removed from under the vehicle.

9 Installation is the reverse of removal.

10 Bleed the brake system (Chapter 9).

Rear axle

11 Raise the rear of the vehicle and support it securely on jackstands. Remove the wheels.

12 Position a jack under the rear axle differential carrier.

13 Disconnect the driveshaft from the rear axle companion flange. Fasten the driveshaft out of the way.

14 Disconnect the shock absorber lower mounts.

15 Disconnect the vent hose from the fitting on the axle housing and fasten it out of the way.

16 Disconnect and plug the brake hose from the axle housing.

17 Remove the brake drums.

18 Disconnect the parking brake cables from the actuating levers and the backing plate (see Chapter 9).

19 On coil spring models, support the trailing arms with floor jacks on each side and remove the trailing arm-to-axle housing bolts.

20 Lower the trailing arms to clear the axle housing and remove the axle.

21 On leaf spring models, disconnect the rear spring U-bolts. Remove the spacers and clamp plates.

22 Lower the jack under the differential and then remove the rear axle assembly from under the vehicle.

23 Installation is the reverse of removal. Lower the vehicle weight onto the wheels and then tighten the U-bolt nuts or trailing arm mounting nuts to the specified torque.

24 Bleed the brake system (Chapter 9).

22.10 Lubricate the lips of the new pinion seal and seat it squarely in the bore, then drive it into the carrier with a seal driver, a large socket (shown) or a block of wood

24.3a Remove the hub retaining ring . .

24.3b . . then remove the key from the hub

24 Hub/drum assembly and wheel bearings (full-floating axle) - removal, installation and adjustment

Removal

Refer to illustrations 24.3a, 24.3b and 24.3c

1 Remove the axleshaft (Section 21).

2 Raise the vehicle, support it securely on jackstands and remove the rear wheels.

3 Remove the retaining ring and key (if equipped) from the end of the axle housing **(see illustrations)**. Use a special hub nut socket to remove the locknut **(see illustration)**.

4 Release the tang of the retainer from the adjusting nut and remove the nut lock.

5 Remove the adjusting nut, using the tool if necessary.

6 Remove the thrust washer.

7 Pull the hub/drum assembly straight off the axle tube. On some models, the drum can be removed separately after unscrewing the countersunk screws.

8 Remove and discard the oil seal.

9 To further disassemble the hub, use a hammer and a long bar or drift to knock out the inner bearing, cup (race) and oil seal.

10 Remove the outer retaining ring and then knock the outer bearing and cup from the hub using a bearing driver or a large piece of pipe.

11 On models on which the brake drum is not mounted by screws, the drum can be detached from the hub once the assembly has been removed from the vehicle by pressing out the wheel studs. It may be necessary to have the studs removed and installed by a machine shop because a press will probably be required. When reassembling this type of hub, make sure the drain holes are lined up and then apply a thin coat of RTV sealant to the hub oil deflector contact surface before installing the screws.

12 Clean the old sealing compound from the oil seal bore in the hub.

13 Use solvent to wash the bearings, hub and axle tube. A small brush may prove useful; make sure no bristles from the brush embed themselves in the bearing rollers. Allow the parts to air dry.

24.3c Rear axle hub locknut tool

14 Carefully inspect the bearings for cracks, wear and damage. Check the axle tube flange, studs, and hub splines for damage and corrosion. Check the bearing cups (races) for pitting or scoring. Worn or damaged components must be replaced with new ones.

15 Inspect the brake drum for scoring or damage (Chapter 9).

16 Lubricate the bearings and the axle tube contact areas with wheel bearing grease. Work the grease completely into the bearings, forcing it between the rollers, cone and cage.

17 Because of the special tools and techniques required, take the hub and bearings to a dealer or repair shop to have the bearings and new cups and oil seals installed in the hub.

Installation

18 Make sure the axle housing oil deflector is in position, place the hub assembly on the axle tube, taking care not to damage the oil seals.

24.23 Torque the adjusting nut while rotating the hub

19 Install the thrust washer with the tang in its inner diameter in the axle housing keyway.

20 Install the adjusting nut and adjust the bearings as described below.

Adjustment

Refer to illustration 24.23

21 Install the hub/wheel, thrust washer and adjusting nut. The tang on the thrust washer must be engaged in the axle keyway.

22 Rotate the wheel, making sure the hub turns freely without the brakes dragging.

23 While rotating the wheel, tighten the adjusting nut to 50 ft-lbs with a torque wrench **(see illustration)**.

24 Back off the adjusting nut until it is slightly loose, then tighten it hand-tight. Tighten the nut further until one of the slots in the nut align with the axle keyway and install the key. On models with a nut lock, bend the square tang over the slot or flat of the locknut.

25 Install the locknut and tighten it securely.

26 Install the retaining ring in the end of the spindle (if equipped).

27 Install the axleshaft and lower the vehicle.

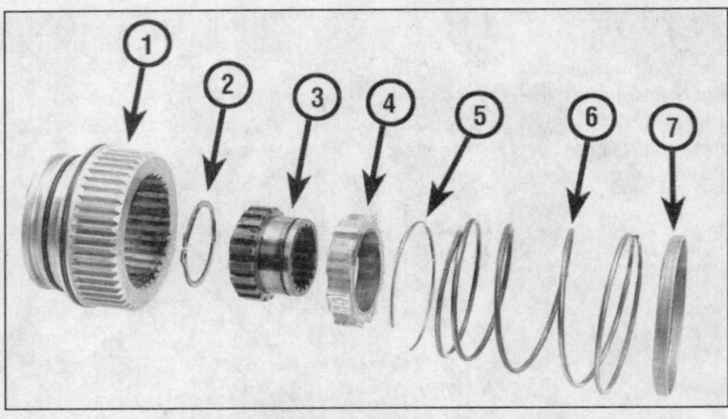

25.3 Typical freewheeling (locking) hub components - exploded view

1 *Drive flange*
2 *Axle shaft snap ring*
3 *Inner drive gear*
4 *Clutch ring*
5 *Internal snap ring*
6 *Spring*
7 *Spring retainer*

25.4 Using a pick or small pliers, remove the internal snap-ring from the hub groove, then remove the outer clutch retaining ring and the actuating cam body (if equipped)

25 Front wheel drive hub - removal and installation

Refer to illustrations 25.3 and 25.4

Full-time four-wheel drive and freewheeling (locking) hub

1 Turn the actuator lever to set the hub in the Lock position.
2 Raise the vehicle and support it securely on jackstands. Rotate the wheel to make sure the hub is locked and remove the wheel.
3 Pry out the hub cover (early models) or remove the six retaining plate bolts. Remove the retaining plate actuating knob and O-ring **(see illustration)**.
4 Remove the internal snap-ring from the hub groove, then remove the outer clutch retaining ring and the actuating cam body (if equipped) **(see illustration)**.
5 Remove the axleshaft snap-ring.
6 Remove the internal snap-ring and hub body assembly.
7 Installation is the reverse of removal.

Automatic locking hub

8 Raise the front of the vehicle, support it securely on jackstands and remove the wheels.
9 Remove the hub retaining screws with a Torx-head tool.
10 Remove the cover, cover plate, rubber seal ring and the bearing assembly.
11 Remove the internal snap-ring with snap-ring pliers and withdraw the hub body, inner drive gear, spring and clutch hub assembly.
12 Installation is the reverse of removal.

26 Front drive axle (K-series) with drum brakes - axleshaft removal and installation

1 Raise the front of the vehicle and support it securely on jackstands.
2 Tap off the hub grease cap.
3 Extract the snap-ring from the axleshaft.
4 On K10 and K1500 models, pull the splined drive flange from the shaft and hub and remove the spacer from the hub.
5 On K20 and K2500 models, remove the stud nuts and washers and withdraw the drive flange and gasket.
6 Remove the bearing locknut, lock ring and adjusting nut.
7 Remove the wheel, hub and brake drum as an assembly.
8 Remove the bolts and lock washers which hold the brake flange plate and spindle to the steering knuckle.
9 Remove the flange plate from the steering knuckle and support the plate to avoid straining the brake hose.
10 Slide the spindle from the shaft and withdraw the axleshaft/joint assembly.
11 Installation is the reverse of removal. When inserting the axleshaft, hold it level to engage the splines with those in the differential side gear.

27 Front drive axle (K-series) with disc brakes - axleshaft removal and installation

1 Raise the front of the vehicle, support it securely on jackstands and remove the front wheel(s).
2 Remove the brake caliper (Chapter 9) and hang it out of the way on a piece of wire so the hose is not stretched.
3 Remove the hub lock mechanism (Section 25).
4 Remove the brake disc/hub assembly.
5 Remove the inner bearing and seal, splash shield, brake bracket and spindle.
6 Withdraw the axleshaft/joint assembly.
7 Installation is the reverse of removal.

28 Front wheel bearings (four-wheel drive) - check, repack and adjustment

Check

1 Raise the vehicle and support it on jackstands.
2 Grasp the front wheel at the sides or top and bottom and move it in-and-out. If it moves less than 0.001-inch or more than 0.010-inch, the bearings should be checked, then repacked with grease or replaced if necessary. **Note:** *A special spanner wrench is necessary for removing, installing and adjusting wheel bearings on these models, as is a torque wrench. Some later models may also require other special spanners or adapters. Read through the procedure for your year and model before deciding to attempt the job.*

Repack

3 To remove the bearings for replacing or repacking, begin by removing the front wheels.
4 Remove the front wheel drive hub (Section 25).
5 Bend the lockwasher ear up from the locknut flat (if equipped).
6 Use a special socket, available at most auto parts stores, to remove the locknut.
7 Remove the lock ring, pin, adjusting nut or drag sleeve retainer, depending on model.
8 On disc brake equipped models, remove the caliper and hang it out of the way on a piece of wire (Chapter 9).
9 Grasp the brake disc or drum securely, pull it out far enough to dislodge the outer wheel bearing and pull the assembly from the spindle.
10 Inspect the bearings for proper lubrication and signs that the grease has been contaminated by dirt (grease will be gritty) or water (grease will have a milky-white appearance).
11 Use a hammer and a long drift or length

of 3/4 inch wood dowel to drive the inner wheel bearing and seal out of the hub (discard the seal).

12 Clean the bearings with solvent and dry them with compressed air (make sure the air does not spin the bearing as this will damage the rollers) or allow them to air dry.

13 Check the bearings for wear, pitting and scoring of the roller and cage. Light discoloration of the bearing surfaces is normal, but if the surfaces are badly worn or damaged, replace the bearings with new ones.

14 Clean the hub with solvent and remove the old grease from the cavity.

15 Inspect the bearing races for wear, signs of overheating, pitting and corrosion. If the races are worn or damaged, drive them out of the hub with a hammer and a drift.

16 Drive the new races in evenly with a section of pipe and a hammer, but be very careful not to damage them or get them cocked in the bore. If you have any doubts about this procedure, take the hubs to a properly equipped machine shop to have the new races pressed in.

17 Clean the spindle with a cloth moistened with solvent, taking care not to get solvent on the brake components.

18 Pack the bearings with high temperature, multi-purpose EP grease prior to installation. Work generous amounts of grease in from the back of the cage so the grease is forced up through the rollers.

19 Add a small amount of grease to the hub cavity and to the center of the spindle. Lubricate the bearing race surfaces in the hub with a light coat of grease.

20 Lubricate the outer edge of the new grease seal, insert the inner bearing and then press the seal into the hub. Make sure the seal is seated completely by tapping it evenly into place using a hammer and a block of wood.

21 Place the hub in position on the spindle, push the assembly into place and install the outer wheel bearing and adjusting nut.

22 If equipped, install the brake caliper.

23 Adjust the wheel bearings as described below.

Adjustment

24 Install the wheels.

25 Rotate the wheel to seat the bearings.

1967 through 1976 models

26 Attach a torque wrench to the special socket and tighten the adjusting nut to 50 ft-lbs while rotating the wheel, then back the nut off 90-degrees.

27 While continuing to rotate the wheel, retighten the nut to 35 ft-lbs and then back it off 3/8-turn.

28 Install the lock ring and pin, turning the adjusting nut only far enough so the pin will engage the nearest notch.

29 Install the locknut. Tighten the locknut to 50 ft-lbs on 1967 through 1971 models or to 80 to 100 ft-lbs on 1972 through 1976 models.

1977 through 1983 models

30 Attach a torque wrench to the special socket.

31 While rotating the wheel, tighten the adjusting nut to 50 ft-lbs and then back it of 90-degrees.

32 While continuing to rotate the wheel, tighten the adjusting nut to 35 ft-lbs and then back it off 3/8-turn.

33 On 1970 through 1980 Series 10, 20, 1500 and 2500 models, install the lock ring and pin and turn the adjusting nut only far enough to engage the nearest notch. Install the locknut. Tighten the locknut to 50 ft-lbs (1977 models) or 80 ft-lbs (1978 through 1980 models).

34 On 1981 through 1983 Series 10, 20, 1500 and 2500 models, install the drag sleeve retainer washer so that it is against the inner nut, with the tang engaged in the spindle keyway. Install the inner nut pin (if equipped) so that it enters one of the retainer washer holes. Install the locknut. Tighten the locknut to between 160 and 205 ft-lbs.

35 On Series 30 and 3500 models, install the lockwasher and outer locknut. Tighten the locknut to 65 ft-lbs. Bend one ear of the lockwasher at least 30-degrees over the inner nut and 60-degrees over the outer nut.

1984 through 1987 models

36 On manual locking hubs, attach a torque wrench to the special socket and rotate the wheel. Tighten the adjusting nut to 70 ft-lbs (Monroe automatic locking hubs) or 50 ft-lbs (all others) and then back the nut off 90-degrees.

37 While continuing to rotate the wheel, retighten the nut to 35 ft-lbs and back it off 3/8-turn.

38 On Monroe automatic locking hubs and manual locking hubs, line up the locknut with the nearest slot and then install the cotter pin or key.

39 On Warner automatic locking hubs, install the drag sleeve washer on the axle and against the inner nut with the tang in the keyway. Install the inner nut pin. Install the outer locknut. Tighten the locknut to between 160 and 205 ft-lbs.

All models

40 Install the front wheel drive hubs and lower the vehicle.

Notes

Chapter 9 Brakes

Contents

	Section
Brake check	See Chapter 1
Brake hoses and lines - check and replacement	20
Brake light switch - removal, installation and adjustment	17
Brake pedal - removal and installation	16
Brake system bleeding	21
Disc brake caliper (Bendix) - removal, overhaul and installation	6
Disc brake caliper (Delco) - removal, overhaul and installation	5
Disc brake pads (Bendix) - replacement	3
Disc brake pads (Delco) - replacement	2
Disc brake rotor - inspection, removal and installation	4
Drum brake shoes - inspection, removal and installation	7
Drum brake wheel cylinder - removal, overhaul and installation	8
Fluid level checks	See Chapter 1
Front wheel bearing (2-wheel drive) check, repack and adjustment	See Chapter 1
General information	1
Hydro-boost system - bleeding	22
Hydro-boost unit - removal and installation	19
Master cylinder (Bendix) - removal, overhaul and installation	10
Master cylinder (Delco) - removal, overhaul and installation	9
Parking brake control and cables (hand lever type) - removal and installation	13
Parking brake (hand lever type) - adjustment	12
Parking brake pedal and cables (pedal type) - removal and installation	15
Parking brake (pedal type) - adjustment	14
Power brake vacuum booster - inspection, removal and installation	18
Pressure regulating combination valve - check, removal and installation	11
Rear Wheel Anti-Lock (RWAL) brake system - general information	23

Specifications

Disc brakes

Pad lining minimum thickness	See Chapter 1
Rotor lateral runout limit	0.004 in
Rotor thickness variation limit	0.0005 in
Rotor minimum thickness	Cast into rotor

Drum brakes

Maximum drum diameter	Cast into drum

Component location

Typical front disc brake assembly

1 *Dust cap*
2 *Brake rotor*
3 *Outer brake pad*
4 *Caliper mounting bolts*
5 *Inner brake pad*
6 *Bleed screw*
7 *Brake hose*
8 *Caliper*

Torque specifications

Torque specifications	Ft-lbs
Master cylinder mounting nuts	25
Vacuum booster mounting nuts	22
Pedal pivot bolt	25
Rear brake anchor pin	140
Caliper mounting bolt	35
Hydro-boost mounting bolts	25
Hydro-boost inlet and outlet unions	25

Component location

Typical rear drum brake assembly

1. *Return spring*
2. *Return spring*
3. *Hold-down spring*
4. *Primary brake shoe*
5. *Adjuster screw assembly*
6. *Actuator lever*
7. *Lever return spring*
8. *Actuator spring*
9. *Hold-down spring*
10. *Secondary brake shoe*

1 General information

Models produced through 1970 are equipped with four-wheel, self-adjusting drum brakes. Later models have disc brakes on the front wheels and self-adjusting drum brakes on the rear.

The parking brake system acts mechanically on the rear wheels only. On earlier models the parking brake is operated by a hand control, while on later vehicles a foot pedal is used.

The master cylinder is a tandem design with separate reservoirs for the two circuits. In the event of a leak or failure in one hydraulic circuit, the other circuit will remain operative.

A visual warning of circuit failure is provided by a warning light, actuated by movement of the combination (pressure differential warning) switch, located in the engine compartment.

Power assisted brakes are available as optional equipment. On most models the power brake system uses a vacuum servo system with engine manifold vacuum and atmospheric pressure to provide assistance. Some heavy duty models are equipped with the Hydro-boost system. On this system the booster operates in conjunction with the power steering system pump and is operated hydraulically. Some manual steering models are equipped with a power steering pump to provide power hydraulic pressure for this system.

2.3 A large C-clamp can be used to push the caliper piston into the bore to provide room for the new, thicker pads (note that the clamp frame is positioned against the flat end of the piston housing and the screw pushes against the outer pad - work slowly and do not apply excessive force!)

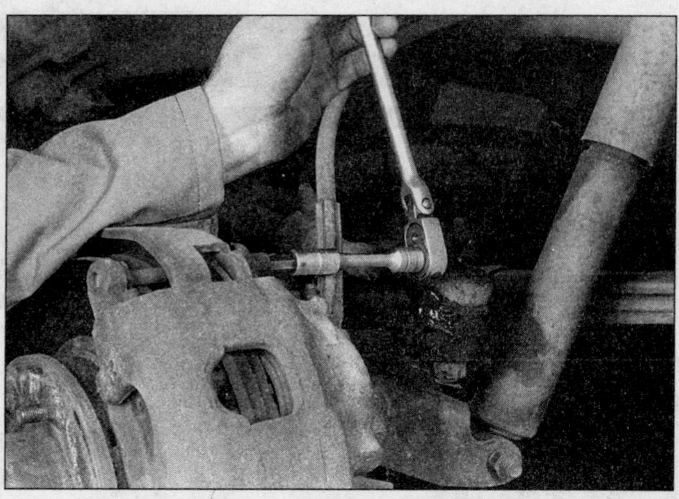

2.5 An Allen head or Torx socket will be necessary to unscrew the caliper mounting bolts

2.6a Lift the caliper and pads straight up off the brake disc . . .

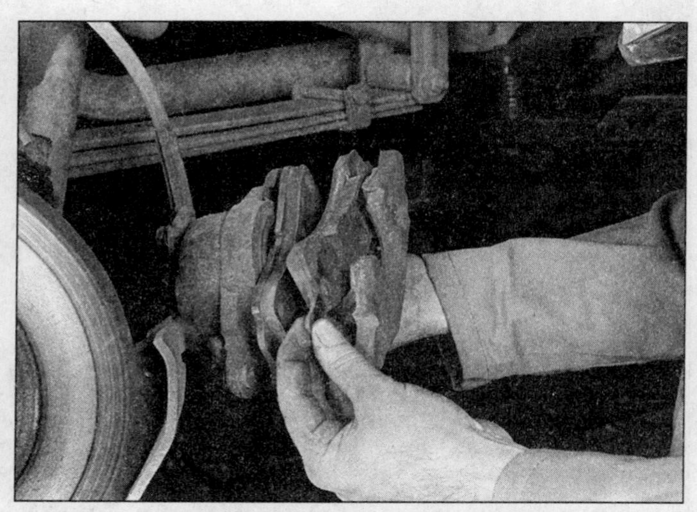

2.6b . . . then remove the outer brake pad from the caliper

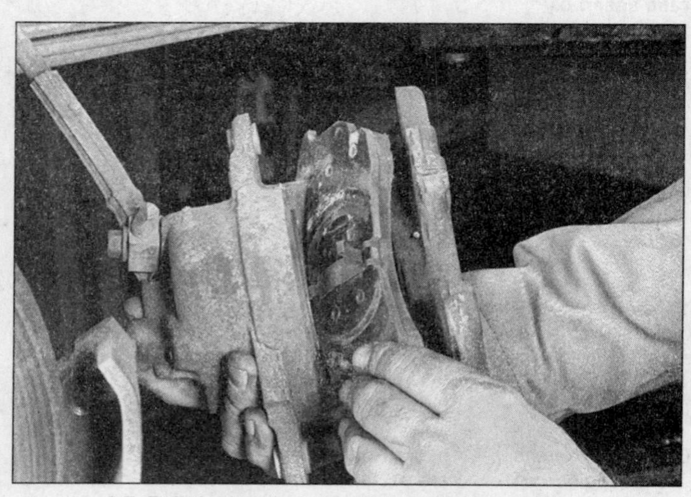

2.7 Pull the inner pad and clip out of the caliper

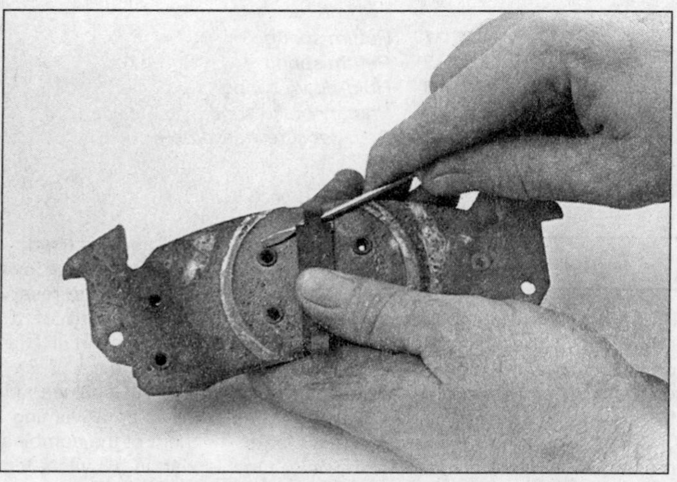

2.12 Pry the clip off the pad with a small screwdriver

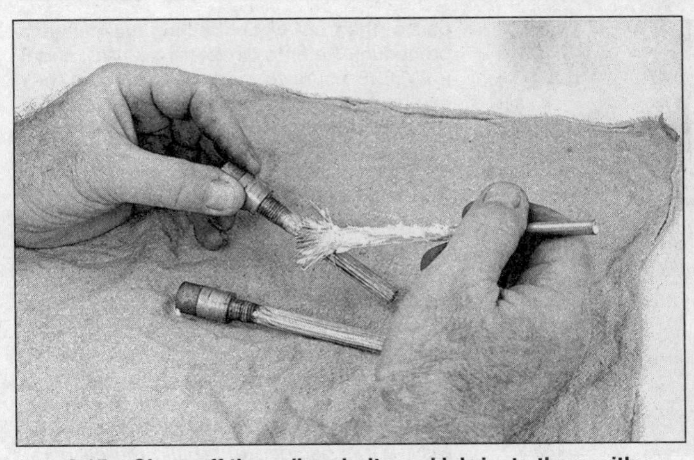

2.15a Clean off the caliper bolts and lubricate them with high-temperature grease

2.15b Install the caliper bolts and tighten them to the torque listed in this Chapter's Specifications

2 Disc brake pads (Delco) - replacement

Refer to illustrations 2.3, 2.5, 2.6a, 2.6b, 2.7, 2.12, 2.15a and 2.15b

Warning: *Disc brake pads must be replaced on both wheels at the same time - never replace the pads on only one wheel. Also, brake system dust may contain asbestos, which is hazardous to your health. Do not blow it out with compressed air and do not inhale it. Do not use gasoline or solvents to remove the dust. Use brake system cleaner only.*

1 Raise the front of the vehicle and support it securely on jackstands. Remove the wheels.

2 Remove about two-thirds of the fluid from the rear compartment of the master cylinder reservoir because as the pistons are pushed in for clearance to allow the pads to be removed, the fluid will be forced back into the reservoir. Set the cover loosely in place to prevent spraying fluid.

3 Use a C-clamp with the screw side resting against the outer pad to compress the caliper piston **(see illustration)**. Tighten the clamp to move the caliper so that the piston will be depressed to the bottom of the bore.

4 Remove the clamp and make sure the pads are now clear of the disc.

5 Remove the two caliper mounting bolts **(see illustration)**.

6 Lift the caliper off the disc and remove the outer pad **(see illustrations)**.

7 Pull the inner pad out of the caliper and remove it **(see illustration)**. Fasten the caliper out of the way with a piece of wire to prevent damage to the brake hose.

8 Extract the pad retaining clip from the cavity in the piston if it didn't come out with the pad.

9 Remove the sleeves from the inboard ears on the caliper and then inspect the rubber bushings at the four caliper ears. If the bushings are damaged or hardened, pry them out and replace them with new ones.

10 Clean the caliper and mounting bolts and check them for corrosion and damage. Replace the bolts with new ones if they are significantly corroded or damaged.

11 Begin reassembly by installing the sleeves in the caliper. Make sure that the sleeves are installed with the end nearest the pads flush with the machined surface of the caliper ear.

12 Pry the inner pad retaining clip off the old pad with a small screwdriver **(see illustration)**. Press the clip onto the new pad so that the single tang end is over the notch in the center of the edge of the pad. Press the two tangs at the end of the inboard shoe spring over the bottom edge of the pad.

13 Position the inner pad complete with clip in the caliper so that the wear indicator is facing the rear of the caliper.

14 Position the outer pad in the caliper so that the tab at the bottom of the pad is engaged in the cutout in the caliper.

15 Hold the caliper in position over the disc and screw the mounting bolts (lubricated with white lithium base grease) through the sleeves in the inboard ears and the mounting bracket **(see illustrations)**. Pass the bolts through the outboard holes and tighten them to the specified torque.

16 Install the brake pads on the opposite wheel, then install the wheels and lower the vehicle. Add brake fluid to the reservoir until it is 1/4-inch from the top (see Chapter 1).

17 Pump the brakes several times to seat the pads against the disc, then check the fluid level again.

18 Check the operation of the brakes before taking the vehicle in traffic. Try to avoid heavy brake applications until the brakes have been applied lightly several times to seat the pads.

3 Disc brake pads (Bendix) - replacement

Refer to illustration 3.4

Warning: *Disc brake pads must be replaced on both wheels at the same time - never replace the pads on only one wheel. Also, brake system dust may contain asbestos, which is hazardous to your health. Do not blow it out with compressed air and do not inhale it. Do not use gasoline or solvents to remove the dust. Use brake system cleaner only.*

1 Raise the front of the vehicle and support it securely on jackstands. Remove the wheels.

2 Remove about two-thirds of the fluid from the master cylinder reservoir.

3 Use a C-clamp with the screw resting against the backing plate of the outer pad to compress the caliper piston. Tighten the clamp to move the caliper so that the piston can be depressed to the bottom of the bore.

4 Remove the locking bolt, then use a brass drift to drive out the caliper support key and spring **(see illustration)**.

5 Lift the caliper from the disc and fasten it out of the way with a piece of wire to the suspension arm so that the brake hose is not stretched.

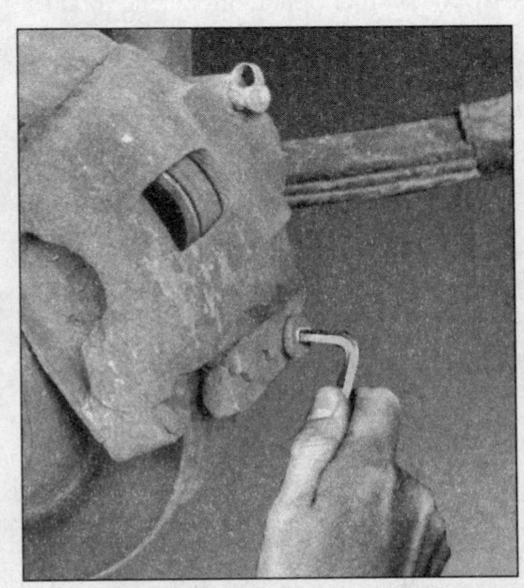

3.4 Remove the locking bolt and drive the caliper support key and spring out using a brass drift and hammer

4.3 Checking the brake disc for runout with a dial indicator

6 Remove the inner pad and discard the pad clip.
7 Remove the outer pad from the caliper.
8 Clean the brake components with brake system solvent, taking care not to inhale any of the dust.
9 Smear silicone grease on the sliding surfaces of the caliper and install a new inner pad clip, making sure that the loop of the spring is away from the disc.
10 Install the inner pad in the groove in the steering knuckle and the outer pad in the caliper.
11 Position the caliper over the disc, making sure that the brake hose is not twisted.
12 Tap the support key and spring into position and install the lock bolt. Make sure that the boss on the bolt engages with the cutout in the key.
13 Install the brake pads on the opposite wheel, then install the wheels and lower the vehicle. Add brake fluid to the reservoir until it is 1/4-inch from the top (see Chapter 1).
14 Pump the brakes several times until the pads seat against the disc, then check the fluid level again.
15 Check the operation of the brakes before taking the vehicle in traffic. Try to avoid heavy brake applications until the brakes have been applied lightly several times to seat the pads.

4 Disc brake rotor - inspection, removal and installation

Inspection

Refer to illustration 4.3

1 Raise the vehicle and support it securely on jackstands. Remove the wheel and hold the rotor in place (if necessary) with two lug nuts.
2 Visually inspect the rotor surface for score marks and other damage. Light scratches and shallow grooves are normal after use and are not detrimental to brake operation. Deep scoring - over 0.015-inch- requires rotor removal and refinishing by an automotive machine shop. Be sure to check both sides of the rotor. **Note:** *Rotors are normally routinely refinished when pads are replaced.*
3 Rotor runout causes the brake pedal to pulsate under hard braking. Refinishing the rotors will correct this problem. If you wish to check rotor runout, place a dial indicator at a point about 1/2-inch from the outer edge of the rotor **(see illustration)**. Set the indicator to zero and turn the rotor. The indicator reading should not exceed 0.006-inch. If it does, the rotor should be refinished by an automotive machine shop.
4 It is absolutely critical that the rotor not be machined to a thickness under the minimum stamped on the rotor itself.

Removal and installation

5 The rotor is an integral part of the front hub. Hub removal and installation is done as part of the front wheel bearing maintenance procedure. Refer to Chapters 1 (2 WD) and 8 (4 WD). Be sure to mark the rotors so they can be returned to the same sides they were removed from.

5 Disc brake caliper (Delco) - removal, overhaul and installation

Refer to illustration 5.5

Note: *If an overhaul is indicated (usually because of fluid leakage) explore all options before beginning this procedure. New and factory rebuilt calipers are available on an exchange basis, which makes this job quite easy. If it is decided to rebuild the calipers, make sure that a rebuild kit is available before proceeding. Always rebuild the calipers in pairs - never rebuild just one of them.*

1 Carry out the operations described in Steps 3 through 7 of Section 2 to the point where the caliper is removed from the disc.
2 Mark the relative position of the brake hose union-to-caliper connection and remove the bolt which secures the hose union to the caliper. Retain the bolt and the two copper sealing washers (one on each side of the union block). Wrap a plastic bag around the end of the hose to prevent fluid loss and contamination.
3 Remove the caliper and clean it with brake system solvent.
4 Place several shop towels or rags in the center of the caliper, then force the piston out

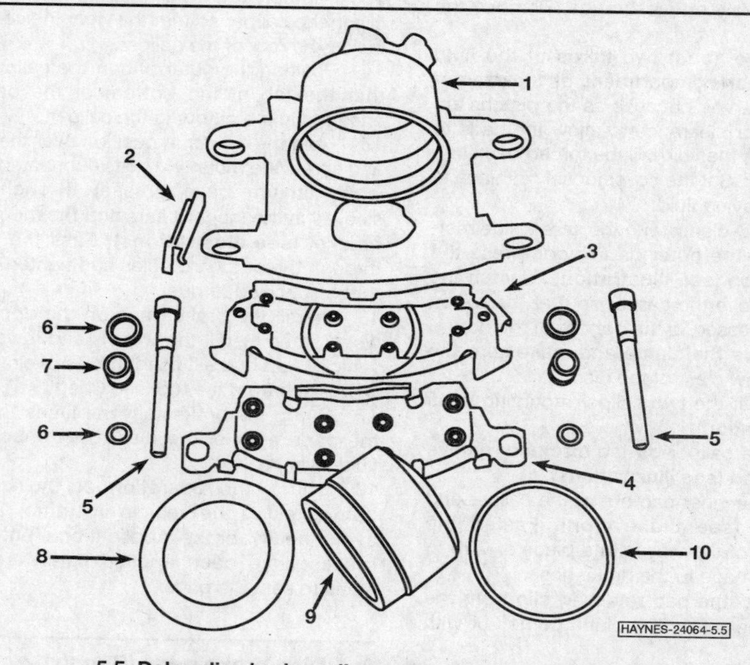

5.5 Delco disc brake caliper components - exploded view

1 *Caliper assembly*
2 *Spring*
3 *Inner brake pad*
4 *Outer brake pad*
5 *Mounting bolts*
6 *Bushings*
7 *Sleeves*
8 *Dust boot*
9 *Piston*
10 *Seal*

7.2a After squirting penetrating oil around the studs and axle flange, tap sharply around the face of the drum to free it

7.2b Once the drum is free, tap around the outer circumference to drive it off the studs - work carefully and don't use excessive force!

7.2c Use a hammer and chisel to knock the plug out of the drum for access to the adjuster

of the bore by directing compressed air into the inlet fitting. Only gentle air pressure is required; a foot or hand operated tire pump is adequate. **Warning:** *Never place your fingers in front of the piston in an attempt to catch or protect it when applying compressed air - serious injury could occur.*

5 With the piston removed, inspect the surfaces of the piston and cylinder **(see illustration)**. If there is any sign of scoring or bright wear areas the caliper must be replaced with a new one.

6 If the components are in good condition, discard the piston seal and dust boot and obtain an overhaul kit, which will contain all the replaceable items.

7 Wash the piston and cylinder bore in clean brake fluid or brake cleaner only. **Warning:** *DO NOT, under any circumstances, clean brake system parts with gasoline or petroleum-based solvents.*

8 Manipulate the new piston seal into the groove in the cylinder using your fingers only.

9 Engage the new dust boot with the groove in the end of the piston. Dip the piston in the clean brake fluid and insert it squarely into the cylinder. Depress the piston to the bottom of the cylinder bore.

10 Seat the boot in the caliper counterbore.

11 Connect the brake hose to the caliper, making sure that the copper washers are in position, and align the union fitting with the marks made before disassembly to ensure that the hose will not rub or twist.

12 Install the caliper as detailed in Section 2, then bleed the front brake circuit (Section 21).

6 Disc brake caliper (Bendix) - removal, overhaul and installation

Note: *If an overhaul is indicated (usually because of fluid leakage) explore all options before beginning this procedure. New and factory rebuilt calipers are available on an exchange basis, which makes this job quite easy. If it is decided to rebuild the calipers, make sure that a rebuild kit is available before proceeding. Always rebuild the calipers in pairs - never rebuild just one of them.*

1 Carry out the operations described in Section 3 to the point where the caliper is removed from the disc.

2 Mark the relative position of the brake hose union-to-caliper connection and remove the bolt which secures the hose union to the caliper. Retain the bolt and the two copper sealing washers (one on each side of the union block). Wrap a plastic bag around the end of the hose to prevent fluid loss and contamination.

3 Remove the caliper and clean it with brake system solvent.

4 Place several shop towels or rags in the center of the caliper, then force the piston out of the bore by directing compressed air into the inlet fitting. Only gentle air pressure is required; a foot or hand operated tire pump is adequate.

5 With the piston removed, inspect the surfaces of the piston and cylinder. If there is any sign of scoring or bright wear areas the caliper must be replaced with a new one.

6 If the components are in good condition, discard the piston seal and dust boot and obtain an overhaul kit, which will contain all the replaceable items.

7 Wash the piston and cylinder bore in clean brake fluid or brake cleaner only. **Warning:** *DO NOT, under any circumstances, clean brake system parts with gasoline or petroleum-based solvents!*

8 Manipulate the new piston seal into its groove in the cylinder using your fingers only.

9 Engage the new dust boot with the groove in the end of the piston. Dip the piston in the clean brake fluid and insert it squarely into the cylinder. Seat the boot in the caliper bore using a seal driver. If you're careful, this can also be done with a drift punch.

10 Install the caliper as detailed in Section 3.

11 Connect the brake hose to the caliper, making sure that the copper washers are in position, and align the union fitting with the marks made before disassembly to ensure that the hose will not rub or twist. Bleed the front brake circuit (Section 21).

7 Drum brake shoes - inspection, removal and installation

Refer to illustrations 7.2a, 7.2b 7.2c, 7.3a, 7.3b, 7.4, 7.5, 7.6, 7.7, 7.8, 7.14 and 7.20

Warning: *Brake shoes must be replaced on both wheels at the same time - never replace the shoes on only one wheel. Also, brake system dust may contain asbestos, which is hazardous to your health. Do not blow it out with compressed air and do not inhale it. Do not use gasoline or solvents to remove the dust. Use brake system cleaner only.*

1 Raise the vehicle, support it securely on jackstands and remove the wheels.

2 Remove the brake drum/hub (see Chapter 8 if necessary). If the brake drums are stuck due to corrosion between the axle flange and the wheel studs and the brake drum, spray penetrating oil around the flange and studs and allow it to soak in. Tap sharply with a hammer around the studs and flange to break the drum loose and then work around the back edge of the drum to remove it **(see illustrations)**. If the drum is locked on the shoes because of severe wear (causing the shoes to be locked in the grooves in the drum wear surface), then the plug must be knocked out of the drum **(see illustration)**. Rotate the drum until the actuator lever can be released from the adjusting screw by pulling it out with a thin rod inserted through the opening **(see illustration 22.12 in Chapter 1)**. The adjuster can then be backed off to release the drum. **Note:** *If the plug is knocked out, make sure that it is removed from the drum. Install a rubber plug afterwards to protect the brakes.*

3 Wash the brake assembly with brake

cleaner before beginning work **(see illustrations)**.

4 Unhook the brake return springs **(see illustration)**.

5 Remove the hold-down springs by depressing the spring and rotating the slotted washer 90° to align the slot with the hold-down pin head **(see illustration)**.

6 Remove the actuator link, hold-down spring, lever pivot, actuator lever, pawl and lever return spring **(see illustration)**.

7 Separate the brake shoes. This is achieved by removing the adjuster assembly and spring **(see illustration)**.

8 Remove the parking brake strut and spring and remove the parking brake lever from the secondary shoe **(see illustration)**.

9 Clean the brake drum and check it for score marks, deep grooves, hard spots (which will appear as small discolored areas) and cracks. If the drum is worn, scored or out-of-round, it can be resurfaced at an automotive machine shop. If this is done, the maximum drum diameter (stamped on the outside) must not be exceeded. Minor imperfections can be removed with fine emery cloth.

10 If there is any evidence of brake fluid leakage at the wheel cylinders, the cylinders must be replaced or overhauled (See Section 8).

7.3a Use brake system cleaner to clean the brake assembly - be very careful not to breathe the dust or fumes!

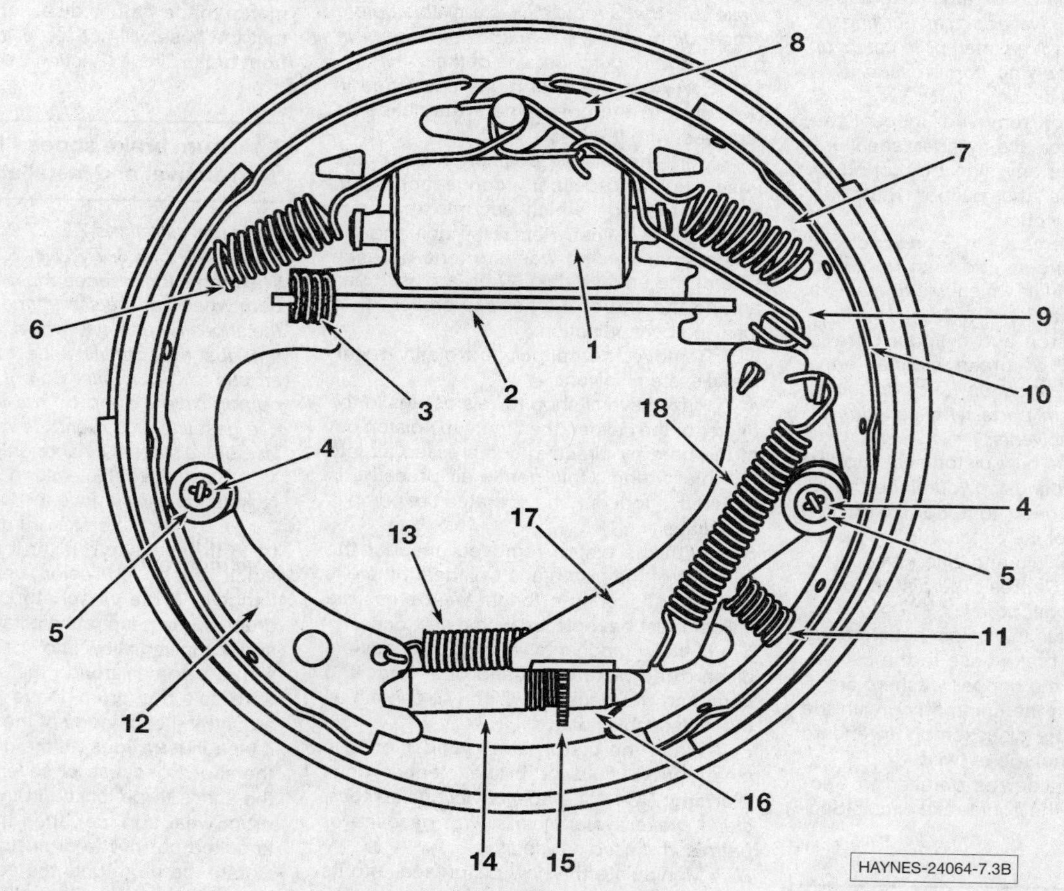

7.3b Typical rear drum brake assembly

1 *Wheel cylinder*
2 *Parking brake strut*
3 *Strut spring*
4 *Hold-down pin*
5 *Hold-down spring and retainer*
6 *Return spring*
7 *Return spring*
8 *Shoe guide*
9 *Actuator link*
10 *Secondary shoe*
11 *Actuator lever return spring*
12 *Primary shoe*
13 *Adjuster screw spring*
14 *Pivot nut*
15 *Adjusting screw*
16 *Adjusting screw socket*
17 *Actuator lever*
18 *Actuator spring*

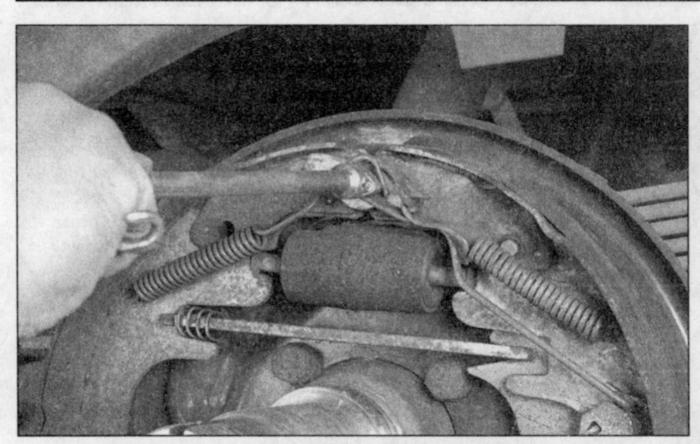

7.4 Remove the shoe return springs - the spring tool shown here is available at most auto parts stores and makes this job much easier and safer

7.5 Remove the front hold-down spring and pin - the hold-down spring tool shown here is available at most auto parts stores

11 Check the backing plate mounting bolts for tightness. Clean all rust and dirt from the shoe contact areas on the backing plate with fine emery cloth.

12 Check the adjuster operation. If the adjusters are seized, badly worn or loose, the adjuster assembly should be replaced with a new one.

13 Lubricate the parking brake cable and fulcrum end of the parking brake cable with brake lube or white lithium-base grease. Attach the lever to the shoe and make sure that it moves freely.

14 Apply brake lube or white lithium-base grease to the shoe contact areas on the backing plate **(see illustration)**. Do not get any on the shoe linings.

15 Connect the brake shoes with the

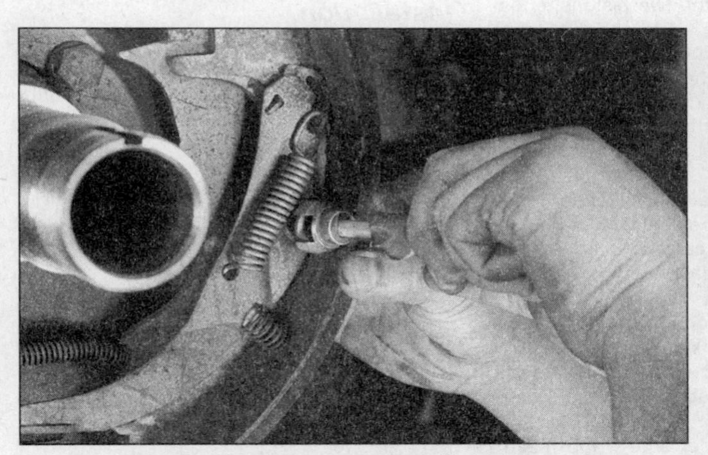

7.6 Remove the rear hold-down spring and actuator lever pivot - be careful not to let the pivot fall out of the lever

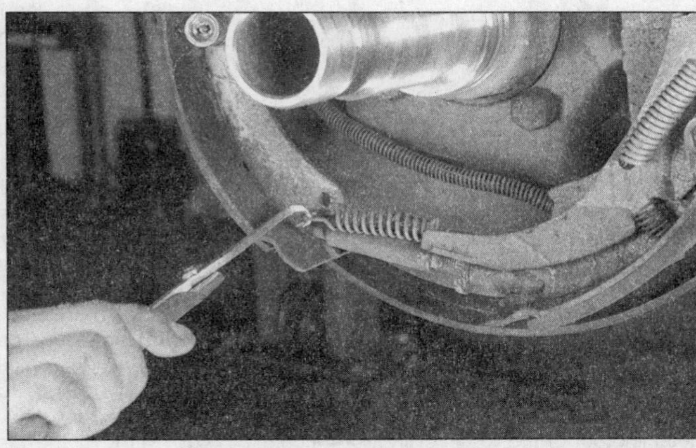

7.7 Remove the adjuster spring - the spring tool shown here is available at most auto parts stores

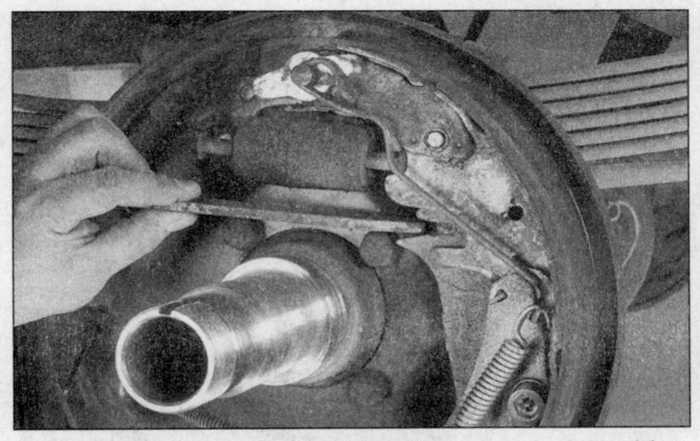

7.8 Remove the parking brake strut (rear drum brakes only)

7.14 Apply brake system lubricant to the areas on the backing plate that support the shoes (don't use too much lubricant or get it on the shoes or drum)

adjusting screw spring, then place the screw assembly in position.

16 Connect the parking brake cable to the lever (rear shoes only).

17 Secure the primary shoe (short lining) with the hold-down pin, spring and cap. Position the parking brake strut and strut spring (rear brakes).

18 Install the actuator assembly and secondary shoe with the hold down pin, spring and cap. Position the parking brake strut and strut spring (rear shoes) between the shoes.

19 Install the shoe guide over the anchor pin, then install the actuator link.

20 Hook the return springs into the shoes, then install the springs over the anchor pin (primary first, then secondary) **(see illustration)**.

21 Ensure that the actuating lever functions by moving it by hand, then turn the star wheel back to retract the shoes.

22 Install the drum and wheel. Use an adjusting tool to adjust the brakes out until the wheel just drags, then turn the adjuster in until the shoes don't drag. Install a metal cap in the adjusting hole in the drum.

23 Lower the vehicle and test the brake pedal position. If the brake pedal goes to the floor, further adjustment of the brakes is required. If the brake pedal is low, back the vehicle up, making repeated stops, to actuate the self-adjusters, which work only when the vehicle is in reverse. Test the brakes for proper operation before driving in traffic.

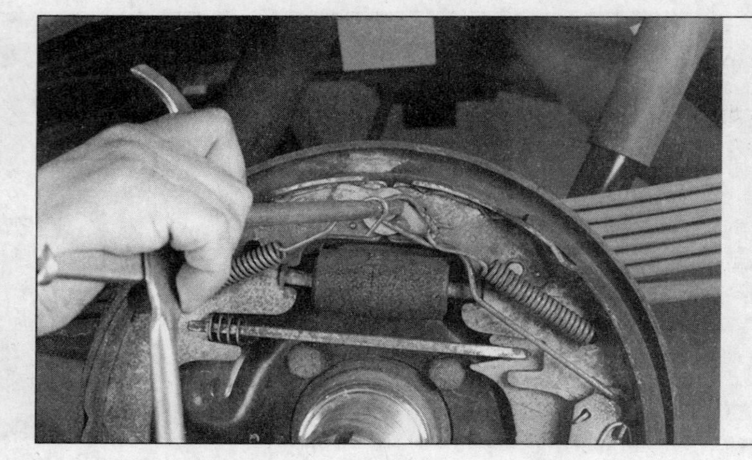

7.20 Install the primary and secondary shoe return springs

8 Drum brake wheel cylinder - removal, overhaul and installation

Removal

1 Refer to Section 7 and remove the brake shoes.

2 Unscrew the brake line fitting from the rear of the wheel cylinder. Don't pull the metal line out of the wheel cylinder - it could bend, making installation difficult.

3 Remove the two bolts securing the wheel cylinder to the brake backing plate.

4 Remove the wheel cylinder.

5 Plug the end of the brake line to prevent the loss of brake fluid and the entry of dirt.

Overhaul

Refer to illustration 8.6

Note: *Purchase two wheel cylinder rebuild kits before beginning this procedure. Never overhaul only one wheel cylinder - always rebuild both of them at the same time.*

6 To disassemble the wheel cylinder, first remove the rubber boot from each end of the cylinder and push out the two pistons, cups (seals) and spring expander **(see illustration)**. Discard the rubber parts and use new ones from the rebuild kit when reassembling the wheel cylinder.

7 Inspect the pistons for scoring and scuff marks. If present, the pistons should be replaced with new ones.

8 Examine the inside of the cylinder bore for score marks and corrosion. If these conditions exist, the cylinder can be honed slightly to restore it, but replacement is recommended.

9 If the cylinder is in good condition, clean it with brake cleaner. **Warning:** *DO NOT, under any circumstances, use gasoline or petroleum-based solvents to clean bake parts!*

10 Remove the bleeder screw and make sure that the hole is clean.

11 Lubricate the cylinder bore with clean brake fluid, then insert one of the new rubber cups into the bore. Make sure the lip on the rubber cup faces in.

12 Place the expander spring in the opposite end of the bore and push it in until it contacts the rear of the rubber cup.

13 Install the remaining cup in the cylinder bore.

14 Attach the rubber boots to the pistons, then install the pistons and boots.

15 The wheel cylinder is now ready for installation.

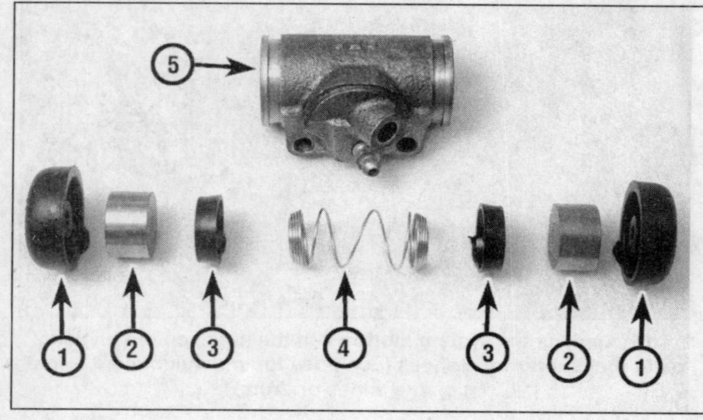

8.6 Wheel cylinder components - exploded view

1 *Dust boot*
2 *Piston*
3 *Cup*
4 *Spring*
5 *Wheel cylinder housing*

Installation

16 Installation is the reverse of removal. Attach the brake line to the wheel cylinder before installing the mounting bolts and tighten the line fitting after the wheel cylinder mountings bolts have been tightened.

17 Bleed the brakes before driving the vehicle.

9 Master cylinder (Delco) - removal, overhaul and installation

Refer to illustrations 9.1, 9.3, 9.4a and 9.4b

Note: *Purchase a master cylinder rebuild kit before beginning this procedure. The kit will include all of the replacement parts necessary for the overhaul. The rubber replacement parts, particularly the seals, are the key to*

9.1 Remove the master cylinder mounting nuts (arrows)

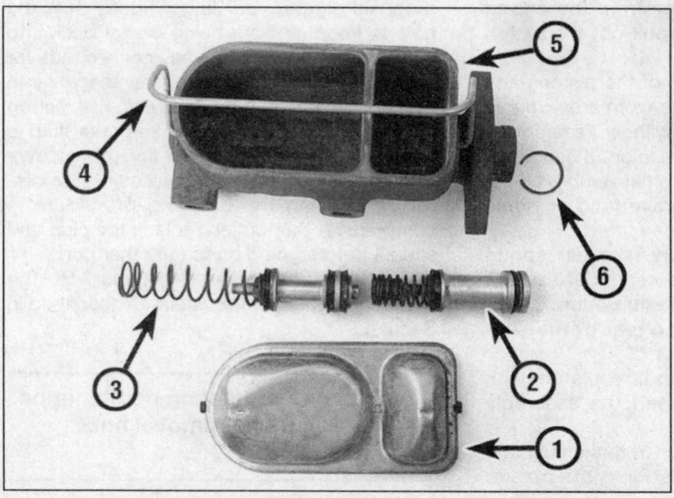

9.3 Early model Delco master cylinder components - exploded view

1 Reservoir cover
2 Primary piston and spring
3 Secondary piston and spring
4 Bail
5 Master cylinder body
6 Lock ring

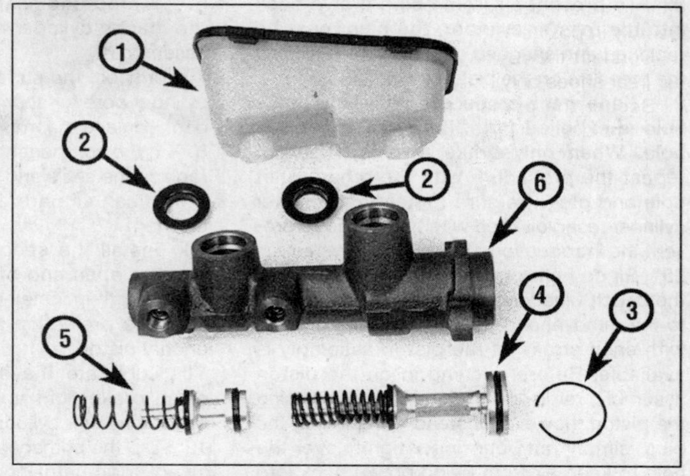

9.4a Later model Delco master cylinder components - exploded view

1 Reservoir assembly (cover, diaphragm and reservoir body)
2 Grommets
3 Lock ring
4 Primary piston and spring
5 Secondary piston and spring
6 Master cylinder body

fluid control in the master cylinder. As such, it's very important to install them correctly. Be very careful not to let them come in contact with petroleum-based solvents or lubricants.

Removal and overhaul

1 Place rags or newspapers under the master cylinder. Using a flare-nut wrench, unscrew the line fittings from the master cylinder. Remove the master cylinder mounting nuts and detach it from the vehicle **(see illustration)**. Drain out the old fluid and discard it.

2 Clean the exterior of the master cylinder with brake system cleaner. **Warning:** *DO NOT, under any circumstances, clean brake parts with gasoline or petroleum-based solvents!*

3 On earlier models, remove the reservoir cover and pour out any remaining brake fluid **(see illustration)**.

4 On later models with a plastic reservoir, remove the reservoir by prying it loose from the master cylinder body with a large screwdriver **(see illustrations)**.

5 Use a wooden dowel to depress the primary piston to eject the hydraulic fluid from the master cylinder.

6 Inspect the bottom of the front fluid reservoir. If a stop screw is visible, remove it.

7 Mount the master cylinder in a vise, using block of wood to protect its surface, then extract the snap-ring from the end of the cylinder.

8 Extract the primary piston and the secondary piston and spring. The latter can be ejected by applying air pressure at the front fluid outlet.

9 Examine the surfaces of the cylinder bore and the secondary piston. If there is evidence of scoring or bright wear areas, replace master cylinder with a new one.

10 If the components are in good condition, wash them in clean brake fluid. Discard all rubber components and the primary piston assembly. The repair kit will contain all necessary replaceable components including a new (completely assembled) primary piston.

11 Install new seals in the grooves of the secondary piston using your fingers only to manipulate them into position. The front seal must have its lip towards the pointed end of the piston. The seal which has the smallest internal diameter is the front seal.

12 The second seal should be installed on the secondary piston so that its lips face the pointed end of the piston. The third seal should be installed in the rear groove of the secondary piston so that its lips face the flat end of the piston.

13 Lubricate the cylinder bores and piston assemblies with clean brake fluid.

14 Install the secondary piston spring over the pointed end of the secondary piston and insert the assembly into the master cylinder bore. Take care not to trap or distort the seal lips.

15 Insert the primary piston and pushrod (complete with retainer) into the master cylinder bore. Exert pressure on the pushrod and install the snap-ring. Install the stop screw if equipped.

16 On models with a detachable plastic reservoir, insert the reservoir grommets into the master cylinder and press it into place until seated.

17 Fill the master cylinder reservoirs with clean brake fluid.

18 Depress the primary piston two or three times with a screwdriver until the cylinder is filled with fluid.

Bench bleeding procedure

19 Before installing the new master cylinder, it should be bench bled. Because it will be necessary to apply pressure to the master cylinder piston and, at the same time, control flow from the brake line outlets, it is recommended that the master cylinder be mounted in a vise. Use caution not to clamp the vise too tightly, or the master cylinder body might be cracked.

9.4b Pry the reservoir out of the master cylinder grommets (Delco master cylinder)

20 Insert threaded plugs into the brake line outlet holes and snug them down so that there will be no air leakage past them, but not so tight that they cannot be easily loosened.

21 Fill the reservoir with brake fluid of the recommended type (see *Recommended lubricants and fluids* in Chapter 1.)

22 Remove one plug and push the piston assembly into the master cylinder bore to expel the air from the master cylinder. A large Phillips screwdriver can be used to push on the piston assembly.

23 To prevent air from being drawn back into the master cylinder, the plug must be replaced and snugged down before releasing the pressure on the piston assembly.

24 Repeat the procedure until only brake fluid is expelled from the brake line outlet hole. When only brake fluid is expelled, repeat the procedure with the other outlet hole and plug. Be sure to keep the master cylinder reservoir filled with brake fluid to prevent the introduction of air into the system.

25 Since high pressure is not involved in the bench bleeding procedure, an alternative to the removal and replacement of the plugs with each stroke of the piston assembly is available. Before pushing in on the piston assembly, remove the plug. Before releasing the piston, however, instead of replacing the plug, simply put your finger tightly over the hole to keep air from being drawn back into the master cylinder. Wait several seconds for brake fluid to be drawn from the reservoir into the piston bore, then depress the piston again, removing your finger as brake fluid is expelled. Be sure to put your finger back over the hole each time before releasing the piston, and when the bleeding procedure is complete for that outlet, replace the plug and snug it before going on to the other port.

26 Attach the master cylinder to the booster and reconnect the brake lines.

27 Bleed the system as described in Section 21.

10 Master cylinder (Bendix) - removal, overhaul and installation

Note: *Purchase a master cylinder rebuild kit before beginning this procedure. The kit will include all of the replacement parts necessary for the overhaul. The rubber replacement parts, particularly the seals, are the key to fluid control in the master cylinder. As such, it's very important to install them correctly. Be very careful not to let them come in contact with petroleum-based solvents or lubricants.*

Removal and overhaul

1 Remove the master cylinder mounting nuts and detach it from the vehicle **(see illustration 9.1)**. Drain out the old fluid and discard it.

2 Clean the exterior of the master cylinder with brake system cleaner and wipe it dry with a lint-free cloth. **Warning:** *DO NOT, under any circumstances, clean brake parts with gasoline or petroleum-based solvents!*

3 Remove the four bolts which secure the reservoir to the body.

4 Remove the small O-ring and the two compensating valve seals from the recesses on the underside of the reservoir. Do not remove the two small filters unless they are damaged and require replacement.

5 Depress the primary piston with a screwdriver, then remove the compensating valve poppets and springs.

6 Extract the snap-ring from the end of the master cylinder and pull out the piston assemblies.

7 Inspect the surfaces of the pistons and cylinder bore for scoring or worn areas. If evident, replace the master cylinder assembly.

8 If the components are in good condition, replace the seals and piston assemblies.

9 Clean all parts in brake fluid or brake cleaner.

10 Install the secondary (shorter) spring into the open end of the secondary piston actuator, then install the piston return spring onto the projection at the rear of the secondary piston.

11 Lubricate the secondary piston with clean brake fluid and insert the assembly into the master cylinder.

12 Dip the primary piston in brake fluid and insert it (actuator end first) into the master cylinder.

13 Depress the pistons with a screwdriver until the snap-ring can be installed.

14 Install the compensating valve seals and the O-ring in the reservoir recesses.

15 Hold the pistons depressed and install the compensating valve springs and poppets, then secure the reservoir. Tighten the bolts securely.

Bench bleeding procedure

16 Before installing the new master cylinder, it should be bench bled. Because it will be necessary to apply pressure to the master cylinder piston and, at the same time, control flow from the brake line outlets, it is recommended that the master cylinder be mounted in a vise. Use caution not to clamp the vise too tightly, or the master cylinder body might be cracked.

17 Insert threaded plugs into the brake line outlet holes and snug them down so that there will be no air leakage past them, but not so tight that they cannot be easily loosened.

18 Fill the reservoir with brake fluid of the recommended type (see *Recommended lubricants and fluids* in Chapter 1).

19 Remove one plug and push the piston assembly into the master cylinder bore to expel the air from the master cylinder. A large Phillips screwdriver can be used to push on the piston assembly.

20 To prevent air from being drawn back into the master cylinder, the plug must be replaced and snugged down before releasing the pressure on the piston assembly.

21 Repeat the procedure until only brake fluid is expelled from the brake line outlet hole. When only brake fluid is expelled, repeat the procedure with the other outlet hole and plug. Be sure to keep the master cylinder reservoir filled with brake fluid to prevent the introduction of air into the system.

22 Since high pressure is not involved in the bench bleeding procedure, an alternative to the removal and replacement of the plugs with each stroke of the piston assembly is available. Before pushing in on the piston assembly, remove the plug. Before releasing the piston, however, instead of replacing the plug, simply put your finger tightly over the hole to keep air from being drawn back into the master cylinder. Wait several seconds for brake fluid to be drawn from the reservoir into the piston bore, then depress the piston again, removing your finger as brake fluid is expelled. Be sure to put your finger back over the hole each time before releasing the piston, and when the bleeding procedure is complete for that outlet, replace the plug and snug it before going on to the other port.

23 Attach the master cylinder to the booster and bleed the system as described in Section 21.

11 Pressure regulating combination valve - check, removal and installation

1 This device is located adjacent to the brake master cylinder and incorporates the dual brake circuit pressure differential valve and the warning light switch.

2 Disconnect the electrical lead from the switch terminal and connect the lead to a good ground.

3 Turn the ignition switch On. The warning light on the instrument panel should light. If it does not, check for a burned out bulb or faulty wiring.

4 To remove the valve, disconnect the wire and the brake lines, then remove the mounting bolts.

5 Installation is the reverse of removal. Bleed the brake system when finished (Section 21).

12 Parking brake (hand lever type) - adjustment

Refer to illustration 12.1

1 The parking brake on early models (through 1970) is a hand lever type. Adjustment may be required due to cable stretch or wear in the linkage or after installing new components **(see illustration)**.

2 To adjust, release the lever, then pull it out two notches.

3 Raise the rear of the vehicle until the wheels are clear of the ground and support the vehicle securely on jackstands.

4 Loosen the locknuts on the cable equalizer and adjust until a slight drag is felt when the rear wheels are turned.

5 Release the hand control and check that the rear wheels now turn without any drag.

6 Tighten the equalizer locknuts.

13 Parking brake control and cables (hand lever type) - removal and installation

1 Release the hand control.

2 Disconnect the front cable from the equalizer by removing the locknut at the rear end of the cable.

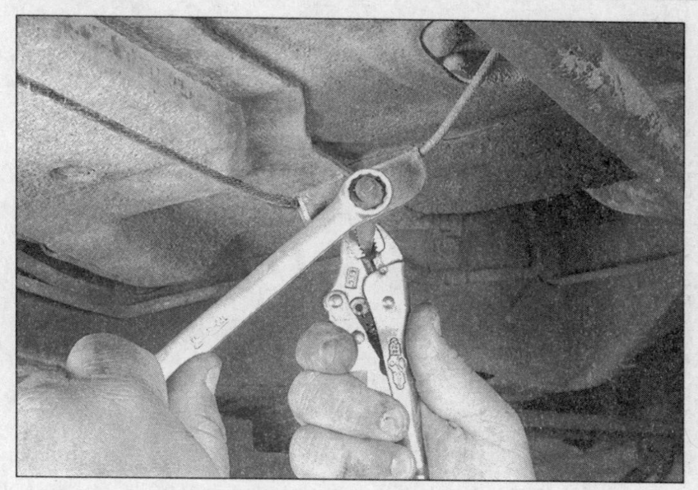

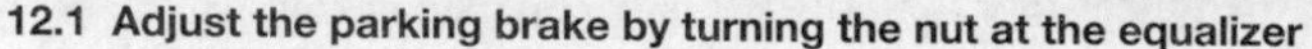

12.1 Adjust the parking brake by turning the nut at the equalizer

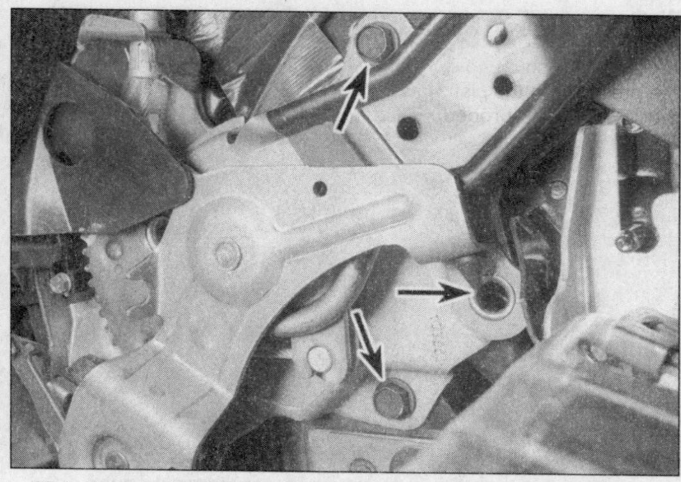

15.2 Remove the parking brake pedal mounting bolts (arrows)

3 Pull up the control lever until the ball end of the cable can be slipped out of the lever.

4 Remove the bolts which attach the lever to the engine compartment wall.

5 To remove the rear cables, release the hand control lever and then remove the equalizer locknut and the front cable connecting clevis so that all tension is removed from the rear cables.

6 Disconnect the rear cable clevis from the equalizer.

7 Extract the U-shaped retainer from the rear cable bracket on the frame. Pull the cable from the bracket.

8 Remove the rear brake drum and shoes as described in Section 7.

9 Disconnect the end of the parking brake cable from the parking brake actuating lever.

10 Compress the fingers which hold the cable conduit into position in the brake backing plate and pull out the cable assembly.

11 Installation of all components is the reverse of removal. Adjust the cables as described in Section 12.

14 Parking brake (pedal type) - adjustment

1 The parking brake on all later models is foot pedal operated and it is normally self-adjusting through the automatic adjusters in the rear brake drums. However, supplementary adjustment may be needed in the event of cable stretch, wear in the linkage or after installation of new components.

2 Raise the rear of the vehicle until the wheels are clear of the ground and support it securely on jackstands. Release the parking brake pedal by pulling on the release lever.

3 Apply the parking brake four notches of its ratchet.

4 Loosen the locknut on the cable equalizer and then tighten the adjuster nut **(see illustration 12.1)** until a slight drag is felt when the rear wheels are turned.

5 Release the parking brake pedal and check that there is no longer any drag when the wheels are turned.

6 Tighten the equalizer locknut and lower the vehicle to the ground.

15 Parking brake pedal and cables (pedal type) - removal and installation

Refer to illustration 15.2

1 Release the parking brake pedal.

2 Remove the pedal mounting nuts and bolts, then lower the assembly **(see illustration)**.

3 Slip the ball at the end of the cable out of the clevis on the parking brake pedal.

4 To remove the front cable, remove the adjusting nut from the equalizer, then extract the clip from the rear section of the front cable and from the lever arm. Disconnect the cable.

5 To remove the center cable, remove the adjusting nut at the equalizer. Unhook the connector at each end of the cable and disengage the hooks and guides.

6 To remove the rear cable, remove the rear brake drum as described in Section 7. Loosen the adjusting nut at the equalizer. Disengage the rear cable at the connector and from the brake shoe operating lever. Compress the conduit fingers and withdraw the cable assembly from the brake backing plate.

7 Installation is the reverse of removal. Adjust the parking brake as described in Section 14.

16 Brake pedal - removal and installation

1 On vehicles with a manual transmission, the clutch and brake pedals operate on a common pivot shaft (see Chapter 7), while on vehicles equipped with an automatic transmission the brake pedal is removed after unscrewing the nut on the end of the pivot bolt and disconnecting the pushrod.

2 On both types the pedal is removed after removing the nut and through-bolt.

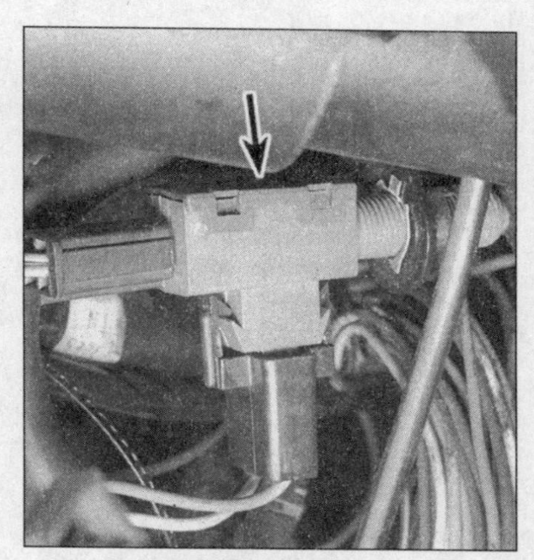

17.2 Later model brake light switch

17 Brake light switch - removal, installation and adjustment

Refer to illustration 17.2

1 On earlier models, unplug the electrical connector and loosen the switch locknut until the switch can be unscrewed from the bracket. The position of the switch should be adjusted by screwing it in or out so that the brake lights are actuated after the pedal has been depressed between 3/8 and 5/8-inch.

2 On later models, unplug the wiring connector(s), depress the brake pedal and pull the switch out of the clip **(see illustration)**. Insert the switch into the tubular clip with the brake pedal depressed until the switch assembly seats on the tube clip. A clicking sound will be heard. Release the brake pedal

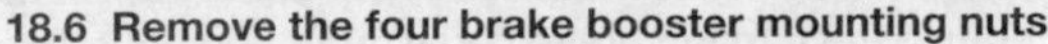

18.6 Remove the four brake booster mounting nuts

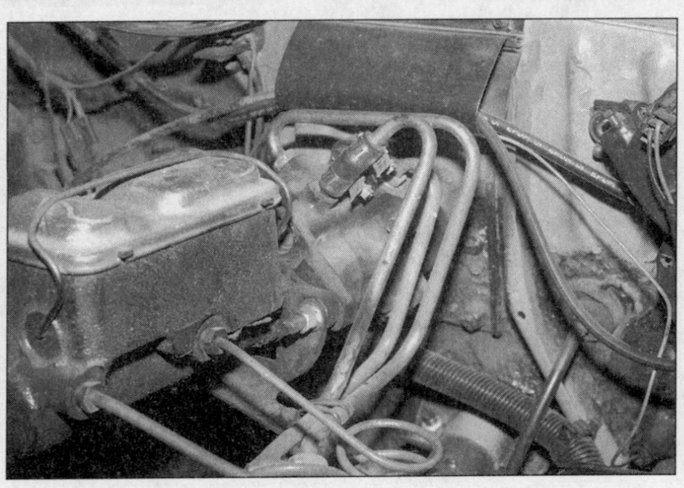

19.3 Hydro-boost unit mounting details

and allow it to return to its "at rest" position. It may be necessary to pull the pedal back slightly. The switch should activate the brake lights after the brake pedal is depressed between 1/2 and 1-inch.

18 Power brake vacuum booster - inspection, removal and installation

Refer to illustration 18.6

1 The power brake booster unit requires no special maintenance apart from periodic inspection of the vacuum hose and the case. Early models will have an in-line filter which should be inspected periodically and replaced if clogged or damaged.

2 Dismantling of the power unit requires special tools and is not ordinarily performed by the home mechanic. If a problem develops, it is recommended that a new or factory rebuilt unit be installed.

3 In the engine compartment, remove the nuts attaching the master cylinder to the booster and carefully pull the master cylinder forward until it clears the mounting studs. Be careful not to bend or kink the brake lines.

4 Disconnect the vacuum hose where it attaches to the power brake booster.

5 In the passenger compartment, disconnect the power brake pushrod from the brake pedal.

6 Remove the nuts attaching the booster to the firewall **(see illustration)**.

7 Carefully lift the booster unit away from the firewall and out of the engine compartment.

8 Upon installation, place the booster into position and tighten the retaining nuts. Connect the brake pedal.

9 Install the master cylinder and vacuum hose.

10 Carefully test the operation of the brakes before placing the vehicle in normal operation.

19 Hydro-boost unit - removal and installation

Refer to illustration 19.3

1 With the engine off, depress and release the brake pedal several times to discharge all pressure from the accumulator. Unbolt the master cylinder from the booster and pull it gently forward without straining the brake lines.

2 Disconnect the booster pushrod from the booster bracket pivot lever and disconnect the brake (steering fluid) lines from the booster unit. Plug all openings.

3 Remove the booster support braces and then remove the support bracket and the booster itself **(see illustration)**.

4 Installation is the reverse of removal. After installation, bleed the system as described in Section 22.

20 Brake hoses and lines - check and replacement

Refer to illustration 20.4

1 About every six months, with the vehicle raised and placed securely on jackstands, the flexible hoses which connect the steel brake lines with the front and rear brake assemblies should be inspected for cracks, chafing of the outer cover, leaks, blisters and other damage. These are important and vulnerable parts of the brake system and inspection should be complete. A light and mirror will be needed for a thorough check. If a hose exhibits any of the above defects, replace it with a new one.

Flexible hose replacement

2 Clean all dirt away from the ends of the hose.

3 Disconnect the brake line from the hose fitting using a back-up wrench on the fitting. Be careful not to bend the frame bracket or line. If necessary, soak the connections with penetrating oil.

4 Remove the U-clip from the female fitting at the bracket **(see illustration)** and remove the hose from the bracket.

5 Disconnect the hose from the caliper, discarding the copper washers on either side of the fitting.

6 Using new copper washers, attach the new brake hose to the caliper.

7 Pass the female fitting through the frame

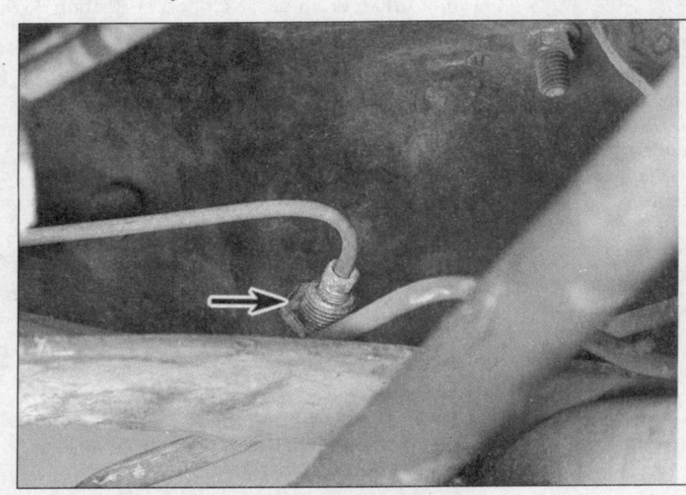

20.4 Remove the U-clip or nut from the female fitting at the bracket (arrow)

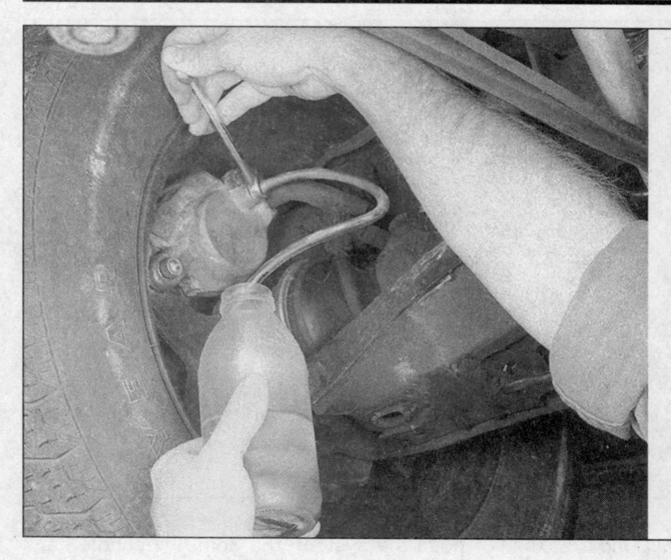

21.8 When bleeding the brakes, a clear piece of tubing is attached to the bleeder screw fitting and submerged in brake fluid - air bubbles can be easily seen in the tube and container (when no more bubbles appear, the air has been purged from the caliper or wheel cylinder)

or frame bracket. With the least amount of twist in the hose, install the fitting in this position. **Note:** *The weight of the vehicle must be on the suspension, so the vehicle should not be raised while positioning the hose.*

8 Install the U-clip in the female fitting at the frame bracket.

9 Attach the brake line to the hose fitting using a back-up wrench on the fitting.

10 Carefully check to make sure the suspension or steering components do not make contact with the hose. Have an assistant push on the vehicle and also turn the steering wheel lock-to-lock during inspection.

11 Bleed the brake system as described in Section 21.

Rigid brake line replacement

12 When replacing brake lines it is important that the proper replacements be purchased. Do not use copper tubing for any brake system connections. Auto parts stores and brake supply houses carry various lengths of prefabricated brake line. These sections can be bent with a tubing bender.

13 When installing the new line make sure it is well supported in the brackets and has plenty of clearance between moving or hot components.

14 After installation, check the master cylinder fluid level and add fluid as necessary. Bleed the brake system as outlined in the next Section and test the brakes carefully before placing the vehicle into normal operation.

21 Brake system bleeding

Refer to illustration 21.8

Note: *Bleeding the brake system is necessary to remove any air that manages to find its way into the system when it has been opened during removal and installation of a hose, line, caliper or master cylinder.*

1 It will probably be necessary to bleed the system at all four brakes if air has entered the system due to low fluid level, or if the brake lines have been disconnected at the master cylinder.

2 If a brake line was disconnected only at a wheel, then only that caliper or wheel cylinder must be bled.

3 If a brake line is disconnected at a fitting located between the master cylinder and any of the brakes, that part of the system served by the disconnected line must be bled.

4 Remove any residual vacuum from the brake power booster by applying the brake several times with the engine off.

5 Remove the master cylinder reservoir cover(s) and fill the reservoir(s) with brake fluid. Reinstall the cover. **Note:** *Check the fluid level often during the bleeding operation and add fluid as necessary to prevent the fluid level from falling low enough to allow air bubbles into the master cylinder.*

6 Have an assistant on hand, as well as a supply of new brake fluid, an empty clear plastic container, a length of plastic, rubber or vinyl tubing to fit over the bleeder valve and a wrench to open and close the bleeder valve.

7 Beginning at the right rear wheel, loosen the bleeder screw slightly, then tighten it to a point where it is snug but can still be loosened quickly and easily.

8 Place one end of the tubing over the bleeder screw fitting and submerge the other end in brake fluid in the container **(see illustration)**.

9 Have the assistant pump the brakes a few times to get pressure in the system, then hold the pedal firmly depressed.

10 While the pedal is held depressed, open the bleeder screw just enough to allow a flow of fluid to leave the valve. Watch for air bubbles to exit the submerged end of the tube. When the fluid flow slows after a couple of seconds, tighten the screw and have your assistant release the pedal.

11 Repeat Steps 9 and 10 until no more air is seen leaving the tube, then tighten the bleeder screw and proceed to the left rear wheel, the right front wheel and the left front wheel, in that order, and perform the same procedure. Be sure to check the fluid in the master cylinder reservoir frequently.

12 Never use old brake fluid. It contains moisture which can boil, rendering the brakes useless.

13 Refill the master cylinder with fluid at the end of the operation.

14 Check the operation of the brakes. The pedal should feel solid when depressed, with no sponginess. If necessary, repeat the entire process. **Warning:** *Do not operate the vehicle if you are in doubt about the effectiveness of the brake system.*

22 Hydro-boost system - bleeding

1 On vehicles equipped with a brake booster which is operated in conjunction with the power steering system or pump, the bleeding of the brake system is carried out as described in the preceding Section. The steering brake booster system, however, should be bled in the following manner whenever lack of power assistance indicates the need for it.

2 If the power steering fluid has foamed due to low fluid level, fill the reservoir and park the vehicle with the reservoir cap removed until the foam has cleared.

3 Top up the power steering reservoir with the specified fluid (Chapter 1). Leave the reservoir cap off.

4 Start the engine and let it run for a few seconds.

5 Repeat the operations described in Steps 3 and 4 until the fluid level remains constant after the engine has been run.

6 Raise the front of the vehicle until the wheels are off the ground.

7 With the engine running at approximately 1500 rpm, turn the steering wheel gently from lock-to-lock, lightly contacting the stops. Check the reservoir level.

8 Lower the vehicle.

9 Turn the wheels from lock-to-lock with the engine running while depressing the brake pedal several times.

10 Turn off the engine and depress the brake pedal four or five times.

11 Check the reservoir fluid level, refilling as necessary.

12 If the fluid is extremely foamy, repeat the procedure.

13 On models equipped with manual steering system and a power steering pump is used generate the necessary brake pressure for the brake booster, use the power steering system bleeding procedure described in Chapter 10.

23 Rear Wheel Anti-Lock (RWAL) brake system - general information

Refer to illustration 23.2

The Rear Wheel Anti-lock (RWAL) brake

system is designed to maintain vehicle maneuverability, directional stability and optimum deceleration under severe braking conditions on most road surfaces. It does so by monitoring the rotational speed of the wheels and controlling the brake line pressure to the rear wheels during braking. This prevents the rear wheels from locking up prematurely during hard braking.

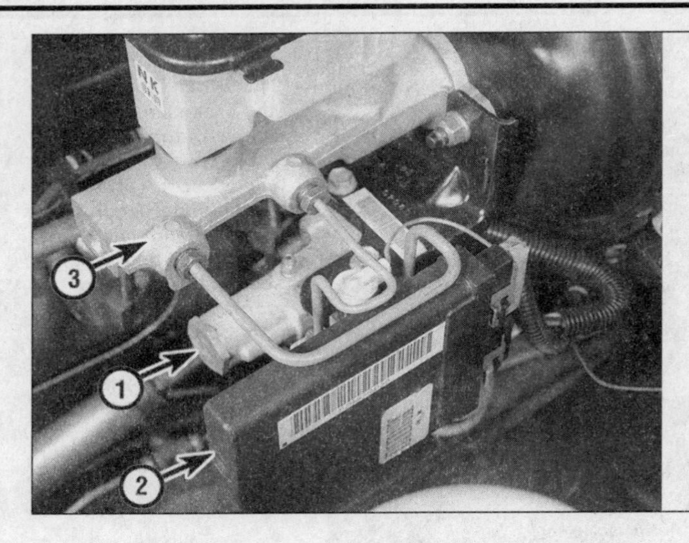

23.2 RWAL system components

1 *Isolation/dump valve*
2 *RWAL control module*
3 *Master cylinder*

Actuator assembly

The actuator assembly includes the master cylinder and a control valve which consists of a dump valve and an isolation valve **(see illustration)**. The valve operates by changing the rear brake fluid pressure in response to signals from the control module.

Control module

The RWAL control module is mounted in the engine compartment below the master cylinder and is the "brain" for the system. The function of the control module is to accept and process information received from the speed sensor and brake light switch to control the hydraulic line pressure, avoiding wheel lock up. The control module also constantly monitors the system, even under normal driving conditions, to find faults within the system.

If a problem develops within the system, the BRAKE warning light will glow on the dashboard. A diagnostic code will also be stored, which, when retrieved by a service technician, will indicate the problem area or component.

Speed sensor

A speed sensor is located in the transmission extension housing on 2WD models and in the transfer case on 4WD models. The speed sensor sends a signal to the control module indicating rear wheel rotational speed.

Brake light switch

The brake light switch signals the control module when the driver steps on the brake pedal. Without this signal the anti-lock system won't activate.

Diagnosis and repair

If the BRAKE warning light on the dashboard comes on and stays on, make sure the parking brake is off and there's no problem with the brake hydraulic system. If neither of these is the cause, the RWAL system is probably malfunctioning. Although a special electronic tester is necessary to properly diagnose the system, the home mechanic can perform a few preliminary checks before taking the vehicle to a dealer service department which is equipped with this tester.

a) *Make sure the brakes, calipers and wheel cylinders are in good condition.*
b) *Check the electrical connectors at the control module assembly.*
c) *Check the fuses.*
d) *Follow the wiring harness to the speed sensor and brake light switch and make sure all connections are secure and the wiring isn't damaged.*

If the above preliminary checks don't rectify the problem, the vehicle should be diagnosed by a dealer service department.

Chapter 10
Steering and suspension systems

Contents

	Section
Balljoints (two-wheel drive) - check and replacement	4
Chassis lubrication	See Chapter 1
Front axle balljoints (four-wheel drive) - check and adjustment	11
Front axle kingpins (four-wheel drive with disc brakes) - overhaul	15
Front axle kingpins (four-wheel drive with drum brakes) - check and adjustment	13
Front axle kingpins (four-wheel drive with drum brakes) - overhaul	14
Front axle steering knuckle and balljoints (four-wheel drive) - removal, overhaul and installation	12
Front coil spring (two-wheel drive) - removal and installation	3
Front leaf spring (four-wheel drive) - removal and installation	10
Front lower control arm (two-wheel drive) - removal and installation	6
Front stabilizer bar - removal and installation	2
Front steering knuckle (two-wheel drive) - removal and installation	8
Front suspension assembly - removal and installation	9
Front upper control arm (two-wheel drive) - removal and installation	5
Front wheel bearing (two-wheel drive) check, repack and adjustment	See Chapter 1
General information	1
Intermediate shaft coupling (early models) - removal, overhaul and installation	29
Manual steering gear - check and adjustment	26
Manual steering gear - removal and installation	32
Pitman arm (four-wheel drive) - removal and installation	25
Pitman arm (two-wheel drive) - removal and installation	24
Pitman shaft oil seal (two-wheel drive) - replacement in vehicle	31
Power steering fluid level check	See Chapter 1
Power steering gear - removal and installation	45
Power steering pump - removal and installation	44
Power steering system - bleeding	43
Rear auxiliary leaf spring (early two-wheel drive) - removal and installation	17
Rear coil spring (early two-wheel drive) - removal and installation	16
Rear control arm (1967 through 1972) - removal and installation	18
Rear leaf spring (1973 on) - removal and installation	19
Shock absorbers - removal and installation	7
Steering angles and wheel alignment - general information	46
Steering column (1967 through 1972) - overhaul	36
Steering column (1973 on) - overhaul	38
Steering column (1967 through 1972) - removal and installation	35
Steering column (1973 on) - removal and installation	37
Steering column lock - removal and installation	42
Steering column lower bearing (1967 through 1972) - replacement in vehicle	34
Steering column (tilt version) - overhaul	39
Steering column turn signal switch (1967 through 1972) - removal and installation	40
Steering column turn signal switch (1973 on) - removal and installation	41
Steering column upper bearing (1967 through 1972) - replacement in vehicle	33
Steering damper - check and replacement	22
Steering idler arm (two-wheel drive) - removal and installation	23
Steering shaft flexible coupling - removal and installation	30
Steering shaft universal joint coupling - removal and installation	28
Steering system - general information	20
Steering wheel - removal and installation	27
Suspension and steering check	See Chapter 1
Tie-rod ends - replacement	21
Tire and tire pressure checks	See Chapter 1
Tire rotation	See Chapter 1

Specifications

Steering system

Steering gear preloads	
Manual steering (1967 through 1971)	
Worm bearing preload	5 to 9 in-lbs
Over center (sector lash) preload	4 to 10 in-lbs
Total preload not to exceed	14 in-lbs
Power steering (1967 through 1971)	
Worm bearing preload	7 to 9 in-lbs
Over center (sector lash) preload	6 to 10 in-lbs
Total preload not to exceed	18 in-lbs
Manual steering (1972 on)	
Worm bearing preload	4 to 6 in-lbs
Over center (sector lash) preload	4 to 10 in-lbs
Total preload not to exceed	14 in-lbs
Power steering (1972 on)	
Steering gear ball drag	3 in-lbs max
Worm bearing preload	1 to 16 in-lbs
Over center (sector lash) preload	3 to 6 in-lbs
Total preload not to exceed	14 in-lbs
Front hub bearing endplay	
two-wheel drive models	0.001 to 0.005 in
four-wheel drive models	0.001 to 0.010 in

Torque specifications

	Ft-lbs
Front suspension (two-wheel drive)	
Lower control arm shaft U-bolt	
C10 (through 1974)	45
C10 (1975 on)	85
C20 and C30	85
Upper control arm shaft nuts	
C10	70
C20 and C30	105
Control arm flexible bushing bolt (C10)	140
Upper control arm steel bushings bolt (C20 and C30)	
new	190
used	115
Lower control arm steel bushing bolt (C20 and C30)	
new	280
used	130
Upper balljoint nut	
C10	50
C20 and C30	90
Crossmember-to-side rail bolt	65
Crossmember-to-bottom rail bolt	90 to 100
Crossmember brake support struts bolt	60
Stabilizer bar-to-control arm bolt	25
Stabilizer bar-to-frame bolt	25
Shock absorber upper mounting bolt	140
Shock absorber lower mounting bolt	60
Caliper mounting bolt	35
Front suspension (four-wheel drive)	
Upper balljoint nut on axle casing yoke	100
Upper balljoint adjusting sleeve nut	50
Lower balljoint nut on axle casing yoke	80
Wheel bearing locknut	
K10 and K20	
Through 1997	50
1978 on	80
K30	65
Knuckle (kingpin) bearing retainer bolts	
K10 and K20	25
K30	70 to 90
Stabilizer bar-to-anchor plate bolt	130
Stabilizer bar-to-frame bolt	55
Spindle-to-knuckle nuts (K10 and K20)	25

	Ft-lbs (unless otherwise indicated)
Spindle-to-knuckle nuts (K30)	60
Shock absorber upper mounting bolt	65
Shock absorber lower mounting bolt	65
Spring front eye bolt	90
Spring rear eye bolt	50
Spring-to-rear shackle nut	50
Spring U-bolt	150
Spring front support-to-frame bolt	25
Suspension bumper bolt	25
Stabilizer-to-spring plate bolt	130
Rear suspension (coil spring type)	
Rear axle-to-control arm U-bolt	145
Rear axle-to-control arm U-bolt (with auxiliary springs)	270
Coil spring-to-axle U-bolt	120
Control arm front pivot bolt	145
Coil spring lower mounting bolt	55
Coil spring upper mounting bolt	50
Shock absorber mounting bolt	75
Transverse tie-rod-to-frame bolt	135
Transverse tie-rod-to-axle bolt	135
Auxiliary leaf spring bolt (C10 and C20)	370
Rear suspension (leaf spring type)	
Spring-to-axle U-bolt	140
Spring front eye bolt	90
Spring rear shackle nut	90
Shock absorber upper mounting bolt	140
Shock absorber lower mounting bolt	115
Rear stabilizer-to-anchor plate bolt	25
Steering (two-wheel drive)	
Tie-rod stud nut	
C10	35
C20 and C30	45
Tie-rod clamp bolt	22
Idler arm mounting bolt	30
Idler arm-to-relay rod nut	60
Pitman arm-to-relay rod nut	60
Pitman arm-to-shaft nut	
manual steering	140
power steering	180
Steering gear mounting bolt	65 to 75
Steering wheel nut	30
Power steering pump pulley nut	60
Power steering pump bracket and support bolt	25
Steering shaft lower coupling clamp bolt	30
Steering gear adjuster plug lock nut	
manual steering	85
power steering	80
Lash adjuster screw lock nut	35
Steering column bracket-to-dash panel	15
Dash panel bracket-to-dash bolt	20
Lock bolt spring screw (tilt column)	36 in-lbs
Bearing housing support screw (tilt column)	60 in-lbs
Steering (four-wheel drive)	
Same as for two-wheel drive where applicable except:	
Tie-rod balljoints	45
Tie-rod clamp bolts	35
Steering connecting rod nut	50
Steering connecting rod clamp nut	40
Pitman arm-to-shaft nut	90

Component location

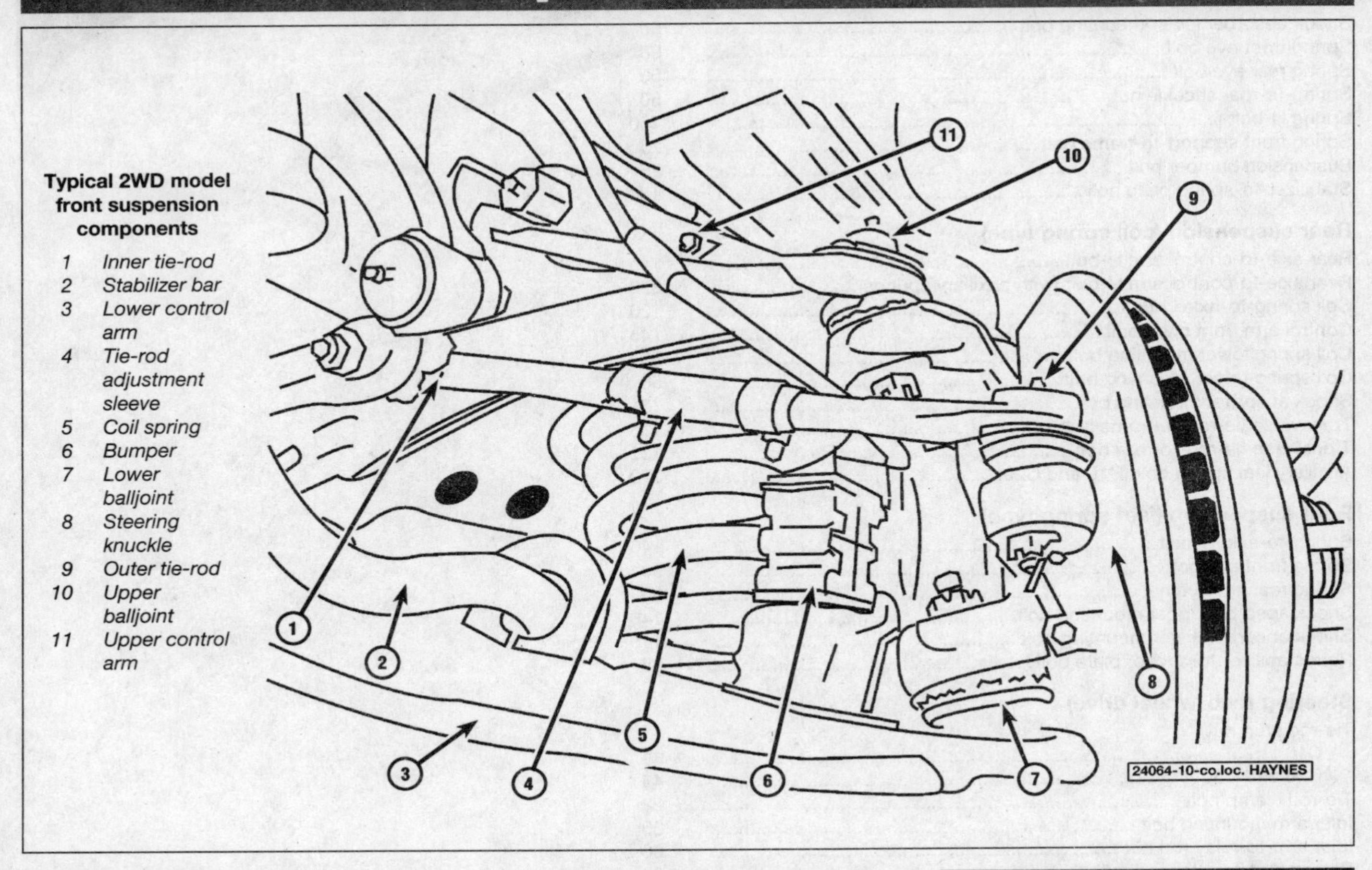

1 General information

Front suspension on two-wheel drive (C-series) vehicles is an independent type using upper and lower wishbone control arms, coil springs and telescopic shock absorbers. On four-wheel drive (K-series) vehicles, leaf springs are used.

Rear suspension on two-wheel drive models built between 1967 and 1972 consists of coil springs, trailing arms and telescopic shock absorbers, while four-wheel drive and heavy duty vehicles use leaf springs. All 1973 and later models have rear leaf springs and telescopic shock absorbers with a stabilizer bar or tie-rod.

2 Front stabilizer bar - removal and installation

Refer to illustrations 2.2, 2.3a and 2.3b

1 Raise the vehicle and support it securely on jackstands.

2 Disconnect the bushings and support

2.2 Remove the stabilizer support bracket-to-frame retaining bolts (arrows) - (2WD models)

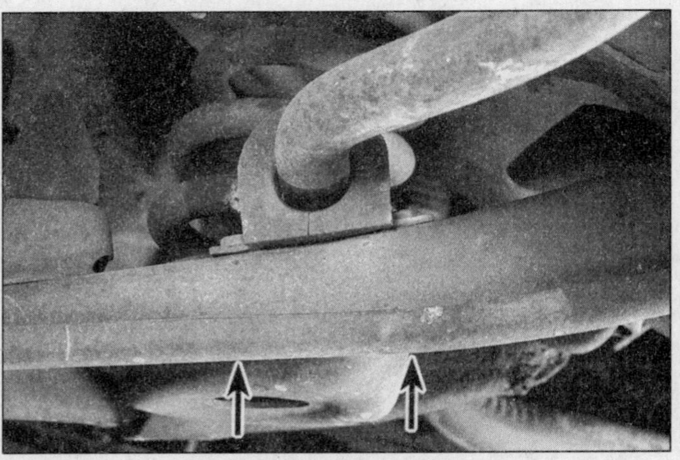

2.3a Remove the stabilizer support bracket-to-control arm retaining bolts (arrows) (2WD models)

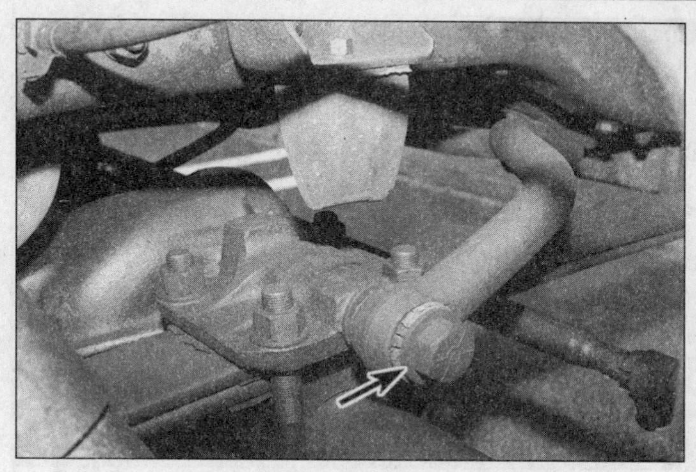

2.3b Remove the stabilizer support bracket retaining bolts (arrows) - (4WD models)

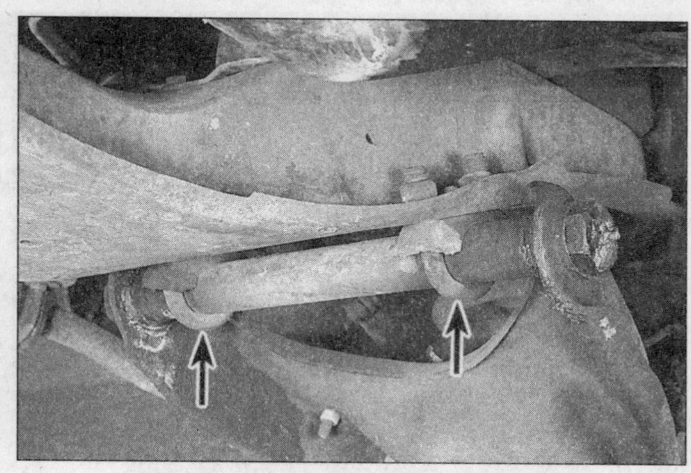

3.6 Lower control arm U-bolts (arrows)

brackets which hold the bar to the frame **(see illustration)**.

3 Disconnect the brackets which hold the bar to the lower control arms or spring anchor plates (four-wheel drive) **(see illustrations)**.

4 Replace any worn bushings and install by reversing the removal operations. Make sure that the slits in the flexible bushings at the frame attachment points are towards the front of the vehicle.

3 Front coil spring (two-wheel drive) - removal and installation

Refer to illustration 3.6

1 Raise the vehicle and support it securely on jackstands so the suspension control arms hang free.

2 Disconnect the shock absorber lower mounting.

3 Disconnect the stabilizer bar from the lower control arm.

4 Install a chain through the spring and around the lower control arm as a safety precaution.

5 A suitable spring compressor that installs through the center of the spring will be required to compress the spring so that its tension can be removed from the lower control arm. **Warning:** *If the proper tool is not available it is recommended that this operation be left to a dealer or properly equipped shop. The spring is under great pressure and injury could result if it should become disengaged from the tool.*

6 Once the spring has been compressed, remove the two U-bolts which secure the lower control arm to the crossmember **(see illustration)**.

7 Lower the control arm until the spring, still compressed by the tool, can be removed.

8 If the original spring is to be reinstalled, there is no need to remove the spring compressor. If a new spring is to be installed carefully remove the compressor from the old spring and attach it to the new spring.

9 Installation is the reverse of removal, making sure that the alignment hole in the pivot cross shaft of the lower control arm is engaged with the stud on the crossmember.

4 Balljoints (two-wheel drive) - check and replacement

Refer to illustrations 4.3a, 4.3b, 4.10 and 4.11

Check

Upper balljoint

1 Raise the vehicle and support it securely on jackstands. Place a floor jack under the lower control arm and raise it enough to raise the upper control arm off its stop. Grasp the top and bottom of the tire and try to wiggle it in and out while as assistant feels for movement at the balljoint. If there is any perceptible movement, the balljoint is worn and must be replaced with a new one.

Lower balljoint

2 Raise the vehicle, support it securely on jackstands and remove the wheels.

3 Support the weight of the suspension with a jack. Measure the distance between the tip of the balljoint stud and the tip of the grease fitting below the balljoint (A) (if equipped with a grease fitting) or between

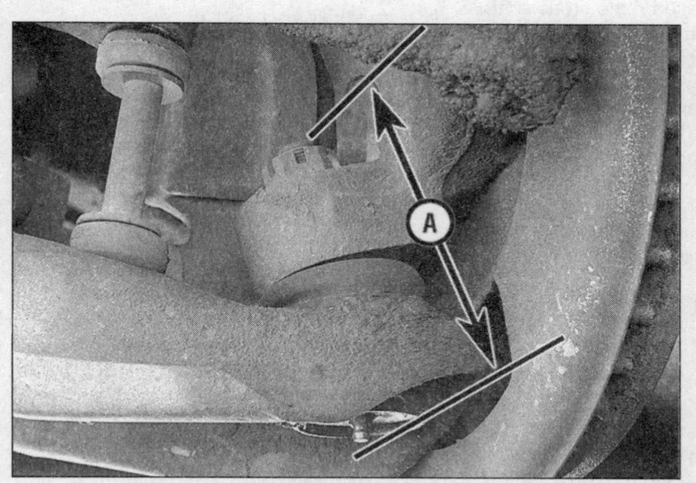

4.3a Measure between the points shown on models with grease fittings to determine balljoint wear

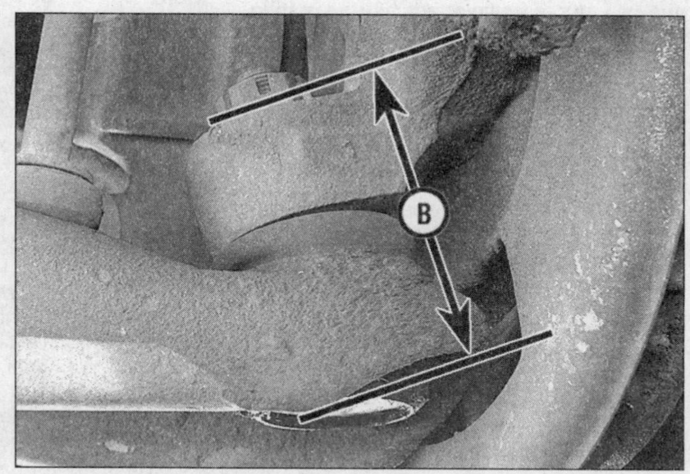

4.3b On models without grease fittings, measure between the points shown to check for balljoint wear

4.10 Balljoint removal requires a special tool which can be fabricated from a large bolt, nut, washer and socket

4.11 Upper balljoint installation

the lower surface of the control arm and the top of the balljoint (B) **(see illustrations)**.

4 Lower the jack to allow the suspension to hang free and repeat the measurement.

5 If the difference between the two measurements exceeds 3/32-inch, the balljoint is worn and must be replaced with a new one.

Replacement

6 Raise the vehicle, support it securely on jackstands and remove the front wheel.

7 Remove the brake caliper and hang it out of the way (Chapter 9).

8 Remove the cotter key from the balljoint and back off the nut two turns.

9 Place a floor jack under the lower control arm. **Note:** *The jack must remain under the control arm during removal and installation of the balljoint to hold the spring and control arm in position.*

10 Separate the balljoint from the steering knuckle using a balljoint separator to press the balljoint out of the steering knuckle **(see illustration)**.

11 To remove the upper balljoint from the control arm, drill out the rivets, remove the balljoints and clean the control arm. Install the new upper balljoint against the mating surface of the control arm and secure it with the supplied nuts and bolts **(see illustration)**.

12 To remove the lower balljoint, remove the control arm (Section 6) and take it to an automotive machine shop to have the balljoint pressed out of the lower control arm and a new one pressed in.

13 Inspect the tapered holes in the steering knuckle, removing any accumulated dirt. If out-of-roundness, deformation or other damage is noted, the knuckle must be replaced with a new one (Section 8).

14 Reconnect the balljoints to the steering knuckle and tighten the nuts to the specified torque.

15 If the cotter key does not line up with the opening in the castellated nut, tighten (never loosen) the nut just enough to allow installation of the cotter pin.

16 Install the lubrication fittings and lubricate the new balljoints.

17 Install the wheels and lower the vehicle.

18 The front end alignment should be checked by a dealer or alignment shop.

5 Front upper control arm (two-wheel drive) - removal and installation

1 Raise the front of the vehicle and support it securely on jackstands under the frame crossmember.

2 Remove the wheel and support the lower control arm with a jack.

3 Disconnect the upper balljoint (Section 4).

4 With the upper control arm free, the pivot cross shaft can be unbolted from the crossmember **(see illustration 9.7a)**. Take great care not to let the shims drop from their location between the cross shaft and crossmember until their position and number have been established. These shims control the front wheel camber and caster.

5 If the cross shaft pivot bushings are worn, they can be replaced if the cross shaft is first withdrawn using a suitable extractor tool. It is recommended that this job be left to your dealer or a properly equipped shop.

6 Place the control arm and cross shaft in position on the crossmember so that the convex and concave faces of the alignment washers meet.

7 Install the securing nuts, and before tightening, insert the camber shims in their original positions. Tighten the nut at the thinner shim pack first.

8 Connect the balljoint to the steering knuckle, tighten to the specified torque and install a new cotter pin. If the cotter key holes are not in alignment, tighten the nut further; never back it off.

9 Install the wheel and then lower the vehicle. Lubricate the balljoint.

10 Have the front end alignment checked by a dealer or alignment shop.

6 Front lower control arm (two-wheel drive) - removal and installation

1 Raise the vehicle and support it securely on jackstands.

2 Remove the front spring (Section 3) and support the inner end of the control arm with a jack.

3 Remove the brake caliper and hang it out of the way on a piece of wire (Chapter 9).

4 Disconnect the lower balljoint (Section 4).

5 The inner cross shaft will already have been disconnected for removal of the coil spring so that the control arm can now be removed from the vehicle .

6 Take the control arm to a dealer or properly equipped shop to have the balljoint and bushings replaced.

7 Installation is the reverse of removal. Tighten all nuts to their specified torque, and lubricate the new balljoint on completion.

7 Shock absorbers - removal, testing and installation

1 Removal of a shock absorber should be carried out with the weight of the vehicle resting normally on its wheels. If this is not possible because of lack of access, raise the vehicle under the axles or suspension lower control arm so that the shock absorber is in its normal attitude, neither fully extended nor retracted.

2 Disconnect the upper and lower mountings and remove the unit.

3 Inspect the shock absorber for signs of fluid leakage. If any is visible, replace the unit.

4 Examine the flexible mounting bushings. If they are worn or have hardened or deteriorated, replace them.

5 On conventional-type shock absorbers, secure the lower mounting eye in the jaws of a vise, and fully extend and compress the shock absorber at least six times. If the oper-

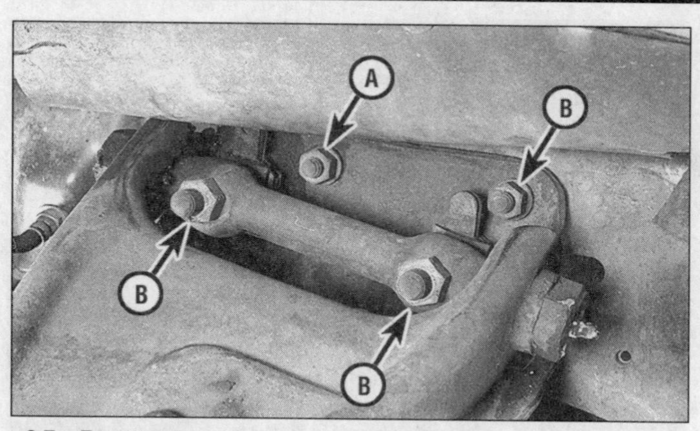

9.7a Remove the upper crossmember-to-frame mounting bolts (A) - bolts (B) hold the cross-shaft to the crossmember

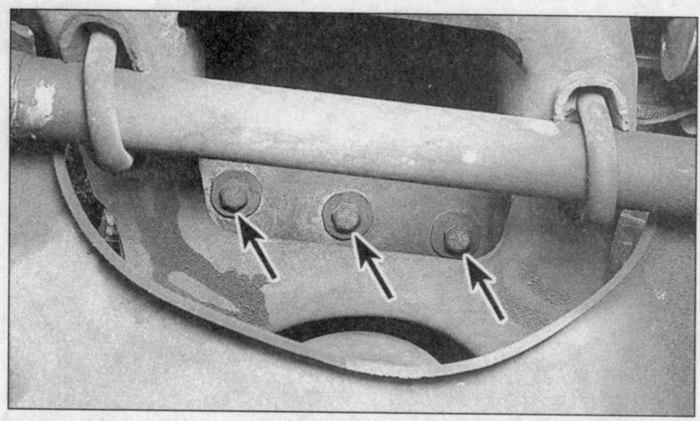

9.7b Remove the lower crossmember-to-frame mounting bolts

ation is noisy, jerky or offers no resistance in either direction then the units must be replaced as a pair.

6 Gas-filled shock absorbers should only be tested with the unit in an inverted position.

7 Installation is the reverse of removal. Tighten the mounting nuts and bolts to the specified torque.

8 Front steering knuckle (two-wheel drive) - removal and installation

1 Raise the front of the vehicle and support it securely on jackstands.

2 Support the lower control arm with a jack so that the coil spring is compressed in its normal ride height.

3 Remove the wheel.

4 Remove the caliper/hub assembly or brake drum (see Chapter 1).

5 Remove the disc splash shield or drum brake backing plate.

6 Disconnect the tie-rod end from the knuckle.

7 Disconnect the balljoints from the steering knuckle (Section 4).

8 Remove the steering knuckle.

9 Installation is the reverse of removal. Adjust the front wheel bearings (Chapter 1 and 8) and have the front wheel alignment checked by a dealer or alignment shop.

9 Front suspension assembly - removal and installation

Refer to illustrations 9.7a and 9.7b

1 Raise the front of the vehicle, support on it securely on jackstands and remove the wheels.

2 Disconnect the front shock absorber lower mounting.

3 Disconnect the idler arm and the Pitman arm from the steering linkage relay rod.

4 Support the engine under the oil pan and remove the engine mount through bolts.

5 Disconnect the brake line from the T-union on the crossmember.

6 Support the crossmember with a jack.

7 Remove the bolts which attach the crossmember to the frame side members **(see illustrations)**, lower the jack and withdraw the complete front axle assembly from the vehicle.

8 Installation is the reverse of removal. After installation bleed the brake system (Chapter 9) and have the front wheel alignment checked by a dealer or alignment shop.

10 Front leaf spring (four-wheel drive) - removal and installation

Refer to illustrations 10.3 and 10.11

1 Raise the vehicle and support it securely on jackstands.

2 Support the axle with a jack so that the leaf springs are not under tension. Remove the wheel.

3 Remove the nuts from the spring U-bolt **(see illustration)**.

4 Remove the pivot bolts from the front and rear shackle spring eyes. Lift the spring from the vehicle.

5 Inspect the spring eye bushings. Later types are of a flexible design while earlier versions have plain bushings with lubricators. Either type can be replaced, using a suitable tool or suitable size bolt, nut, washers and tubular spacers to draw out the old bushing.

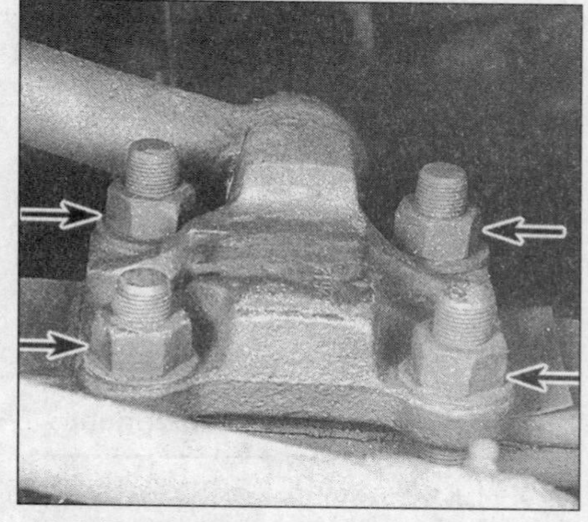

10.3 Remove the spring U-bolt retaining nuts (arrows)

6 Use a wire brush to clean the spring and inspect for damage and broken leaves.

7 Install the spring bushings.

8 Place the spring upper cushion in position on the spring.

9 Install the front of the spring into the frame and install the bolt and nut finger tight.

10 Attach the rear of the spring to the shackle and tighten the bolt and nut finger tight.

11 Install the lower spring pad and retainer plate. Tighten the bolts to the specified torque **(see illustration)** in the order shown in the illustration.

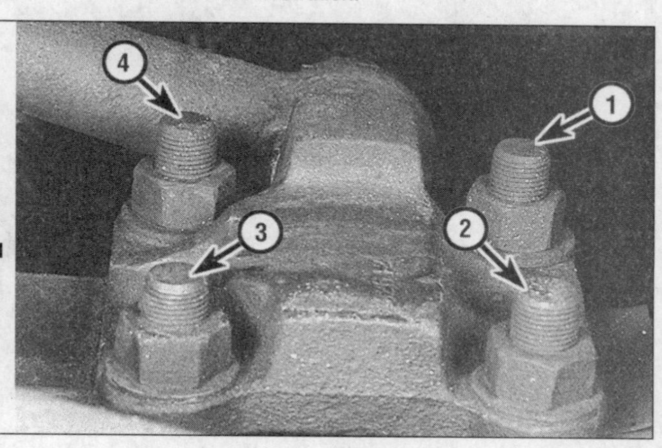

10.11 Four-wheel drive leaf spring U-bolt nut tightening sequence; tighten all nuts uniformly, then tighten nuts 1 and 3 to 25 ft-lbs before fully tightening all of the nuts to the specified torque in a 2-4-1-3 sequence

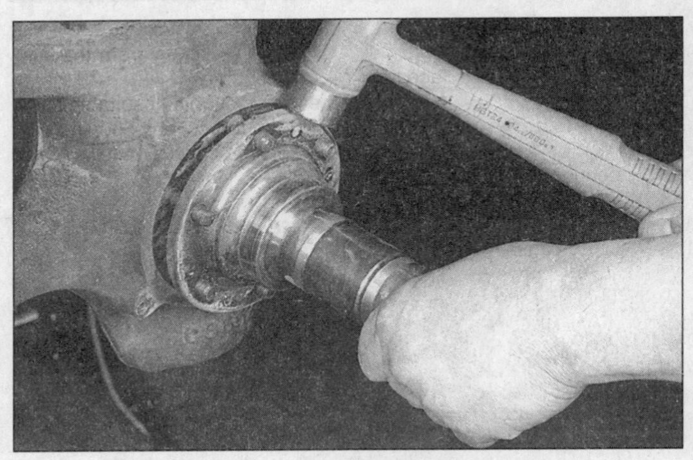

12.3 Remove the front spindle by tapping on the spindle with a hammer

12.5 Removing the steering arm nuts (new nuts must be used for reassembly)

12 Tighten the shackle and pivot bolts and nuts to the specified torque.
13 Lower the vehicle.
14 On some earlier vehicles, a stop plate is installed at the rear of the rear spring shackle. This should be adjusted to give a 1/8-inch clearance when the vehicle weight is resting on the suspension.

11 Front axle balljoints (four-wheel drive) - check and adjustment

Refer to illustration 11.4

1 This adjustment is normally only required if excessive play in the steering, severe or irregular tire wear, or persistent loosening of the tie-rod is observed.

Check

2 Raise the front of the vehicle, support it securely and remove the wheels.
3 Disconnect the steering tie-rod from the steering arm (Section 21).
4 Attach a spring scale to the hole in the steering arm and record the pull required to keep the knuckle assembly turning after starting from the initial straight ahead position. The pull should be between 20 and 25 pounds in either direction.

Adjustment

5 Where necessary, adjust the balljoint by removing the upper balljoint stud nut and turning the adjusting sleeve. Tighten the balljoint nut to the specified torque on completion.
6 Install the wheels and connect the steering linkage.

12 Front axle steering knuckle and balljoints (four-wheel drive) - removal, overhaul and installation

Refer to illustrations 12.3, 12.5, 12.7 and 12.16

1 Raise the vehicle, support it securely on jackstands and remove the front wheels.
2 Remove the front hubs (Chapter 8).
3 Remove the spindle retaining bolts and tap off the spindle and thrust washer **(see illustration)**.
4 Disconnect the steering linkage tie-rod from the knuckle arm.
5 If the steering arm is to be removed, new nuts must be used when it is replaced **(see illustration)**.
6 Remove the cotter key from the upper balljoint nut.
7 Remove the upper and lower balljoint retaining nuts **(see illustration)**.
8 Disconnect the balljoints (Section 4).
9 Do not extract the adjusting sleeve from the knuckle unless completely new assemblies are being installed. Never release the adjusting sleeve by more than two threads during removal of the knuckle and also take great care when removing the knuckle, otherwise the hardened threads of the sleeve can damage the softer threads in the knuckle.
10 It is recommended that new balljoints are installed by a dealer or properly equipped shop. If a suitable press or clamp and the necessary suitable size pieces or pipe or sockets are available you can carry out the work yourself, observing the following:

a) *Extract the snap-ring from the lower balljoint.*

12.7 Unscrew the balljoint retaining nut

12.16 Install the spindle/driveshaft with the chamfer toward the seal

b) Remove the lower balljoint first, followed by the upper one.

11 Insert the lower balljoint (the one without a cotter key hole in the stud) into the knuckle and install a new snap-ring.

12 Install the upper balljoint until it is fully seated.

13 Engage the balljoints with the knuckle yokes. Tighten the lower balljoint nut, the upper balljoint adjusting sleeve and the upper balljoint nut to the specified torque. Insert a new cotter key to secure the upper balljoint nut. If the key holes are not in alignment, tighten the nut further - do not back it off.

14 Install the steering arm (new nuts) and tighten the nuts to the specified torque.

15 Connect the steering tie-rod and tighten the nut to the specified torque.

16 Install the bronze thrust washer over the axleshaft (chamfer towards slinger) and install the spindle **(see illustration)**.

17 Assemble the spindle to the knuckle and tighten the nuts to the specified torque.

18 Install the hubs and adjust the bearings (Chapter 8).

13 Front axle kingpins (four-wheel drive with drum brakes) - check and adjustment

1 This adjustment is normally only required if excessive play in the steering, severe tire wear or persistent loosening of the tie-rod is noted.

Check

2 Raise the front of the vehicle, support it securely on jackstands and remove the wheels.

3 Disconnect the steering tie-rods from the steering knuckle.

4 Extract the kingpin seals and retainers.

5 Using a spring scale or torque wrench, check the turning torque of each knuckle. This should be between 5 and 10 in-lbs for axles series FH033 and FH035 and 15 to 35 ft-lbs for axle FHD035.

Adjustment

6 Adjust if necessary by removing or adding shims to the steering knuckle. On FH033 and FH035 axles the shims are located under the steering arm and on FHD035 axles they are located between the lower trunnion cap and the knuckle.

14 Front axle kingpins (four-wheel drive with drum brakes) - overhaul

1 Remove the axleshaft and joint assembly as described in Chapter 8.

2 Disconnect the steering tie-rod from the steering arm on the knuckle.

3 Extract the seal retainers and seals from the steering knuckle.

4 Remove the upper and lower bearing retainers and shims from the knuckle. Note the location of the shims for exact installation in their original positions.

5 Remove the knuckle and the bearing cups from the axle housing.

6 To reassemble, press new bearing cups into the axle housing.

7 Mate the steering knuckle to the axle housing yoke and install the bearing retainers. Install the original or equivalent thickness shims and tighten the bearing retainer bolts to the specified torque. On K2500 models, position the key in the kingpin and install the bushing by aligning the kingpin and key with the keyway.

8 Check the bearing preload as described in Section 13.

9 Locate the seal ring, rubber seal and felt seal against the inboard face of the knuckle and align the bolt holes. Install the steel retainer.

10 Connect the steering tie-rod and install the axleshaft as described in Chapter 8.

15 Front axle kingpins (four-wheel drive with disc brakes) overhaul

1 Remove the hub assembly as described in Chapter 8.

2 Remove the spindle retaining nuts.

3 Tap on the spindle with lightly with a plastic hammer to dislodge it and then remove the spindle.

4 Inspect the bronze spacer which is located between the axleshaft joint assembly and the bearing. If it is worn, replace it.

5 Carefully remove the four nuts from the upper kingpin cap, working alternately in a criss-cross pattern. Work slowly because the spring under the cap will force the cap up with considerable pressure. Remove the spring, the gasket and the cap.

6 From the underside of the knuckle, remove the four cap screws from the lower kingpin bearing cap. Extract the bearing cap and lower the kingpin.

7 Extract the upper kingpin tapered bushing and knuckle from the yoke. Remove the kingpin felt seal, followed by the knuckle.

8 Remove the upper kingpin from the yoke. A special tool may be required to remove this and, since the new kingpin must be installed with a tightening torque of between 500 and 600 ft-lbs, it is recommended that the removal and installation operations be left to a dealer or properly equipped shop.

9 Remove the lower kingpin bearing cup, cover and seal by driving the assembly out with a hammer and suitable piece of pipe or socket. Discard the seal and obtain a new one. Check the grease retainer for distortion and replace if necessary.

10 Begin reassembly by driving in the lower kingpin bearing cup and new grease retainer. Pack the retainer with grease, apply grease to the bearing cover and install it.

11 Install the new grease seal. When fully installed it will protrude slightly from the surface of the yoke flange.

12 Install the upper kingpin. If suitable tools are available, tighten to between 500 and 600 ft-lbs. Otherwise have the job done by a dealer or properly equipped shop.

13 Assemble the felt seal to the kingpin. Install the knuckle and the tapered bushing over the kingpin.

14 Install the lower bearing cap and kingpin with the four cap screws. Tighten the screws alternately in a criss-cross pattern to the specified torque.

15 Assemble the compression spring onto the upper kingpin bushing. Install the bearing cap along with a new gasket and install the nuts. Tighten the nuts alternately in a criss-cross pattern to the specified torque.

16 Rear coil spring (early two-wheel drive) - removal and installation

Refer to illustration 16.4

1 Disconnect the shock absorber lower mounting from the control arm.

2 Raise the rear of the vehicle and support it securely on jackstands. Support the rear axle with a jack so that the coil spring is not under tension or compression.

3 Unscrew the spring clamp bolts, the cover bolt from under the control arm and the upper bolt from inside the spring coils.

4 Lower the jack until the control arm is sufficiently low to enable the spring to be withdrawn **(see illustration)**.

5 Installation is the reverse of removal, making sure that the end of the coil spring seats in the notch of the spring mount. Tighten the bolts to the specified torque.

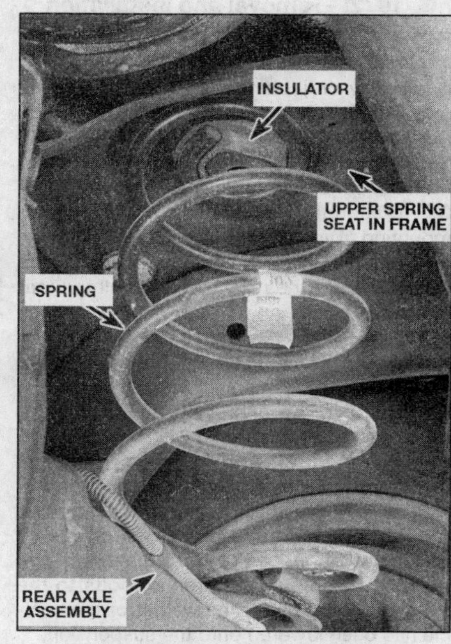

16.4 Rear coil spring details (1967 through 1972 models)

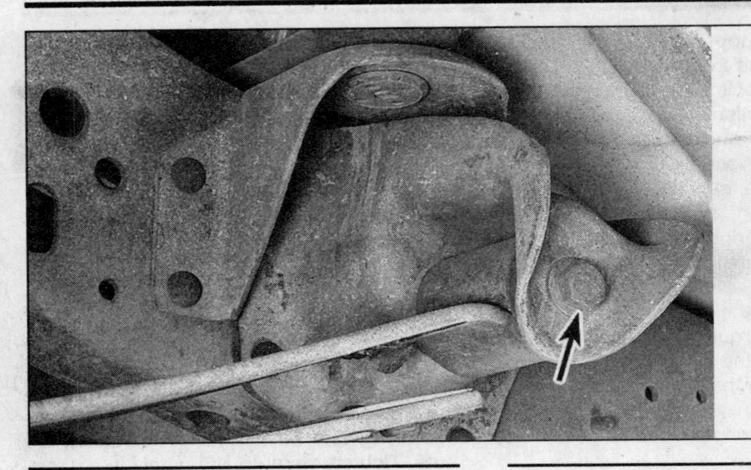

19.4 Remove the rear leaf spring through bolt (arrow)

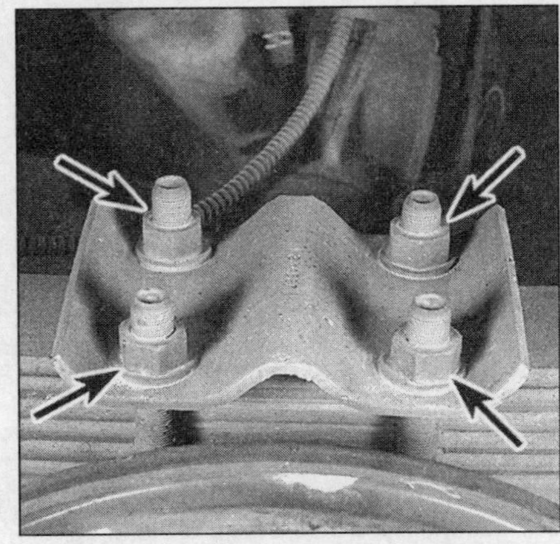

19.5 Remove the U-bolt retaining nuts (arrows) and withdraw the spring plate

17 Rear auxiliary leaf spring (early two-wheel drive) - removal and installation

1 Raise the rear of the vehicle, support it securely on jackstands and support the rear axle with a jack so that there is no load on the spring and the leaf spring does not contact the control arm bumper.
2 Extract the cotter key from the spring retaining bolt, unscrew the spring retaining bolt and withdraw the spring from the frame bracket.
3 The spring bumper on the lower control arm can be removed if the rear axle is supported and the U-bolts which hold the arm to the axle are disconnected.
4 Installation is the reverse of removal. Tighten the bolts to the specified torque.

18 Rear control arm (1967 through 1972) - removal and installation

1 Raise the rear of the vehicle and support it securely on jackstands. Support the rear axle with a jack to remove the load from the rear coil springs.
2 Disconnect the lower end of the spring from the control arm.
3 Disconnect the shock absorber lower mounting from the control arm (Section 16).
4 Release the U-bolt nuts which hold the control arm to the rear axle and pull the rear of the arm down.
5 Disconnect the parking brake lever at the clip on the bracket attachment on the control arm.
6 Remove the pivot bolt at the forward end of the control arm and withdraw the arm.
7 The control arm bushings can be replaced by using a press or a bolt and nut with suitable pieces of tubing to draw out the old bushings and to install the new ones.
8 Installation of the control arm is the reverse of removal. Tighten all bolts to the specified torque after the weight of the vehicle has been lowered onto the suspension.

19 Rear leaf spring (1973 on) - removal and installation

Refer to illustrations 19.4 and 19.5

1 Raise the rear of the vehicle and support it securely on jackstands.
2 Support the rear axle with a jack so that tension is removed from the spring.
3 Loosen, but do not remove, the shackle-to-spring hanger bolt.
4 Remove the spring-to-front hanger through-bolt, then remove the shackle bolt **(see illustration)**.
5 Remove the U-bolts and withdraw the spring plate **(see illustration)**.
6 Remove the spring.
7 New spring bushings can be installed using a suitable tool or equivalent such as a bolt, nuts and suitable pieces of tubing to draw out the old bushings and pull the new ones into position. On some heavy duty springs fitted to C20 and C30 vehicles, the front bushing is staked in position. Relieve the staking before pressing out the bushing and stake the new bushing after installation.
8 Clean the spring with a wire brush and inspect for cracks. If any are found, have a new leaf installed by a dealer or properly equipped shop or obtain a new spring.
9 Installation is the reverse of removal. Adjust the height of the axle with the jack to align the spring eyes with the hanger. Tighten the U-bolt nuts diagonally in a criss-cross pattern. Lower the vehicle weight onto the suspension before tightening the bolts and nuts to the specified torque.

20 Steering system - general information

Steering is a recirculating ball type with power steering as an option. On later models an impact absorbing steering column design and a tilt column are available.

21 Tie-rod ends - replacement

Refer to illustrations 21.10a and 21.10b

1 Wear in the steering gear and linkage is indicated when there is considerable movement in the steering wheel without corresponding movement at the wheels. Wear is also indicated when the vehicle tends to "wander".
2 Raise the front of the vehicle and support it on jackstands.
3 Remove the cotter keys and loosen the nuts on the tie-rod ends. Ideally, a tie-rod end separator tool should now be used to disconnect the tie-rod end from the steering arm. If this is not available, place a heavy hammer against one side of the eye and then strike the opposite side with a light hammer. This will momentarily distort the eye so that the

21.10a Position of the tie-rod ends on the connecting rod of a four-wheel drive steering linkage

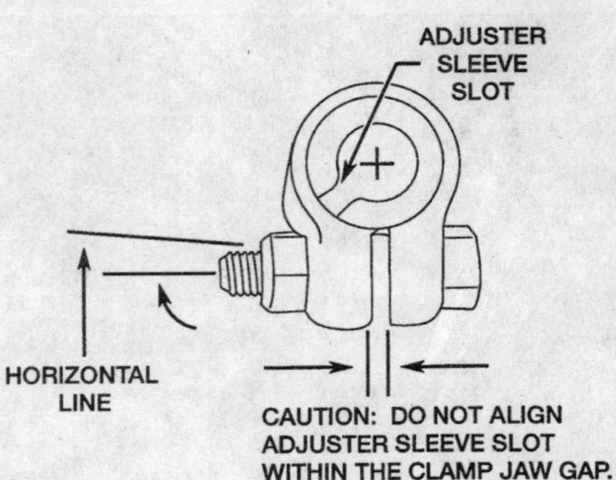

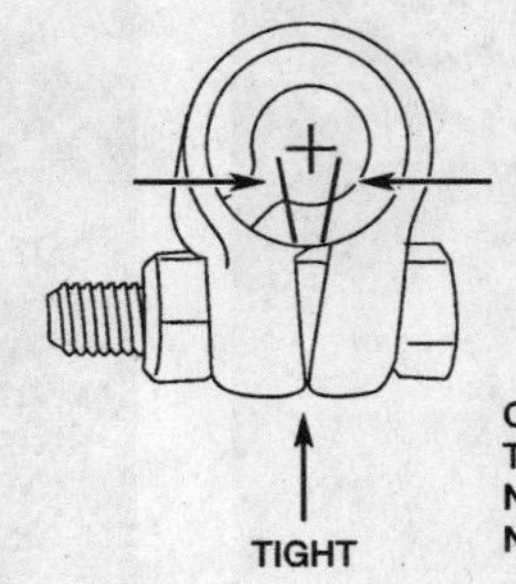

21.10b Typical two-wheel drive tie-rod assembly details

tapered stud drops free. Before beginning, squirt penetrating oil into the tie-rod connection and allow it to penetrate.

4 Remove the tie-rod end nut and disconnect the tie-rod end.

5 Count the number of exposed threads at the point where the tie-rod end enters the tie-rod and record this figure.

6 Release the clamp pinch bolt and remove the tie-rod end assembly.

7 If the inner tie-rod end is worn, it should be removed from the relay rod in the same manner.

8 Clean and lubricate the threads in the tie-rod so that the new tie-rod ends can be easily screwed into position.

9 Screw the tie-rod end assembly into position so that the same number of threads is exposed as was originally recorded. If both balljoints were removed from the tie-rod, then an equal number of threads should be exposed with a difference not to exceed three threads.

10 Position the clamps as shown, set the tie-rod ends in their correct attitudes and reconnect their studs to the steering arm eye or relay rod **(see illustrations)**.

11 Tighten the tie-rod end nuts. If the cotter key hole and stop are not in alignment, tighten the nut further, never back it off.

12 Lubricate the tie-rods on completion and then have the front end alignment (toe-in) checked by a dealer or alignment shop.

22 Steering damper - check and replacement

Refer to illustration 22.4

1 The steering damper is basically a shock absorber located in the steering linkage designed to damp out road and steering shocks.

Check

2 Whenever the vehicle is raised for lubrication or inspection (Chapter 1), inspect the damper for loose mounting nut, worn bushings and fluid leakage. Disconnect one end of the damper and extend and compress it several times making sure the action is smooth, with no skips or noise. The damper is a sealed unit which must be replaced with a new unit if a fault develops.

Replacement

3 Raise the front of the vehicle and support it securely on jackstands.

4 Remove the cotter keys and nuts and withdraw the damper **(see illustration)**.

5 Installation is the reverse of removal.

23 Steering idler arm (two-wheel drive) - removal and installation

1 Raise the front of the vehicle and support it securely on jackstands.

2 Extract the cotter key and nut from the end of the idler arm.

3 Using a suitable puller tool, disconnect the relay rod stud from the eye of the idler arm.

4 Remove the retaining bolts and lower the idler arm from the frame.

5 Installation is the reverse of removal. Tighten the nuts to the specified torque and install a new cotter pin.

24 Pitman arm (two-wheel drive) - removal and installation

Refer to illustration 24.5

1 Raise the front of the vehicle and support it securely on jackstands.

2 Disconnect the relay rod stud from the eye in the Pitman arm.

3 Mark the Pitman arm-to-shaft relationship.

4 Remove the Pitman arm nut from the shaft.

5 Using a puller, draw the arm from the shaft **(see illustration)**.

6 Installation is the reverse of removal, making sure to align the marks on shaft and arm and tighten the nuts to the specified torque.

22.4 Remove the steering damper retaining bolts (arrows)

24.5 A special puller is required to detach the Pitman arm from the shaft

26.7 Loosen the adjuster screw locknut and then back off the adjuster screw 1/4 turn before measuring steering gear preload

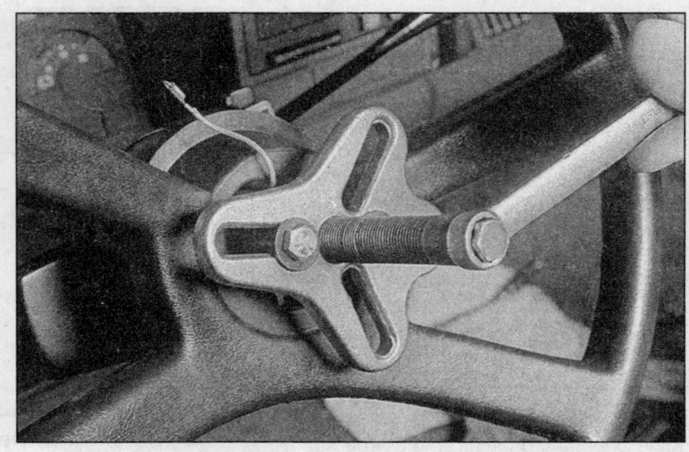

27.5 A puller is required for removing the steering wheel

25 Pitman arm (four-wheel drive) - removal and installation

1 Raise the front of the vehicle and support it securely on jackstands.
2 Remove the main securing nut and the pinch bolt.
3 Mark the Pitman arm-to-shaft relationship.
4 Draw the arm from the shaft using a puller (Section 24).
5 When installing the arm, spread the clamp just enough with a pry bar or large screwdriver to enable the arm to slide onto the splined shaft. Do not attempt to drive it onto the shaft or the steering gear will be damaged inside the housing.
6 Tighten the shaft nut and pinch bolt to the specified torque.

26 Manual steering gear - check and adjustment

Refer to illustration 26.7

1 If lost motion, slackness and vibration are noted in the manual steering mechanism itself, several adjustments are possible.
2 Inspect the steering linkage to make sure it is not worn and that all attaching bolts are secure (Chapter 1).

Worm bearing adjustment

3 Disconnect the negative cable at the battery. Place the cable out of the way so it cannot accidentally come in contact with the negative terminal of the battery, as this would once again allow power into the electrical system of the vehicle.
4 Raise the front of the vehicle and support securely on jackstands.
5 Remove the Pitman arm nut and washer.
6 Mark the relationship of the Pitman arm to the shaft and then, using a suitable puller, remove the arm.
7 Loosen the adjuster screw locknut on the steering gear and then back off the adjuster screw 1/4-turn **(see illustration)**.
8 Remove the horn button or shroud.
9 Turn the steering in one direction gently until it comes up against its stop. Now measure the number of turns to the opposite stop, then divide the number of turns in two in order that the steering can then be set in its center position. Mark the rim of the steering wheel with a piece of tape and count the number of turns as the tape passes a fixed point on the instrument panel.
10 The steering wheel must now be rotated through 90-degrees to measure bearing drag. To do this, either attach a correctly calibrated torque wrench and socket to the steering wheel nut or attach a cord and spring balance to the handle of a socket wrench and note the force required to move the steering wheel. If the latter is used, make sure that the cord is attached one-inch from the center of the nut down the wrench handle as the worm bearing preload (see Specifications) is measured in in-lbs.
11 Turn the adjuster screw as necessary and then repeat the turning check until preload is to specification, then tighten the adjuster screw locknut.

Pitman shaft lash adjustment (over center preload)

12 Set the steering to the center position as described in Step 9.
13 Release the lash adjuster locknut and turn the adjuster screw clockwise until any backlash is eliminated. Retighten the locknut.
14 Check the turning torque as described in Step 10, taking the highest reading as the wheel passes through the center point. This should be the specified value more than the figure to which the worm bearing preload was finally set. If necessary release the locknut and turn the adjuster screw until the correct over center torque is obtained.
15 Install the Pitman arm, connect the battery and install the horn button.

Early model steering column lower bearing adjustment

16 Loosen the clamp at the lower end of the steering shaft.
17 While an assistant pushes down with moderate pressure on the steering wheel, set the position of the clamp as shown.
18 Tighten the clamp bolt to the specified torque without moving the position of the clamp.

Shift tube adjustment

19 Set the steering column shift tube lever in the Neutral or Drive position.
20 Loosen the adjusting ring clamp screws and turn the shift tube adjusting ring to obtain between 0.033 and 0.036-inch clearance using a feeler gauge between the shift tube lever and the adjusting ring.
21 When the adjustment is correct, tighten the clamp ring screws.

27 Steering wheel - removal and installation

Refer to illustration 27.5

1 Disconnect the negative cable at the battery. Place the cable out of the way so it cannot accidentally come in contact with the negative terminal of the battery, as this would once again allow power into the electrical system of the vehicle.
2 Remove the horn button or shroud, cup, Belleville spring and bushing.
3 Set the front wheels in the straight ahead position.
4 Remove the steering wheel retaining nut and extract the washer. Mark the relationship of the steering wheel to the upper end of the shaft.
5 A steering wheel puller will now be required to remove the steering wheel from the shaft **(see illustration)**. Do not attempt to remove the wheel by hitting it from the rear side as damage to the column may result.
6 Installation is the reverse of removal.

29.1 Steering shaft coupling

30.2 Use a 12-point socket to remove the coupling clamp pinch bolt (arrow)

Make sure that the turn signal switch is in the Neutral position before pushing the wheel into place.

28 Steering shaft universal joint coupling - removal and installation

1 This type of joint is used on earlier models which have an intermediate shaft with a splined slip joint.
2 Mark the relationship of the steering shaft to the worm shaft. Extract the bolt, nut and washer from the clamp on the steering gear worm shaft.
3 From the upper yoke of the universal joint, remove the two snap-rings, the two yoke bushings, and the cork washers by tapping the yoke shaft with a plastic mallet.
4 Move the center cross of the universal joint to separate the upper and lower sections of the coupling.
5 The lower sections of the coupling can be dismantled in the same way once it is removed from the vehicle.
6 Reassembly and installation are the reverse of the removal procedure.

29 Intermediate shaft coupling (early models) - removal, overhaul and installation

Refer to illustration 29.1

1 This type of coupling cannot be removed until either the steering gear has been lowered (see Section 36) or the steering column has been pulled up (see Section 35) **(see illustration)**.
2 Remove the intermediate shaft-to-flexible coupling bolt and withdraw the lower shaft and coupling assembly.
3 Pry off the snap-ring and slide the cover from the shaft.
4 Remove the bearing blocks, spring and tension spring from the pivot pin.
5 Wipe the grease from the pin and the end of the shaft. Scribe a locating mark on the pin and the end of the shaft. Scribe a locating mark on the pin on the same side of the shaft chamfer.
6 Support the shaft assembly (chamfer up) and use a press to remove the pin from the shaft.
7 Slide the washer and seal from the end of the shaft.
8 Clean all the components and slide the seal and washer onto the steering shaft so that the lip of the washer is towards the upper end of the shaft.
9 Install the pin from the chamfered side, aligning the previously made scribe marks. Make sure that the pin is centered or the coupling will bind.
10 Apply wheel bearing grease liberally to the bearing blocks and to the inside of the cover.
11 Locate the tension spring and bearing blocks on the pin.
12 Slide the cover over the bearing blocks, aligning the cover reference mark with the one on the shaft. Install the washer and seal into the end of the cover and secure with the snap-ring.
13 Install the lower shaft and coupling assembly onto the intermediate shaft and lower flexible coupling. Install the coupling bolts.

30 Steering shaft flexible coupling - removal and installation

Refer to illustration 30.2

1 Remove the nuts from the flange bolts.
2 Use a 12-point socket to remove the coupling clamp pinch bolt **(see illustration)**.
3 Remove the bolts which retain the steering box to the frame and then lower the steering box just far enough to allow withdrawal of the coupling.
4 Installation is the reverse of removal, making sure to attach the coupling to the worm shaft so that the flat on the shaft is in alignment with the flat on the coupling. Push the coupling onto the worm shaft until the coupling reinforcement bottoms.

31 Pitman shaft oil seal (two-wheel drive) - replacement in vehicle

1 A leaking Pitman shaft (sector shaft) oil seal can be replaced without having to remove or dismantle the steering gear.
2 Raise the vehicle and support it securely on jackstands.
3 Remove the Pitman arm.
4 Center the steering gear by setting the front wheels in the straight ahead position and then checking that the worm shaft flat is in the uppermost position.
5 Remove the side cover from the steering box together with the Pitman shaft.
6 Pry the defective Pitman shaft oil seal from its seat and tap a new one into position.
7 Remove the lash adjuster screw locknut and then separate the side cover from the Pitman shaft by turning the lash adjuster screw clockwise and unscrewing from the cover.
8 Install the Pitman shaft into the steering box so that the center tooth of the sector in the shaft enters the center tooth cutout of the ball nut.
9 Fill the steering gear with the specified type of lubricant and then place a new side cover gasket in position.
10 Install the side cover onto the lash adjuster screw by passing a thin screwdriver through the hole in the side cover and turning the screw counterclockwise until it bottoms. Turn the screw back 1/4-turn.
11 Install the side cover.
12 Carry out the adjustments described in Section 26.
13 Install the Pitman arm.

32 Manual steering gear - removal and installation

1 Raise the vehicle and support it securely

on jackstands.

2 Set the front wheels in the straight ahead position. Mark the relationship of the Pitman arm-to-shaft and the coupling to worm shaft relationship.

3 Remove the nut or pinch bolt which retains the Pitman arm to the Pitman shaft. Remove the arm with a suitable puller.

4 Remove the mounting bolts and lift the gear from the vehicle.

5 Because of the special tools and techniques required to overhaul the steering gear, it is recommended that a worn assembly be replaced with a new or rebuilt unit.

6 Installation is reverse of removal, but observe the following points:

a) *Make sure the coupling bolt passes easily through the shaft undercut.*
b) *Make sure that the flexible coupling pins are centered in the slots in the steering shaft flange.*

33 Steering column upper bearing (1967 through 1972) - replacement in vehicle

1 Remove the steering wheel (Section 27).

2 Remove the directional signal cam.

3 Pry out the upper bearing. On tilt column models, the upper bearings are part of the bearing housing and if they are worn, the complete assembly will have to be replaced.

4 Installation is the reversal of removal, but before installing the steering wheel make sure that the signal switch is in the Neutral position.

34 Steering column lower bearing (1967 through 1972) - replacement in vehicle

1 Remove the intermediate shaft and coupling as described in Section 29.

2 Remove the preload spring washer and spring from the end of the steering shaft.

3 On tilt column models, pry off the lower bearing reinforcement clip and remove the lower bearing and adapter as a complete assembly. Press the bearing from the adapter.

4 Installation is the reverse of removal except on tilt columns, make sure that the wide tab of the adapter engages in the open slot in the steering column. Make sure that when the reinforcement clip is installed, the three tabs are fully engaged both in the reinforcement and in the column.

35 Steering column (1967 through 1972) - removal and installation

1 Remove the clamp bolt from the coupling just above the steering gear.

2 On 3-speed and automatic transmission models, disconnect the shifter rods.

3 Release the clamp on the engine side of the firewall and slide the clamp down the column (mast jacket).

4 Remove the steering wheel (Section 27).

5 Disconnect the column wiring connector.

6 Remove the dash panel trim plate, parking brake lever and support assembly.

7 Remove the column upper clamp screws from the dash panel.

8 Lower the column and withdraw it, at the same time rotating it so that the shift levers clear the opening in the firewall. Draw the column into the vehicle interior.

9 Installation is the reverse of removal, taking care to adjust the lower bearing as described in Section 34.

36 Steering column (1967 through 1972) - overhaul

1 If the column has a shifter tube, slide the rubber grommet back from the shift lever support housing, drive out the selector lever pivot pin and remove the shift lever.

2 Remove the steering wheel retaining nut.

3 Slide the steering shaft out of the mast jacket and remove the backup light switch.

4 Extract the three locking screws from the directional signal housing, rotate the housing and remove it.

5 Remove the spacer and thrust washer from the shifter lever housing and bushing seat from the upper end of the mast jacket.

6 Remove the bushing and bushing seat from the other end of the mast jacket.

7 Remove the bolt and screws from the adjuster ring clamp at the lower end of the mast jacket and withdraw the clamp, lower bearing and adjuster ring.

8 On three speed models, remove the first and reverse shift lever and spacer.

9 On automatic transmission models, remove the three screws from the selector plate clamping ring.

10 Place the mast jacket upright on two pieces of wood and push down on the shift lever. Place a block of wood on the upper end of the shifter tube and tap the tube out of the mast jacket.

11 Remove the felt seal from the shifter tube.

12 Remove the firewall clamp, seal and panel seal from the mast jacket.

13 Reassembly the reverse of disassembly, but on automatic transmission vehicles adjust the shifter tube as described in Section 26.

37 Steering column (1973 on) - removal and installation

Refer to illustration 37.8

1 Disconnect the negative cable at the battery. Place the cable out of the way so it cannot accidentally come in contact with the negative terminal of the battery, as this would once again allow power into the electrical system of the vehicle.

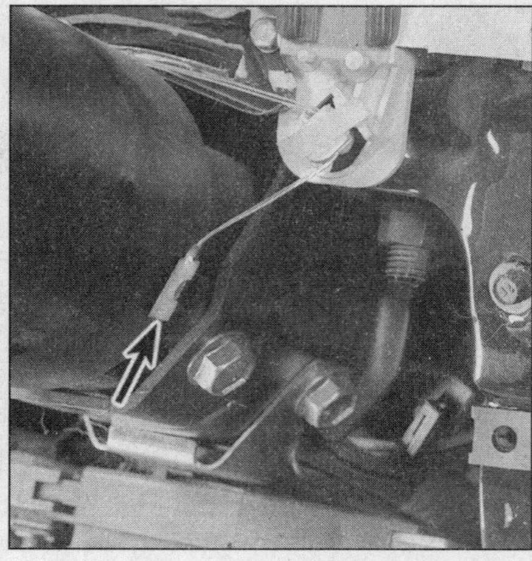

37.8 Typical automatic transmission indicator cable (arrow)

2 Remove the steering wheel (Section 27).

3 Remove the nuts and washers which secure the flanged end of the steering shaft to the flexible coupling.

4 Disconnect the transmission control linkage from the column shift tube levers.

5 Disconnect the wiring connectors on the side of the column.

6 Extract the floor pan trim cover screws and remove the cover.

7 Remove the screws which secure the two halves of the floor pan cover. Extract the screws which secure the halves and the seal to the floor pan and remove them.

8 Disconnect the shift position indicator cable (automatic transmission models) **(see illustration)**.

9 Move the front seat as far to the rear as possible to provide maximum clearance. Remove the two column bracket-to-instrument panel nuts. Lower the column and remove it.

10 Begin installation by assembling the lower dash cover and the upper dash cover to the seal. Use the rubber dowel plug attached to the seal to do this.

11 Attach the bracket to the column jacket and tighten the four bolts to the specified torque.

12 Install the column and position the flange to the flexible joint and then fit the lockwashers and nuts. The coupling on manual steering assemblies must be installed before the column itself.

13 Loosen the two nuts from the screws at the instrument panel bracket.

14 Position the lower clamp and tighten the attaching nuts to the specified torque.

15 Tighten the two nuts to the specified torque.

16 Install the seal and the covers.

17 Install the screws and tighten to the specified torque.

18 Tighten the two nuts to the specified torque.

19 If plastic spacers are used to center the flexible coupling pins (these spacers are supplied specially for the purpose), remove them at this time.
20 Install the indicator cable (automatic transmission).
21 Install the instrument panel trim cover.
22 Connect the transmission control linkage at the shift tube levers.
23 Install the steering wheel.
24 Connect the battery ground cable.

38 Steering column (1973 on) - overhaul

1 Disconnect the negative cable at the battery. Place the cable out of the way so it cannot accidentally come in contact with the negative terminal of the battery, as this would once again allow power into the electrical system of the vehicle.
2 With the column removed from the vehicle, remove the dash seal.

Disassembly

3 Secure the column in a vise by gripping one set of the weld nuts attached to the column.
4 Remove the directional signal switch and lock cylinder and the ignition key warning switch and the ignition switch.
5 On column shift models, drive out the upper shift lever pivot pin and remove the shift lever.
6 Remove the upper bearing and thrust washer.
7 Remove the four screws which attach the turn signal switch and ignition lock housing to the jacket. Remove the housing assembly.
8 Take out the thrust cap from the lower side of the housing and lift the ignition switch actuating rod and rack assembly together with the shift lock bolt and spring assembly from the housing.
9 Remove the shift gate.
10 Remove the ignition switch actuator sector through the lock cylinder hole by pushing on the block tooth sector with a rod or punch.
11 Remove the gearshift lever hosing and shroud, or the transmission control lock tube housing and shroud, as applicable.
12 Remove the shift lever spring from the gearshift housing, or the lock tube spring, as applicable.
13 Pull the steering shaft from the lower end of the jacket assembly.
14 Remove the back-up switch or Neutral safety switch (2 screws).
15 Remove the lower bearing retainer clip.

Automatic and floor shift models

16 Remove the lower bearing retainer adapter assembly, shift tube spring and washer. Press out the lower bearing by applying pressure to the outer race, then slide out the shift tube assembly.

Column shift (manual)

17 Remove the lower bearing adapter, bearing and first/reverse shift lever. Press out the lower bearing by applying pressure to the outer race. Remove the three screws from the lower end bearing and slide out the shift tube assembly.
18 From the upper end of the mast jacket, remove the gearshift housing lower bearing.

Reassembly

19 Replace any worn components and begin reassembly by applying a thin coat of lithium base grease to all friction surfaces and then installing the sector into the turn signal housing. To do this, reach through the lock cylinder hole, place the sector onto the shaft with the tang end to the outside of the hole. Press the sector onto the shaft using a blunt tool.
20 Install the shift gate onto the housing.
21 Insert the rack preload spring into the housing from the lower end so that both ends of the spring are attached to the housing.
22 Assemble the locking bolt to the crossover arm on the rack.
23 Insert the rack and lock bolt assembly into the housing (teeth up). Align the first tooth on the sector with the first tooth on the rack so that the block teeth will line up when the rack assembly is pushed in.
24 Install the thrust cup into the housing.
25 Install the gearshift housing lower bearing, aligning the indentations with the projections in the jacket.
26 Install the shift lever spring into the housing.
27 Install the housing and shroud assemblies onto the mast jacket, rotating slightly to ensure proper seating in the bearing.
28 With the shift lever housing in position, and the gearshift housing at Park, pull the rack down and install the directional signal switch and lock cylinder housing onto the jacket. When seated, install the four screws.
29 Press the lower bearing fully into the adapter assembly.

Automatic and floor shift models

30 Assemble the spring, and the lower bearing and adapter assembly into the bottom of the jacket. Hold the adapter in place then install the lower bearing reinforcement and retainer. Ensure that the retainer snaps into the slots.

Column shift (manual)

31 Loosely install the three screws in the jacket and shift tube bearing. Assemble the first/reverse lever and lower the bearing and adapter assembly into the bottom of the jacket. Hold the adapter in place and install the bearing reinforcement and retainer. Ensure that the retainer snaps into the slots. Place a 0.005-inch feeler gauge blade between the second/third lever and spacer then turn the upper shift tube bearing down and tighten the three screws. Remove the feeler gauge.
32 Install the neutral safety or back up switch. Refer to Chapter 7 for further information.
33 Slide the steering shaft into the column then install the upper bearing thrust washer.
34 Install the ignition key warning switch, directional signal switch, lock cylinder assembly and ignition switch, as described previously.
35 Install the shift lever and shift lever pivot pin, then remove the assembly from the vise.
36 Install the four dash bracket-to-column screws and tighten to the specified torque.

39 Steering column (tilt version) - overhaul

Refer to illustrations 39.3, 39.4, 39.7, 39.26 and 39.39

1 Follow the standard column dismantling procedures in Steps 1 through 3 in Section 36.
2 Remove the tilt release lever then drive out the shift lever pivot pin and remove the shift lever from the housing.
3 Remove the three directional signal housing retaining screws **(see illustration)**.
4 Install the tilt release lever and move the column to the highest position. Use a screwdriver to remove the tilt lever spring retainer

39.3 Remove the three screws (arrows) which retain the switch cover assembly

39.4 To release the tilt mechanism retainer, push it in and turn it counterclockwise

39.7 Use a slide hammer and a 8/32 machine screw to remove the bearing housing pivot pins (tilt column)

by pressing in approximately 3/16-inch then turning 1/8-turn (45-degrees) counterclockwise until the ears align with the grooves in the housing **(see illustration)**.

5 Remove the pot joint-to-steering shaft clamp bolt, then remove the intermediate shaft and pot joint assembly.

6 Push the upper shaft in sufficiently to remove the upper bearing inner race and seat. Pry off the lower bearing retainer and remove the bearing reinforcement, bearing and bearing adapter assembly from the lower end of the mast jacket.

7 Withdraw the upper bearing housing pivot pins **(see illustration)**.

8 Install the tilt release lever and disengage the lock shoes, then remove the bearing housing by pulling up to extend the rack fully down. Move the housing to the left to disengage the ignition switch rack from the actuator rod.

9 Remove the steering shaft assembly from the upper end of the column, then the upper bearing seat and inner race.

10 Dismantle the shaft by removing the centering balls and the anti-lash spring.

11 Remove the automatic transmission shift position indicator wire, where applicable.

12 Remove the four screws retaining the shaft bearing housing support, followed by the housing support. Remove the ignition switch actuator rod.

13 Use a puller to remove the shift tube (or transmission control lock tube on floor shift models) from the lower end of the mast jacket.

14 Remove the bearing housing support lockplate by sliding it out of the jacket notches and tipping it down towards the hub at the 12 o'clock position. Slide it under the jacket opening and remove the wave washer.

15 Remove the shift lever housing or lock tube housing from the mast jacket. Remove the shift lever spring by winding it up with pliers, then pulling it out. On floor shift models, the spring plunger has to be removed.

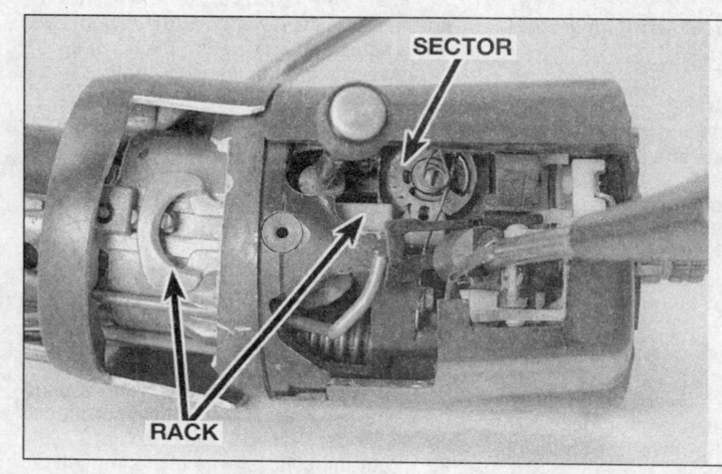

39.26 Lock bolt and rack assembly installation details (tilt column)

16 To dismantle the bearing housing, remove the tilt lever opening shield then take out the lock bolt spring by removing the retaining screw and moving the spring clockwise.

17 Remove the snap-ring from the sector driveshaft, then use a hammer a small punch to lightly tap the driveshaft from the sector. Remove the driveshaft, sector, lock bolt, rack and rack spring.

18 Drive out the tilt release lever, then remove the lever and spring. To relieve the load on the release lever, hold the shoes in and wedge a block between the top of the shoes (over the slots) and the bearing housing.

19 Drive out the lock shoe retaining pin then the lock shoe and springs. **Note:** *With the tilt lever opening on the left and the shoe uppermost, the four-slot shoe is on the left.*

20 If the bearings are to be replaced, remove the separator and balls. Carefully drive out the race from the housing, followed by the second race.

21 During the assembly procedure, all friction surfaces should be lightly lubricated with white lithium base grease.

22 Where dismantled, carefully press the bearings into the housing using a suitable sized socket.

23 Install the lock shoe springs, shoes and shoe pin, using a suitable rod (approximately 0.180-inch diameter) for locating purposes.

24 Install the shoe release lever, spring and pin. To relieve the release lever load, hold the shoes in and wedge a block between the top of the shoes (over the slots) and bearing housing.

25 Install the sector driveshaft. Lightly tap it on until the snap-ring can be installed.

26 Install the lock bolt and engage it with the sector cam surface, then install the rack and spring. The block tooth on the rack must engage correctly in the sector **(see illustration)**. Install the tilt release lever.

27 Install the lock bolt spring and tighten the retaining screw securely.

28 Wind up the shift lever spring with pliers and install (push) it onto the housing. On floor shift models the plunger has to be installed.

29 Slide the gearshift lever housing onto the steering mast jacket.

30 Install the wave washer for the bearing support lockplate.

31 Install the lockplate, working it into the notches in the jacket by tipping towards the

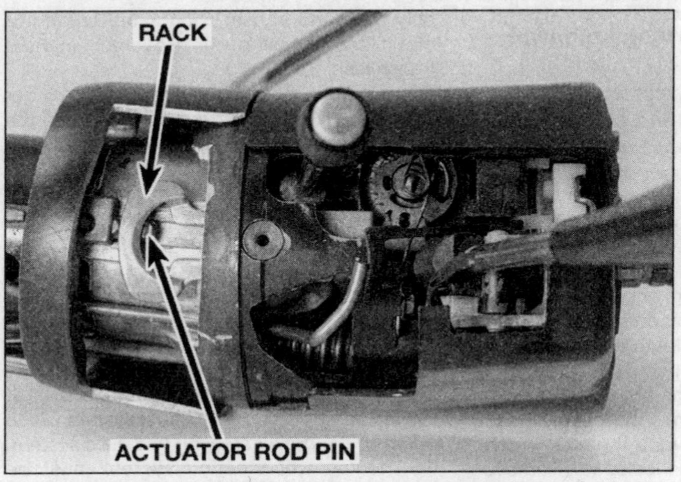

39.39 Bearing housing installation details (tilt column)

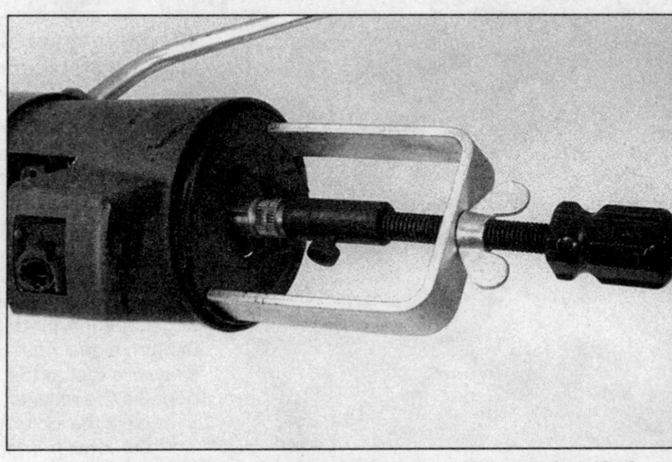

41.5 A special tool is required to depress the steering shaft lockplate so the retaining ring can be removed

housing hub at the 12 o'clock position and sliding it under the jacket opening. The lockplate can then be slid into the notches in the jacket.

32 Carefully install the shift tube into the lower end of the mast jacket, aligning the keyway in the tube with the key in the shift lever housing. The next part of the operation ideally requires the use of a special tool, but suitable spacers, washers and a long bolt can be used as an alternative. Install the tube as shown and pull the shift tube into the housing by rotating the outer nut. Do not exert any load on the end of the shift tube and ensure that the shift tube lever is aligned with the slotted opening at the lower end of the mast jacket.

33 Install the bearing support thrust washer and retaining ring by pulling the shift lever housing up to compress the wave washer.

34 Install the bearing support, aligning the "V" in the support with the "V" in the jacket. Insert the support to lockplate screws and tighten to the specified torque.

35 Align the lower bearing adapter with the notches in the jacket, then push the adapter into the lower end. Install the lower bearing, bearing reinforcement and retainer, ensuring that the clip is aligned with the slots in the reinforcement, jacket and adapter.

36 Install the centering balls and anti-lash spring in the upper shaft, then install the lower shaft from the same side of the balls as the spring ends protrude.

37 Install the steering shaft assembly into the shift tube from the upper end, guiding the shaft carefully through the tube and bearing.

38 Install the ignition switch actuator rod through the shift lever housing and insert it in the bearing support slot. Extend the rack down from the housing.

39 Assemble the bearing housing over the steering shaft, engaging the rack over the end of the actuator rod **(see illustration)**.

40 Install the external release lever, then hold the lock shoes in the disengaged position and assemble the bearing housing over the steering shaft until the pivot pin holes align. Now install the pivot pins.

41 Place the bearing housing in the fully up position, then install the tilt lever spring guide, spring and spring retainer. Using a screwdriver, push in the retainer and turn clockwise to engage in the housing.

42 Install the upper bearing inner race and seat.

43 Install the tilt lever opening shield.

44 Remove the tilt release lever, install the directional signal housing and tighten the three retaining screws.

45 Install the tilt release lever and the shift lever, then drive in the shift lever pin.

46 Install the ignition key warning switch, lock, cylinder, directional switch and ignition switch.

47 Align the grooves across the upper end of the pot joint with the steering shaft flat and assemble the intermediate shaft assembly to the upper shaft. Install the clamp and bolt and tighten to the specified torque.

48 Install the neutral safety or back-up switch. Refer to Chapter 8 for further information.

49 Install the four dash panel bracket-to-column screws and tighten to the specified torque. **Note:** *Ensure that the slotted openings in the bracket face the upper end of the column.*

40 Steering column turn signal switch (1967 through 1972) - removal and installation

1 Disconnect the negative cable at the battery. Place the cable out of the way so it cannot accidentally come in contact with the negative terminal of the battery, as this would once again allow power into the electrical system of the vehicle.

2 Remove the steering wheel (Section 27).

3 Extract the turn signal canceling cam and spring.

4 Remove the column-to-instrument panel trim plate.

5 Disconnect the signal switch wiring harness at the half-moon shaped connector.

6 Pry the wiring harness protector from the slots in the steering column.

7 Mark the position of each wire and then push each one out of the connector using a tool similar to the one shown to depress the wire retainer tabs.

8 Remove the turn signal lever screw and withdraw the lever.

9 Depress the hazard warning lamp knob, then unscrew and remove the knob.

10 If a tilt column is installed unscrew and remove the tilt release lever. If the vehicle is equipped with automatic transmission, remove the speed selector position indicator dial screws and withdraw the dial, needle, cap and bulb from the housing cover. Pull off the turn signal housing cover using a small two-legged puller with jaws that face out. The turn signal switch can be removed after extracting the three mounting screws and guiding the wiring harness through the opening in the shift level housing.

11 Installation of the turn signal switch to both types of columns is the reverse of removal.

41 Steering column turn signal switch (1973 on) - removal and installation

Refer to illustration 41.5

1 Disconnect the negative cable at the battery. Place the cable out of the way so it cannot accidentally come in contact with the negative terminal of the battery, as this would once again allow power into the electrical system of the vehicle.

2 Remove the steering wheel (Section 27).

3 Remove the column-to-instrument panel trim cover.

4 Using a screwdriver, pry the cover from the lockplate.

5 A compressor tool will now be required to compress the lockplate so that the snap-ring can be pried out of its groove **(see illustration)**. Remove the lockplate.

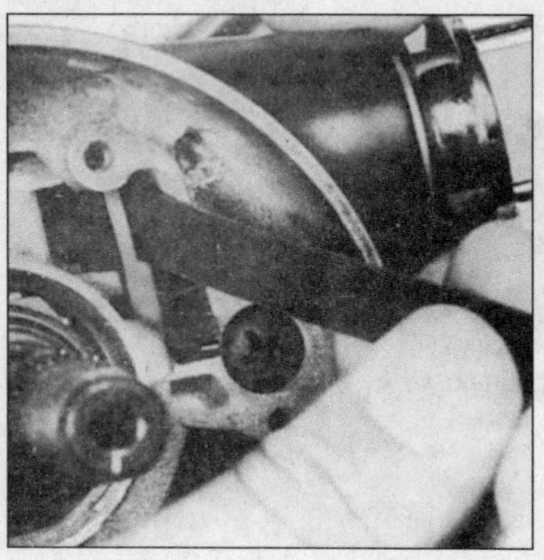

42.3 **Ignition lock cylinder removal on early models requires a narrow flat rod to depress the lock retainer**

6 From the end of the shaft, slide off the signal canceling cam, the bearing preload spring and the thrust washer.
7 Remove the directional signal lever screw and the lever.
8 Push the hazard warning knob in and then unscrew the knob.
9 Remove the three switch mounting screws.
10 Pull the switch connector from the jacket bracket. Feed the connector through the column support bracket and then pull the switch straight up.
11 Remove the turn signal wire protector by gripping its tab with a pair of pliers and pulling down.
12 On tilt column models, set the signal and shifter in Low. Remove the harness cover by pulling it down. Remove the three switch mounting screws and pull the switch up to remove.
13 Installation is the reverse of removal, but use a new lockplate snap-ring.

42 Steering column lock - removal and installation

Refer to illustrations 42.3 and 42.4

1 Disconnect the negative cable at the battery. Place the cable out of the way so it cannot accidentally come in contact with the negative terminal of the battery, as this would once again allow power into the electrical system of the vehicle.
2 Remove the steering wheel (Section 27).
3 On early models, insert a thin rod into the turn signal housing slot **(see illustration)**. Keep the tool to the right side of the slot and depress the retainer at the bottom of the slot to release the lock. Remove the lock cylinder.
4 On later models, place the lock in the Run position. Remove the lockplate. Remove the directional signal switch just far enough to slip it over the end of the shaft without pulling the harness out of the column. Remove the lock retaining screw and the lock cylinder **(see illustration)**. **Note:** *Be very careful when removing the screw (use a magnetic screwdriver, if available). If the retaining screw is dropped during removal, it could fall into the steering column. The column would have to be completely disassembled to remove the screw.*
5 On early models, install the lock by holding the cylinder sleeve and turning the knob clockwise against its stop.
6 Insert the cylinder into the housing so that the key on the cylinder sleeve is aligned with the keyway in the housing. Push the cylinder into the abutment of the cylinder and the sector.
7 Hold a 0.070-inch diameter drill bit between the lock bezel and the housing. Rotate the cylinder counterclockwise while maintaining a gentle pressure until the drive section of the cylinder mates with the sector.
8 Push in until the snap-ring locks in the grooves. Remove the drill and check the operation of the lock, which should engage only when the transmission is in Park (automatic transmission) or Reverse (manual transmission).
9 On later models, hold the lock cylinder in your hand and, with the key in, rotate the knob clockwise to the stop. Align the cylinder key with the keyway in the housing and push the lock cylinder all the way in. Install the lock retaining screw and tighten it securely.
10 Install the turn signal switch and the steering wheel.

43 Power steering system - bleeding

Note: *Refer to Chapter 9 for details of bleeding the brake assist Hydroboost system which relies on the power steering hydraulic system.*

1 The power steering system must be bled whenever a line is disconnected. Bubbles can be seen in power steering fluid which has air in it and the fluid can often have a tan or milky appearance. Low fluid level can cause air to mix in with the fluid resulting in a noisy pump as well as foaming of the fluid.
2 Open the hood and check the fluid level in the reservoir, adding the specified fluid necessary to bring it up to the proper level.
3 Raise the front wheels off the ground.
4 With the power steering pump and fluid at normal operating temperature, start the engine and slowly turn the steering wheel several times from left-to-right and back again, lightly contacting the stops. Check the fluid level, topping it up as necessary until it remains steady and no more bubbles appear in the reservoir.

44 Power steering pump - removal and installation

Refer to illustration 44.4

1 Disconnect the fluid hoses from the power steering pump. Cap or tape the open ends of the hoses to prevent loss of fluid and

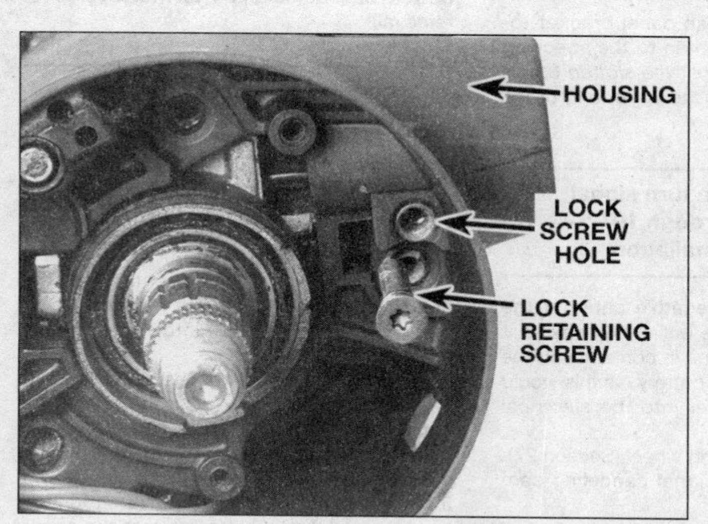

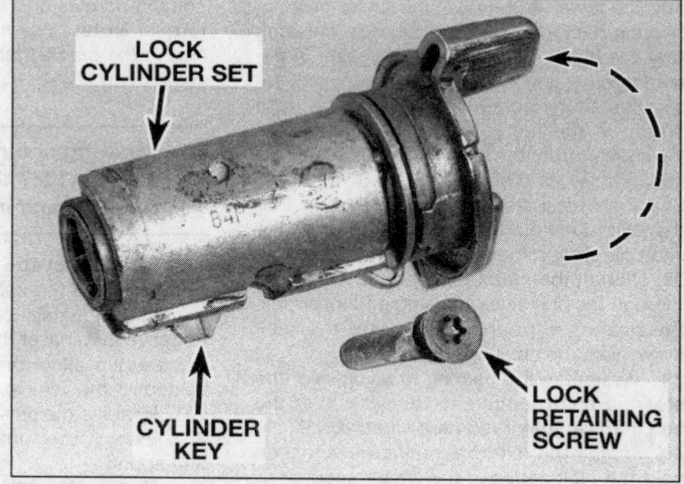

42.4 Later model ignition lock cylinder , the lock cylinder is held in place by a screw

44.4 Typical V8 power steering pump installation details

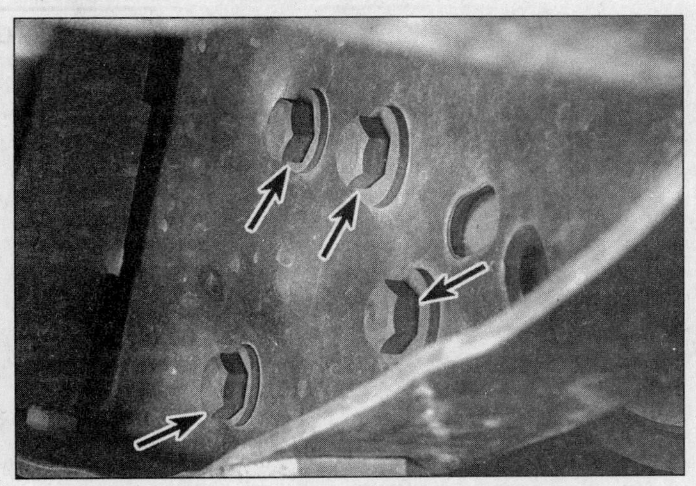

45.5 Remove the four steering gear mounting bolts (arrows)

fasten them out of the way.
2 Catch any fluid which drains from the pump in a suitable container.
3 Loosen the pump-to-bracket mounting nuts, push the pump in toward the engine and remove the drivebelt.
4 Unbolt and remove the pump from the vehicle **(see illustration)**.
5 Installation is the reverse of removal.
6 Adjust the drivebelt tension and bleed the system.

45 Power steering gear - removal and installation

Refer to illustration 45.5

1 Disconnect the flexible hoses from the steering gear. Plug the ends of the hoses and fasten them out of the way.
2 Tape over the holes in the steering gear case to prevent the entry of dirt.
3 Disconnect the steering shaft flexible coupling.
4 Mark the Pitman arm-to-shaft relationship and remove the Pitman arm (Section 31).
5 Remove the bolts retaining the steering gear to the frame and lower the gear from the vehicle **(see illustration)**.
6 If the gear is faulty or worn, it should be replaced with a new or factory reconditioned unit. Do not dismantle the original unit.
7 Installation is the reverse of removal. Take care when connecting the flexible coupling and make sure that the pinch bolt passes through the notch in the steering shaft.
8 Make sure that the coupling alignment pins are centered in the slots in the steering shaft flange. Special plastic screws are available from your dealer for this job, but they must be removed before tightening the coupling bolts to the specified torque.
9 Bleed the system as described in Section 43.

46 Steering angles and wheel alignment - general information

Correct wheel alignment is essential to proper steering and even tire wear. Symptoms of alignment problems are pulling of the steering to one side or the other and uneven tire wear.

If these symptoms are present, check for the following before having the alignment adjusted:

Loose steering gear mounting bolts
Misadjusted steering gear
Worn or damaged wheel bearings
Bent tie-rods
Worn tie-rod ends
Improper tire pressure
Mixing tires of different construction

Front wheel alignment should be left to a dealer service department or an alignment shop.

Notes

Chapter 11 Body

Contents

	Section
Body and frame repair - major damage	5
Body repair - minor damage	4
Bumpers - removal and installation	19
Carburetor outside air intakes - removal and installation	25
Door inside handle - removal and installation	15
Door lock assembly - removal and installation	12
Door lock cylinder - removal and installation	14
Door outside handle - removal and installation	13
Door - removal, installation and adjustment	18
Door striker - adjustment	17
Door trim panel - removal and installation	8
Door vent window assembly - removal and installation	9
Door window glass and regulator - removal and installation	10
Front fender - removal and installation	26
General information	1
Hood lock - removal, installation and adjustment	23
Hood release cable (later models) - removal and installation	24
Hood - removal and installation	22
Maintenance - body and frame	2
Maintenance - hinges and locks	6
Maintenance - upholstery and carpets	3
Power-operated door lock - removal and installation	16
Power-operated window glass and regulator - removal and installation	11
Radiator grille (1967 through 1971 models) - removal and installation	20
Radiator grille (1972 and later models) - removal and installation	21
Rear view mirrors - removal and installation	28
Tailgate - removal and installation	27
Windshield and fixed glass - removal and installation	7

Specifications

Torque specifications

	Ft-lbs
Front bumper bolt	35
Front bumper-to-bracket bolt	70
Rear bumper-to-outer bracket	35
Rear bumper outer bracket and brace	50
Step bumper-to-bracket or frame	40
Lock striker-to-body pillar	45
Door hinge bolt	35
Tailgate hinge bolt	35
Hood hinge bolt	18
Fender-to-cowl bolt	35
Tailgate anchor plate	18
Tailgate support cable bolts	25
Tailgate trunnion bolts	18
Tailgate trunnion bolts (with chain supports)	35

1 General information

These models were available in a wide variety of body styles, including pick-up, crew cab pickup and station wagon (Suburban). The vehicles covered in this manual have a separate frame and body.

As with other parts of the vehicle, proper maintenance of body components plays an important part in retention of the vehicle's market value. It is far less costly to handle small problems before they grow into larger ones. Information in this Chapter will tell you all you need to know to keep seals sealing, body panels aligned and general appearance up to par.

Major body components which are particularly vulnerable in accidents are removable. These include the hood, fenders, grille, doors, box and tailgate. It is often cheaper and less time consuming to replace an entire panel than it is to attempt a restoration of the old one. However, this must be decided on a case-by-case basis.

2 Maintenance - body and frame

1 The condition of your vehicle's body is very important, because it is on this that the second hand value will mainly depend. It is much more difficult to repair a neglected or damaged body than it is to repair mechanical components. The hidden areas of the body, such as the fender wells, the frame, and the engine compartment, are equally important, although they obviously do not require as frequent attention as the rest of the body.

2 Once a year, or every 12,000 miles, it's a good idea to have the underside of the body and the frame steam cleaned. All traces of dirt and oil will be removed and the underside can then be inspected carefully for rust, damaged brake lines, frayed electrical wiring, damaged cables and other problems. The front suspension components should be greased after completion of this job.

3 At the same time, clean the engine and the engine compartment using either a steam cleaner or a water soluble degreaser.

4 The fender wells should be given particular attention, as undercoating can peel away and stones and dirt thrown up by the tires can cause the paint to chip and flake, allowing rust to set in. If rust is found, clean down to the bare metal and apply an anti-rust paint.

5 The body should be washed as needed. Wet the vehicle thoroughly to soften the dirt, then wash it down with a soft sponge and plenty of clean soapy water. If the surplus dirt is not washed off very carefully, it will in time wear down the paint.

6 Spots of tar or asphalt coating thrown up from the road should be removed with a cloth soaked in solvent.

7 Once every six months, give the body and chrome trim a thorough waxing. If a chrome cleaner is used to remove rust from any of the vehicle's plated parts, remember that the cleaner also removes part of the chrome, so use it sparingly.

3 Maintenance - upholstery and carpets

1 Every three months remove the carpets or mats and clean the interior of the vehicle (more frequently if necessary). Vacuum the upholstery and carpets to remove loose dirt and dust.

2 If the upholstery is soiled, apply upholstery cleaner with a damp sponge and wipe it off with a clean, dry cloth.

4 Body repair - minor damage

See photo sequence

Repair of minor scratches

1 If the scratch is superficial and does not penetrate to the metal of the body, repair is very simple. Lightly rub the scratched area with a fine rubbing compound to remove loose paint and built up wax. Rinse the area with clean water.

2 Apply touch-up paint to the scratch, using a small brush. Continue to apply thin layers of paint until the surface of the paint in the scratch is level with the surrounding paint. Allow the new paint at least two weeks to harden, then blend it into the surrounding paint by rubbing with a very fine rubbing compound. Finally, apply a coat of wax to the scratch area.

3 If the scratch has penetrated the paint and exposed the metal of the body, causing the metal to rust, a different repair technique is required. Remove all loose rust from the bottom of the scratch with a pocket knife, then apply rust inhibiting paint to prevent the formation of rust in the future. Using a rubber or nylon applicator, coat the scratched area with glaze-type filler. If required, the filler can be mixed with thinner to provide a very thin paste, which is ideal for filling narrow scratches. Before the glaze filler in the scratch hardens, wrap a piece of smooth cotton cloth around the tip of a finger. Dip the cloth in thinner and then quickly wipe it along the surface of the scratch. This will ensure that the surface of the filler is slightly hollow. The scratch can now be painted over as described earlier in this section.

Repair of dents

4 When repairing dents, the first job is to pull the dent out until the affected area is as close as possible to its original shape. There is no point in trying to restore the original shape completely as the metal in the damaged area will have stretched on impact and cannot be restored to its original contours. It is better to bring the level of the dent up to a point which is about 1/8-inch below the level of the surrounding metal. In cases where the dent is very shallow, it is not worth trying to pull it out at all.

5 If the back side of the dent is accessible, it can be hammered out gently from behind using a soft-face hammer. While doing this, hold a block of wood firmly against the opposite side of the metal to absorb the hammer blows and prevent the metal from being stretched.

6 If the dent is in a section of the body which has double layers, or some other factor makes it inaccessible from behind, a different technique is required. Drill several small holes through the metal inside the damaged area, particularly in the deeper sections. Screw long, self tapping screws into the holes just enough for them to get a good grip in the metal. Now the dent can be pulled out by pulling on the protruding heads of the screws with locking pliers.

7 The next stage of repair is the removal of paint from the damaged area and from an inch or so of the surrounding metal. This is easily done with a wire brush or sanding disk in a drill motor, although it can be done just as effectively by hand with sandpaper. To complete the preparation for filling, score the surface of the bare metal with a screwdriver or the tang of a file or drill small holes in the affected area. This will provide a good grip for the filler material. To complete the repair, see the Section on filling and painting.

Repair of rust holes or gashes

8 Remove all paint from the affected area and from an inch or so of the surrounding metal using a sanding disk or wire brush mounted in a drill motor. If these are not available, a few sheets of sandpaper will do the job just as effectively.

9 With the paint removed, you will be able to determine the severity of the corrosion and decide whether to replace the whole panel, if possible, or repair the affected area. New body panels are not as expensive as most people think and it is often quicker to install a new panel than to repair large areas of rust.

10 Remove all trim pieces from the affected area except those which will act as a guide to the original shape of the damaged body, such as headlight shells, etc. Using metal snips or a hacksaw blade, remove all loose metal and any other metal that is badly affected by rust. Hammer the edges of the hole inward to create a slight depression for the filler material.

11 Wire brush the affected area to remove the powdery rust from the surface of the metal. If the back of the rusted area is accessible, treat it with rust-inhibiting paint.

12 Before filling is done, block the hole in some way. This can be done with sheet metal riveted or screwed into place, or by stuffing the hole with wire mesh.

13 Once the hole is blocked off, the affected area can be filled and painted. See the following sub-section on filling and painting.

Filling and painting

14 Many types of body fillers are available, but generally speaking, body repair kits which contain filler paste and a tube of resin hardener are best for this type of repair work. A wide, flexible plastic or nylon applicator will be necessary for imparting a smooth and contoured finish to the surface of the filler material. Mix up a small amount of filler on a clean piece of wood or cardboard (use the hardener sparingly). Follow the manufacturer's instructions on the package, otherwise the filler will set incorrectly.

15 Using the applicator, apply the filler paste to the prepared area. Draw the applicator across the surface of the filler to achieve the desired contour and to level the filler surface. As soon as a contour that approximates the original one is achieved, stop working the paste. If you continue, the paste will begin to stick to the applicator. Continue to add thin layers of paste at 20-minute intervals until the level of the filler is just above the surrounding metal.

16 Once the filler has hardened, the excess can be removed with a body file. From then

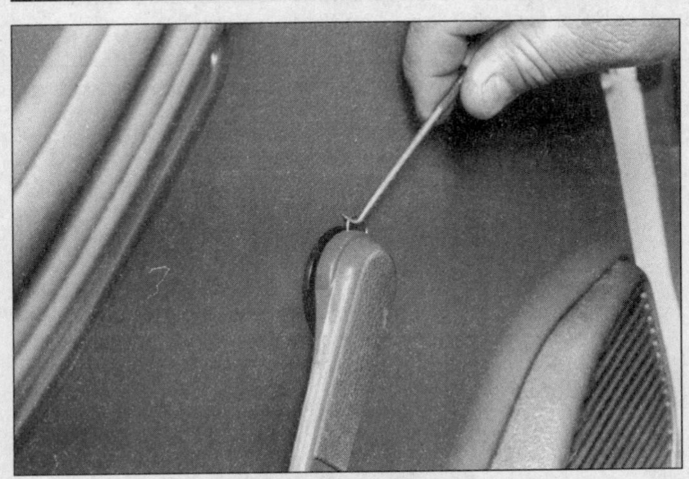

8.1 If a window crank clip removal tool is not available, press the door trim in, then extract the clip with a small hooked tool

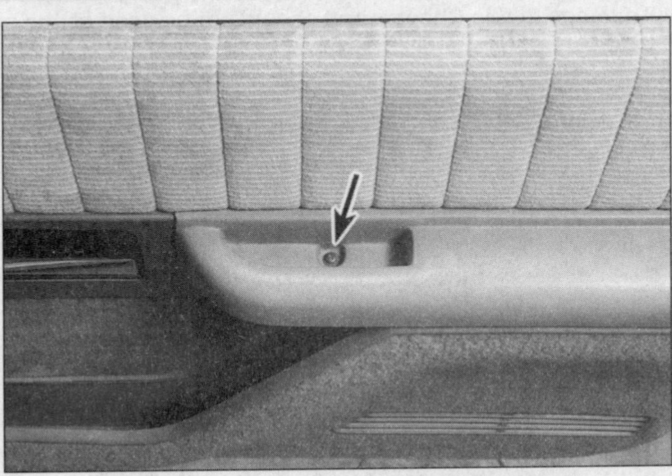

8.2 Remove the armrest retaining screw (arrow)

on, progressively finer grades of sandpaper should be used, starting with a 180-grit paper and finishing with 600-grit wet-or-dry paper. Always wrap the sandpaper around a flat rubber or wooden block, otherwise the surface of the filler will not be completely flat. During the sanding of the filler surface, the wet-or-dry paper should be periodically rinsed in water. This will ensure that a very smooth finish is produced in the final stage.

17 At this point, the repair area should be surrounded by a ring of bare metal, which in turn should be encircled by the finely feathered edge of good paint. Rinse the repair area with clean water until all of the dust produced by the sanding operation is gone.

18 Spray the entire area with a light coat of primer. This will reveal any imperfections in the surface of the filler. Repair the imperfections with fresh filler paste or glaze filler and once more smooth the surface with sandpaper. Repeat this spray-and-repair procedure until you are satisfied that the surface of the filler and the feathered edge of the paint are perfect. Rinse the area with clean water and allow it to dry completely.

19 The repair area is now ready for painting. Spray painting must be carried out in a warm, dry, windless and dust free atmosphere. These conditions can be created if you have access to a large indoor work area, but if you are forced to work in the open, you will have to pick the day very carefully. If you are working indoors, dousing the floor in the work area with water will help settle the dust which would otherwise be in the air. If the repair area is confined to one body panel, mask off the surrounding panels. This will help minimize the effects of a slight mismatch in paint color. Trim pieces such as chrome strips, door handles, etc., will also need to be masked off or removed. Use masking tape and several thicknesses of newspaper for the masking operations.

20 Before spraying, shake the paint can thoroughly, then spray a test area until the spray painting technique is mastered. Cover the repair area with a thick coat of primer. The thickness should be built up using several thin layers of primer rather than one thick one. Using 600-grit wet-or-dry sandpaper, rub down the surface of the primer until it is very smooth. While doing this, the work area should be thoroughly rinsed with water and the wet-or-dry sandpaper periodically rinsed as well. Allow the primer to dry before spraying additional coats.

21 Spray on the top coat, again building up the thickness by using several thin layers of paint. Begin spraying in the center of the repair area and then, using a circular motion, work out until the whole repair area and about two inches of the surrounding original paint is covered. Remove all masking material 10 to 15 minutes after spraying on the final coat of paint. Allow the new paint at least two weeks to harden, then use a very fine rubbing compound to blend the edges of the new paint into the existing paint. Finally, apply a coat of wax.

5 Body and frame repair - major damage

1 Major damage must be repaired by an auto body/frame repair shop with the necessary welding and hydraulic straightening equipment.

2 If the damage has been serious, it is vital that the frame be checked for proper alignment or the vehicle's handling characteristics may be adversely affected. Other problems, such as excessive tire wear and wear in the driveline and steering may occur.

3 Due to the fact that all of the major body components (hood, fenders, etc.) are separate and replaceable units, any seriously damaged components should be replaced rather than repaired. Sometimes these components can be found in a wrecking yard that specializes in used vehicle components, often at considerable savings over the cost of new parts.

6 Maintenance - hinges and locks

Every 3000 miles or three months, the door, hood and tailgate hinges and locks should be lubricated with a few drops of oil. The door striker plates should also be given a thin coat of white lithium-base grease to reduce wear and ensure free movement.

7 Windshield and fixed glass - removal and installation

1 Replacement of the windshield and fixed glass requires the use of special fast-setting adhesive/caulk materials. These operations should be left to a dealer or a shop specializing in glass work.

2 Windshield-mounted rear view mirror support removal is also best left to experts, as the bond to the glass also requires special tools and adhesives.

8 Door trim panel - removal and installation

Refer to illustrations 8.1 and 8.2

1 Remove the window crank and door handle. On early models, the crank and door handle are retained by screws, while on later models a clip retainer is used. A special tool, available at auto parts stores, makes clip removal easier **(see illustration)**. If the tool is not available, press the door trim in, then extract the clip with a length of wire with a small hook at the end.

2 Remove the armrest screws and the door pocket screws **(see illustration)**.

3 Remove the trim panel from the door by inserting a trim removal tool under the edge and prying the clips out.

4 Installation is the reverse of removal. Replace any panel retaining clips broken during removal.

These photos illustrate a method of repairing simple dents. They are intended to supplement *Body repair - minor damage* in this Chapter and should not be used as the sole instructions for body repair on these vehicles.

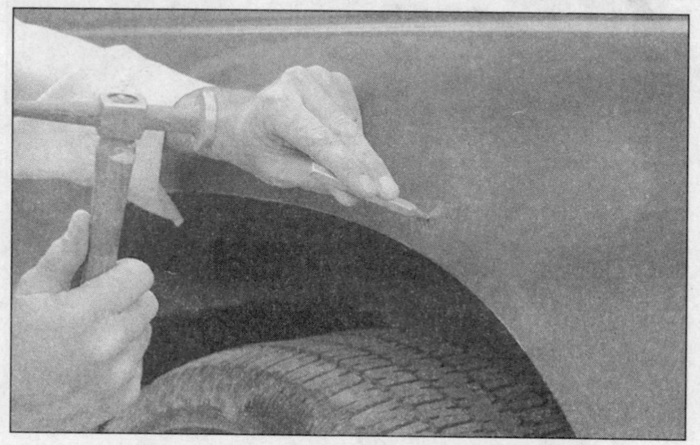

1 If you can't access the backside of the body panel to hammer out the dent, pull it out with a slide-hammer-type dent puller. In the deepest portion of the dent or along the crease line, drill or punch hole(s) at least one inch apart . . .

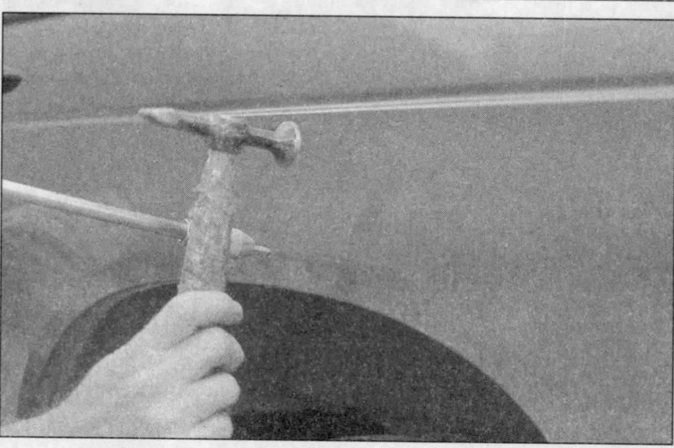

2 . . . then screw the slide-hammer into the hole and operate it. Tap with a hammer near the edge of the dent to help 'pop' the metal back to its original shape. When you're finished, the dent area should be close to its original contour and about 1/8-inch below the surface of the surrounding metal

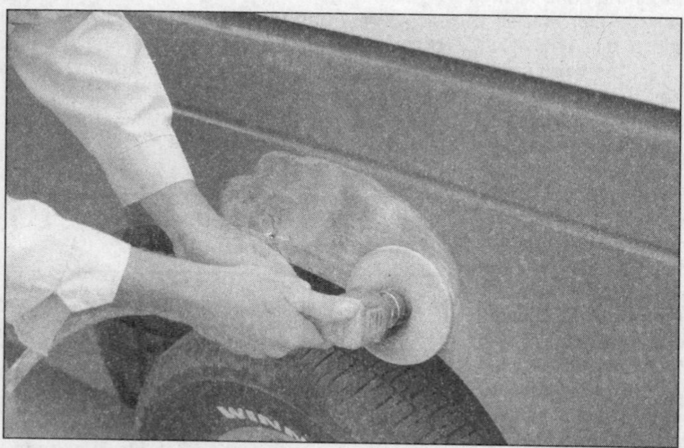

3 Using coarse-grit sandpaper, remove the paint down to the bare metal. Hand sanding works fine, but the disc sander shown here makes the job faster. Use finer (about 320-grit) sandpaper to feather-edge the paint at least one inch around the dent area

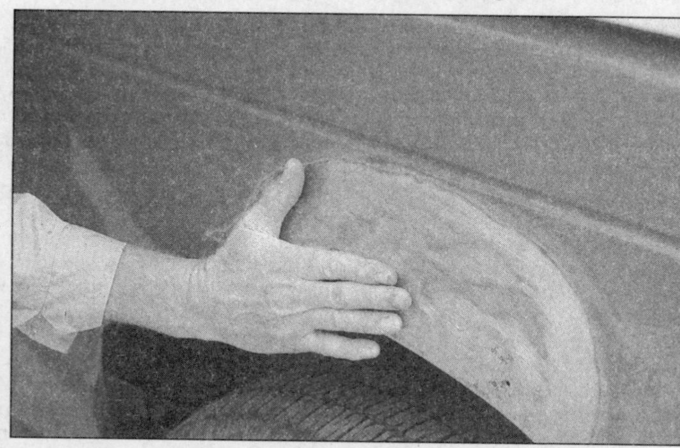

4 When the paint is removed, touch will probably be more helpful than sight for telling if the metal is straight. Hammer down the high spots or raise the low spots as necessary. Clean the repair area with wax/silicone remover

5 Following label instructions, mix up a batch of plastic filler and hardener. The ratio of filler to hardener is critical, and, if you mix it incorrectly, it will either not cure properly or cure too quickly (you won't have time to file and sand it into shape)

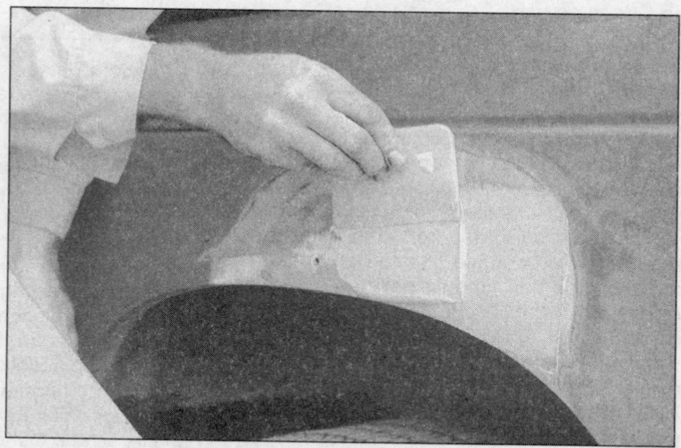

6 Working quickly so the filler doesn't harden, use a plastic applicator to press the body filler firmly into the metal, assuring it bonds completely. Work the filler until it matches the original contour and is slightly above the surrounding metal

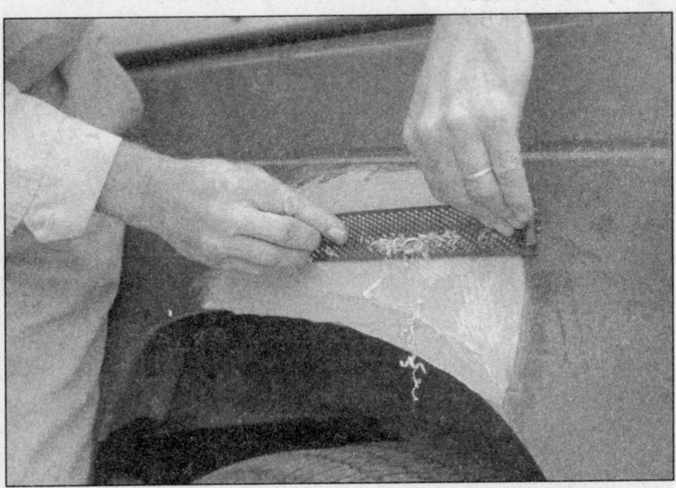

7 Let the filler harden until you can just dent it with your fingernail. Use a body file or Surform tool (shown here) to rough-shape the filler

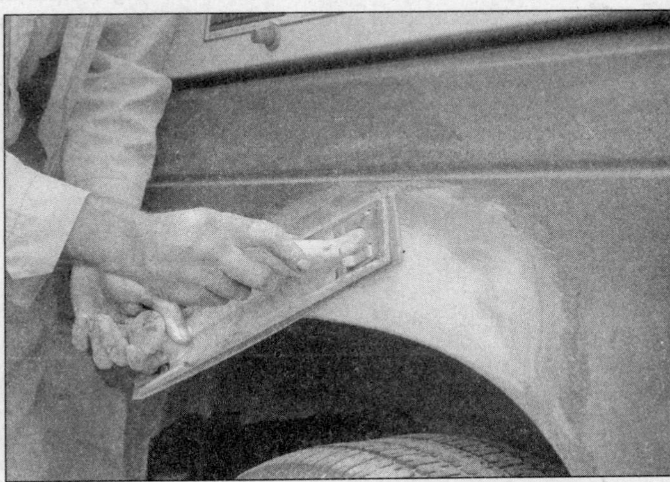

8 Use coarse-grit sandpaper and a sanding board or block to work the filler down until it's smooth and even. Work down to finer grits of sandpaper - always using a board or block - ending up with 360 or 400 grit

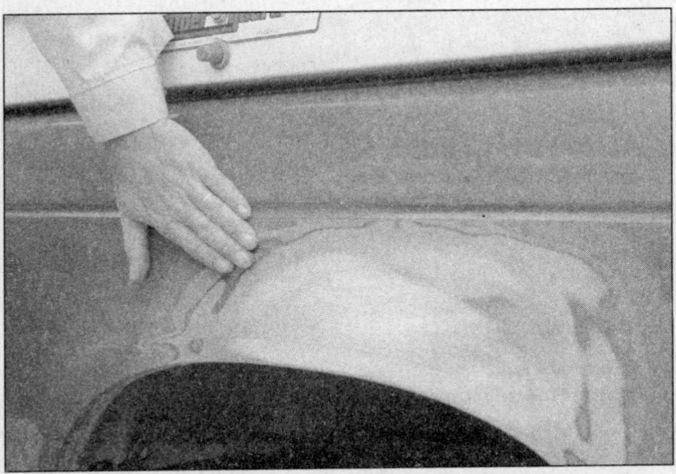

9 You shouldn't be able to feel any ridge at the transition from the filler to the bare metal or from the bare metal to the old paint. As soon as the repair is flat and uniform, remove the dust and mask off the adjacent panels or trim pieces

10 Apply several layers of primer to the area. Don't spray the primer on too heavy, so it sags or runs, and make sure each coat is dry before you spray on the next one. A professional-type spray gun is being used here, but aerosol spray primer is available inexpensively from auto parts stores

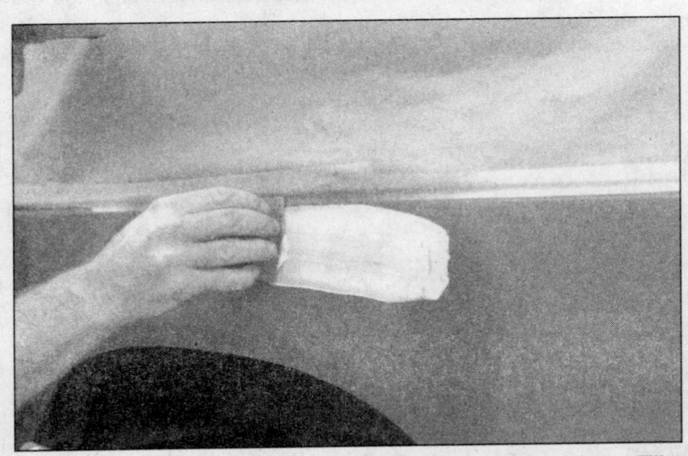

11 The primer will help reveal imperfections or scratches. Fill these with glazing compound. Follow the label instructions and sand it with 360 or 400-grit sandpaper until it's smooth. Repeat the glazing, sanding and respraying until the primer reveals a perfectly smooth surface

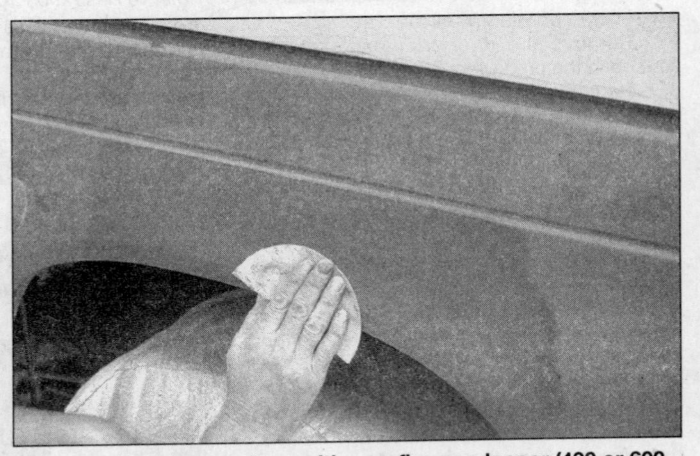

12 Finish sand the primer with very fine sandpaper (400 or 600-grit) to remove the primer overspray. Clean the area with water and allow it to dry. Use a tack rag to remove any dust, then apply the finish coat. Don't attempt to rub out or wax the repair area until the paint has dried completely (at least two weeks)

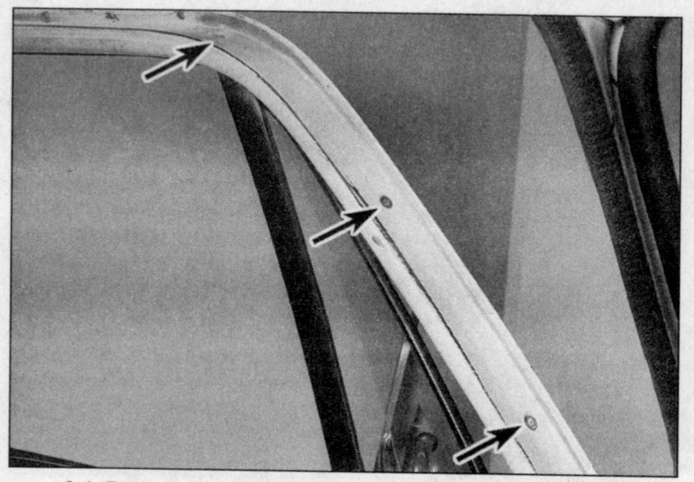

9.4 Remove the window vent retaining screws (arrows)

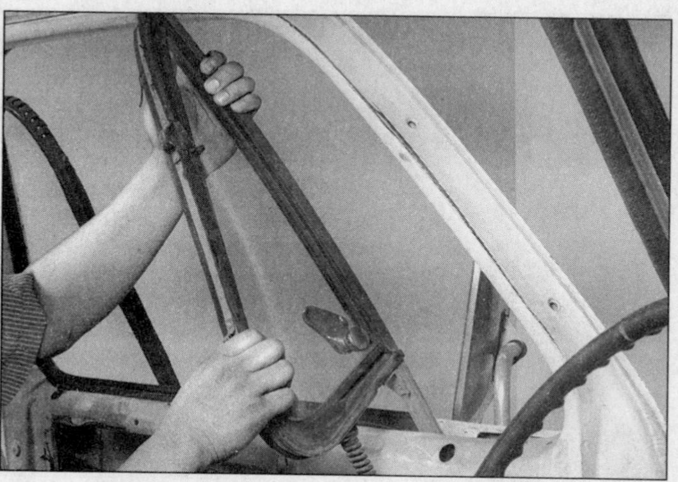

9.7 Lift the vent window up and out of the door

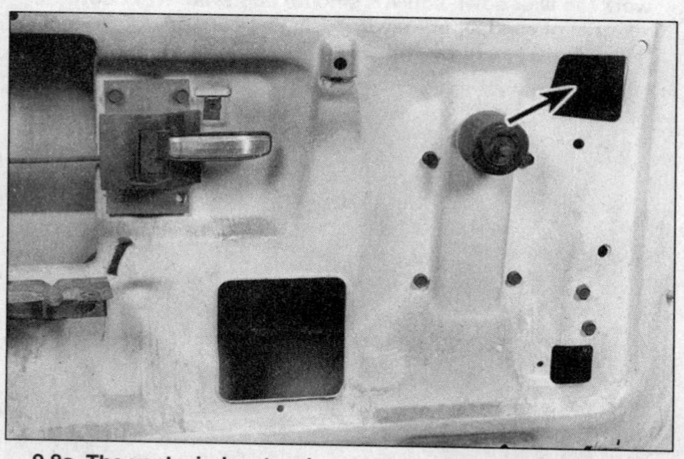

9.8a The vent window tension can be adjusted with a socket wrench inserted through the access hole in the door (arrow)

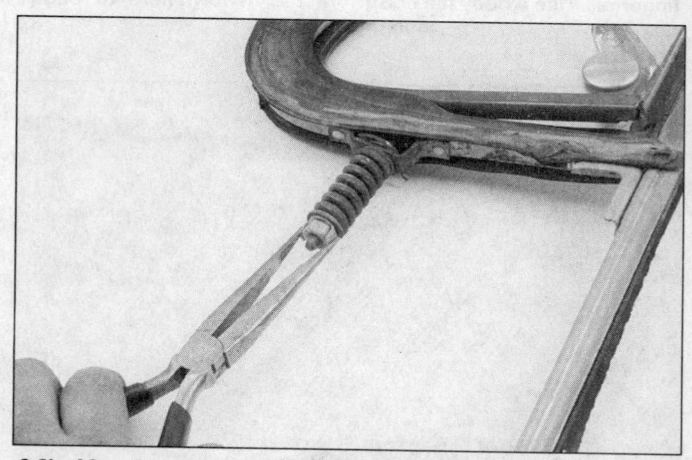

9.8b After the adjustment is made bend the vent window tension nut lock tabs down

9 Door vent window assembly - removal and installation

Refer to illustrations 9.4, 9.7, 9.8a and 9.8b

1 Lower the window all the way.

2 Remove the door trim panel as described in the preceding Section.

3 Remove the door lock knob.

4 Remove the screws which attach the vent window lower assembly to the door panel **(see illustration)**.

5 Loosen the inner and outer panel screws which are accessible through the hole to the rear of the lower vent pivot.

6 Slide the main glass to the rear, away from the vent window.

7 Turn the complete assembly 90-degrees and guide it up and out of the door **(see illustration)**.

8 Installation is the reverse of removal. Once installed, the vent window swivel action can be adjusted by tightening the nut at the base to compress the spring. Access to the nut is through the small hole **(see illustration)**. Bend the nut locking tab over after adjustment **(see illustration)**.

10 Door window glass and regulator - removal and installation

Refer to illustration 10.7

1 Lower the glass completely, remove the window regulator handle and the door lock knob.

2 Remove the door armrest and trim panel.

3 Remove the ventilator assembly as described in the preceding Section.

4 Slide the main window glass forward until the front roller is in alignment with the notch in the sash channel. Disengage the roller.

5 Push the window forward and tilt the front edge up until the rear roller is disengaged.

6 Return the window glass to the level

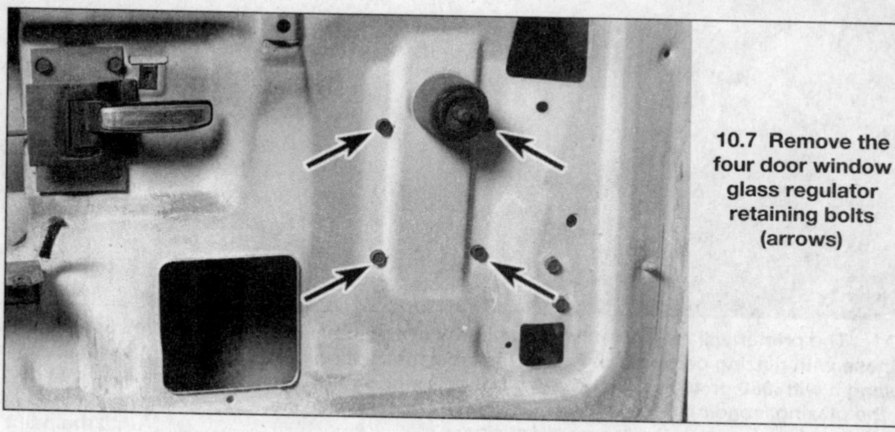

10.7 Remove the four door window glass regulator retaining bolts (arrows)

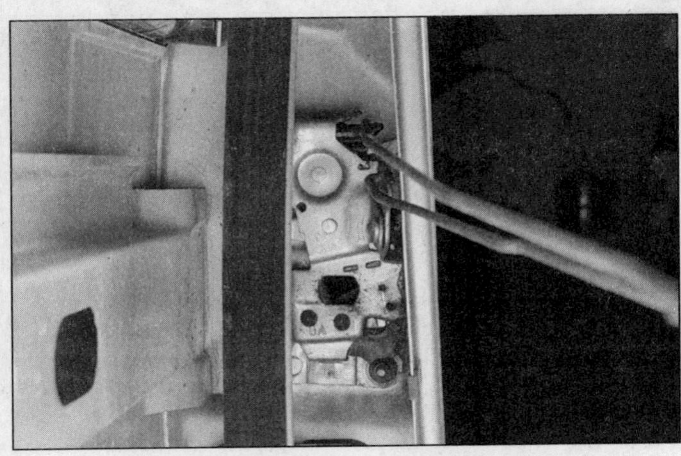
12.4 Typical door lock component layout

13.3 Remove the outside handle mounting bolts (arrows)

position and withdraw it from the door.

7 If necessary, remove the mounting screws and withdraw the regulator assembly through the lower opening in the door **(see illustration)**.

8 Prior to installation, lubricate the window regulator mechanism with chassis grease. Installation of the glass and regulator is the reverse of removal.

11 Power-operated window glass and regulator - removal and installation

1 Disconnect the negative cable from the battery. Place the cable out of the way so it cannot accidentally come in contact with the negative terminal of the battery, as this would once again allow power into the electrical system of the vehicle.

2 Remove the door trim panel.

3 Remove the armrest bracket and the pull assist handle bracket.

4 Remove the bolts from the remote control assembly and position it to one side.

5 Remove the outer glass seal and the rear run glass channel.

6 Remove the regulator-to-door panel nuts and screws. Pull the rear of the glass up as far as possible and rotate it clockwise about 90-degrees to remove the glass from the door.

7 Disconnect the electrical harness from the regulator and turn the motor through 90-degrees. At this point, the following operation must be carried out to relieve the tension from the counterbalance spring, otherwise personal injury can occur during the final stages of removal.

8 Drill a hole through the regulator sector gear and backplate. **Note:** *Do not drill closer than 1/2-inch to the edge of the gear or plate. Install a self-tapping sheet metal screw (10 - 12 x 3/4) into the hole to lock the sector gear.*

9 Remove the screws which hold the motor to the regulator and withdraw the motor.

15.2 Remove the interior door handle mounting bolts (arrows)

10 Installation is the reverse of removal. Be sure to lubricate the motor drive gear and regulator sector teeth. Do not remove the temporary locking screw until the motor has been installed and bolted securely in place.

12 Door lock assembly - removal and installation

Refer to illustration 12.4

1 Raise the window all the way, then remove the regulator handle, remote door lock and trim panel.

2 Extract the clip from the interior rod.

3 Extract the clip from the outside handle rod. This is done by lowering the window and inserting a long thin screwdriver through the glass slot. Raise the glass again when finished.

4 Remove the lock mounting screws and withdraw the lock and rod as an assembly **(see illustration)**.

5 Installation is the reverse of removal.

13 Door outside handle - removal and installation

Refer to illustration 13.3

1 Raise the window glass all the way and remove the trim panel and regulator handle.

2 Remove the clip which attaches the control rod to the lock (Section 12, Step 3).

3 Extract the screws which hold the outside handle to the door panel. Remove the handle and the control rod **(see illustration)**.

4 Installation is the reverse of removal.

14 Door lock cylinder - removal and installation

1 Raise the window glass and remove the trim panel and regulator handle.

2 Detach the lock rod from the lock cylinder.

3 Using a thin screwdriver, pry out the lock cylinder retaining clip. Remove the cylinder.

4 Installation is the reverse of removal.

15 Door inside handle - removal and installation

Refer to illustration 15.2

1 Remove the trim panel, the remote control lock knob and the regulator handle.

2 Remove the bolts which hold the inside handle to the door and detach the handle **(see illustration)**.

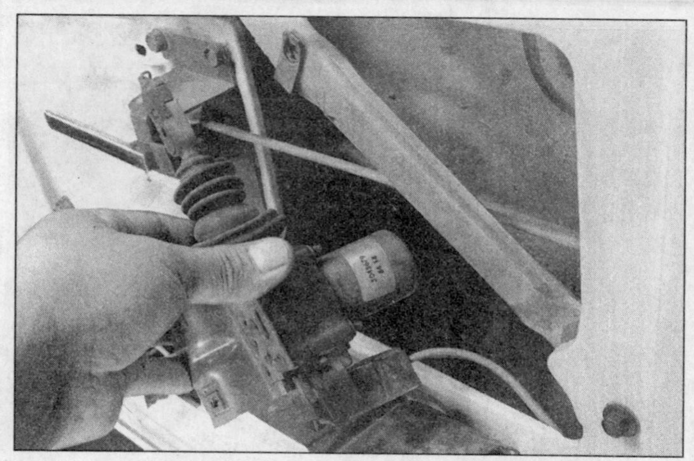

16.4 Remove the power door lock motor through the opening in the door

17.2 Adjusting the door striker using the special spline-drive tool

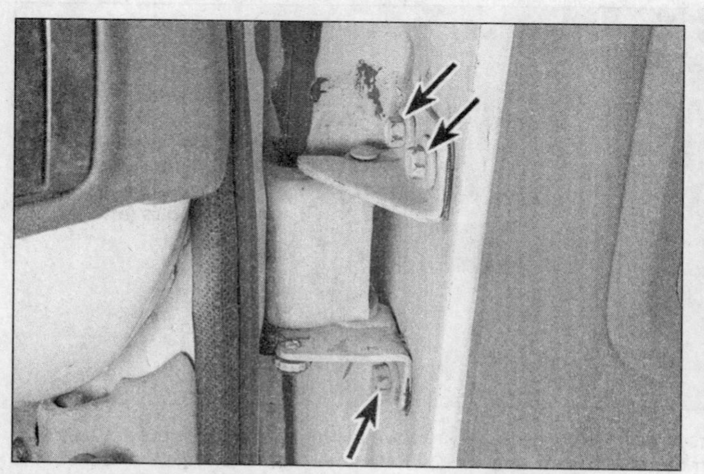

18.3 Door hinge-to-body bolts (arrows) - loosening these bolts will allow fore-and-aft adjustment of the door; loosening the hinge-to-door bolts will allow in-and-out adjustment

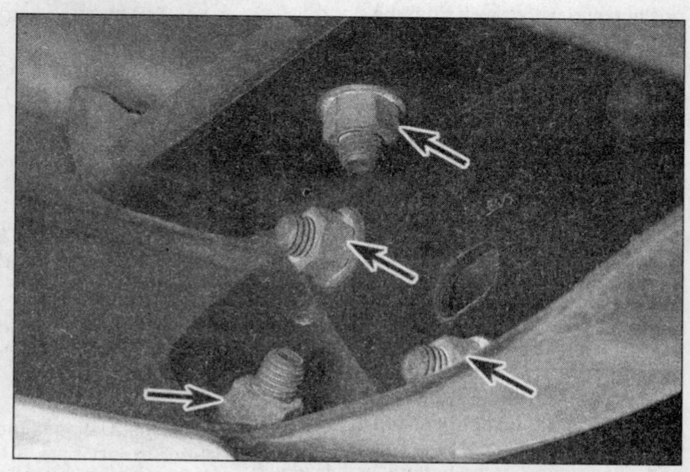

19.2a Remove the front bumper bracket mounting bolts (arrows)

3 Disconnect the control rod from the inside handle.

4 Installation is the reverse of removal.

16 Power-operated door lock - removal and installation

Refer to illustration 16.4

1 Disconnect the battery ground cable.

2 Remove the door trim panel (Section 8).

3 Disconnect the electrical wiring from the motor.

4 Unbolt the motor. Remove the door lock lever from the rubber mount at the top of the motor actuator and withdraw the motor through the opening in the door **(see illustration)**.

17 Door striker - adjustment

Refer to illustrations 17.2a and 17.2b

1 Proper adjustment of the door striker is important because the striker ensures positive closing of the door and supports part of its weight.

2 Loosen the striker with a wrench, reposition it and tighten it. On later models a special door striker wrench, available at most auto parts stores, is required for unscrewing the door striker bolt **(see illustration)**.

3 Adjust the striker until when the door is closed, the door panel is flush with the adjacent body panels and the striker is not so high that it tends to force the door up as it closes.

18 Door - removal, installation and adjustment

Refer to illustration 18.3

1 Open the door all the way and support it on jacks or blocks covered with cloth or pads to prevent damage to the paint.

2 Scribe around the hinges to ensure correct realignment during installation, then unbolt the hinges from the door. Lift the door away.

3 The door check on later models is part of the upper hinge. To remove the hinges from the body pillar, again mark the position of the hinge plates and unbolt them **(see illustration)**. An offset wrench will make the job of removing the bolts easier.

4 Install the door by reversing the removal procedure. Align the door by moving it after loosening the hinge bolts. Adjust the lock striker.

19 Bumpers - removal and installation

Refer to illustrations 19.2a, 19.2b and 19.2c

1 These models were equipped with a wide variety of bumper designs.

2 Removal on all models is simply a matter of unbolting the bumper from the frame, brackets or brace **(see illustrations)**. Some models require disconnection of the rear license plate light wire.

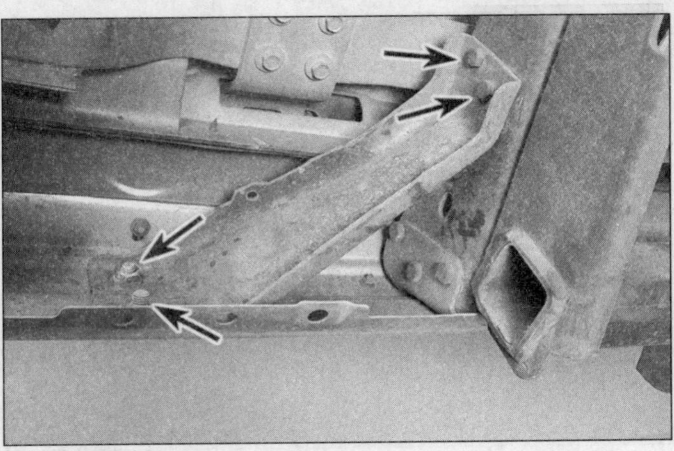

19.2b Remove the rear bumper bracket mounting bolts (arrows)

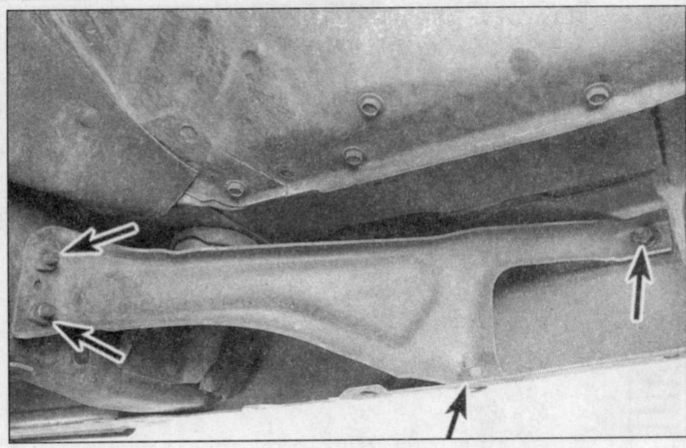

19.2c Later model front bumper installation details

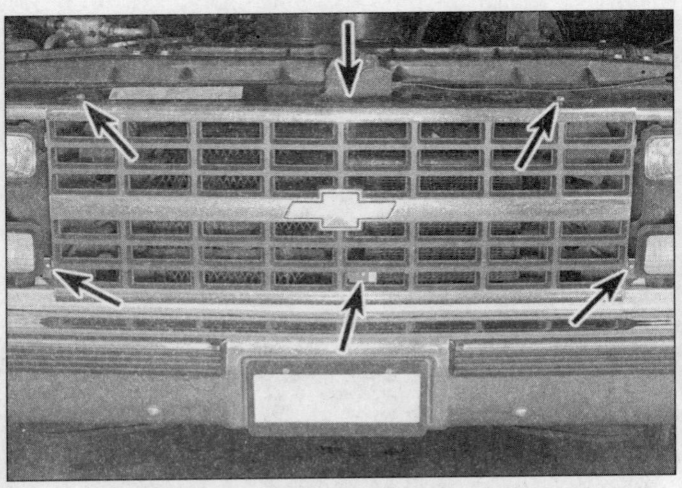

21.2 Remove the grille mounting screws (arrows)

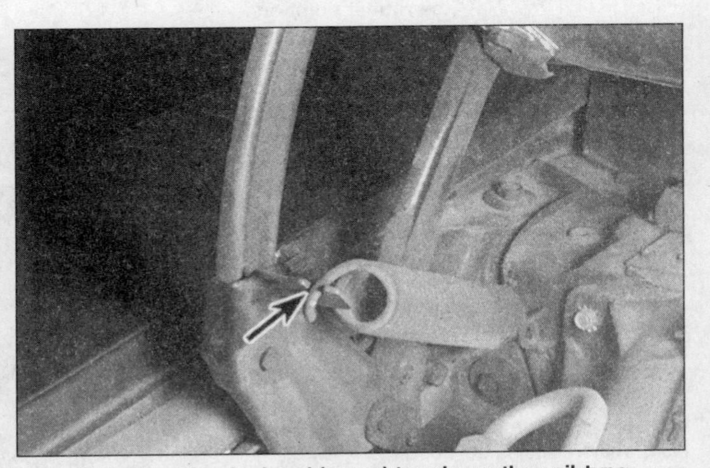

22.3 Use a hooked tool (arrow) to release the coil-type hood spring

20 Radiator grille (1967 through 1971 models) - removal and installation

1 Disconnect the negative cable from the battery. Place the cable out of the way so it cannot accidentally come in contact with the negative terminal of the battery, as this would once again allow power into the electrical system of the vehicle.
2 Remove the front bumper.
3 Remove the headlights.
4 Remove the baffle from the bottom of the grille and radiator support.
5 Unbolt the hood latch from the radiator support.
6 Remove the screw which holds the bottom of the center support to the radiator support.
7 Unplug the parking lights from the radiator support.
8 Remove the four screws from each side of the grille and detach the grille.
9 Installation is the reverse of removal.

21 Radiator grille (1972 and later models) - removal and installation

Refer to illustration 21.2

1 Remove the headlight bezels.
2 Remove the grille screws, followed by the grille **(see illustration)**.
3 Installation is the reverse of removal.

22 Hood - removal and installation

Refer to illustration 22.3

1 Prop the hood open as far as possible.
2 Scribe around the hinges with a pencil or awl.
3 On coil spring-equipped hood mechanisms, disconnect the hinge spring using a hooked tool **(see illustration)**.
4 Remove the hinge bolts and, with the help of an assistant, detach the hood from the vehicle.
5 Installation is the reverse of removal.

23.1 Scribe around the hood lock before removing the lock

23 Hood lock - removal, installation and adjustment

Refer to illustrations 23.1 and 23.5

1 Open the hood and scribe around the lock to ensure correct realignment during installation. Remove the mounting bolts and detach the lock catch and bolt **(see illustration)**.

23.5 Adjust the hood flush with the fenders by raising or lowering the rubber bumpers

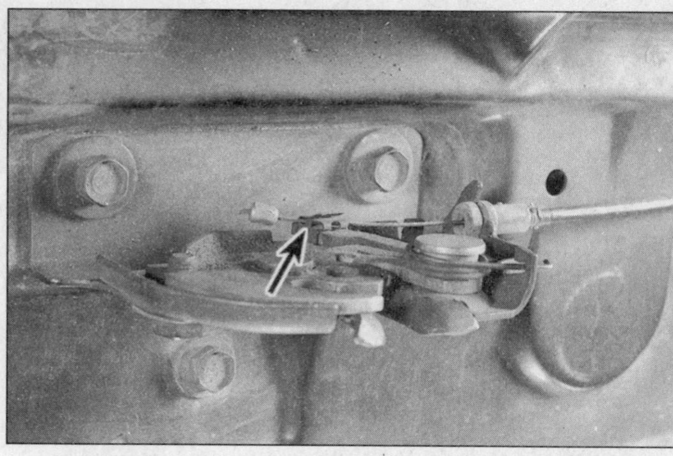

24.2 Remove the hood release cable from the clip (arrow)

2 If new components are being installed, release the locknut on the lock-bolt and adjust the lock-bolt (using a screwdriver in the end slot to screw it in or out) until the distance between the tip of the bolt and the face of the mounting support plate is approximately 2-7/16 inches.
3 Install the support screws and tighten them finger tight. Close the hood and move the support plate until the lock ball just enters the center of the elongated guide in the lock assembly. Open the hood and tighten the lock-bolt support plate screws.
4 Turn the lock-bolt in or out as necessary to obtain positive closure without having to use too much force on the release mechanism when releasing the hood.
5 The hood bumpers should be adjusted to ensure that the hood upper surface is flush with the adjacent surfaces of the fenders **(see illustration)**.

24 Hood release cable (later models) - removal and installation

Refer to illustration 24.2

1 The cable is accessible after removal of the radiator grille.
2 Disconnect the cable from the latch release, unbolt the hand control lever and pull the cable through into the passenger compartment **(see illustration)**.
3 Installation is the reverse of removal.

25 Carburetor outside air intakes - removal and installation

1 Raise the hood, then remove the carburetor air duct from the air cleaner intake (snorkel) by sliding the duct to the rear.
2 Remove the two screws which attach the air intake to the radiator support and detach it.
3 Installation is the reverse of removal.

27.10 Tailgate torque rod (arrow), mounting bolts and bracket (arrows)

26 Front fender - removal and installation

1 Remove the hood.
2 Remove the headlight bezel and disconnect the wiring harness and any other components at the front fender.
3 Remove the screws from the upper and lower radiator grille panels.
4 Remove the screws which retain the fender (wheel opening) flange to the skirt (liner).
5 Remove the skirt-to-fender bolts located inboard on the underside of the skirt.
6 Remove the two screws which attach the support bracket to the fender.
7 Remove the screws which secure the radiator support to the front fender.
8 Remove the bolt and shim which attach the trailing edge of the fender to the hinge pillar.
9 Remove the two bolts which secure the top rear edge of the fender to the cowl.
10 Lift the fender from the vehicle.
11 Installation is the reverse of removal. Be sure to insert new sealing tape between the filler panel and the fender.

27 Tailgate - removal and installation

Refer to illustration 27.10

Suburban models

1 Lower the tailgate and remove the hinge access covers.
2 Remove the tailgate-to-hinge bolts.
3 Remove the left-hand torque rod bracket.
4 On power window models, disconnect the wiring harness.
5 Disconnect the support cables and remove the tailgate and torque rod.
6 Access to the internal components of the tailgate is obtained by removing the large access cover.

Blazer/Jimmy models

7 Lower the tailgate and remove the hinge-to-body bolts on each side.
8 Disconnect the torque rod anchor plate on each side. Do this by removing only the lower bolts and then allowing the plates to swing down.
9 Raise the tailgate enough to disconnect

the support cables.

10 Remove the tailgate by pulling the disconnected hinge from the body, gripping the torque rod with one hand and lifting it over the gravel deflector and bumper **(see illustration)**.

11 Remove the large panel for access to internal components.

Pick-up/Crew cab models

12 Two types of tailgates are used on these models.

13 Removal of either type can be carried out by removing the support chains or links and disconnecting the lower trunnions.

All models

14 Installation of tailgates on all models is the reverse of removal.

28 Rear view mirrors - removal and installation

Interior mirror

1 Remove the retaining screw and lift the mirror up off the mounting plate.

2 Installation is the reverse of removal.

Exterior mirrors

3 Several types of exterior mirrors are used on these models. Remove the retaining nuts and bolts to detach the mirror assembly from the vehicle.

4 Installation is the reverse of removal.

Notes

Chapter 12
Chassis electrical system

Contents

	Section
Battery check and maintenance	See Chapter 1
Battery - removal and installation	See Chapter 5
Bulb replacement	9
Circuit breakers - general information	5
Electrical troubleshooting - general information	2
Fuses - general information	3
Fusible links - general information	4
General information	1
Headlights - adjustment	8
Headlights - removal and installation	7
Instrument cluster (1967 through 1972 models) - removal and installation	11
Instrument cluster (1973 and later models) - removal and installation	12
Radio antenna - removal and installation	17
Radio and speakers - removal and installation	16
Switches - removal and installation	10
Turn signal and hazard flashers - check and replacement	6
Windshield washer pump (later models) - removal and installation	15
Windshield wiper arms - removal and installation	13
Windshield wiper/washer motor - removal and installation	14

Specifications

Bulb application

1967 through 1971	Number
Interior light	211
Indicator and warning lights (instrument panel)	194 or 1895
Brake/tail lights	1157
Rear license plate light	67
Front parking/turn signal light	1157
Headlights	6014
Clearance lights	1155
Side marker light	194
Transmission control light	1445
Back-up light	1156
Heater indicator light	1445

Bulb application

	Number
1972 through 1974	
Same as 1967 through 1971, except for:	
Interior light	212
Corner marker light	67
Cargo area light	1142
Radio dial light	293
Cruise control light	53
Courtesy light	1003
1975	
Interior lights	1003
Interior lights (Jimmy and Suburban)	211
Indicator and warning lights	168
Brake/tail lights	1157
Rear license plate light	67
Front park/turn signal lights	1157
Headlights	6014
Clearance lights	168
Roof marker lights	194
Transmission control light	1445
Back-up light	156
Heater indicator light	1445
Corner marker light	67
Cargo area light	1142
Radio dial (AM) light	1816
Radio dial (AM/FM) light	216
Cruise control light	53
Courtesy light	1003
Windshield wiper switch light	161
Clock light	168
Underhood light	93
1976 on	
Same as 1975, except for:	
Seat belt warning light	168
Cargo area light	211-2

1 General information

The electrical system is a 12-volt, negative ground type. Power for the lights and all electrical accessories is supplied by a lead/acid-type battery which is charged by the alternator.

This Chapter covers repair and service procedures for the various electrical components not associated with the engine. Information on the battery, alternator, distributor and starter motor can be found in Chapter 5.

When working on electrical system components, disconnect the negative battery cable from the battery to prevent electrical shorts and/or fires.

2 Electrical troubleshooting - general information

A typical electrical circuit consists of an electrical component, any switches, relays, motors, etc. related to that component and the wiring and connectors that connect the component to both the battery and the chassis. To aid in locating a problem in any electrical circuit, wiring diagrams are included at the end of this book.

Before tackling any troublesome electrical circuit, first study the appropriate diagrams to get a complete understanding of what makes up that individual circuit. Trouble spots, for instance, can often be narrowed down by noting if other components related to that circuit are operating properly or not. If several components or circuits fail at one time, chances are the problem lies in the fuse or ground connection, as several circuits are often routed through the same fuse and ground connections.

Electrical problems often stem from simple causes, such as loose or corroded connections, a blown fuse or a melted fusible link. Always visually inspect the condition of the fuse, wires and connections in a problem circuit before troubleshooting it.

If testing instruments are going to be utilized, use the diagrams to plan ahead of time where you will make the necessary connections in order to accurately pinpoint the trouble spot.

The basic tools needed for electrical troubleshooting include a circuit tester or voltmeter (a 12-volt bulb with a set of test leads can also be used), a continuity tester, which includes a bulb, battery and set of test leads, and a jumper wire, preferably with a circuit breaker incorporated, which can be used to bypass electrical components.

Voltage checks should be performed if a circuit is not functioning properly.

Connect one lead of a circuit tester to either the negative battery terminal or a known good ground. Connect the other lead to a connector in the circuit being tested, preferably nearest to the battery or fuse. If the bulb of the tester lights up, voltage is present, which means that the part of the circuit between the connector and the battery is problem free. Continue checking the rest of the circuit in the same fashion. When you reach a point at which no voltage is present, the problem lies between that point and the last test point with voltage. Most of the time the problem can be traced to a loose connection. **Note:** *Keep in mind that some circuits receive voltage only when the ignition key is in the Accessory or Run position.*

One method of finding shorts in a circuit is to remove the fuse and connect a test light or voltmeter in its place to the fuse terminals. There should be no voltage present in the circuit.

Move the wiring harness from side-to-side while watching the test light. If the bulb goes on, there is a short to ground somewhere in that area, probably where the insula-

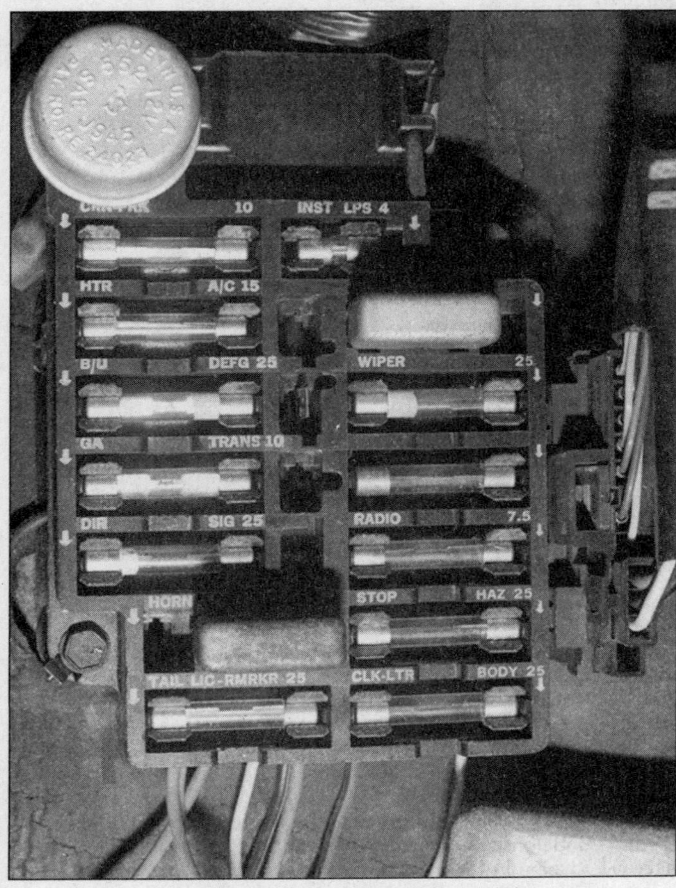

3.2a Typical early fuse block using glass body fuses

3.2b Typical later model fuse block with miniaturized fuses

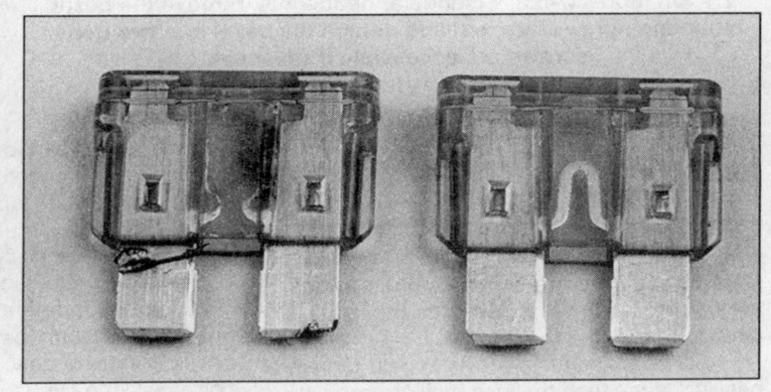

3.4 When a fuse blows, the element between the terminals melts - the fuse on the left is blown, the fuse on the right is good

tion has rubbed through.

The same test can be performed on each component in the circuit, even a switch.

Perform a ground test to check whether a component is properly grounded. Disconnect the battery and connect one lead of a self-powered test light, known as a "continuity tester," to a known good ground. Connect the other lead to the wire or ground connection being tested. If the bulb goes on, the ground is good.

If the bulb does not go on, the ground is not good.

A continuity check determines if there are any breaks in a circuit - if it is conducting electricity properly. With the circuit off (no power in the circuit), a self-powered continuity tester can be used to check the circuit. Connect the test leads to both ends of the circuit (or to the "power" end and a good ground), and if the test light comes on the circuit is passing current properly. If the light doesn't come on, there is a break somewhere in the circuit. The same procedure can be used to test a switch, by connecting the continuity tester to the power in and power out sides of the switch. With the switch turned on, the test light should come on.

When diagnosing for possible open circuits, it is often difficult to locate them by sight because oxidation or terminal misalignment are hidden by the connectors. Merely wiggling a connector on a sensor or in the wiring harness may correct the open circuit condition. Remember this if an open circuit is indicated when troubleshooting a circuit. Intermittent problems may also be caused by oxidized or loose connections.

Electrical troubleshooting is simple if you keep in mind that all electrical circuits are basically electricity running from the battery, through the wires, switches, relays, fuses and fusible links to each electrical component (light bulb, motor, etc.) and back to ground, from which it is passed back to the battery. Any electrical problem is an interruption in the flow of electricity to and from the battery.

3 Fuses - general information

Refer to illustrations 3.2a, 3.2b and 3.4

The electrical circuits of the vehicle are protected by a combination of fuses, circuit breakers and fusible links. The fuse block is located under the dash on the driver's side.

Each fuse protects one or more circuits. The protected circuit is identified on the fuse panel, directly above each fuse **(see illustrations)**. On later models, miniaturized fuses are installed in the fuse block. These compact fuses, with blade terminal design, allow fingertip removal and installation.

If an electrical component fails, always check the fuse first.

A blown fuse, which is nothing more than a broken element, is easily identified through the clear glass (early models) or plastic body (later models) **(see illustrations)**. The easiest way to check fuses is with a test light. Check for power at each end of the fuse. If power is present on one side of the fuse but not the other, the fuse is blown.

Be sure to replace blown fuses with the correct type. Fuses of different ratings are physically interchangeable, but only fuses of the proper rating should be used. Replacing a fuse with one of a higher or lower value than specified is not recommended. Each electrical circuit needs a specific amount of protection. The amperage value of each fuse is molded into the fuse body. **Caution:** *Never bypass a fuse with pieces of metal or foil. Serious damage to the electrical system could result.*

If the replacement fuse immediately fails, do not replace it again until the cause of the problem is isolated and corrected.

7.1 On models with rectangular headlights, remove the bezel mounting screws (arrows) and detach the bezel (the two upper screws are accessible from above)

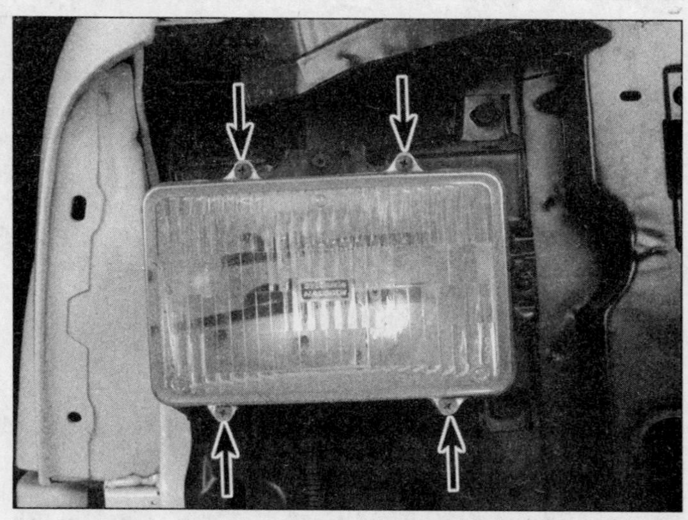

7.2 On rectangular sealed headlights, the bezel is held in place by four screws (arrows) on each side of the housing (on round headlights there are only three screws)

4 Fusible links - general information

Some circuits are protected by fusible links. These links are used in circuits which are not ordinarily fused, such as the ignition circuit.

Although the fusible links appear to be a heavier gauge than the wire they are protecting, the appearance is due to the thick insulation. All fusible links are several wire gauges smaller than the wire they are designed to protect.

The fusible links are generally located near the starter or attached to the positive battery terminal.

Fusible links cannot be repaired, but a new link of the same size wire can be installed as follows:

a) Disconnect the negative cable at the battery.

b) Disconnect the fusible link from the wiring harness.

c) Cut the damaged fusible link out of the wiring just behind the connector.

d) Strip the insulation back approximately 1/2-inch.

e) Position the connector on the new fusible link and crimp it into place.

f) Use rosin core solder at each end of the new link to obtain a good solder joint.

g) Use plenty of electrical tape around the soldered joint. No wires should be exposed.

h) Connect the battery ground cable. Test the circuit for proper operation.

5 Circuit breakers - general information

Circuit breakers protect components such as power windows, power door locks, rear window defoggers and headlights. Some circuit breakers are located in the fuse box.

Because a circuit breaker resets itself automatically, an electrical overload in a circuit breaker protected system will cause the circuit to fail momentarily, then come back on. If the circuit does not come back on, check it immediately. Once the condition is corrected, the circuit breaker will resume its normal function.

6 Turn signal and hazard flashers - check and replacement

Turn signal flasher

1 The turn signal flasher, a small canister shaped unit located in the fuse block or adjacent to it, flashes the turn signals.

2 When the flasher unit is functioning properly, an audible click can be heard during its operation. If the turn signals fail on one side or the other and the flasher unit does not make its characteristic clicking sound, a faulty turn signal bulb is indicated.

3 If both turn signals fail to blink, the problem many be due to a blown fuse, a faulty flasher unit, a broken switch or a loose or open connection. If a quick check of the fuse box indicates that the turn signal fuse has blown, check the wiring for a short before installing a new fuse.

4 To replace the flasher, simply pull it out of the fuse block or wiring harness.

5 Make sure that the replacement unit is identical to the original. Compare the old one to the new one before installing it.

6 Installation is the reverse of removal.

Hazard flasher

7 The hazard flasher, a small canister shaped unit located in the fuse block, flashes all four turn signals simultaneously when activated.

8 The hazard flasher is checked in a fashion similar to the turn signal flasher (see Steps 2 and 3).

9 To replace the hazard flasher, pull it from the fuse block.

10 Make sure the replacement unit is identical to the one it replaces. Compare the old one to the new one before installing it.

11 Installation is the reverse of removal.

7 Headlights - removal and installation

Refer to illustrations 7.1 and 7.2

1 Remove the mounting screws and detach the bezel **(see illustration)**.

2 Remove the headlight retaining ring screws **(see illustration)**. Be sure to remove only the screws which secure the headlight retaining ring - DO NOT disturb the headlight beam adjustment (upper) screws.

3 Pull the sealed beam unit forward and disconnect the plug from the rear.

4 Installation is the reverse of removal. Make sure the number molded into the face of the lens is at the top.

8 Headlights - adjustment

Refer to illustrations 8.1a, 8.1b and 8.5

Note: *Headlight aim is very important! If adjusted incorrectly, they could blind the driver of an oncoming vehicle and cause a serious accident or seriously reduce your ability to see the road. The headlights should be checked for proper aim every 12 months and any time a new sealed beam headlight is installed or front end body work is performed. It should be emphasized that the following*

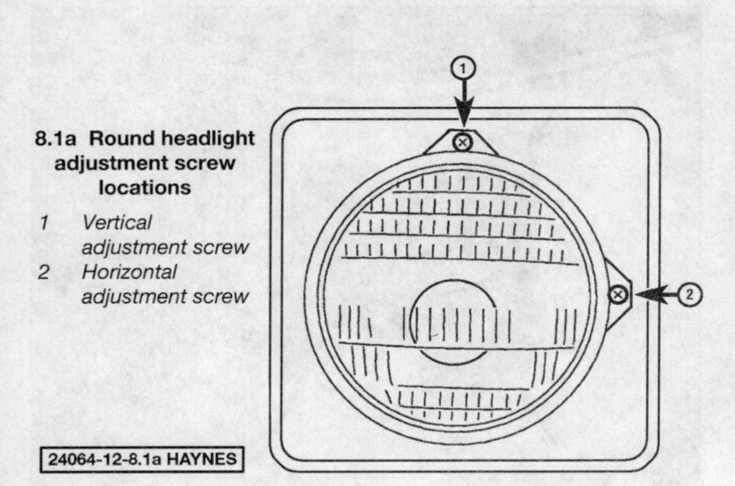

8.1a Round headlight adjustment screw locations

1 *Vertical adjustment screw*
2 *Horizontal adjustment screw*

8.1b Rectangular headlight adjustment screw locations

1 *Horizontal adjustment screw* 2 *Vertical adjustment screw*

8.5 Headlight aiming details

procedure is only intended as a temporary adjustment until the headlights can be adjusted by a properly equipped shop.

1 Headlights have two spring loaded adjusting screws - one on the top, controlling up-and-down movement, and one or two on the side, controlling left-and-right movement **(see illustrations)**.

2 There are several methods of adjusting the headlights. The simplest method requires a blank wall 25-feet in front of the vehicle and a level floor.

3 Park the vehicle 25-feet from the wall.

4 Position masking tape vertically on the wall in reference to the vehicle centerline and the centerlines of both headlights.

5 Position a horizontal tape line in reference to the centerline of all the headlights **(see illustrations)**. **Note:** *It may be easier to position the tape on the wall with the vehicle parked only a few inches away.*

6 Adjustment should be made with the vehicle sitting level, the gas tank half-full and no unusually heavy load in the vehicle.

7 Starting with the low beam adjustment, position the high intensity zone so it is two-inches below the horizontal line and two-inches to the right of the headlight vertical line. Adjustment is made by turning the top adjusting screw clockwise to raise the beam and counterclockwise to lower the beam. The adjusting screw on the side should be used in the same manner to move the beam left-or-right.

8 With the high beams on, the high intensity zone should be vertically centered with the exact center just below the horizontal line. **Note:** *It may not be possible to position the headlight aim exactly for both high and low beams. If a compromise must be made, keep in mind that the low beams are the most used and have the greatest effect on driver safety.*

9 Have the headlights adjusted by a dealer service department or service station at the earliest opportunity.

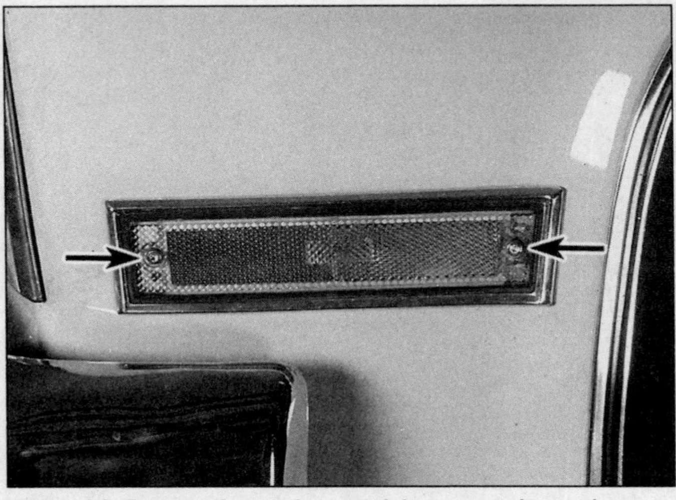

9.5 Remove the two lens retaining screws (arrows)

9.8 Typical tail light lens retaining screws (arrows)

9 Bulb replacement

Refer to illustrations 9.5 and 9.8

1 Disconnect the negative cable at the battery. Place the cable out of the way so it cannot accidentally come in contact with the negative terminal of the battery, as this would once again allow power into the electrical system of the vehicle.

Front parking and turn signal lights

2 Remove the screws and detach the lens.

3 Remove the defective bulb, then install the new one followed by the lens.

Front side marker light

4 If the left side light is being replaced, the hood will first have to be raised and then the bulb holder twisted 90-degrees counter-clockwise and removed from the rear of the light.

5 Access to the bulb holder on the right hand side is possible after removing the mounting screws and pulling the light assembly out **(see illustration)**.

6 Replace the bulb, then reverse the removal procedure.

Rear side marker light

7 The bulb can be replaced after removing the light assembly mounting screws (see Step 5 above).

Rear light cluster

8 Access to the bulbs is obtained by removing the lens screws **(see illustration)**.

9 The bulbs for the tail lights, brake lights, back-up lights or turn signals can then be replaced as necessary.

Rear license plate light

10 Remove the two screws attaching the light assembly to the license plate frame. Remove the lens and replace the bulb.

Instrument panel lights

11 The indicator and illumination bulbs can be replaced by reaching up under the instrument panel and turning the bulb holder counterclockwise.

12 Pull the bulb straight out of the holder, insert the new bulb and press it in securely to lock it in position.

13 Install the holder in the instrument panel. Make sure that the lugs engage in the notches, then turn the holder clockwise to lock it in position.

10 Switches - removal and installation

Refer to illustrations 10.2, 10.8 and 10.40

1 Disconnect the negative cable at the battery. Place the cable out of the way so it cannot accidentally come in contact with the negative terminal of the battery, as this would once again allow power into the electrical system of the vehicle.

Light switch

2 Reach up behind the instrument panel and depress the switch shaft retaining plunger, then remove the switch knob shaft **(see illustration)**. Remove the left side instrument panel trim.

3 Unscrew the nut from the front of the dash, then unplug the electrical connector and remove the switch. Unscrew the ring nut and separate the switch from the holder/rheostat assembly.

4 Install the switch by reversing the removal procedure. Make sure that the grounding ring is installed on the switch.

Headlight dimmer switch - floor mounted

5 Peel back the upper left corner of the floor mat and remove the two screws which hold the switch to the floor pan.

6 Disconnect the plug from the switch terminals.

7 Installation is the reverse of removal.

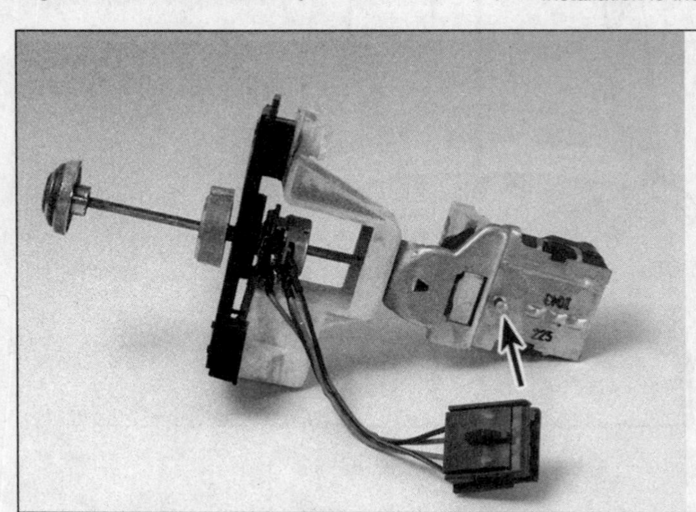

10.2 Depress the headlight switch shaft retaining plunger (arrow) - switch removed for clarity

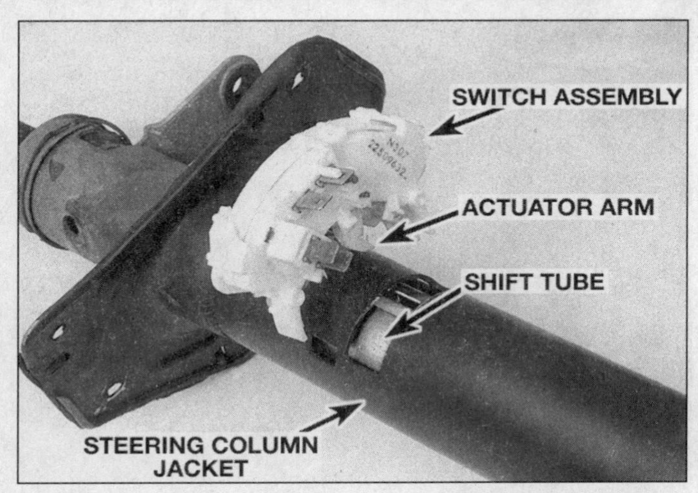

10.8 Neutral start/back-up light switch installation details

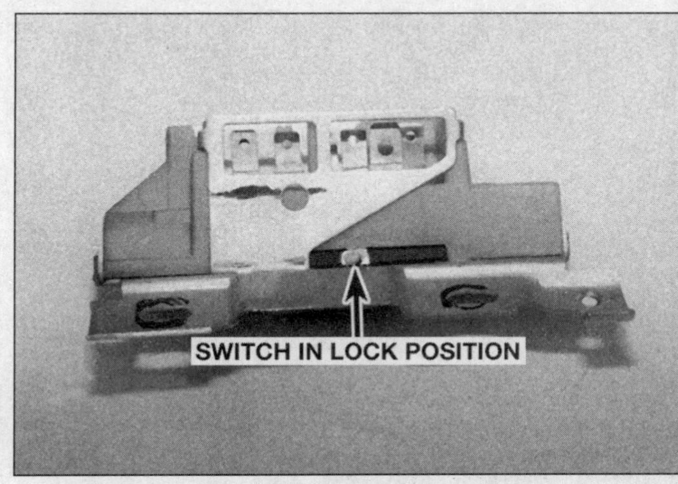

10.40 Set the ignition switch in the locked position

Neutral start switch/back-up light switch (automatic transmission)

8 Disconnect the wire harness from the switch (which is mounted at the upper end of the steering column) **(see illustration)**.

9 Remove the mounting screws and detach the switch.

10 Position the shift lever in Neutral.

11 Insert a 3/32-inch drill bit into the switch gauge hole to a depth of 3/8-inch. New switches are supplied set in the Neutral mode by a plastic shear pin.

12 Place the switch in position on the steering column by engaging the switch carrier tang in the shift tube slot, then install the mounting screws.

13 Withdraw the guide pin.

14 Move the shift lever from Neutral to Park. This will cause the plastic pin to shear off. Return the shift lever to Neutral.

15 Check to make sure that a 5/64-inch drill bit will freely enter the switch hole to a depth of 3/8-inch. If it won't, reset the switch as follows.

16 Make sure that the shift lever is in Neutral.

17 Loosen the switch mounting screws and repeat the procedure described in Step 11.

18 Retighten the switch mounting screws and then repeat the procedure described in Steps 13 through 15. Connect the battery lead and switch wires.

19 The engine should start only in Neutral and Park and the back-up lights should come on when the transmission is shifted to Reverse.

Back-up light switch (manual transmission)

20 On 3-speed transmission equipped vehicles, the switch is mounted on the steering column.

21 Disconnect the negative cable at the battery. Place the cable out of the way so it cannot accidentally come in contact with the negative terminal of the battery, as this would once again allow power into the electrical system of the vehicle.

22 Unplug the switch wiring harness connector.

23 Remove the switch mounting screws and detach the switch.

24 Installation is the reverse of removal.

25 On 4-speed transmission equipped models, the switch is mounted on the transmission.

26 Raise the vehicle and support it securely on jackstands. Working under the vehicle, disconnect the wires from the switch.

27 Remove the switch from the transmission.

28 Installation is the reverse of removal.

Windshield wiper/washer switch (dash mounted)

Note: *Later models with the windshield wiper/washer switch in the steering column require partial disassembly of the column. This difficult procedure should be left to a properly trained and equipped professional technician.*

29 Disconnect the negative cable at the battery. Place the cable out of the way so it cannot accidentally come in contact with the negative terminal of the battery, as this would once again allow power into the electrical system of the vehicle.

30 Remove the instrument panel bezel screws and detach the bezel.

31 Remove the switch mounting screws, pull out the switch and disconnect the wire harness.

32 Installation is the reverse of removal.

Ignition switch

1967 through 1970 models

33 Disconnect the negative cable at the battery. Place the cable out of the way so it cannot accidentally come in contact with the negative terminal of the battery, as this would once again allow power into the electrical system of the vehicle.

34 Turn the ignition switch to the Off position, then insert a stiff piece of wire (such as a welding rod) into the small hole in the lock cylinder face. Push in on the wire and at the same time turn the ignition key counterclockwise until the lock cylinder can be withdrawn.

35 Remove the switch mounting nut, retrieve the switch from under the dash and disconnect the wires from it.

36 The anti-theft connector can be removed from the switch by unsnapping the locking tangs.

37 Installation is the reverse of removal.

1971 and later models

38 The switch is mounted on top of the column jacket, inside the channel section of the brake pedal support.

39 To gain access to the switch, the steering column must be lowered. To do this, carry out the initial operations for removal of the column described in Chapter 10, although it is not necessary to withdraw the steering column.

40 Set the switch in the Lock position, remove the two mounting screws and detach the switch **(see illustration)**.

41 Installation is the reverse of removal. Refer to Chapter 10 for the steering column installation procedure.

Steering column switches (turn signal and lock cylinder)

42 Refer to Chapter 10 for removal and installation procedures.

Headlight dimmer switch - column mounted

43 Remove the trim panel under the steering column, lower and support the steering column, if necessary.

44 Disconnect the wiring connector from the dimmer switch.

45 Remove the screws securing the dimmer switch to the column and remove the switch.

46 Install the new dimmer switch, removing all slack between the switch and the actuator rod.

47 Remaining installation is reverse of removal.

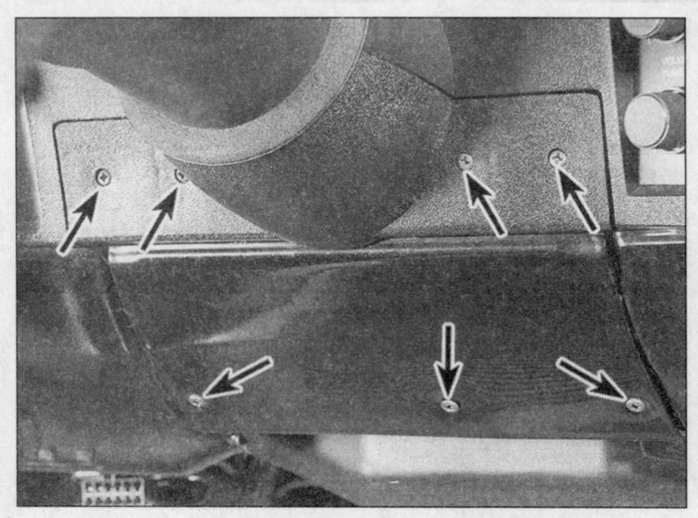

12.2a Remove the lower trim cover retaining screws (arrows)

12.2b Remove the instrument cluster bezel retaining screws (arrows)

11 Instrument cluster (1967 through 1970 models) - removal and installation

1 Disconnect the negative cable at the battery. Place the cable out of the way so it cannot accidentally come in contact with the negative terminal of the battery, as this would once again allow power into the electrical system of the vehicle.
2 Remove the choke control knob.
3 Remove the windshield wiper switch knob and bezel nut.
4 Remove the lighting switch knob/rod and bezel nut.
5 Reach up behind the instrument panel and disconnect the speedometer cable and wire harness connectors.
6 Remove the instrument cluster mounting screws and withdraw the cluster.
7 Installation is the reverse of removal.

12 Instrument cluster (1971 and later models) - removal and installation

Refer to illustration 12.2a and 12.2b

1 Disconnect the negative cable at the battery. Place the cable out of the way so it cannot accidentally come in contact with the negative terminal of the battery, as this would once again allow power into the electrical system of the vehicle.
2 Reach up behind the dash, depress the headlight knob retaining button and remove the knob. Remove the instrument cluster bezel and steering column cover **(see illustrations)**.
3 Remove the clock knob (if equipped).
4 Remove the screws which hold the instrument lens in position and detach the lens.
5 On automatic transmission equipped models, remove the shift indicator.
6 Remove the cluster retainer.
7 Reach up behind the dash panel and disconnect the speedometer cable by depressing the retaining spring clip. Disconnect and plug the oil pressure gauge line (if equipped).
8 Disconnect the wiring harness.
9 Remove the cluster screws and detach the cluster from the instrument panel.
10 The individual instruments and the printed circuit can be removed as necessary.
11 Installation is the reverse of removal.

13 Windshield wiper arms - removal and installation

Refer to illustration 13.3

1 Turn the wipers off in the low speed mode so the wiper arms are in the parked position.

Removal

Early models

2 Using a hooked tool or a small screwdriver, pull on the small spring tang which holds the wiper arm to the splined shaft while simultaneously pulling the arm from the shaft.

Later models

3 Pull the outer end of the wiper arm away from the windshield, which will trip the lock spring from the pivot shaft, hold the arm in this position, pull out on the cap section at the base of the arm and lift the assembly off **(see illustration)**.

Installation

Early models

4 Installation is the reverse of removal. Don't push the arm all the way onto the shaft until the alignment of the arm has been checked. If necessary, the arm can be pulled off again and moved one or two splines on the shaft to correct the alignment without having to pull the spring tang aside. Press the arm onto the shaft until it seats.

13.3 Wiper arm installation details

Later models

5 Place the arm assembly on the splined shaft, making sure there is approximately a 2-1/4 inch clearance between the arm and the windshield molding **(see illustration 13.3)**, and press the assembly on until it's seated on the shaft.

14 Windshield wiper/washer motor - removal and installation

Refer to illustration 14.10

1 Disconnect the negative cable at the battery. Place the cable out of the way so it cannot accidentally come in contact with the negative terminal of the battery, as this would once again allow power into the electrical system of the vehicle.
2 Remove the wiper arms and blades.
3 Remove the heater air intake grille from the plenum chamber.

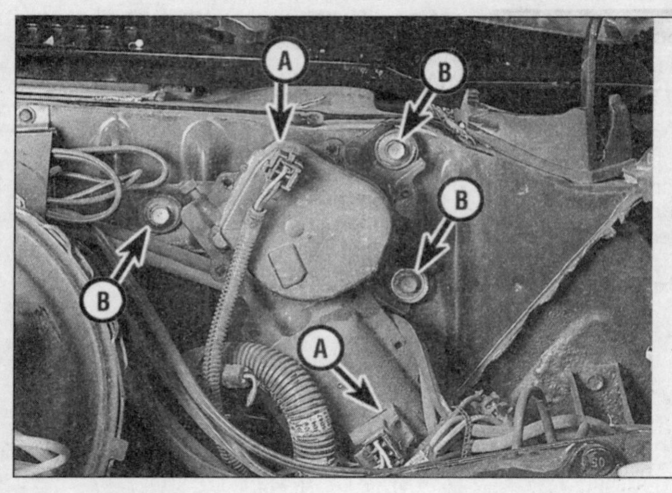

14.10 Remove the wiper electrical connections and mounting bolts - Later model windshield wiper motor

A Electrical connections
B Mounting bolts

16.3 Remove the nuts and washers from the control shafts (arrow) - right side shown

Early models

4 Disconnect the linkage from the wiper motor crank arm and remove the nut and crank arm from the motor shaft.

5 Working under the instrument panel, disconnect all wires and tubes from the wiper/washer motor assembly.

6 Remove the wiper motor mounting screws.

7 Detach the motor from the mounting bracket.

8 Installation is the reverse of removal.

Later models

9 Make sure that the motor was turned off with the wipers in the parked position.

10 Disconnect all wires and hoses from the wiper motor assembly **(see illustration)**.

11 Reach through the access hole in the plenum chamber and loosen the clamp screws at the end of the wiper linkage rod. Pull the rod from the motor crank arm.

12 Remove the wiper motor mounting screws and detach the motor from the firewall.

13 Installation is the reverse of removal.

15 Windshield washer pump (later models) - removal and installation

1 On models built through 1985, the washer pump is mounted on the wiper motor and can be removed without having to remove the complete motor assembly.

2 On 1986 and later models, the washer pump is mounted on the washer fluid reservoir.

Wiper motor mounted pump

3 Disconnect the washer hoses and the wire harness.

4 Remove the three screws which secure the washer pump to the wiper motor and detach the pump.

5 Once the pump has been removed, mark the position of the valve body in relationship to the pump, remove the four screws and detach the valve body.

6 Replace any worn or damaged components and reassemble the pump. Make sure that the valve plate gasket is correctly installed and that the triple O-ring is installed between the valve and pipe assembly.

7 Installation of the pump is the reverse of removal.

Reservoir mounted pump

8 Remove the washer reservoir and disconnect the tubes and wires from the pump.

9 Remove the pump from the reservoir.

10 Installation is the reverse of removal.

16 Radio and speakers - removal and installation

Refer to illustrations 16.3 and 16.14

Radio

1 Disconnect the negative cable at the battery. Place the cable out of the way so it cannot accidentally come in contact with the negative terminal of the battery, as this would once again allow power into the electrical system of the vehicle.

Early models

2 Pull off the radio control knobs and remove the bezels.

3 Remove the nuts and washers from the control shafts **(see illustration)**.

4 Remove the radio support bracket, lift the rear of the radio and push the radio towards the front of the vehicle until the control shafts clear the instrument panel.

5 Pull the radio out, unplug the electrical connectors and lower it from the dash panel.

6 Installation is reverse of removal.

Later models

7 Remove the instrument panel trim plate and lower cover.

8 Remove the receiver rear brace screw.

9 Remove the receiver front mounting screws.

10 Pull the receiver out (towards the rear of the vehicle) and remove the electrical connectors and antenna cable.

11 Installation is the reverse of removal.

Front speaker

12 Remove the four screws from the instrument panel upper bezel.

13 Remove the screws and detach the instrument panel pad.

14 Remove the screws which hold the speaker to the dash panel **(see illustration)**.

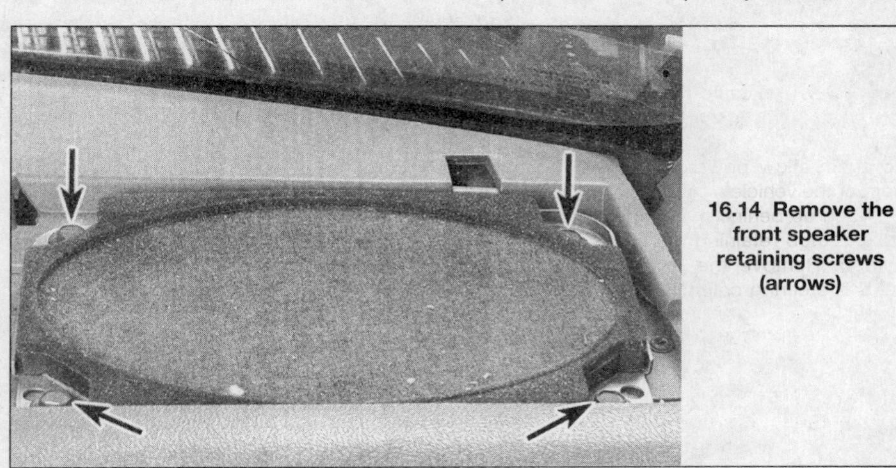

16.14 Remove the front speaker retaining screws (arrows)

15 Raise the speaker and disconnect the wires.
16 Installation is the reverse of removal.

Rear speaker

17 Remove the screws which attach the speaker grill to the trim panel.
18 Disconnect the speaker wires and remove the speaker mounting screws, then detach the speaker.
19 Installation is the reverse of removal.

17 Radio antenna - removal and installation

Body mounted antenna

1 Unscrew the mast nut. Use two wrenches to keep the antenna from turning.
2 Lower the assembly to remove it from the vehicle. Note the installed positions of the seals and spacers.
3 If the assembly is to be completely removed, unplug the antenna lead from the socket in the radio.

Windshield glass mounted antenna

4 Because this type of antenna is imbedded in the glass, windshield replacement is the only cure for a faulty antenna. Consequently, be sure to check the antenna-to-windshield connector plug and wire before concluding that the windshield antenna is defective.

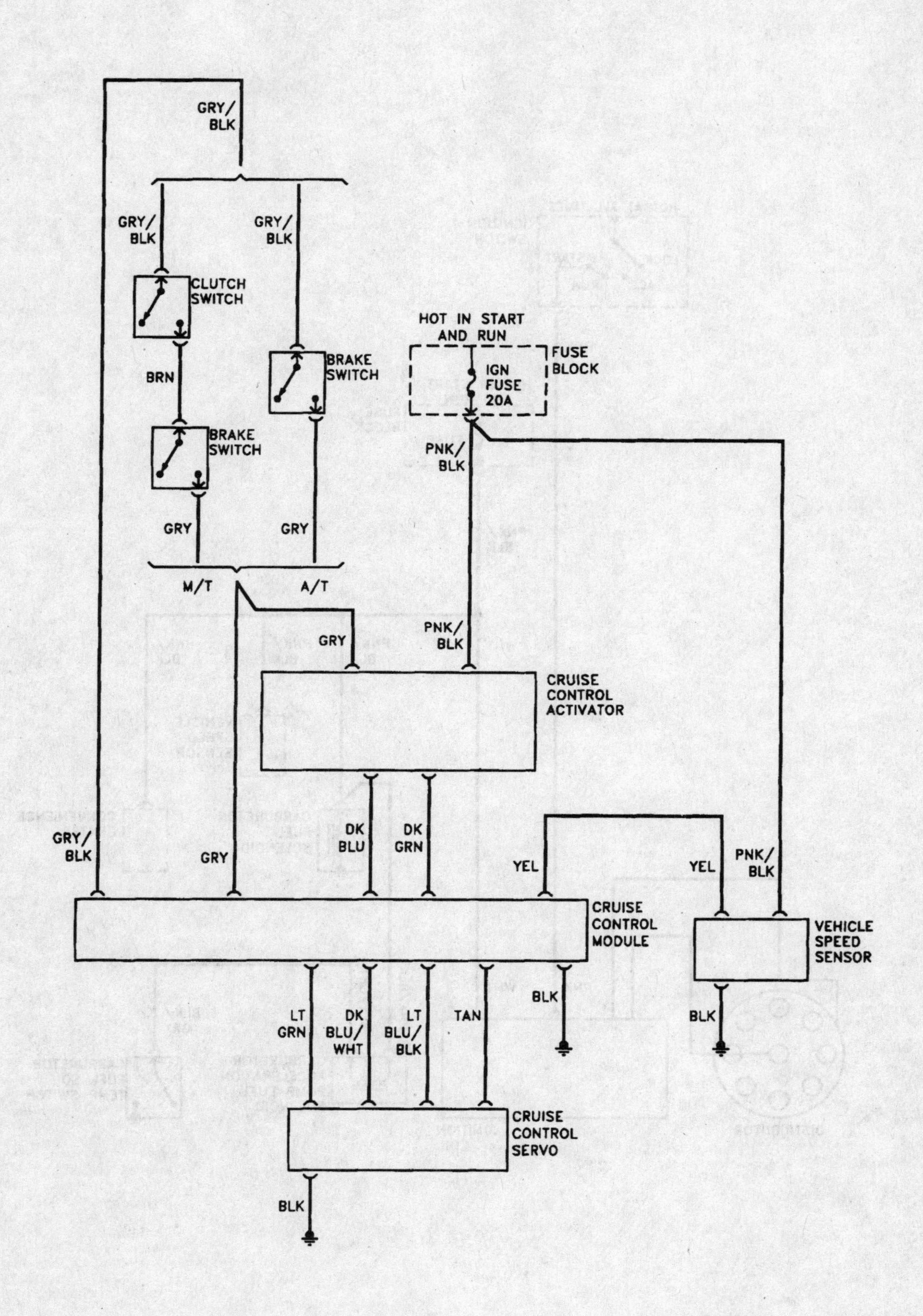

Typical cruise control system

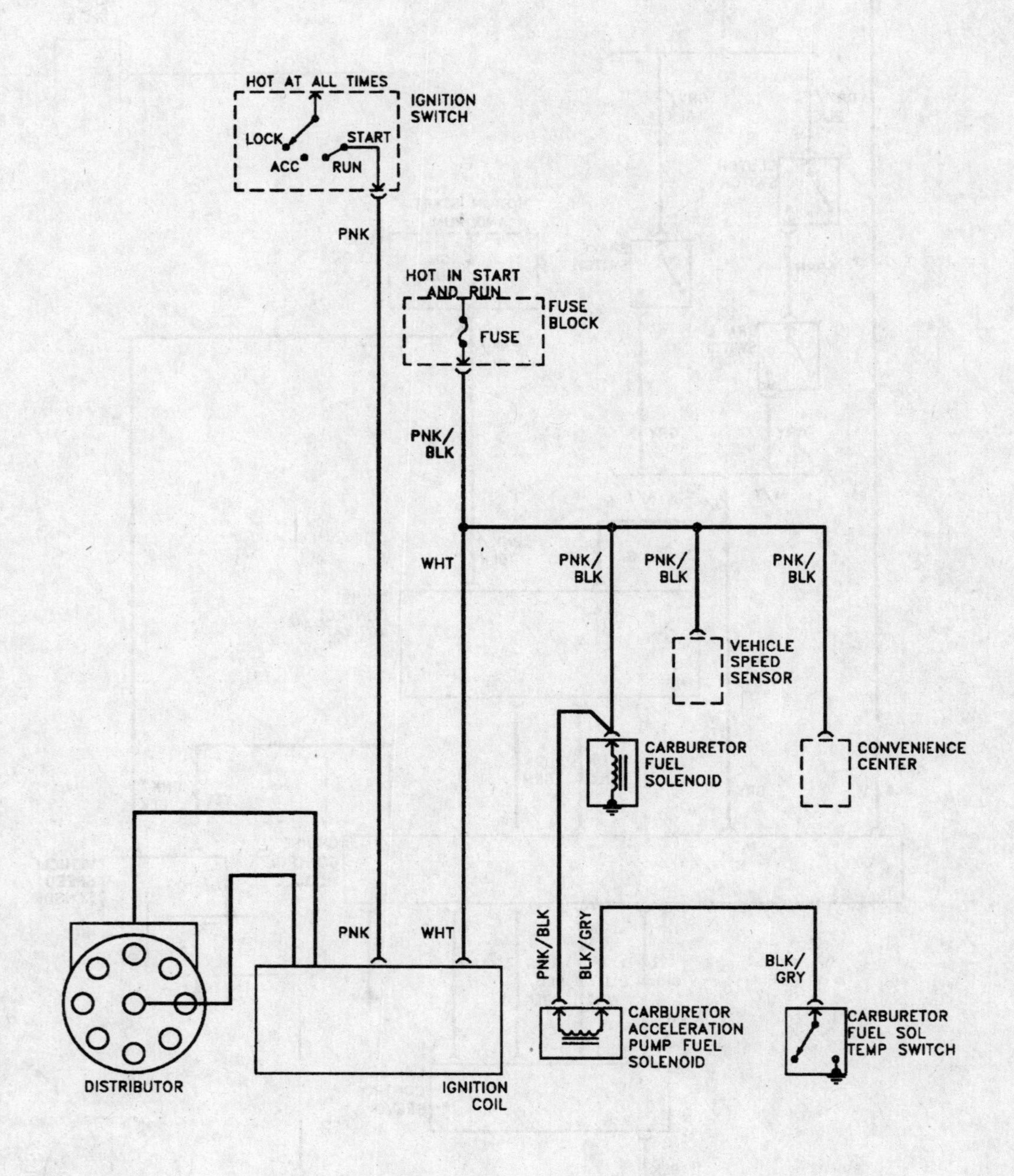

Typical 1981 through 1987 V8 engine ignition and emissions control systems (carbureted models)

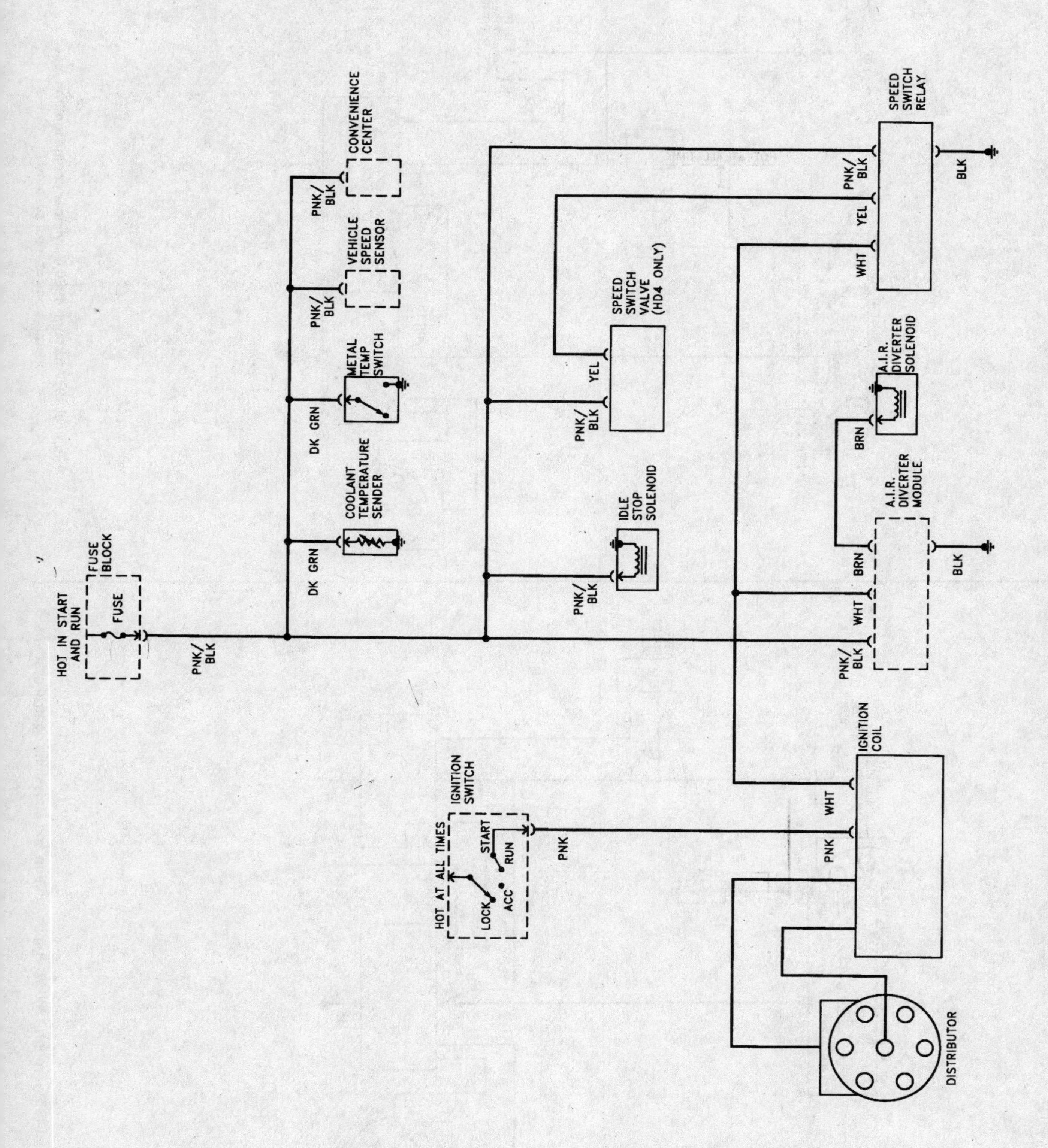

Typical 1981 through 1987 6-cylinder engine ignition and emissions control systems (carbureted models)

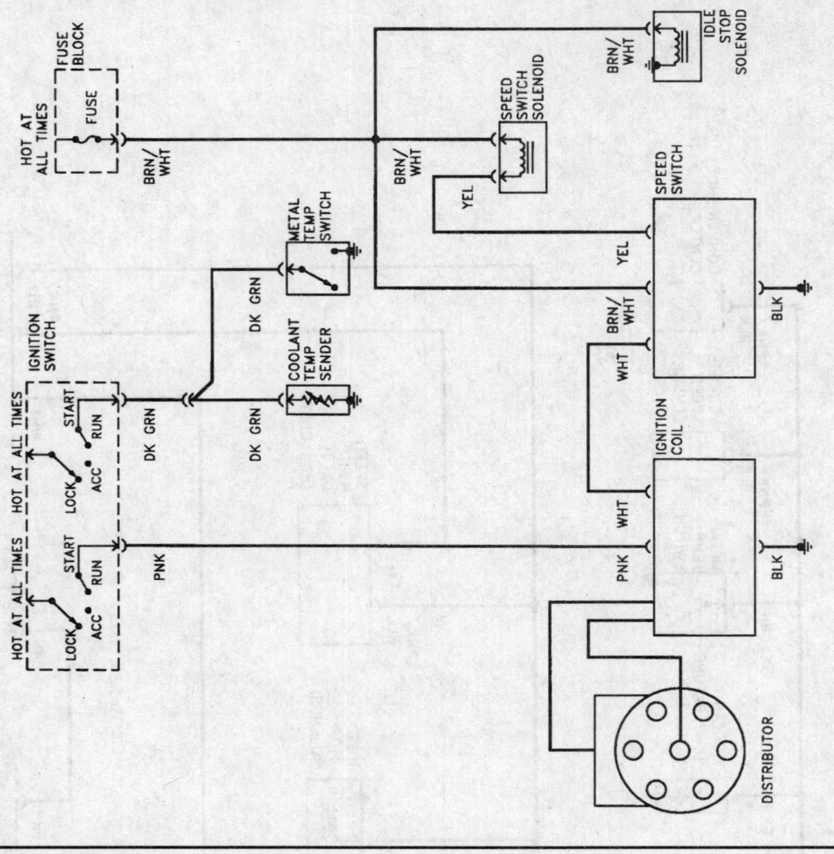

Typical 1977 through 1980 6-cylinder engine ignition and emissions control systems

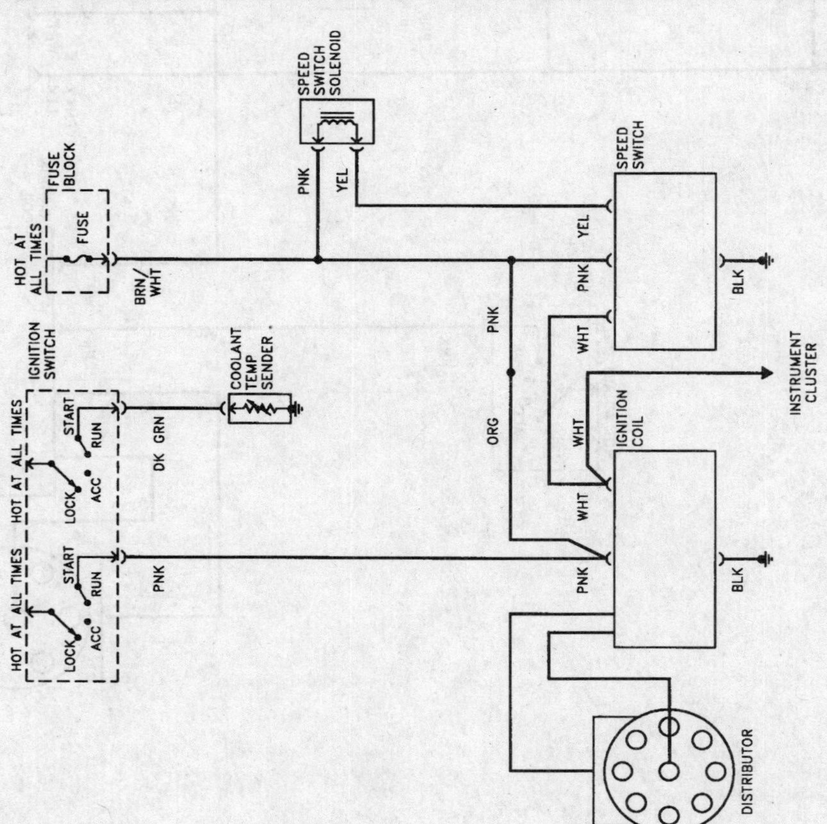

Typical 1977 through 1980 V8 engine ignition and emissions control systems

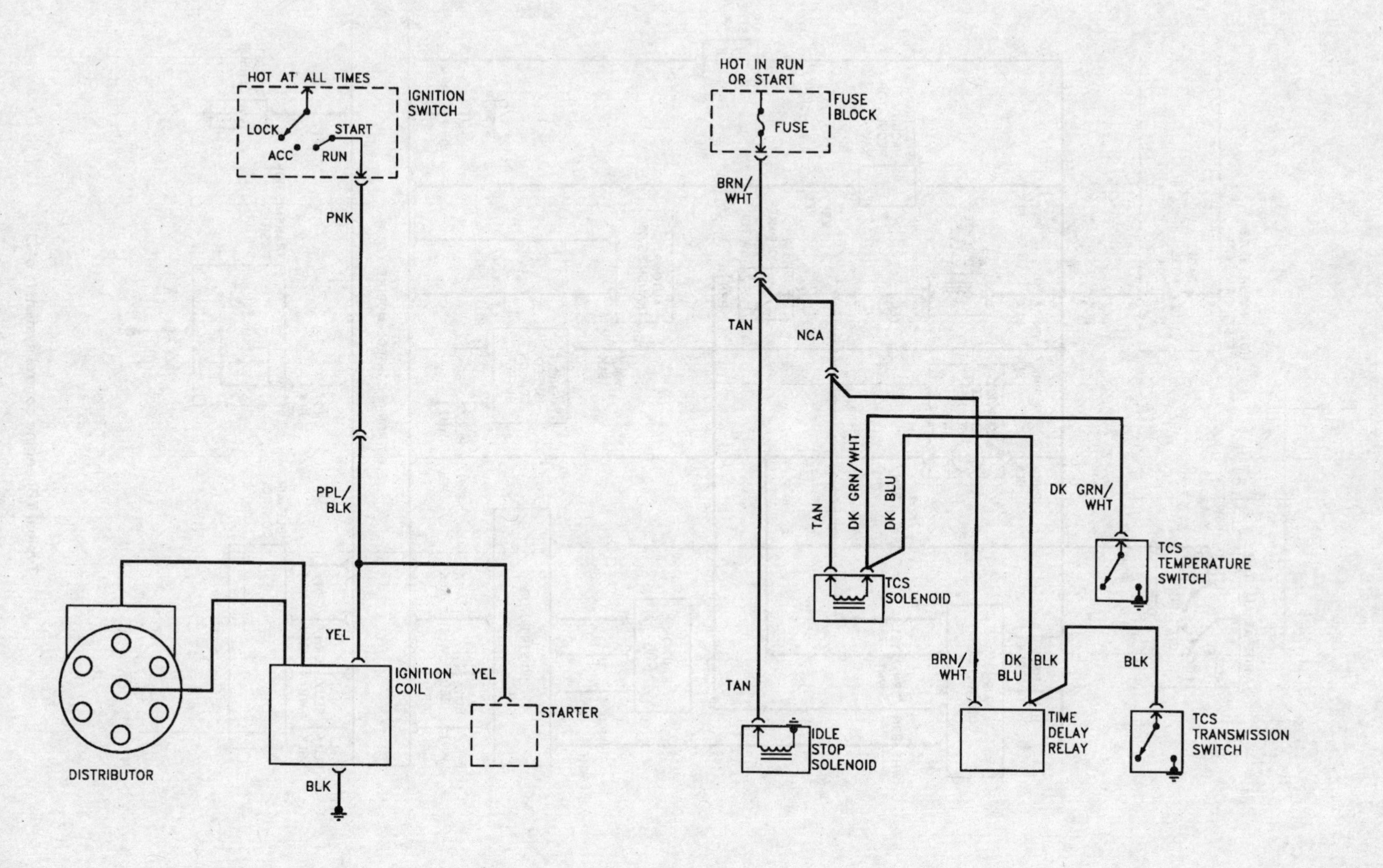

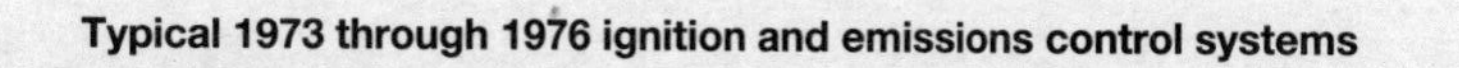
Typical 1973 through 1976 ignition and emissions control systems

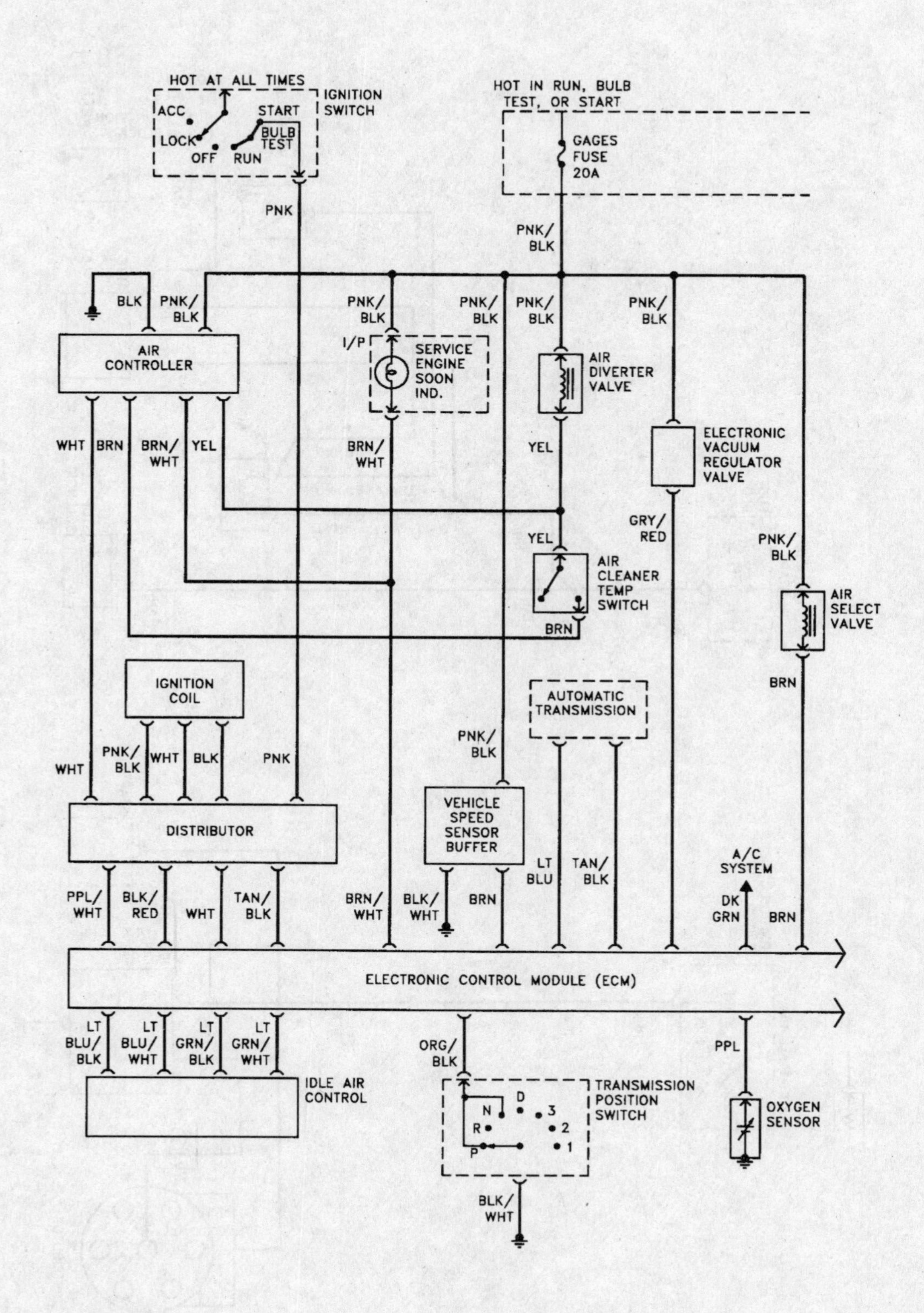

Typical TBI engine control system (1 of 2)

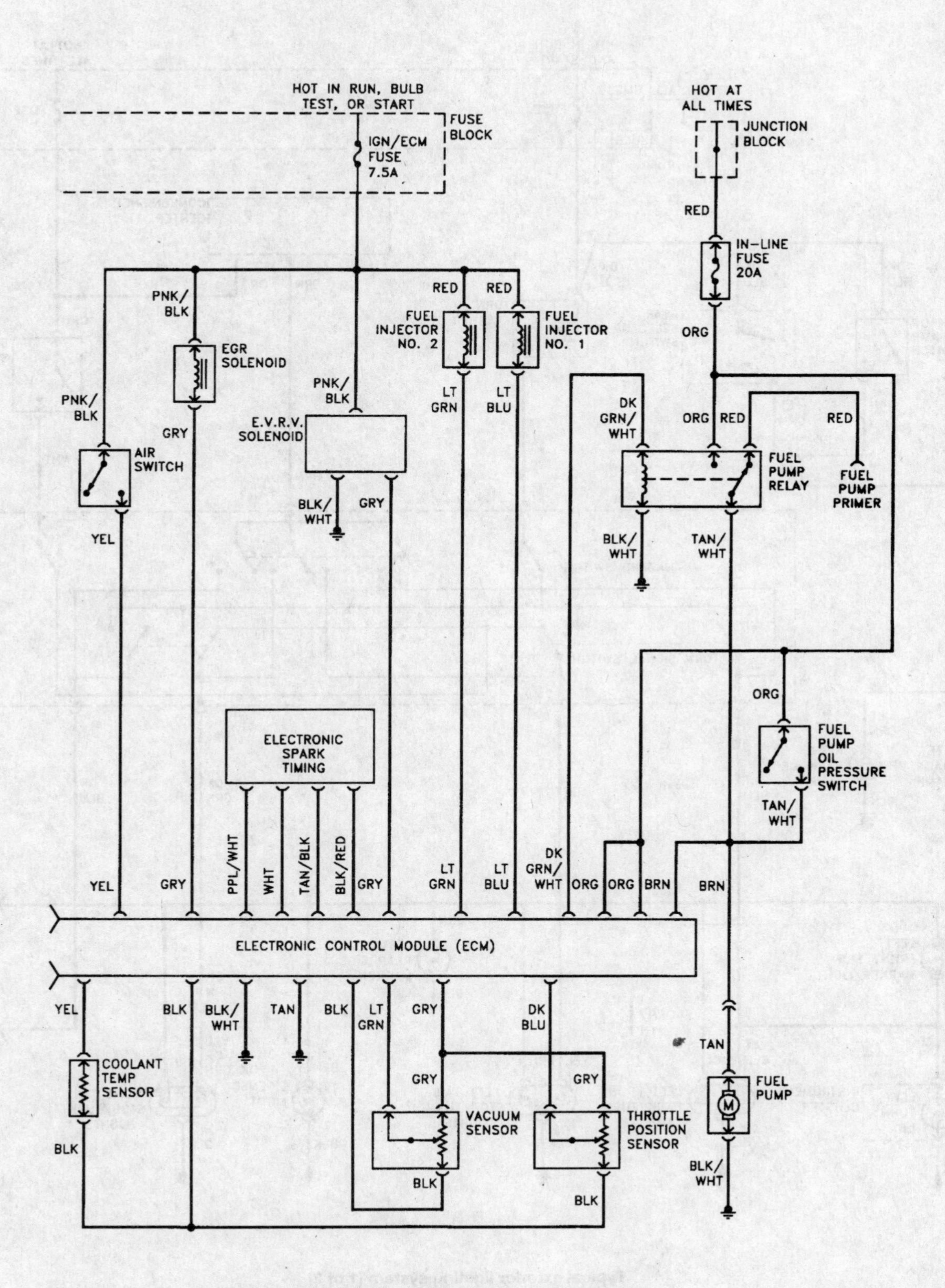

Typical TBI engine control system (2 of 2)

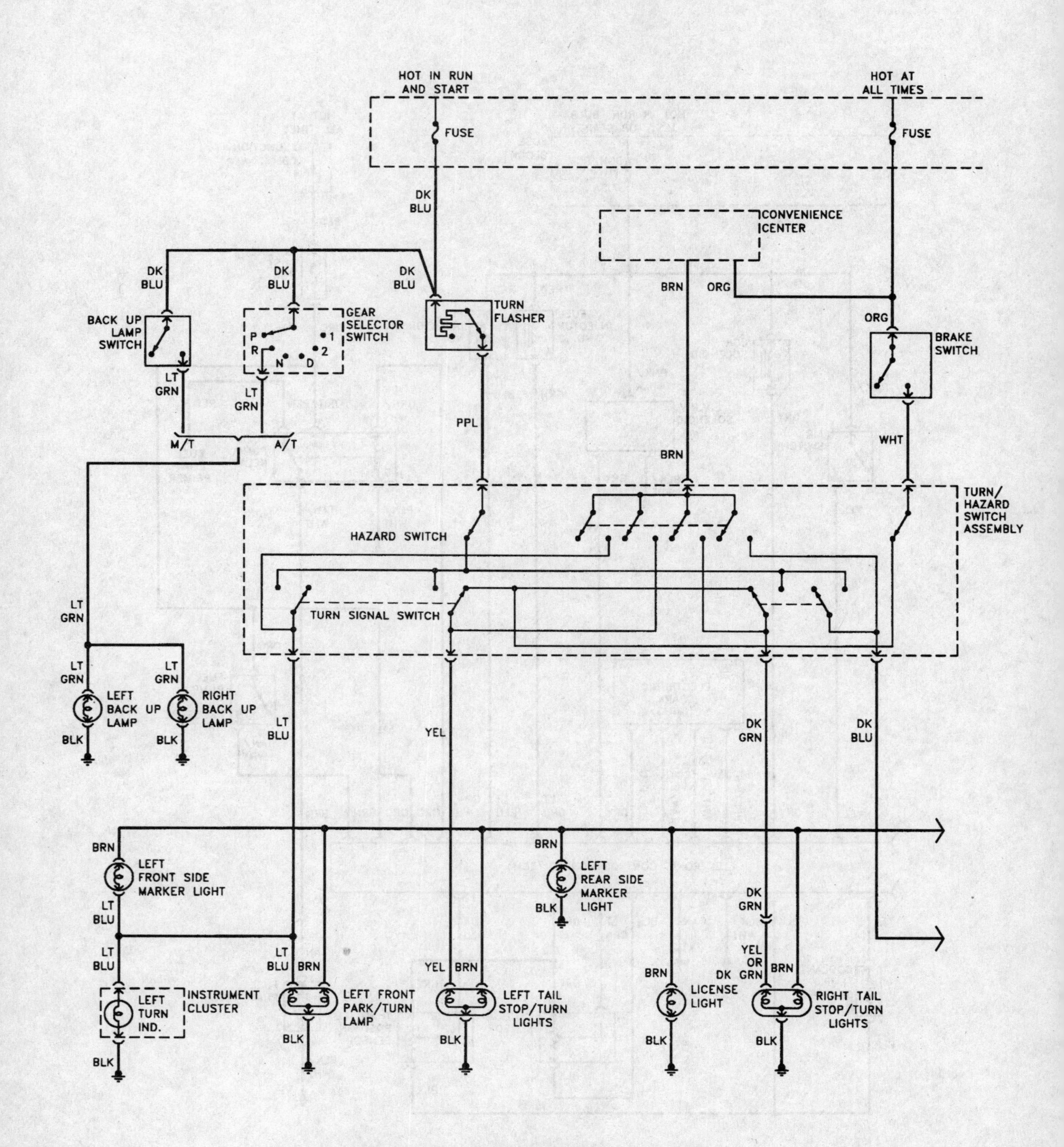

Typical exterior lighting system (1 of 2)

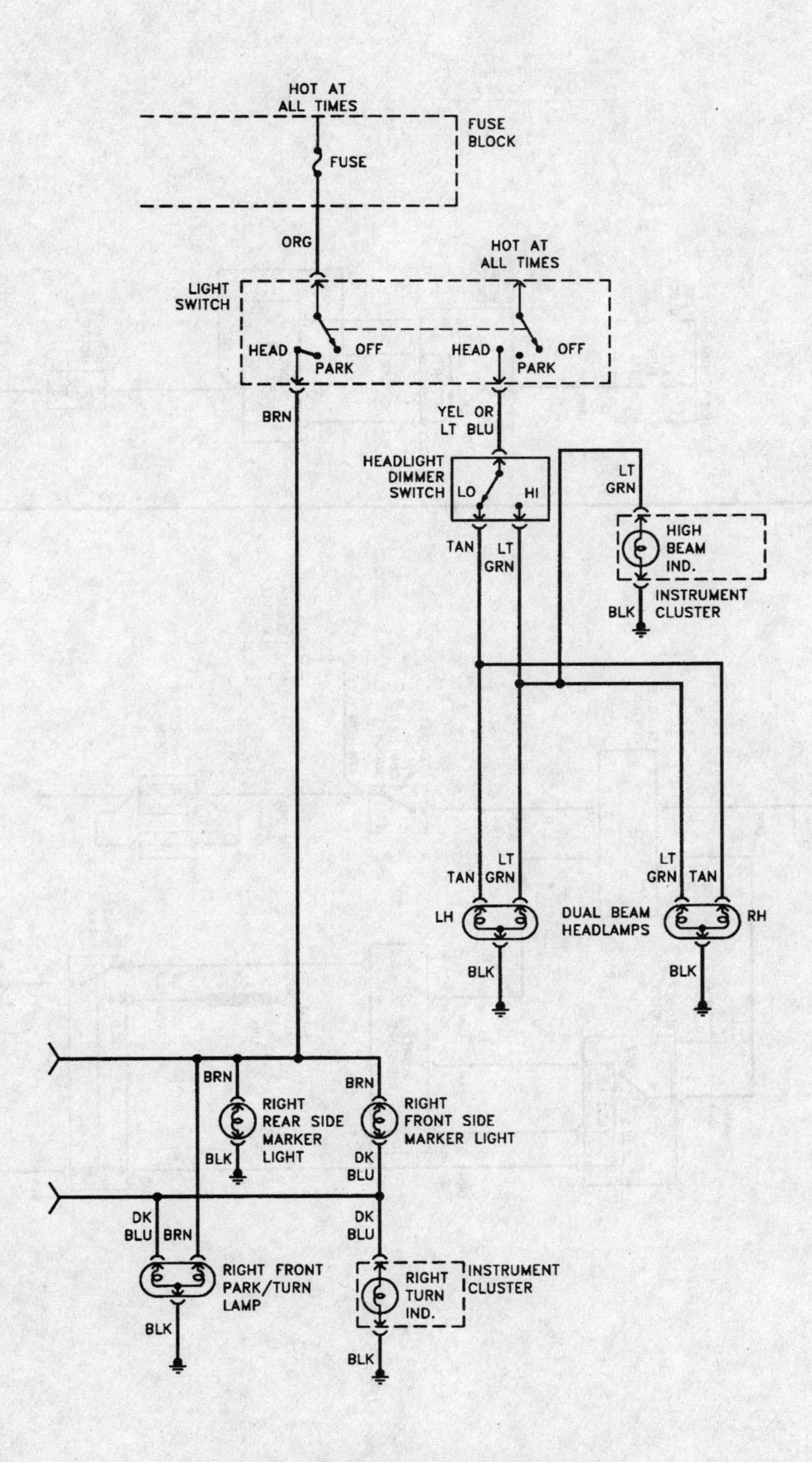

Typical exterior lighting system (2 of 2)

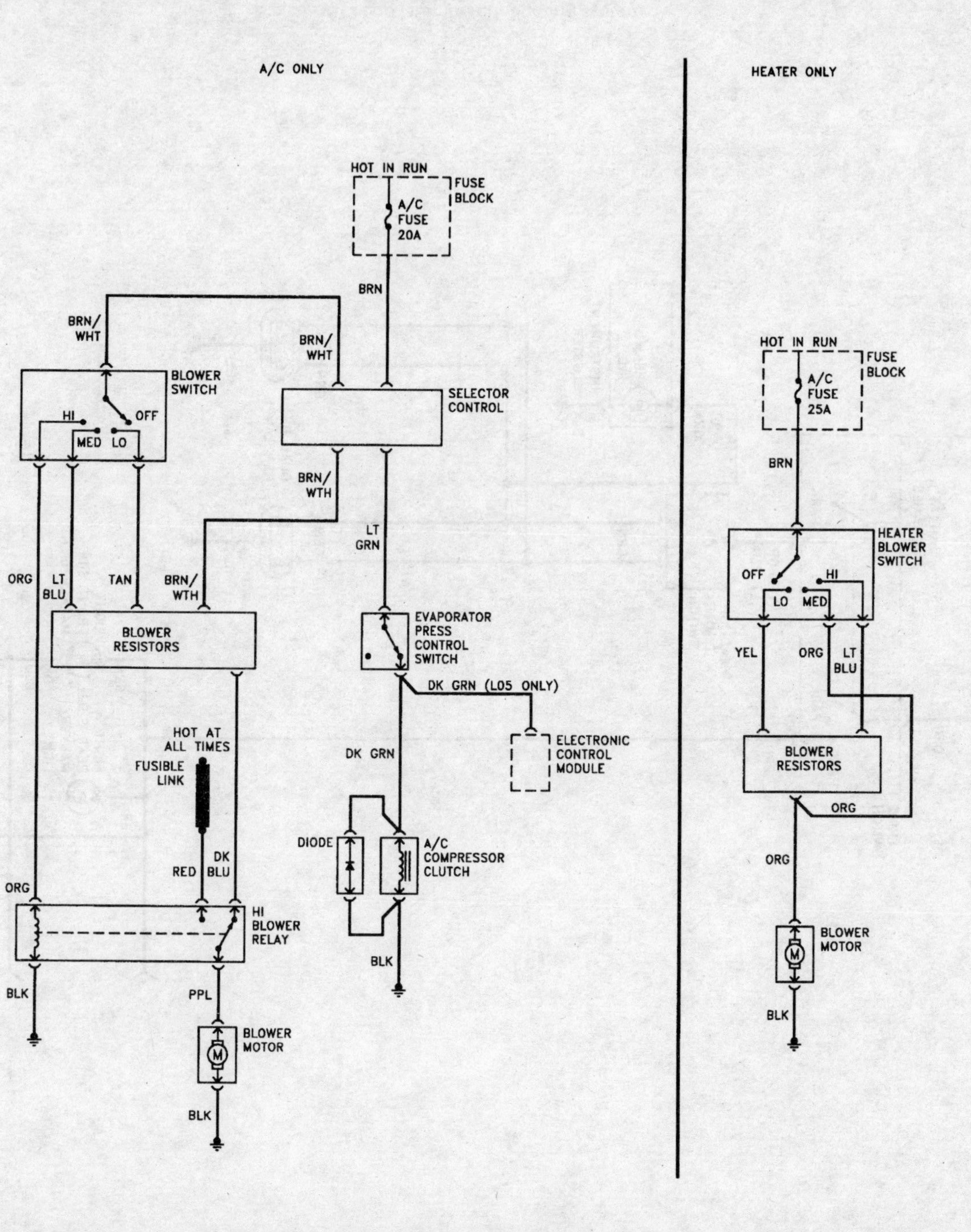

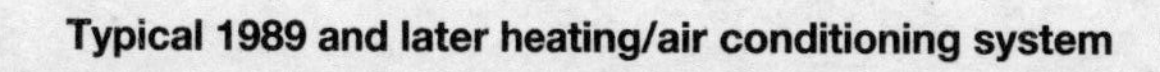
Typical 1989 and later heating/air conditioning system

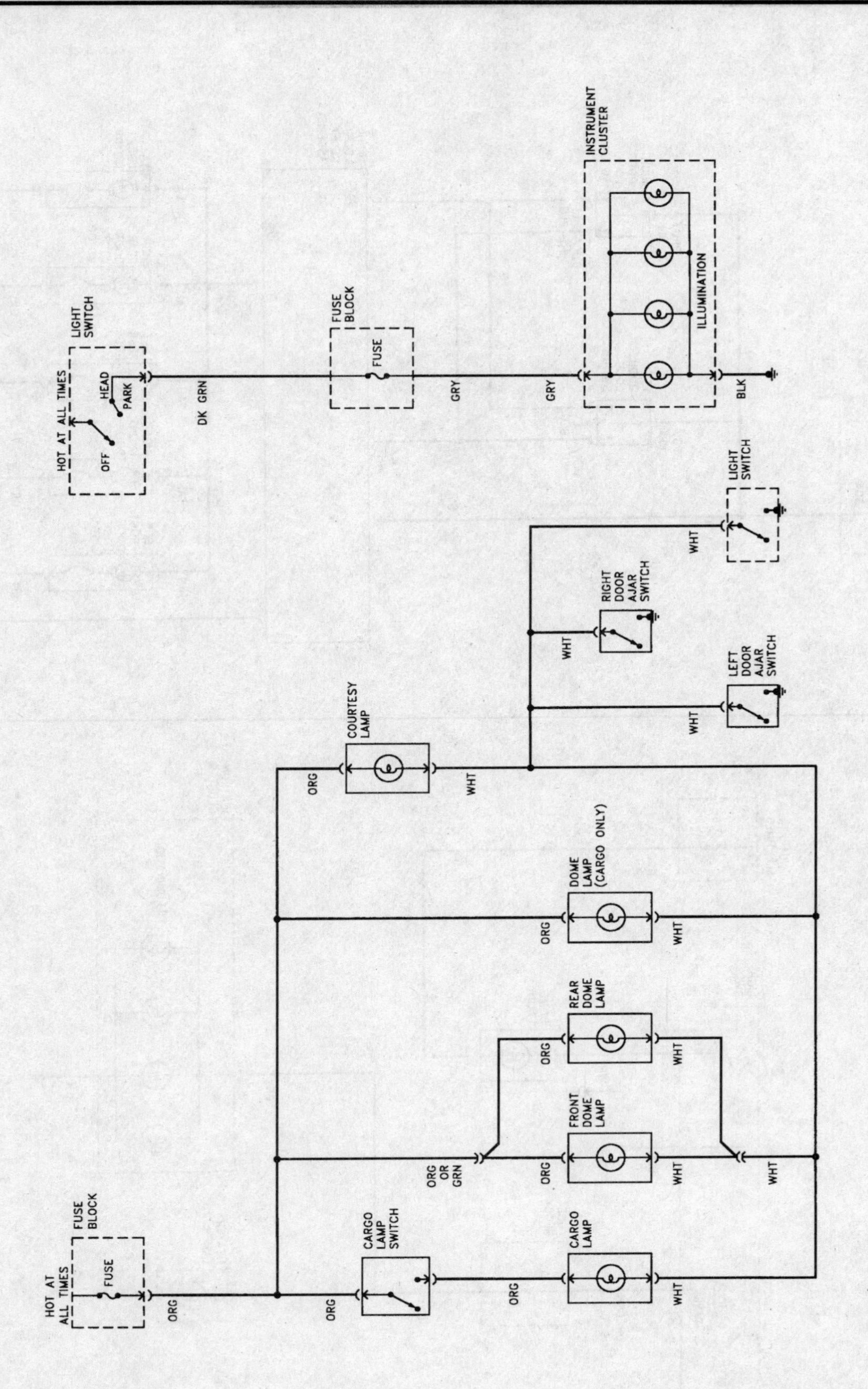

Typical 1973 and later interior lighting system

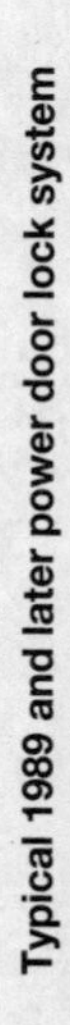
Typical 1989 and later power door lock system

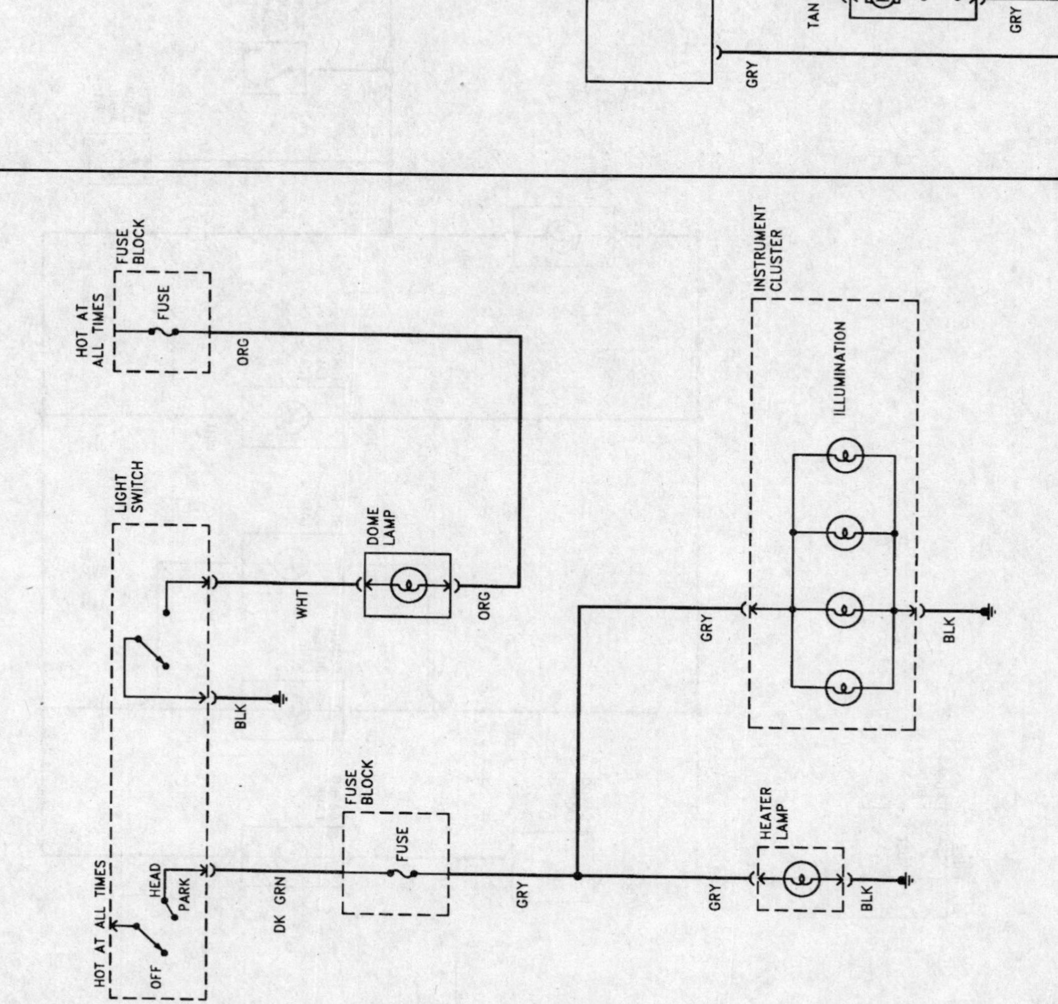

Typical 1972 and earlier interior lighting system

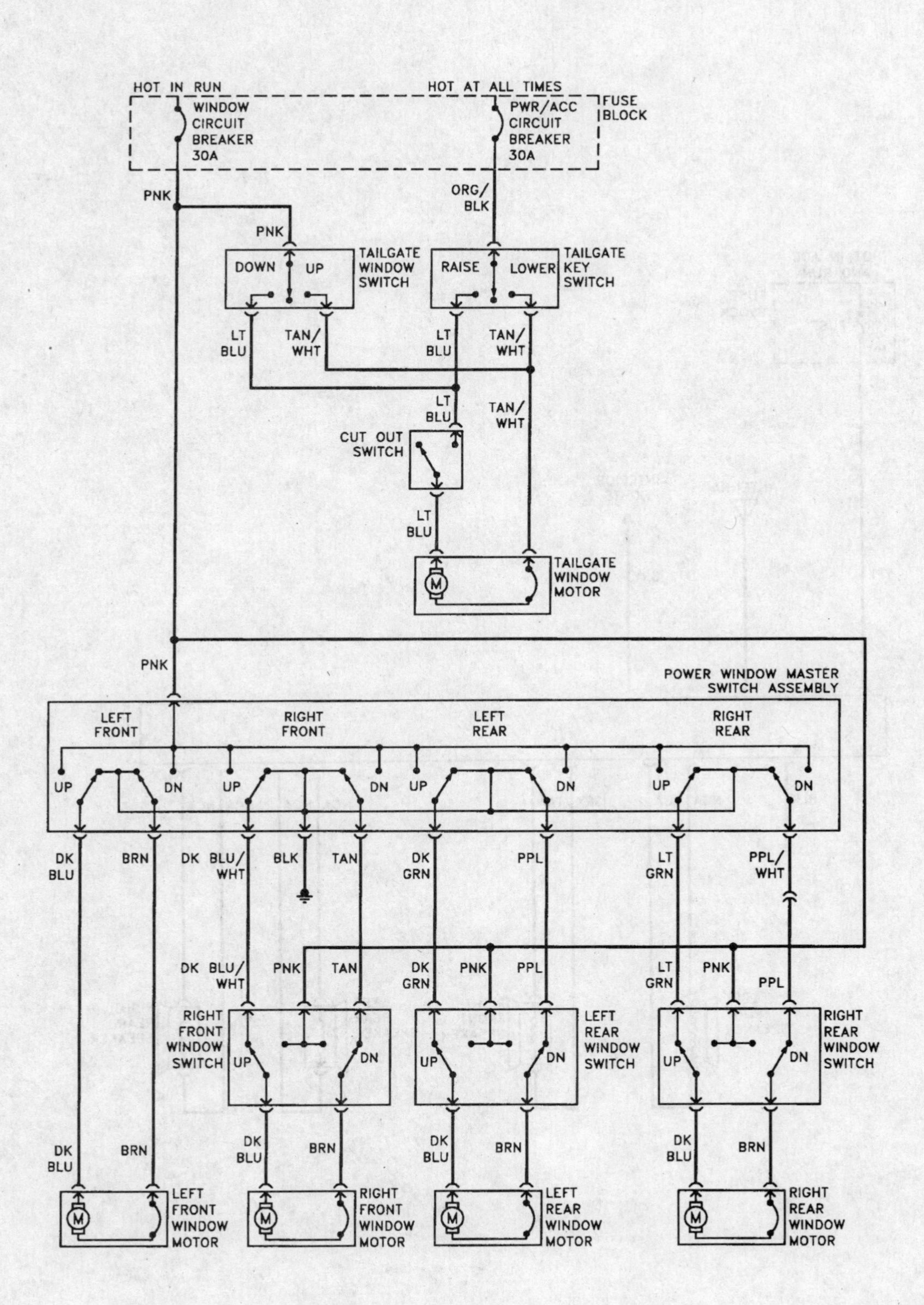

Typical 1989 and later power window system

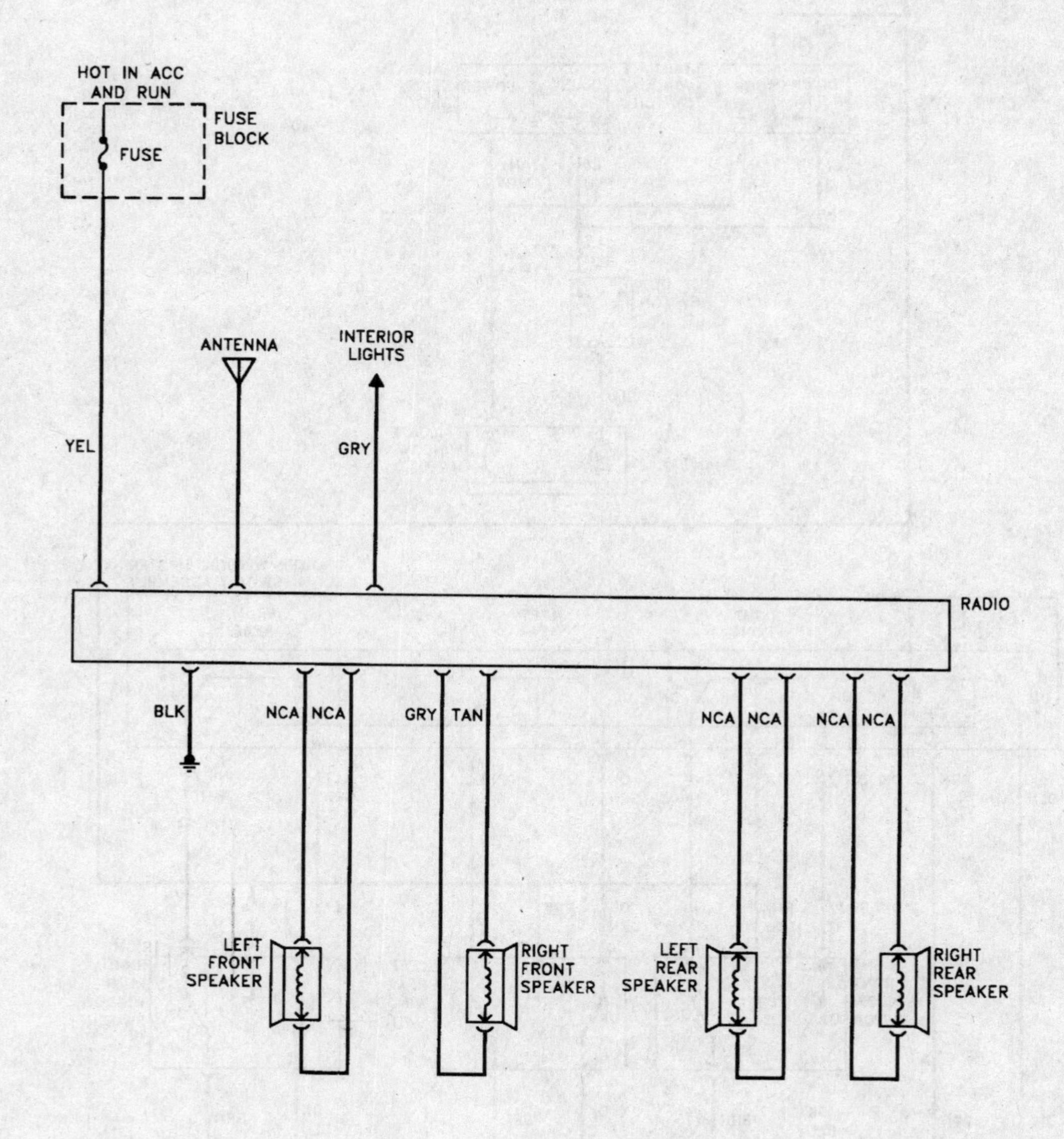

Typical 1977 through 1987 audio system

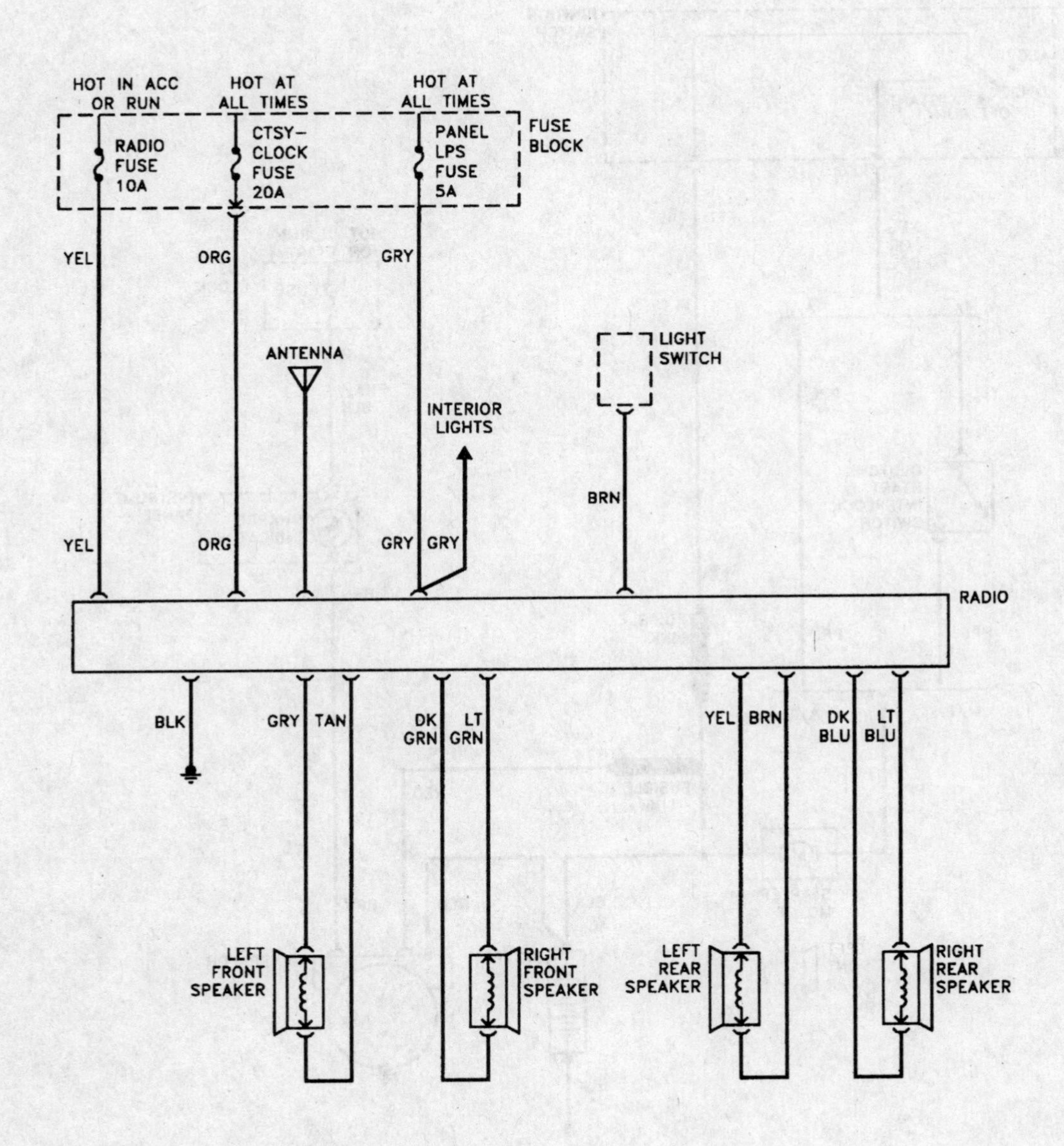

Typical 1989 and later audio system

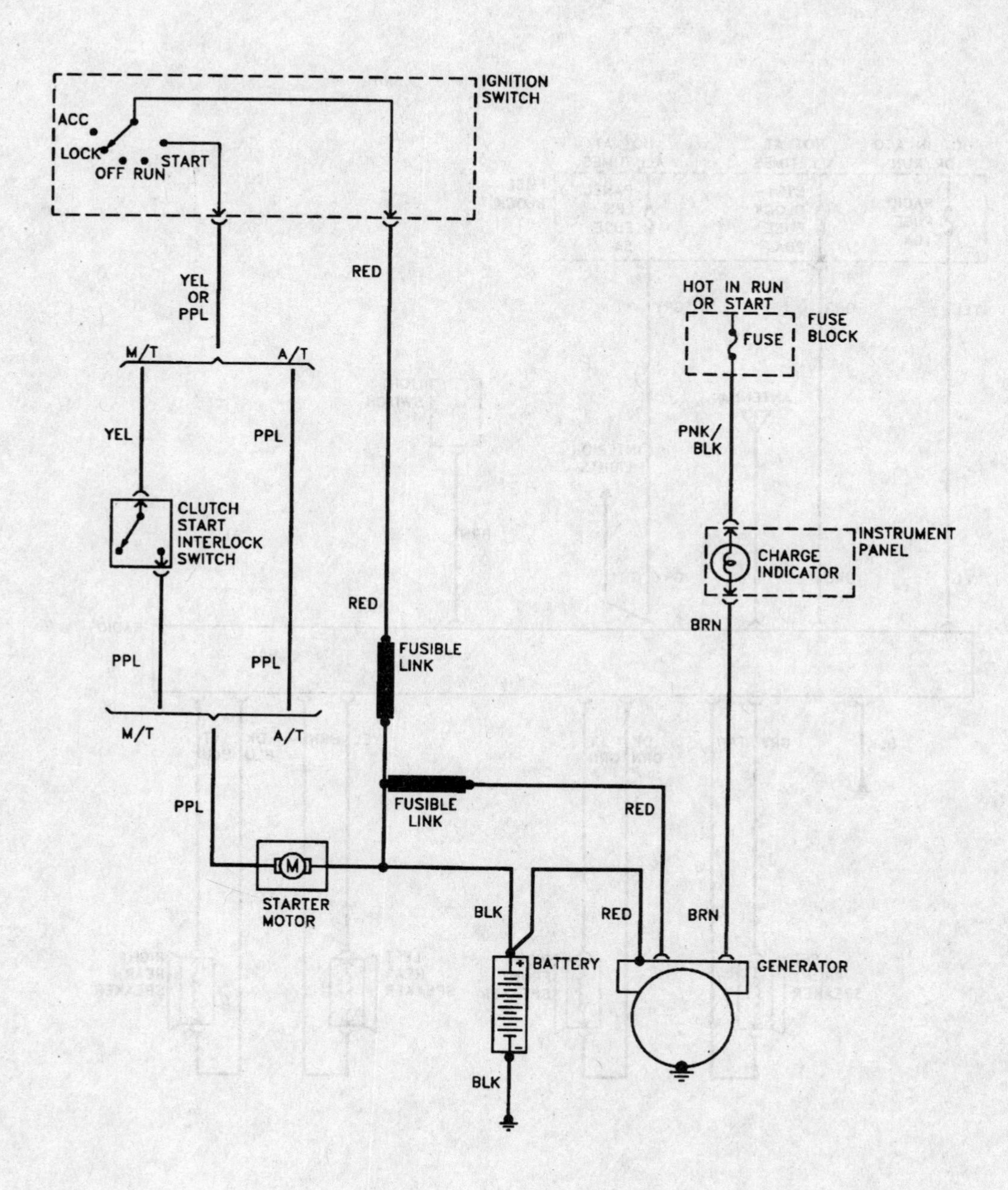

Typical 1977 through 1987 starting and charging systems

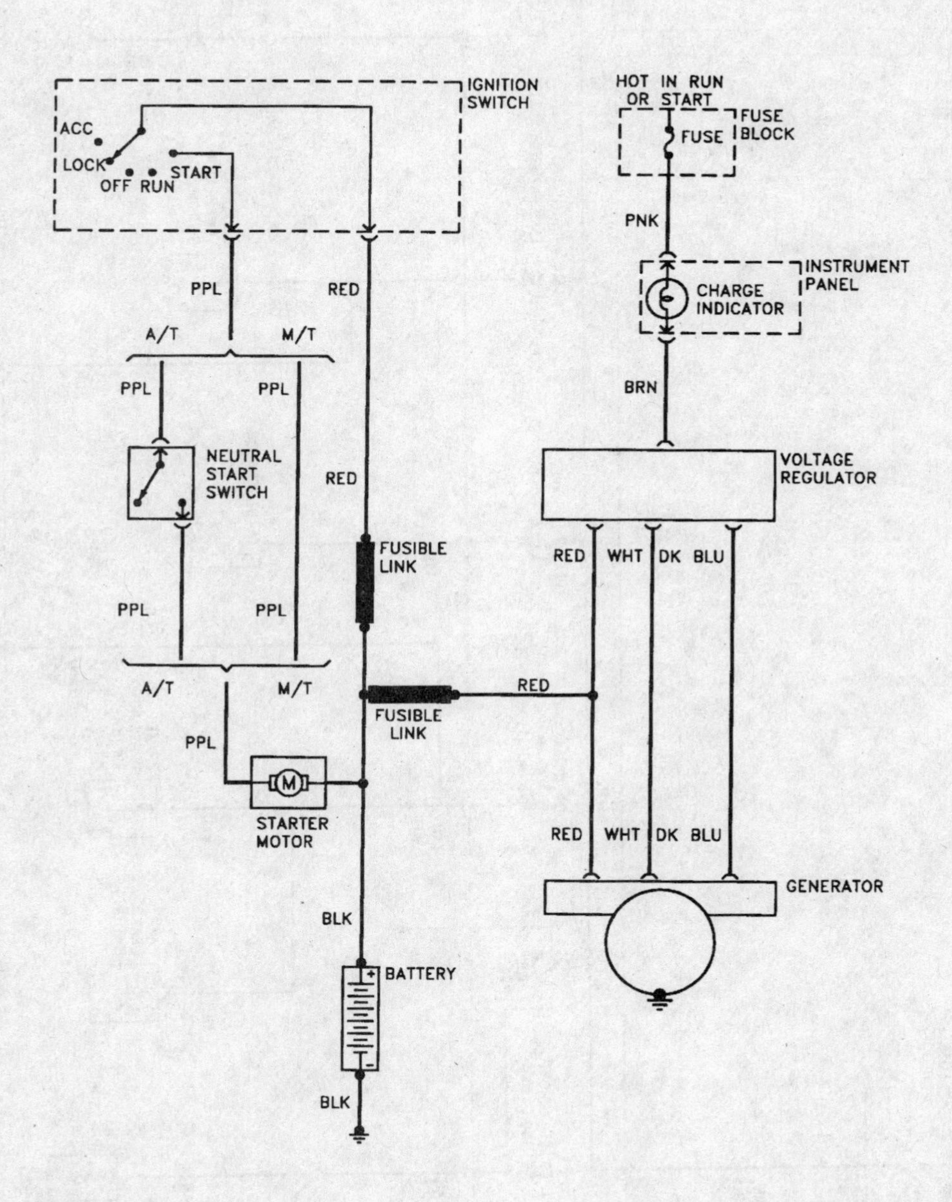

Typical 1967 through 1976 starting and charging systems

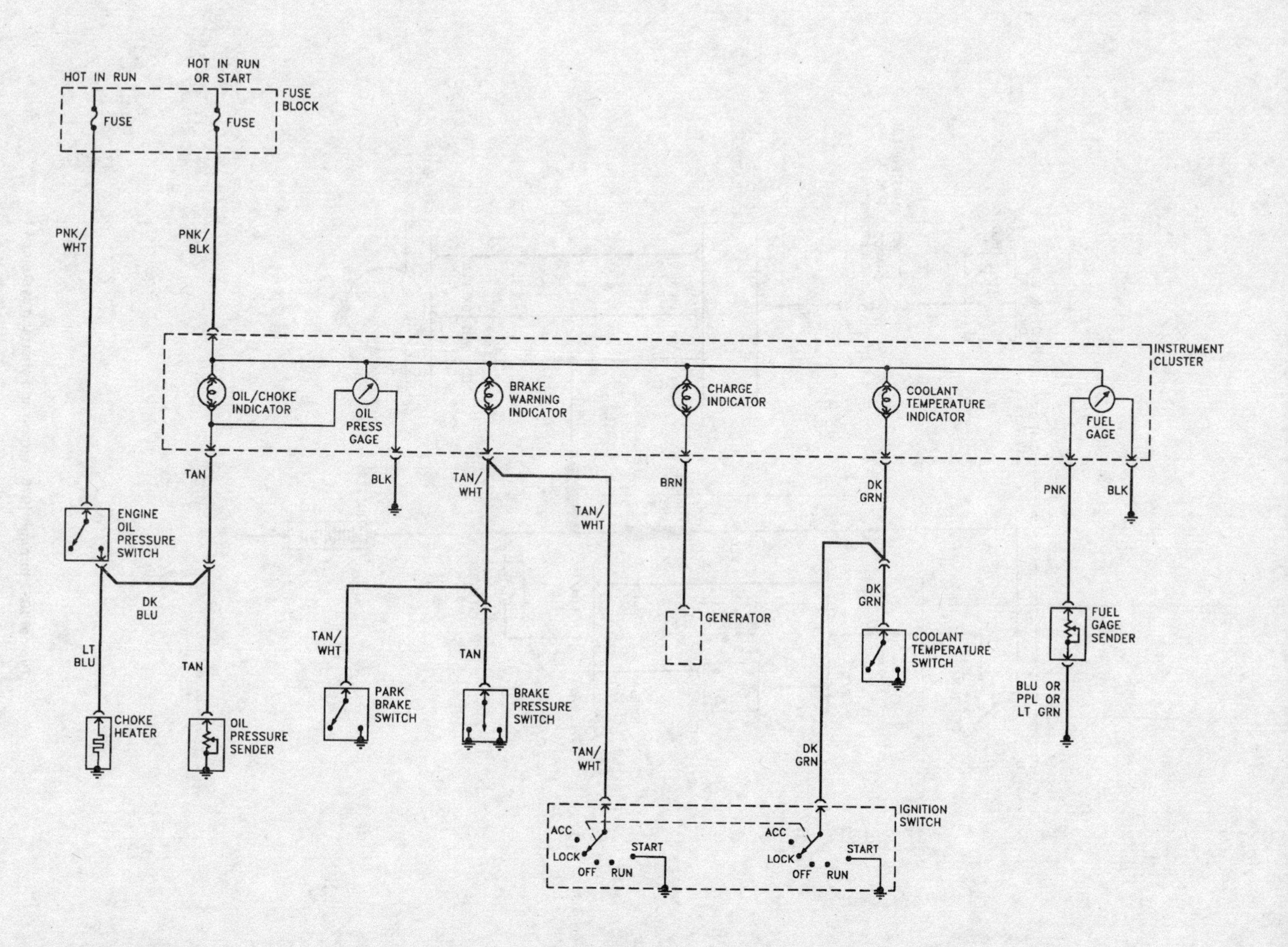

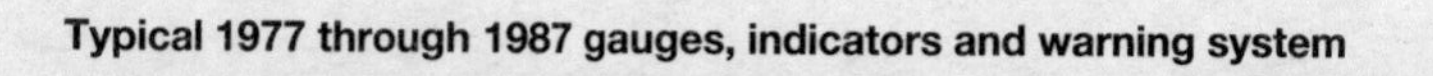
Typical 1977 through 1987 gauges, indicators and warning system

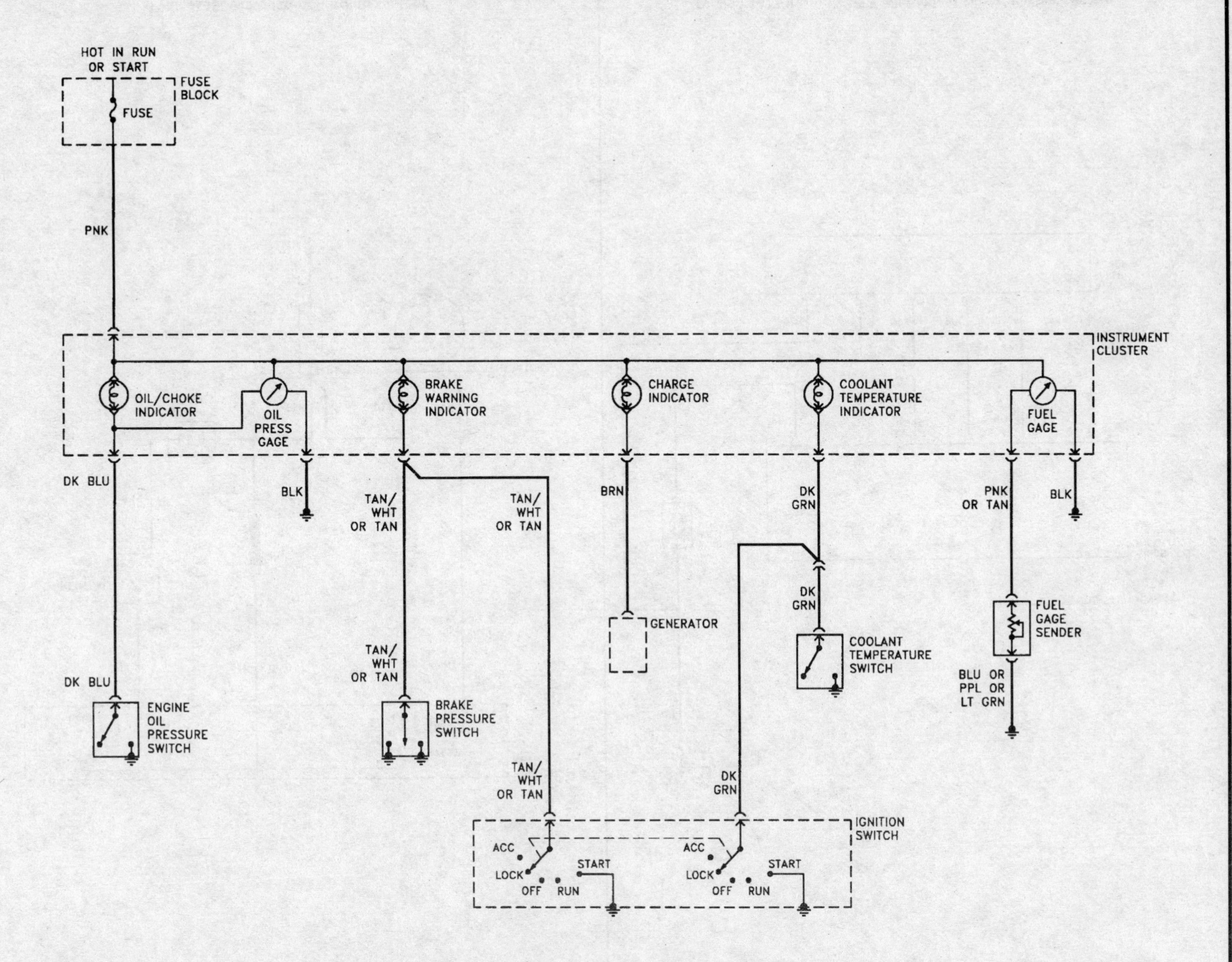

Typical 1967 through 1976 gauges, indicators and warning system

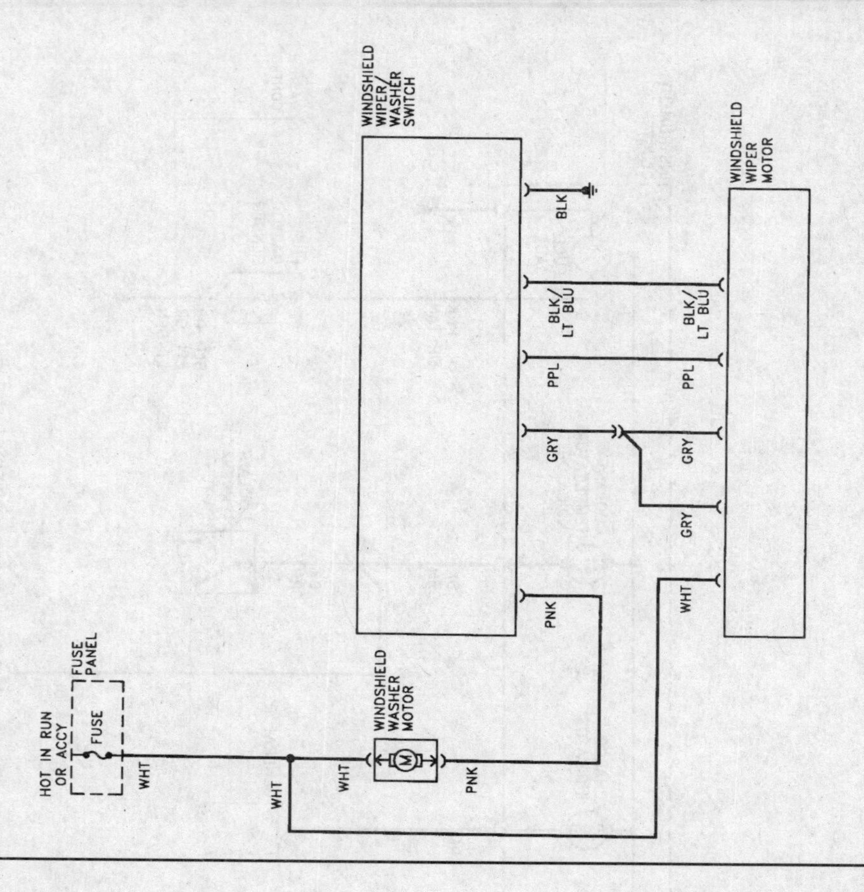

Typical 1977 through 1987 windshield wiper/washer system

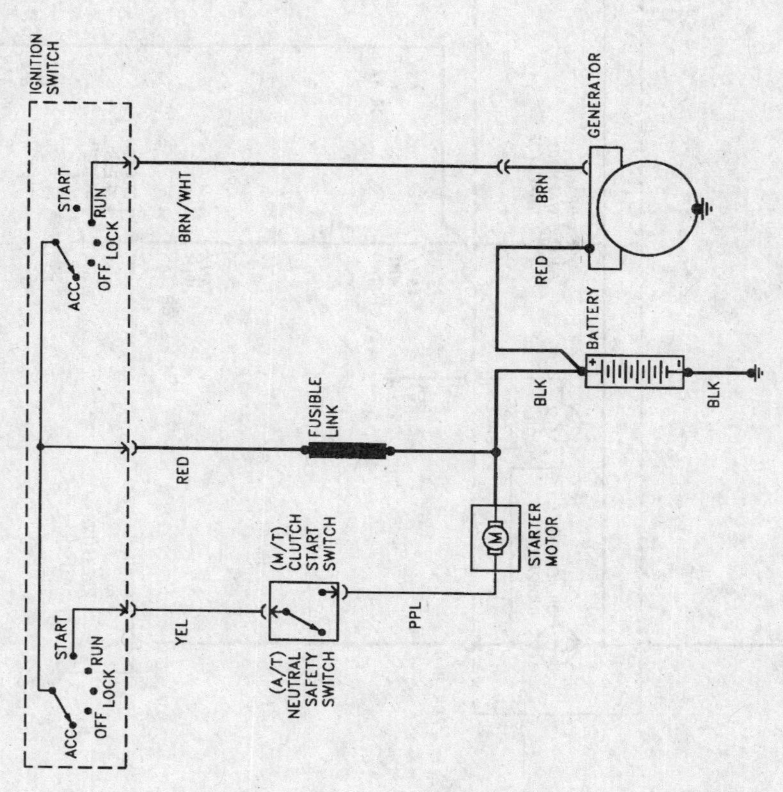

Typical 1989 and later charging system

Typical 1989 and later windshield wiper/washer system

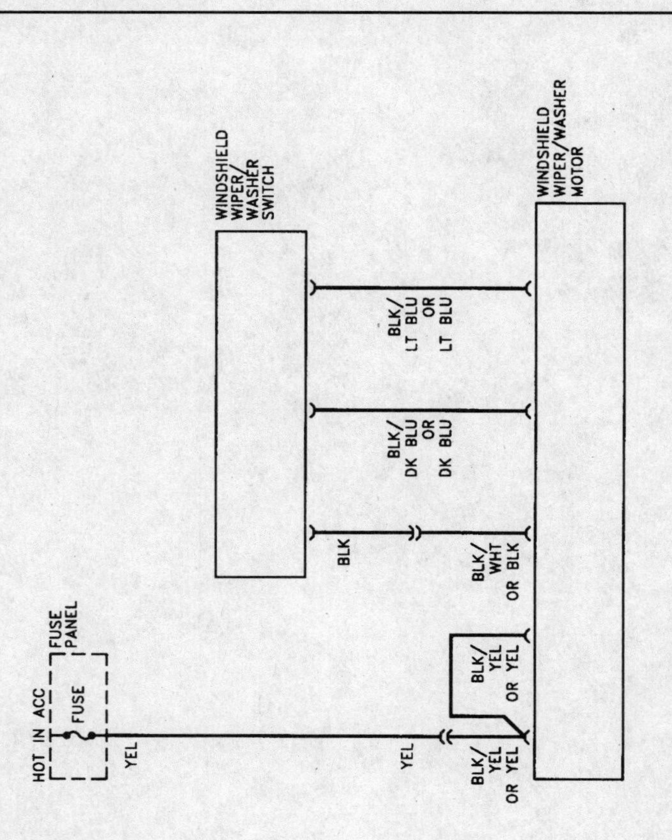

Typical 1967 through 1976 windshield wiper/washer system

Notes

Index

A

About this manual, 0-6
Acknowledgments, 0-2
Air conditioning system
accumulator, removal and installation, 3-8
check and maintenance, 3-7
compressor, removal and installation, 3-8
condenser, removal and installation, 3-8
engine oil cooler, general information, 3-9
Air filter and PCV filter replacement, 1-19
Air Injection Reaction (AIR) system, 6-17
Alignment, steering angles and wheel, general information, 10-19
Alternator
brushes, replacement, 5-11
removal and installation, 5-11
Antifreeze, general information, 3-3
Automatic transmission, 7B-1 through 7B-6
detent cable, adjustment, 7B-2
detent switch, adjustment, 7B-3
diagnosis, general, 7B-2
fluid and filter change, 1-25
fluid level check, 1-9
general information, 7B-1
neutral start switch, adjustment, 7B-4
rear seal, replacement, 7B-4
removal and installation
2-wheel drive models, 7B-5
4-wheel drive models, 7B-4
shift linkage, check and adjustment, 7B-2
speedometer gear, replacement, 7B-4
throttle valve (TV) cable, check and adjustment, 7B-3
throttle valve (TV) cable, replacement, 7B-4
Automotive chemicals and lubricants, 0-19
Auxiliary leaf spring, rear, (early two-wheel drive), removal and installation, 10-10
Axle
assembly, removal and installation, 8-12
balljoints, front, (four-wheel drive), check and adjustment, 10-8
bearing (semi-floating axle), replacement, 8-11
general information, 8-9
kingpins, front, (four-wheel drive with disc brakes) overhaul, 10-9
kingpins, front, (four-wheel drive with drum brakes)
check and adjustment, 10-9
overhaul, 10-9
steering knuckle and balljoints, front, (four-wheel drive), removal, overhaul and installation, 10-8

B

Back-up light switch (manual transmission), 12-7
Balljoints (two-wheel drive), check and replacement, 10-5
Battery cables, check and replacement, 5-5
Battery check and maintenance, 1-11
Battery, removal and installation, 5-4
Body, 11-1 through 11-12
general information, 11-1
maintenance, 11-2
repair
major damage, 11-3
minor damage, 11-2
Booster battery (jump) starting, 0-17
Brake light switch, removal, installation and adjustment, 9-13
Brakes, 9-1 through 9-16
check, 1-18
disc brake caliper, removal, overhaul and installation
Bendix, 9-7
Delco, 9-6

disc brake pads, replacement
Bendix, 9-5
Delco, 9-5
disc brake rotor, inspection, removal and installation, 9-6
drum brake shoes, inspection, removal and installation, 9-7
drum brake wheel cylinder, removal, overhaul and installation, 9-10
general information, 9-3
hoses and lines, check and replacement, 9-14
hydro-boost
system, bleeding, 9-15
unit, removal and installation, 9-14
master cylinder, removal, overhaul and installation
Bendix, 9-12
Delco, 9-10
parking brake
adjustment
hand lever type, 9-12
pedal type, 9-13
control and cables (hand lever type), removal and installation, 9-12
pedal and cables (pedal type), removal and installation, 9-13
pedal, removal and installation, 9-13
power brake vacuum booster, inspection, removal and installation, 9-14
pressure regulating combination valve, check, removal and installation, 9-12
rear wheel anti-lock (RWAL) brake system, general information, 9-16
system bleeding, 9-15
Bulb replacement, 12-6
Bumpers, removal and installation, 11-8
Buying parts, 0-9

C

Carburetor
adjustments, 4-38 through 4-48
choke check, 1-21
general information, 4-31
outside air intakes, removal and installation, 11-10
removal and installation, 4-37
servicing, 4-37
Carburetor/throttle body mounting nut torque check, 1-21
Catalytic converter, 6-23
Charging system
check, 5-10
general information and precautions, 5-9
Chassis electrical system, 12-1 through 12-32
general information, 12-2
Chassis lubrication, 1-14
Circuit breakers, general information, 12-4
Clutch and driveline, 8-1 through 8-16
general information, 8-2
Clutch
adjustment, 8-2
components, removal, inspection and installation, 8-2
cross shaft (later mechanical linkage equipped vehicles), removal and installation, 8-4
description and check, 8-2
hydraulic system, bleeding, 8-5
idler lever and shaft assembly (early models), removal and installation, 8-4
master cylinder, removal and installation, 8-5
operated Neutral start switch, replacement, 8-5
pedal, removal and installation, 8-4
early models, 8-4
later models, pedal free play check and adjustment, 1-17
slave cylinder, removal and installation, 8-5
Coil spring
front, (two-wheel drive), removal and installation, 10-5
rear, (early two-wheel drive), removal and installation, 10-9
Computer Command Control (CCC) system and trouble codes, 6-6
Control arm, rear, (1967 through 1972), removal and installation, 10-10
Conversion factors, 0-20
Coolant reservoir, removal and installation, 3-6
Cooling and heating systems
antifreeze, general information, 3-3
coolant reservoir, removal and installation, 3-6
cooling fan and fan clutch, removal and installation, 3-4
heater (1967 through 1970 models), removal and installation, 3-6
heater (1971 through 1987 models), removal and installation, 3-6
radiator, removal and installation, 3-3
thermostat, removal and installation, 3-3
water pump, check, 3-4
water pump, removal and installation, 3-4
Cooling fan and fan clutch, removal and installation, 3-4
Cooling system
check, 1-12
servicing (draining, flushing and refilling), 1-27
Cooling, heating and air conditioning systems, 3-1 through 3-10
general information, 3-2
Crankshaft
inspection, 2C-16
installation and main bearing oil clearance check, 2C-19

oil seals, six- cylinder inline engines, replacement
front seal, 2B-9
rear seal, 2B-9
removal, 2C-12
Cylinder compression check, 2C-6
Cylinder head
cleaning and inspection, 2C-9
disassembly, 2C-8
reassembly, 2C-11
Cylinder honing, 2C-15

D

Detent cable, adjustment, 7B-2
Detent switch, adjustment, 7B-3
Differential
oil change, 1-25
oil level check, 1-17
pressure sensor, 6-13
Disc brake
caliper, removal, overhaul and installation
Bendix, 9-7
Delco, 9-6
pads, replacement
Bendix, 9-5
Delco, 9-5
rotor, inspection, removal and installation, 9-6
Distributor
cap and rotor check and replacement, 1-30
reference signal, 6-14
removal and installation, 5-6
Door
inside handle, removal and installation, 11-7
lock
assembly, removal and installation, 11-7
cylinder, removal and installation, 11-7
outside handle, removal and installation, 11-7
removal, installation and adjustment, 11-8
striker, adjustment, 11-8
trim panel, removal and installation, 11-3
vent window assembly, removal and installation, 11-6
window glass and regulator, removal and installation, 11-6
Drivebelt check and adjustment, 1-23
Driveshaft
and universal joints, description and check, 8-5
center support bearing, removal and installation, 8-7
front, (four-wheel drive), removal and installation, 8-6
rear, removal and installation, 8-6
removal and installation, 8-6
Drum brake
shoes, inspection, removal and installation, 9-7
wheel cylinder, removal, overhaul and installation, 9-10

E

Early Fuel Evaporation (EFE) system, 6-22
EFE (heat riser) system check, 1-16
Electrical troubleshooting, general information, 12-2
Electronic Control Module/PROM removal and installation, 6-11
Electronic Spark Control (ESC) system, 6-15
Electronic Spark Timing (EST) system, 6-14
Emissions control systems, 6-1 through 6-24
Air Injection Reaction (AIR) system, 6-17
catalytic converter, 6-23
Computer Command Control (CCC) system and trouble codes, 6-6
Early Fuel Evaporation (EFE) system, 6-22
Electronic Control Module/PROM removal and installation, 6-11
Electronic Spark Control (ESC) system, 6-15
Electronic Spark Timing (EST) system, 6-14
Evaporative Emission Control System (EECS), 6-19
Exhaust Gas Recirculation (EGR) system, 6-15
general information, 6-1
information sensors, 6-13
oxygen sensor, 6-14
Positive Crankcase Ventilation (PCV) system, 6-19
Pulse Air Injection Reaction (PAIR) system, 6-21
Thermostatic Air Cleaner (THERMAC), 6-20
Throttle Return Control (TRC) system, 6-21
Transmission Controlled Spark (TCS) system, 6-23
Engine
block
cleaning, 2C-13
inspection, 2C-14
coolant temperature sensor, 6-13
oil and filter change, 1-13
oil cooler, general information, 3-9
overhaul procedures
disassembly sequence, 2C-8
general information, 2C-5
reassembly sequence, 2C-17
rebuilding alternatives, 2C-6
removal and installation
inline six-cylinder engine, 2C-6
V6 engine, 2C-7
V8 engines, 2C-7
removal methods and precautions, 2C-5
six-cylinder inline engines, 2B-1 through 2B-10
crankshaft oil seals, replacement, 2B-9
cylinder head, removal and installation, 2B-6
exhaust manifold (integrated head), removal and installation, 2B-6
general information, 2B-2
hydraulic lifters, removal and installation, 2B-3
intake and exhaust manifold (non-integrated head), removal and installation, 2B-5

mounts, check and replacement, 2B-9
oil pan and oil pump, removal and installation, 2B-9
pushrod cover, removal and installation, 2B-3
repair operations possible with the engine in the vehicle, 2B-2
rocker arm cover, removal and installation, 2B-2
rocker arms and pushrods, removal, inspection and installation, 2B-3
timing cover, gears and camshaft, removal, inspection and installation, 2B-7
Top Dead Center (TDC) for number 1 piston, locating, 2B-7
valve springs, retainers and seals, replacement in vehicle, 2B-4
V8 and V6
camshaft, bearings and lifters, removal, inspection and installation, 2A-11
crankshaft oil seals
replacement, front seal, 2A-13
replacement, rear seal, 2A-13
crankshaft oil seals, replacement, 2A-13
cylinder heads, removal and installation, 2A-8
engine mounts, check and replacement, 2A-15
exhaust manifolds, removal and installation, 2A-7
intake manifold, removal and installation, 2A-6
oil pan, removal and installation, 2A-12
oil pump, removal and installation, 2A-13
repair operations possible with the engine in the vehicle, 2A-3
rocker arm covers, removal and installation, 2A-3
rocker arms and pushrods, removal, inspection and installation, 2A-4
timing cover, chain and sprockets, removal and installation, 2A-10
Top Dead Center (TDC) for number 1 piston, locating, 2A-9
valve springs, retainers and seals, replacement in vehicle, 2A-5
Engine electrical systems, 5-1 through 5-14
Evaporative Emission Control System (EECS), 6-19
Evaporative emissions control system check, 1-28
Exhaust Gas Recirculation (EGR) system, 6-15
check, 1-28
Exhaust system
check, 1-16
component replacement, 4-56
External voltage regulator, check and replacement, 5-10

F

Floor shift linkage, removal, installation and adjustment, 7A-3
Fluid level checks, 1-7
battery electrolyte, 1-8
brake and clutch fluid, 1-8
engine coolant, 1-7
engine oil, 1-7
hydro boost pump fluid, 1-9
windshield washer fluid, 1-8
Front drive axle (K-series), axleshaft removal and installation
with disc brakes, 8-14
with drum brakes, 8-14
Front fender, removal and installation, 11-10
Front suspension assembly, removal and installation, 10-7
Front upper control arm (two-wheel drive), removal and installation, 10-6
Front wheel
bearings (2-wheel drive) check, repack and adjustment, 1-26
drive hub, removal and installation, 8-14
Fuel and exhaust systems, 4-1 through 4-56
general information, 4-2
Fuel filter replacement, 1-20
Fuel injection system
Throttle Body Injection (TBI) unit, disassembly and reassembly, 4-53
throttle linkage, replacement, 4-54
Fuel injection system, general information, 4-52
Fuel system
carburetor
adjustments
2SE/E2SE, 4-48
Rochester 1ME, 4-41
Rochester 1MV, 4-39
Rochester 2G series, 4-41
Rochester 4MV, 4-43
Rochester E4ME/E4MC, 4-50
Rochester M and MV, 4-38
Rochester M2M, 4-49
Rochester M4MC/M4MCA, 4-44
Rochester M4ME, 4-47
removal and installation, 4-37
servicing, 4-37
general information, 4-31
fuel lines, check and replacement, 4-38
fuel pump
check, 4-4
removal and installation, 4-30
Fuel system check, 1-19
Fuel tank
cleaning and repair, 4-55
removal and installation, 4-54
Full-floating axleshaft, removal and installation, 8-11
Fuses, general information, 12-3
Fusible links, general information, 12-4

G

General engine overhaul procedures, 2C-1 through 23-22

H

Headlight dimmer switch, column mounted, 12-7
Headlights
adjustment, 12-4
removal and installation, 12-4
Heater, removal and installation, 3-6
1967 through 1970 models, 3-6
1971 through 1987 models, 3-6
Hinges and locks, maintenance, 11-3
Hood
lock, removal, installation and adjustment, 11-9
release cable (later models), removal and installation, 11-10
removal and installation, 11-9
Hub/drum assembly and wheel bearings (full-floating axle), removal, installation and adjustment, 8-13
Hydro-boost system, bleeding, 9-15
Hydro-boost unit, removal and installation, 9-14

I

Idle speed check and adjustment, 1-22
Ignition
check and replacement, 5-8
module (HEI ignition), replacement, 5-6
pick-up coil (HEI ignition), check and replacement, 5-7
point replacement, 1-31
all models, 1-31
six-cylinder engine, 1-33
V8 engines, 1-32
switch, 12-7
system, check, 5-5
system, general information and precautions, 5-3
timing check and adjustment, 1-34
Information sensors, 6-13
distributor reference signal, 6-14
engine coolant temperature sensor, 6-13
Manifold Absolute Pressure (MAP) sensor, 6-13
oxygen sensor, 6-13
Park/Neutral switch (automatic transmission-equipped vehicles only), 6-14
Throttle Position Sensor (TPS), 6-13
vehicle speed sensor, 6-14
Initial start-up and break-in after overhaul, 2C-22
Instrument cluster, removal and installation
1967 through 1970 models, 12-8
1971 and later models, 12-8
Intermediate shaft coupling (early models), removal, overhaul and installation, 10-13
Introduction to the Chevrolet and GMC pick-ups, 0-6

J

Jacking and towing, 0-18

L

Leaf spring, front, (four-wheel drive), removal and installation, 10-7
Leaf spring, rear, (1973 on), removal and installation, 10-10
Lower control arm, front, (two-wheel drive), removal and installation, 10-6

M

Main and connecting rod bearings, inspection, 2C-17
Maintenance
schedule, 1-6
techniques, tools and working facilities, 0-9
Manifold Absolute Pressure (MAP) sensor, 6-13
Manual steering gear, check and adjustment, 10-12
Manual transmission, 7A-1 through 7A-10
disassembly, overhaul and reassembly
3-speed 76 mm, 7A-4
3-speed 77 mm, 7A-5
4-speed 117 mm, 7A-8
4-speed 89 mm, 7A-7
floor shift linkage, removal, installation and adjustment, 7A-3
general information, 7A-2
oil
change, 1-24
level check, 1-17
rear oil seal, replacement, 7A-3
removal and installation
2-wheel drive models, 7A-3
4-wheel drive models, 7A-3
speedometer driven gear, replacement, 7A-3
steering column shift linkage, adjustment, 7A-2
shift effort, diagnosis, 7A-2
Master cylinder removal, overhaul and installation
Bendix, 9-12
Delco, 9-10

N

Neutral start switch, adjustment, 7B-4
Neutral start switch/back-up light switch (automatic transmission), 12-7

O

Oil seal, replacement
rear, 7A-3
semi-floating axle, 8-10
Oxygen sensor, 6-14

P

Park/Neutral switch (automatic transmission-equipped vehicles only), 6-14
Parking brake
adjustment
hand lever type, 9-12
pedal type, 9-13
control and cables (hand lever type), removal and installation, 9-12
pedal and cables (pedal type), removal and installation, 9-13
Pinion oil seal (all models), replacement, 8-11
Piston rings, installation, 2C-18
Piston/connecting rod assembly
inspection, 2C-15
installation and rod bearing oil clearance check, 2C-20
removal, 2C-11
Pitman arm, removal and installation
four-wheel drive, 10-12
two-wheel drive, 10-11
Pitman shaft oil seal (two-wheel drive), replacement in vehicle, 10-13
Positive Crankcase Ventilation (PCV) system, 6-19
Positive Crankcase Ventilation (PCV) valve check and replacement, 1-27
Power brake vacuum booster, inspection, removal and installation, 9-14
Power steering fluid level check, 1-11
Power-operated door lock, removal and installation, 11-8
Power-operated window glass and regulator, removal and installation, 11-7
Pre-oiling engine after overhaul, 2C-22
Pressure regulating combination valve, check, removal and installation, 9-12
Pulse Air Injection Reaction (PAIR) system, 6-21

R

Radiator grille, removal and installation
1967 through 1971 models, 11-9
and support (1972 and later models), removal and installation, 11-9
Radiator, removal and installation, 3-3
Radio
and speakers, removal and installation, 12-9
antenna, removal and installation, 12-10
Rear view mirrors, removal and installation, 11-11
Rear wheel anti-lock (RWAL) brake system, general information, 9-16

S

Safety first!, 0-21
Seat back latch check, 1-24
Seat belt check, 1-24
Semi-floating rear axleshaft, removal and installation, 8-9
Shift linkage, check and adjustment, 7B-2
Shock absorbers, removal, testing and installation, 10-6
Spark plug
replacement, 1-28
wire check and replacement, 1-29
Speedometer
driven gear, replacement, 7A-3
gear, replacement, 7B-4
Stabilizer bar, front, removal and installation, 10-4
Starter
motor
in-vehicle check, 5-13
removal and installation, 5-13
safety switch check, 1-24
solenoid, removal and installation, 5-13
Starting system, general information, 5-12
Steering and suspension systems, 10-1 through 10-20
general information, 10-4
Steering
column
lower bearing (1967 through 1972), replacement in vehicle, 10-14
overhaul, 10-14
removal and installation, 10-14
shift linkage, adjustment, 7A-2
turn signal switch, removal and installation, 10-17
upper bearing (1967 through 1972), replacement in vehicle, 10-14
damper, check and replacement, 10-11
idler arm (two-wheel drive), removal and installation, 10-11

knuckle and balljoints (four-wheel drive), removal, overhaul and installation, 10-8
knuckle, front, (two-wheel drive), removal and installation, 10-7
shaft universal joint coupling, removal and installation, 10-13

Steering system
damper, check and replacement, 10-11
general information, 10-10
idler arm (two-wheel drive), removal and installation, 10-11
intermediate shaft coupling (early models), removal, overhaul and installation, 10-13
manual steering gear
check and adjustment, 10-12
removal and installation, 10-13
Pitman arm, removal and installation
four-wheel drive, 10-12
two-wheel drive, 10-11
Pitman shaft oil seal (two-wheel drive), replacement in vehicle, 10-13
power steering
gear, removal and installation, 10-19
pump, removal and installation, 10-18
system, bleeding, 10-18
steering shaft flexible coupling, removal and installation, 10-13
tie-rod ends, replacement, 10-10
wheel, removal and installation, 10-12

Suspension
assembly, front, removal and installation, 10-7
auxiliary leaf spring, rear (early two-wheel drive), removal and installation, 10-10
balljoints (two-wheel drive), check and replacement, 10-5
coil spring
front (two-wheel drive), removal and installation, 10-5
rear (early two-wheel drive), removal and installation, 10-9
control arm, rear (1967 through 1972), removal and installation, 10-10
front axle
balljoints (four-wheel drive), check and adjustment, 10-8
kingpins
four-wheel drive
with disc brakes, overhaul, 10-9
with drum brakes, check, adjustment and overhaul, 10-9
leaf spring
front (four-wheel drive), removal and installation, 10-7
rear (1973 on), removal and installation, 10-10
lower control arm, front (two-wheel drive), removal and installation, 10-6
stabilizer bar, front, removal and installation, 10-4
steering knuckle and balljoints (four-wheel drive), removal, overhaul and installation, 10-8
steering knuckle, front (two-wheel drive), removal and installation, 10-7
shock absorbers, removal, testing and installation, 10-6

Suspension and steering check, 1-16
Switches, removal and installation, 12-6
back-up light switch (manual transmission), 12-7
headlight dimmer switch, column mounted, 12-7
Neutral start switch/back-up light switch (automatic transmission), 12-6
windshield wiper/washer switch, 12-7

T

Tailgate, removal and installation, 11-10
Thermostat, removal and installation, 3-3
Thermostatic Air Cleaner (THERMAC), 6-20
Thermostatic air cleaner check, 1-21
Throttle
linkage
inspection, 1-21
replacement, 4-54
Throttle Body Injection (TBI) unit, disassembly and reassembly, 4-53
Throttle Return Control (TRC) system, 6-21
Throttle valve (TV) cable
check and adjustment, 7B-3
replacement, 7B-4
Tie-rod ends, replacement, 10-10
Tire and tire pressure checks, 1-9
Tire rotation, 1-17
Tools, 0-12
Transfer case, 7C-1 through 7C-14
general information, 7C-2
oil change, 1-24
oil level check, 1-17
removal and installation, 7C-3
shift linkage, check and adjustment, 7C-2
Transmission Controlled Spark (TCS) system, 6-23
Troubleshooting, 0-22
Tune-up and routine maintenance, 1-1 through 1-36
Tune-up general information, 1-7
Turn signal and hazard flashers, check and replacement, 12-4

U

Underhood hose check and replacement, 1-12
Universal joints, removal, overhaul and installation, 8-7
Upholstery and carpets, maintenance, 11-2

V8 and V6 engines, 2A-1 through 21-16
Valves, servicing, 2C-11
Vehicle identification numbers, 0-7
Vehicle speed sensor, 6-14

Water pump
check, 3-4
removal and installation, 3-4
Wheel bearings, front (four-wheel drive), check, repack and adjustment, 8-14
Windshield
and fixed glass, removal and installation, 11-3
washer pump (later models), removal and installation, 12-9
wiper arms, removal and installation, 12-8
wiper/washer
motor, removal and installation, 12-8
switch, 12-7
Wiper blade inspection and replacement, 1-13
Working facilities, 0-16

Haynes Automotive Manuals

NOTE: If you do not see a listing for your vehicle, consult your local Haynes dealer for the latest product information.

JRA

20 **Integra** '86 thru '89 & **Legend** '86 thru '90
21 **Integra** '90 thru '93 & **Legend** '91 thru '95
50 **Acura TL** all models '99 thru '08

C

Jeep CJ - *see JEEP (50020)*
20 **Mid-size models** '70 thru '83
25 **(Renault) Alliance & Encore** '83 thru '87

DI

20 **4000** all models '80 thru '87
25 **5000** all models '77 thru '83
26 **5000** all models '84 thru '88
30 **Audi A4** '02 thru '08

STIN-HEALEY

Sprite - *see MG Midget (66015)*

W

20 **3/5 Series** '82 thru '92
21 **3-Series** incl. Z3 models '92 thru '98
22 **3-Series**, '99 thru '05, Z4 models
25 **320i** all 4 cyl models '75 thru '83
50 **1500 thru 2002** except Turbo '59 thru '77

CK

10 **Buick Century** '97 thru '05
Century (front-wheel drive) - *see GM (38005)*
20 **Buick, Oldsmobile & Pontiac Full-size (Front-wheel drive)** '85 thru '05
Buick Electra, LeSabre and Park Avenue; **Oldsmobile** Delta 88 Royale, Ninety Eight and Regency; **Pontiac** Bonneville
25 **Buick, Oldsmobile & Pontiac Full-size (Rear wheel drive)** '70 thru '90
Buick Estate, Electra, LeSabre, Limited, **Oldsmobile** Custom Cruiser, Delta 88, Ninety-eight, **Pontiac** Bonneville, Catalina, Grandville, Parisienne
30 **Mid-size Regal & Century** all rear-drive models with V6, V8 and Turbo '74 thru '87
Regal - *see GENERAL MOTORS (38010)*
Riviera - *see GENERAL MOTORS (38030)*
Roadmaster - *see CHEVROLET (24046)*
Skyhawk - *see GENERAL MOTORS (38015)*
Skylark - *see GM (38020, 38025)*
Somerset - *see GENERAL MOTORS (38025)*

DILLAC

30 **Cadillac Rear Wheel Drive** '70 thru '93
Cimarron - *see GENERAL MOTORS (38015)*
DeVille - *see GM (38031 & 38032)*
Eldorado - *see GM (38030 & 38031)*
Fleetwood - *see GM (38031)*
Seville - *see GM (38030, 38031 & 38032)*

EVROLET

05 **Chevrolet Engine Overhaul Manual**
10 **Astro & GMC Safari Mini-vans** '85 thru '05
15 **Camaro V8** all models '70 thru '81
16 **Camaro** all models '82 thru '92
17 **Camaro & Firebird** '93 thru '02
Cavalier - *see GENERAL MOTORS (38016)*
Celebrity - *see GENERAL MOTORS (38005)*
20 **Chevelle, Malibu & El Camino** '69 thru '87
24 **Chevette & Pontiac T1000** '76 thru '87
Citation - *see GENERAL MOTORS (38020)*
27 **Colorado & GMC Canyon** '04 thru '08
32 **Corsica/Beretta** all models '87 thru '96
40 **Corvette** all V8 models '68 thru '82
41 **Corvette** all models '84 thru '96
45 **Full-size Sedans** Caprice, Impala, Biscayne, Bel Air & Wagons '69 thru '90
46 **Impala SS & Caprice and Buick Roadmaster** '91 thru '96, **Impala** - *see LUMINA (24048)*
Lumina '90 thru '94 - *see GM (38010)*
47 **Impala & Monte Carlo** all models '06 thru '08
48 **Lumina & Monte Carlo** '95 thru '05
Lumina APV - *see GM (38035)*
50 **Luv Pick-up** all 2WD & 4WD '72 thru '82
Malibu '97 thru '00 - *see GM (38026)*
55 **Monte Carlo** all models '70 thru '88
Monte Carlo '95 thru '01 - *see LUMINA (24048)*
59 **Nova** all V8 models '69 thru '79
60 **Nova and Geo Prizm** '85 thru '92
64 **Pick-ups '67 thru '87** - Chevrolet & GMC, all V8 & in-line 6 cyl, 2WD & 4WD '67 thru '87; Suburbans, Blazers & Jimmys '67 thru '91
65 **Pick-ups '88 thru '98** - Chevrolet & GMC, full-size pick-ups '88 thru '98, C/K Classic '99 & '00, Blazer & Jimmy '92 thru '94; Suburban '92 thru '99; Tahoe & Yukon '95 thru '99

24066 **Pick-ups '99 thru '06** - Chevrolet Silverado & GMC Sierra '99 thru '06, Suburban/Tahoe/ Yukon/Yukon XL/Avalanche '00 thru '06
24067 **Chevrolet Silverado & GMC Sierra** '07 thru '09
24070 **S-10 & S-15 Pick-ups** '82 thru '93, **Blazer & Jimmy** '83 thru '94,
24071 **S-10 & Sonoma Pick-ups** '94 thru '04, including **Blazer, Jimmy & Hombre**
24072 **Chevrolet TrailBlazer, GMC Envoy & Oldsmobile Bravada** '02 thru '09
24075 **Sprint** '85 thru '88 & **Geo Metro** '89 thru '01
24080 **Vans - Chevrolet & GMC** '68 thru '96
24081 **Chevrolet Express & GMC Savana** Full-size Vans '96 thru '07

CHRYSLER

10310 **Chrysler Engine Overhaul Manual**
25015 **Chrysler Cirrus, Dodge Stratus, Plymouth Breeze** '95 thru '00
25020 **Full-size Front-Wheel Drive** '88 thru '93
K-Cars - *see DODGE Aries (30008)*
Laser - *see DODGE Daytona (30030)*
25025 **Chrysler LHS, Concorde, New Yorker, Dodge** Intrepid, **Eagle** Vision, '93 thru '97
25026 **Chrysler LHS, Concorde, 300M, Dodge** Intrepid, '98 thru '04
25027 **Chrysler 300, Dodge Charger & Magnum** '05 thru '09
25030 **Chrysler & Plymouth Mid-size** front wheel drive '82 thru '95
Rear-wheel Drive - *see Dodge (30050)*
25035 **PT Cruiser** all models '01 thru '09
25040 **Chrysler** Sebring, **Dodge** Avenger '95 thru '05
Dodge Stratus '01 thru 05

DATSUN

28005 **200SX** all models '80 thru '83
28007 **B-210** all models '73 thru '78
28009 **210** all models '79 thru '82
28012 **240Z, 260Z & 280Z** Coupe '70 thru '78
28014 **280ZX** Coupe & 2+2 '79 thru '83
300ZX - *see NISSAN (72010)*
28018 **510 & PL521 Pick-up** '68 thru '73
28020 **510** all models '78 thru '81
28022 **620 Series Pick-up** all models '73 thru '79
720 Series Pick-up - *see NISSAN (72030)*
28025 **810/Maxima** all gasoline models, '77 thru '84

DODGE

400 & 600 - *see CHRYSLER (25030)*
30008 **Aries & Plymouth Reliant** '81 thru '89
30010 **Caravan & Plymouth Voyager** '84 thru '95
30011 **Caravan & Plymouth Voyager** '96 thru '02
30012 **Challenger/Plymouth Saporro** '78 thru '83
30013 **Caravan, Chrysler Voyager, Town & Country** '03 thru '07
30016 **Colt & Plymouth Champ** '78 thru '87
30020 **Dakota Pick-ups** all models '87 thru '96
30021 **Durango** '98 & '99, **Dakota** '97 thru '99
30022 **Durango** '00 thru '03 **Dakota** '00 thru '04
30023 **Durango** '04 thru '06, **Dakota** '05 and '06
30025 **Dart, Demon, Plymouth Barracuda, Duster & Valiant** 6 cyl models '67 thru '76
30030 **Daytona & Chrysler Laser** '84 thru '89
Intrepid - *see CHRYSLER (25025, 25026)*
30034 **Neon** all models '95 thru '99
30035 **Omni & Plymouth Horizon** '78 thru '90
30036 **Dodge and Plymouth Neon** '00 thru'05
30040 **Pick-ups** all full-size models '74 thru '93
30041 **Pick-ups** all full-size models '94 thru '01
30042 **Pick-ups** full-size models '02 thru '08
30045 **Ram 50/D50 Pick-ups & Raider and Plymouth Arrow Pick-ups** '79 thru '93
30050 **Dodge/Plymouth/Chrysler** RWD '71 thru '89
30055 **Shadow & Plymouth Sundance** '87 thru '94
30060 **Spirit & Plymouth Acclaim** '89 thru '95
30065 **Vans - Dodge & Plymouth** '71 thru '03

EAGLE

Talon - *see MITSUBISHI (68030, 68031)*
Vision - *see CHRYSLER (25025)*

FIAT

34010 **124 Sport Coupe & Spider** '68 thru '78
34025 **X1/9** all models '74 thru '80

FORD

10320 **Ford Engine Overhaul Manual**
10355 **Ford Automatic Transmission Overhaul**
36004 **Aerostar Mini-vans** all models '86 thru '97
36006 **Contour & Mercury Mystique** '95 thru '00
36008 **Courier Pick-up** all models '72 thru '82
36012 **Crown Victoria & Mercury Grand Marquis** '88 thru '10
36016 **Escort/Mercury Lynx** all models '81 thru '90
36020 **Escort/Mercury Tracer** '91 thru '02
36022 **Escape & Mazda Tribute** '01 thru '07
36024 **Explorer & Mazda Navajo** '91 thru '01
36025 **Explorer/Mercury Mountaineer** '02 thru '10
36028 **Fairmont & Mercury Zephyr** '78 thru '83
36030 **Festiva & Aspire** '88 thru '97
36032 **Fiesta** all models '77 thru '80
36034 **Focus** all models '00 thru '07
36036 **Ford & Mercury Full-size** '75 thru '87
36044 **Ford & Mercury Mid-size** '75 thru '86
36048 **Mustang V8** all models '64-1/2 thru '73
36049 **Mustang II** 4 cyl, V6 & V8 models '74 thru '78
36050 **Mustang & Mercury Capri** '79 thru '86
36051 **Mustang** all models '94 thru '04
36052 **Mustang** '05 thru '07
36054 **Pick-ups & Bronco** '73 thru '79
36058 **Pick-ups & Bronco** '80 thru '96
36059 **F-150 & Expedition** '97 thru '09, **F-250** '97 thru '99 & **Lincoln Navigator** '98 thru '09
36060 **Super Duty Pick-ups, Excursion** '99 thru '10
36061 **F-150** full-size '04 thru '09
36062 **Pinto & Mercury Bobcat** '75 thru '80
36066 **Probe** all models '89 thru '92
36070 **Ranger/Bronco II** gasoline models '83 thru '92
36071 **Ranger** '93 thru '10 & **Mazda Pick-ups** '94 thru '09
36074 **Taurus & Mercury Sable** '86 thru '95
36075 **Taurus & Mercury Sable** '96 thru '05
36078 **Tempo & Mercury Topaz** '84 thru '94
36082 **Thunderbird/Mercury Cougar** '83 thru '88
36086 **Thunderbird/Mercury Cougar** '89 and '97
36090 **Vans** all V8 Econoline models '69 thru '91
36094 **Vans** full size '92 thru '05
36097 **Windstar Mini-van** '95 thru '07

GENERAL MOTORS

10360 **GM Automatic Transmission Overhaul**
38005 **Buick Century, Chevrolet Celebrity, Oldsmobile Cutlass Ciera & Pontiac 6000** all models '82 thru '96
38010 **Buick Regal, Chevrolet Lumina, Oldsmobile Cutlass Supreme & Pontiac Grand Prix** (FWD) '88 thru '07
38015 **Buick Skyhawk, Cadillac Cimarron, Chevrolet Cavalier, Oldsmobile Firenza & Pontiac J-2000 & Sunbird** '82 thru '94
38016 **Chevrolet Cavalier & Pontiac Sunfire** '95 thru '05
38017 **Chevrolet Cobalt & Pontiac G5** '05 thru '09
38020 **Buick Skylark, Chevrolet Citation, Olds Omega, Pontiac Phoenix** '80 thru '85
38025 **Buick Skylark & Somerset, Oldsmobile Achieva & Calais and Pontiac Grand Am** all models '85 thru '98
38026 **Chevrolet Malibu, Olds Alero & Cutlass, Pontiac Grand Am** '97 thru '03
38027 **Chevrolet Malibu** '04 thru '07
38030 **Cadillac Eldorado, Seville, Oldsmobile** Toronado, **Buick Riviera** '71 thru '85
38031 **Cadillac Eldorado & Seville, DeVille, Fleetwood & Olds Toronado, Buick Riviera** '86 thru '93
38032 **Cadillac DeVille** '94 thru '05 & **Seville** '92 thru '04 **Cadillac DTS** '06 thru '10
38035 **Chevrolet Lumina APV, Olds Silhouette & Pontiac Trans Sport** all models '90 thru '96
38036 **Chevrolet Venture, Olds Silhouette, Pontiac Trans Sport & Montana** '97 thru '05
General Motors Full-size Rear-wheel Drive - see BUICK (19025)
38040 **Chevrolet Equinox** '05 thru '09 **Pontiac Torrent** '06 thru '09

GEO

Metro - *see CHEVROLET Sprint (24075)*
Prizm - *'85 thru '92 see CHEVY (24060), '93 thru '02 see TOYOTA Corolla (92036)*
40030 **Storm** all models '90 thru '93
Tracker - *see SUZUKI Samurai (90010)*

GMC

Vans & Pick-ups - *see CHEVROLET*

HONDA

42010 **Accord CVCC** all models '76 thru '83
42011 **Accord** all models '84 thru '89
42012 **Accord** all models '90 thru '93

(Continued on other side)

Haynes Automotive Manuals (continued)

NOTE: If you do not see a listing for your vehicle, consult your local Haynes dealer for the latest product information.

42013 **Accord** all models '94 thru '97
42014 **Accord** all models '98 thru '02
42015 **Accord** models '03 thru '07
42020 **Civic 1200** all models '73 thru '79
42021 **Civic 1300 & 1500 CVCC** '80 thru '83
42022 **Civic 1500 CVCC** all models '75 thru '79
42023 **Civic** all models '84 thru '91
42024 **Civic & del Sol** '92 thru '95
42025 **Civic** '96 thru '00, **CR-V** '97 thru '01, **Acura Integra** '94 thru '00
42026 **Civic** '01 thru '10, **CR-V** '02 thru '09
42035 **Odyssey** all models '99 thru '04
42037 **Honda Pilot** '03 thru '07, **Acura MDX** '01 thru '07
42040 **Prelude CVCC** all models '79 thru '89

HYUNDAI

43010 **Elantra** all models '96 thru '06
43015 **Excel & Accent** all models '86 thru '09
43050 **Santa Fe** all models '01 thru '06
43055 **Sonata** all models '99 thru '08

ISUZU

Hombre - *see CHEVROLET S-10 (24071)*
47017 **Rodeo, Amigo & Honda Passport** '89 thru '02
47020 **Trooper & Pick-up** '81 thru '93

JAGUAR

49010 **XJ6** all 6 cyl models '68 thru '86
49011 **XJ6** all models '88 thru '94
49015 **XJ12 & XJS** all 12 cyl models '72 thru '85

JEEP

50010 **Cherokee, Comanche & Wagoneer Limited** all models '84 thru '01
50020 **CJ** all models '49 thru '86
50025 **Grand Cherokee** all models '93 thru '04
50026 **Grand Cherokee** '05 thru '09
50029 **Grand Wagoneer & Pick-up** '72 thru '91 Grand Wagoneer '84 thru '91, Cherokee & Wagoneer '72 thru '83, Pick-up '72 thru '88
50030 **Wrangler** all models '87 thru '08
50035 **Liberty** '02 thru '07

KIA

54070 **Sephia** '94 thru '01, **Spectra** '00 thru '09

LEXUS

ES 300 - *see TOYOTA Camry (92007)*

LINCOLN

Navigator - *see FORD Pick-up (36059)*
59010 **Rear-Wheel Drive** all models '70 thru '10

MAZDA

61010 **GLC Hatchback (rear-wheel drive)** '77 thru '83
61011 **GLC (front-wheel drive)** '81 thru '85
61015 **323 & Protogé** '90 thru '00
61016 **MX-5 Miata** '90 thru '09
61020 **MPV** all models '89 thru '98
Navajo - *see Ford Explorer (36024)*
61030 **Pick-ups** '72 thru '93
Pick-ups '94 thru '00 - *see Ford Ranger (36071)*
61035 **RX-7** all models '79 thru '85
61036 **RX-7** all models '86 thru '91
61040 **626** (rear-wheel drive) all models '79 thru '82
61041 **626/MX-6 (front-wheel drive)** '83 thru '92
61042 **626, MX-6/Ford Probe** '93 thru '01

MERCEDES-BENZ

63012 **123 Series Diesel** '76 thru '85
63015 **190 Series** four-cyl gas models, '84 thru '88
63020 **230/250/280** 6 cyl sohc models '68 thru '72
63025 **280 123 Series** gasoline models '77 thru '81
63030 **350 & 450** all models '71 thru '80
63040 **C-Class:** C230/C240/C280/C320/C350 '01 thru '07

MERCURY

64200 **Villager & Nissan Quest** '93 thru '01
All other titles, see FORD Listing.

MG

66010 **MGB** Roadster & GT Coupe '62 thru '80
66015 **MG Midget, Austin Healey Sprite** '58 thru '80

MITSUBISHI

68020 **Cordia, Tredia, Galant, Precis & Mirage** '83 thru '93
68030 **Eclipse, Eagle Talon & Ply. Laser** '90 thru '94
68031 **Eclipse** '95 thru '05, **Eagle Talon** '95 thru '98
68035 **Galant** '94 thru '03
68040 **Pick-up** '83 thru '96 & **Montero** '83 thru '93

NISSAN

72010 **300ZX** all models including Turbo '84 thru '89
72011 **350Z & Infiniti G35** all models '03 thru '08
72015 **Altima** all models '93 thru '06
72020 **Maxima** all models '85 thru '92
72021 **Maxima** all models '93 thru '04
72030 **Pick-ups** '80 thru '97 **Pathfinder** '87 thru '95
72031 **Frontier Pick-up, Xterra, Pathfinder** '96 thru '04
72032 **Frontier & Xterra** '05 thru '08
72040 **Pulsar** all models '83 thru '86
Quest - *see MERCURY Villager (64200)*
72050 **Sentra** all models '82 thru '94
72051 **Sentra & 200SX** all models '95 thru '06
72060 **Stanza** all models '82 thru '90
72070 **Titan pick-ups** '04 thru '09
Armada '05 thru '10

OLDSMOBILE

73015 **Cutlass** V6 & V8 gas models '74 thru '88
For other OLDSMOBILE titles, see BUICK, CHEVROLET or GENERAL MOTORS listing.

PLYMOUTH

For PLYMOUTH titles, see DODGE listing.

PONTIAC

79008 **Fiero** all models '84 thru '88
79018 **Firebird** V8 models except Turbo '70 thru '81
79019 **Firebird** all models '82 thru '92
79025 **G6** all models '05 thru '09
79040 **Mid-size Rear-wheel Drive** '70 thru '87
For other PONTIAC titles, see BUICK, CHEVROLET or GENERAL MOTORS listing.

PORSCHE

80020 **911** except Turbo & Carrera 4 '65 thru '89
80025 **914** all 4 cyl models '69 thru '76
80030 **924** all models including Turbo '76 thru '82
80035 **944** all models including Turbo '83 thru '89

RENAULT

Alliance & Encore - *see AMC (14020)*

SAAB

84010 **900** all models including Turbo '79 thru '88

SATURN

87010 **Saturn** all S-series models '91 thru '02
87011 **Saturn Ion** '03 thru '07
87020 **Saturn** all L-series models '00 thru '04
87040 **Saturn VUE** '02 thru '07

SUBARU

89002 **1100, 1300, 1400 & 1600** '71 thru '79
89003 **1600 & 1800** 2WD & 4WD '80 thru '94
89100 **Legacy** all models '90 thru '99
89101 **Legacy & Forester** '00 thru '06

SUZUKI

90010 **Samurai/Sidekick & Geo Tracker** '86 thru '01

TOYOTA

92005 **Camry** all models '83 thru '91
92006 **Camry** all models '92 thru '96
92007 **Camry, Avalon, Solara, Lexus ES 300** '97 thru '01
92008 **Toyota Camry, Avalon and Solara and Lexus ES 300/330** all models '02 thru '06
92015 **Celica Rear Wheel Drive** '71 thru '85
92020 **Celica Front Wheel Drive** '86 thru '99
92025 **Celica Supra** all models '79 thru '92
92030 **Corolla** all models '75 thru '79
92032 **Corolla** all rear wheel drive models '80 thru '87
92035 **Corolla** all front wheel drive models '84 thru '92
92036 **Corolla & Geo Prizm** '93 thru '02
92037 **Corolla** models '03 thru '08
92040 **Corolla Tercel** all models '80 thru '82
92045 **Corona** all models '74 thru '82
92050 **Cressida** all models '78 thru '82
92055 **Land Cruiser** FJ40, 43, 45, 55 '68 thru '82
92056 **Land Cruiser** FJ60, 62, 80, FZJ80 '80 thru '96
92060 **Matrix & Pontiac Vibe** '03 thru '08
92065 **MR2** all models '85 thru '87
92070 **Pick-up** all models '69 thru '78
92075 **Pick-up** all models '79 thru '95
92076 **Tacoma, 4Runner, & T100** '93 thru '04
92077 **Tacoma** all models '05 thru '09
92078 **Tundra** '00 thru '06 & **Sequoia** '01 thru '07
92079 **4Runner** all models '03 thru '09
92080 **Previa** all models '91 thru '95
92081 **Prius** all models '01 thru '08
92082 **RAV4** all models '96 thru '05
92085 **Tercel** all models '87 thru '94
92090 **Sienna** all models '98 thru '09
92095 **Highlander & Lexus RX-330** '99 thru '06

TRIUMPH

94007 **Spitfire** all models '62 thru '81
94010 **TR7** all models '75 thru '81

VW

96008 **Beetle & Karmann Ghia** '54 thru '79
96009 **New Beetle** '98 thru '05
96016 **Rabbit, Jetta, Scirocco & Pick-up** gas models '75 thru '92 & Convertible '80 thru '9
96017 **Golf, GTI & Jetta** '93 thru '98 **& Cabrio** '95 thru '02
96018 **Golf, GTI, Jetta** '99 thru '05
96020 **Rabbit, Jetta & Pick-up** diesel '77 thru '8
96023 **Passat** '98 thru '05, **Audi A4** '96 thru '01
96030 **Transporter 1600** all models '68 thru '79
96035 **Transporter 1700, 1800 & 2000** '72 thru '7
96040 **Type 3 1500 & 1600** all models '63 thru '7
96045 **Vanagon** all air-cooled models '80 thru '83

VOLVO

97010 **120, 130 Series & 1800 Sports** '61 thru '7
97015 **140 Series** all models '66 thru '74
97020 **240 Series** all models '76 thru '93
97040 **740 & 760 Series** all models '82 thru '88
97050 **850 Series** all models '93 thru '97

TECHBOOK MANUALS

10205 **Automotive Computer Codes**
10206 **OBD-II & Electronic Engine Manageme**
10210 **Automotive Emissions Control Manual**
10215 **Fuel Injection Manual, 1978 thru 1985**
10220 **Fuel Injection Manual, 1986 thru 1999**
10225 **Holley Carburetor Manual**
10230 **Rochester Carburetor Manual**
10240 **Weber/Zenith/Stromberg/SU Carburetor**
10305 **Chevrolet Engine Overhaul Manual**
10310 **Chrysler Engine Overhaul Manual**
10320 **Ford Engine Overhaul Manual**
10330 **GM and Ford Diesel Engine Repair Manu**
10333 **Engine Performance Manual**
10340 **Small Engine Repair Manual,** 5 HP & Les
10341 **Small Engine Repair Manual,** 5.5 - 20 HP
10345 **Suspension, Steering & Driveline Manu**
10355 **Ford Automatic Transmission Overhaul**
10360 **GM Automatic Transmission Overhaul**
10405 **Automotive Body Repair & Painting**
10410 **Automotive Brake Manual**
10411 **Automotive Anti-lock Brake (ABS) System**
10415 **Automotive Detaiing Manual**
10420 **Automotive Electrical Manual**
10425 **Automotive Heating & Air Conditioning**
10430 **Automotive Reference Manual & Dictionary**
10435 **Automotive Tools Manual**
10440 **Used Car Buying Guide**
10445 **Welding Manual**
10450 **ATV Basics**
10452 **Scooters 50cc to 250cc**

SPANISH MANUALS

98903 **Reparación de Carrocería & Pintura**
98904 **Carburadores para los modelos Holley & Rochester**
98905 **Códigos Automotrices de la Computado**
98910 **Frenos Automotriz**
98913 **Electricidad Automotriz**
98915 **Inyección de Combustible 1986 al 1999**
99040 **Chevrolet & GMC Camionetas** '67 al '87
99041 **Chevrolet & GMC Camionetas** '88 al '98
99042 **Chevrolet & GMC Camionetas Cerradas** '68 al '95
99043 **Chevrolet/GMC Camionetas** '94 thru '04
99055 **Dodge Caravan & Plymouth Voyager** '84 al '9
99075 **Ford Camionetas y Bronco** '80 al '94
99077 **Ford Camionetas Cerradas** '69 al '91
99088 **Ford Modelos de Tamaño Mediano** '75 al '86
99091 **Ford Taurus & Mercury Sable** '86 al '95
99095 **GM Modelos de Tamaño Grande** '70 al '
99100 **GM Modelos de Tamaño Mediano** '70 al '
99106 **Jeep Cherokee, Wagoneer & Comanch** '84 al '00
99110 **Nissan Camioneta** '80 al '96, **Pathfinder** '87 al '
99118 **Nissan Sentra** '82 al '94
99125 **Toyota Camionetas y 4Runner** '79 al '95

Over 100 Haynes motorcycle manuals also available

Haynes North America, Inc., 861 Lawrence Drive, Newbury Park, CA 91320-1514 • (805) 498-6703 • http://www.haynes.co